Abiotic Stress-Mediated Sensing and Signaling in Plants: An Omics Perspective

Sajad Majeed Zargar
Mohammad Yousuf Zargar
Editors

Abiotic Stress-Mediated Sensing and Signaling in Plants: An Omics Perspective

Springer

Editors
Sajad Majeed Zargar
Division of Plant Biotechnology
Sher-e-Kashmir University of Agricultural
Sciences & Technology of Kashmir
Srinagar, Jammu & Kashmir, India

Mohammad Yousuf Zargar
Directorate of Research
Sher-e-Kashmir University of Agricultural
Sciences & Technology of Kashmir
Srinagar, Jammu & Kashmir, India

ISBN 978-981-10-7478-3 ISBN 978-981-10-7479-0 (eBook)
https://doi.org/10.1007/978-981-10-7479-0

Library of Congress Control Number: 2018930876

Printed on acid-free paper

This Springer imprint is published by the registered company Springer Nature Singapore Pte Ltd. part of Springer Nature.
The registered company address is: 152 Beach Road, #21-01/04 Gateway East, Singapore 189721, Singapore

We dedicate this book to our beloved parents.

Preface

Present climatic changes lead to decrease in timely precipitation, increase in temperature and have drastic effect on sustainability. Water, food, health, land and environment that are the essential needs of life are all threatened by the changing climate. There will be a tremendous decline in crop yield as well as food quality that will have a major impact on all other aspects of life. The negative effects of the changing climate are most apparent on genetic resources and biodiversity that represent one of the major limiting factors for crop improvement. As such, there is a need to understand molecular mechanisms and the biology behind tolerance to these abiotic stresses.

Plants have evolved a wide range of mechanisms to cope with various abiotic stresses. In various crop plants, molecular mechanisms involved in a single stress tolerance have been revealed comparatively and independently; however, in order to have a holistic understanding of major and common events happening among various abiotic stresses, there is a need to elucidate the signalling pathways involved. So far, several molecules, like transcription factors and kinases, have been identified as promising candidates that are involved in crosstalk between stress signalling pathways. There are various other pathways that are involved in a crosstalk for abiotic stress tolerance. However, there is a need to better understand the tolerance mechanisms for different abiotic stresses through elucidation of signalling and sensing mechanisms involved.

Here, we have covered various topics that include the impact of different abiotic stresses on plants, the molecular mechanisms leading to tolerance for different abiotic stresses, signalling cascades revealing crosstalk among various abiotic stresses and elucidation of major candidate molecules that may provide abiotic stress tolerance in plants.

Srinagar, Jammu & Kashmir, India

Sajad Majeed Zargar
Mohammad Yousuf Zargar

Acknowledgements

A lot of effort were put in the completion of this book, and without the support of people around us, it would not have been possible.

We would like to thank our organisation (SKUAST-Kashmir), particularly Honourable Vice Chancellor Prof. Nazeer Ahmed, for their support, guidance and encouragement. We would also like to thank Extension Director Prof. M A Teli, Education Director Prof. S A Wani and Dean Prof. F A Zaki of the Faculty of Horticulture for their support and guidance.

SMZ acknowledges the support and encouragement of the scientific staff of the Division of Plant Biotechnology, especially Prof. Riaz Ahmad Shah, Prof. Shafiq A. Wani, Prof. Nazir A. Ganai and Prof. F A Nehvi.

SMZ also acknowledges the support of students especially Dr. Reetika and collaborators especially Prof. Randeep Rakwal (University of Tsukuba, Japan) for his generosity. SMZ also thanks his family for their support and best wishes.

Contents

Contributors

Aayushi Agrawal Department of Biotechnology, Sharda University, Greater Noida, India

Aejaz Ahmed Dar School of Biotechnology, Sher-e-Kashmir University of Agricultural Sciences & Technology of Jammu, Jammu, Jammu & Kashmir, India

Taufika Islam Anee Department of Agronomy, Faculty of Agriculture, Sher-e-Bangla Agricultural University, Dhaka, Bangladesh

Sreshti Bagati School of Biotechnology, Sher-e-Kashmir University of Agricultural Sciences & Technology of Jammu, Jammu, Jammu & Kashmir, India

Shamim Ara Bagum Stress Breeding Lab, Bangladesh Agricultural Research Institute, Gazipur, Bangladesh

Ajay Bhardwaj Kerala Agricultural University, Thrissur, Kerala, India

N. Bharat Kumar Central Silk Board (CSB), Central Sericultural Research &Training Institute (CSR&TI), Pampore, Jammu & Kashmir, India

Masum Billah Stress Breeding Lab, Bangladesh Agricultural Research Institute, Gazipur, Bangladesh

Chetan Chauhan Department of Botany, University of Delhi, Delhi, India

S. S. Chauhan Central Silk Board (CSB), Central Sericultural Research &Training Institute (CSR&TI), Pampore, Jammu & Kashmir, India

Samira Chugh Department of Botany, Gargi College, University of Delhi, New Delhi, India

Sandip Das Department of Botany, University of Delhi, New Delhi, India

Priyanka Deveshwar Department of Plant Molecular Biology, University of Delhi, New Delhi, India

Rose Mary Francies Kerala Agricultural University, Thrissur, Kerala, India

Mudasir Gani Central Silk Board (CSB), Central Sericultural Research &Training Institute (CSR&TI), Pampore, Jammu & Kashmir, India

R. Geeta Department of Botany, University of Delhi, New Delhi, India

M. K. Ghosh Central Silk Board (CSB), Central Sericultural Research &Training Institute (CSR&TI), Pampore, Jammu & Kashmir, India

Lal Chand Godara Indian Council of Agricultural Research (ICAR), Central Agroforestry Research Institute (CAFRI), Jhansi, Uttar Pradesh, India

Parul Goel Department of Biotechnology, CSIR-Institute of Himalayan Bioresource Technology, Palampur, Himachal Pradesh, India

Mirza Hasanuzzaman Department of Agronomy, Faculty of Agriculture, Sher-e-Bangla Agricultural University, Dhaka, Bangladesh

Neelima Hossain Stress Breeding Lab, Bangladesh Agricultural Research Institute, Gazipur, Bangladesh

Aditi Jain Department of Botany, University of Delhi, New Delhi, India

Muskan Jain Laboratory of Plant Molecular Biology, Faculty of Life Sciences and Biotechnology, South Asian University, New Delhi, India

Gauri Joshi Department of Botany, University of Delhi, New Delhi, India

Jashon Jot Kaur Punjab Agricultural University, Ludhiana, Punjab, India

Deepak Kumar Department of Plant Sciences, Central University of Jammu, Jammu, Jammu & Kashmir, India

Pratibha Kumari Leibniz Institute of Plant Biochemistry, Martin Luther University, Halle (Saale), Germany

Sumita Kumari School of Biotechnology, Sher-e-Kashmir University of Agricultural Sciences and Technology, Jammu, India

Reetika Mahajan School of Biotechnology, Sher-e-Kashmir University of Agricultural Sciences & Technology of Jammu, Jammu, Jammu & Kashmir, India

Jubayer Al Mahmud Department of Agroforestry and Environmental Science, Faculty of Agriculture, Sher-e-Bangla Agricultural University, Dhaka, Bangladesh

S. J. Mirajkar Department of Agricultural Botany, Dr. Panjabrao Deshmukh Krishi Vidyapith, Akola, Maharashtra, India

Ananda Mustafiz Laboratory of Plant Molecular Biology, Faculty of Life Sciences and Biotechnology, South Asian University, New Delhi, India

Preeti Nagar Laboratory of Plant Molecular Biology, Faculty of Life Sciences and Biotechnology, South Asian University, New Delhi, India

Kamrun Nahar Department of Agricultural Botany, Faculty of Agriculture, Sher-e-Bangla Agricultural University, Dhaka, Bangladesh

Muslima Nazir Division of Plant Biotechnology, Sher-e-Kashmir University of Agricultural Sciences & Technology of Kashmir, Srinagar, Jammu & Kashmir, India

G. C. Nikalje Department of Botany, R. K. Talreja College of Arts, Science and Commerce, Ulhasnagar, Thane, Maharashtra, India

Nuclear Agriculture and Biotechnology Division, Bhabha Atomic Research Centre, Mumbai, Maharashtra, India

T. D. Nikam Department of Botany, Savitribai Phule Pune University, Pune, India

Sheezan Rasool Division of Plant Biotechnology, Sher-e-Kashmir University of Agricultural Sciences & Technology of Kashmir, Srinagar, Jammu & Kashmir, India

Anjana Rustagi Department of Botany, Gargi College, University of Delhi, New Delhi, India

Pawan Saini Central Silk Board (CSB), Central Sericultural Research &Training Institute (CSR&TI), Pampore, Jammu & Kashmir, India

Devyani Sharma Laboratory of Plant Molecular Biology, Faculty of Life Sciences and Biotechnology, South Asian University, New Delhi, India

Shweta Sharma Department of Botany, Gargi College, University of Delhi, New Delhi, India

Anil Kumar Singh ICAR-Indian Institute of Agricultural Biotechnology, PDU Campus, IINRG, Ranchi, Jharkhand, India

Charan Singh Indian Council of Agricultural Research (ICAR), Indian Institute of Wheat & Barley Research (IIWBR), Karnal, Haryana, India

Swati Singh Department of Botany, University of Delhi, Delhi, India

Jebi Sudan School of Biotechnology, Sher-e-Kashmir University of Agricultural Sciences and Technology, Jammu, India

P. Suprasanna Nuclear Agriculture and Biotechnology Division, Bhabha Atomic Research Centre, Mumbai, India

M. Tofazzal Islam Department of Biotechnology, Bangabandhu Sheikh Mujibur Rahman Agricultural University, Gazipur, Bangladesh

Pushpika Udawat Department of Plant Molecular Biology, University of Delhi, New Delhi, India

Mohammed Shalim Uddin Stress Breeding Lab, Bangladesh Agricultural Research Institute, Gazipur, Bangladesh

Uneeb Urwat Division of Plant Biotechnology, Sher-e-Kashmir University of Agricultural Sciences & Technology of Kashmir, Srinagar, Jammu & Kashmir, India

M. Y. Zargar Directorate of Research, Sher-e-Kashmir University of Agricultural Sciences & Technology of Kashmir, Srinagar, India

Sajad Majeed Zargar Division of Plant Biotechnology, Sher-e-Kashmir University of Agricultural Sciences & Technology of Kashmir, Srinagar, Jammu & Kashmir, India

About the Editors

Sajad Majeed Zargar, PhD is currently an assistant professor at Sher-e-Kashmir University of Agricultural Sciences and Technology of Kashmir (SKUAST-Kashmir) in India. He was previously a visiting professor at the Nara Institute of Science and Technology, Japan. He has worked as an assistant professor at SKUAST-Jammu, Baba Ghulam Shah Badshah University, Rajouri (BGSB), in India. He has also worked as scientist at Advanta India Limited, Hyderabad, India, and TERI (The Energy and Resources Institute), New Delhi, India. Dr. Zargar is recipient of the CREST overseas fellowship from DBT and is also recipient of the Goho grant from the Govt. of Japan. He has received several awards for his work and research. He is also a member and representative of INPPO (International Plant Proteomics Organization). His editorial activities and scientific memberships include publishing research and review articles in international journals and as a reviewer. He has been affiliated with several internationally reputed journals and is also reviewer of reputed journals, *Frontiers in Plant Science*, *3 Biotech*, *Scientia Horticulturae*, *Methods in Ecology and Evolution*, *Australian Journal of Crop Science* and many others. Dr. Zargar has been invited to give many lectures at professional meetings and workshops and has received grants for research projects under his supervision.

Mohammad Yousuf Zargar, PhD is currently research director at Sher-e-Kashmir University of Agricultural Sciences and Technology of Kashmir (SKUAST-Kashmir) in India. He was previously dean of the Faculty of Agriculture, dean of the Faculty of Forestry and associate research director and associate dean of the Faculty of Agriculture, SKUAST-Kashmir. Prior to that, he has worked as professor-cum-chief scientist (microbiology) at SKUAST-Kashmir. Dr. Zargar has published more than 160 research papers in scientific journals of national and international repute. He has guided eight students for the doctoral programme and six for the master's programme. He has received many honours and awards for his contribution to research work. His major contribution has been on fermentation technology for mass multiplication of biofertilisers and biocontrol agents. Dr. Zargar has received appreciations for his product "Shalimar microbes", a consortium of microbes that decompose waste in a short period. Dr. Zargar has been actively pursuing research on cold-tolerant microbes for solid/liquid waste decomposition and nutrient mobilisation. He has also served as associate editor of several scientific journals. Presently, Dr. Zargar is guiding research programmes of the university in agriculture, animal husbandry and other allied sections.

"Omics": A Gateway Towards Abiotic Stress Tolerance

1

Sreshti Bagati, Reetika Mahajan, Muslima Nazir, Aejaz Ahmed Dar, and Sajad Majeed Zargar

Abstract

Abiotic stresses like temperature, drought, salinity, etc. bring about severe changes in the growth and development of plants which lead to excessive amassing of secondary metabolites. These environmental stresses pose as a major constraint to the yield and quality of crop plants on field. Abiotic stress response in plants is an intricate process involving multitude of stress-related genes. At the molecular level, plants respond to abiotic stresses by altering the transcriptional activity of the stress response-related genes. Activation of the stress-inducible genes initiates enzyme synthesis responsible for the synthesis of various osmoprotectants. Although many efforts have been made to elucidate the mechanism of adaptation of plants to stressful conditions, yet the utilization of the acquired knowledge for improving crop performance and productivity under abiotic stress hasn't been achieved because of the intricate genetics involved in stress tolerance mechanisms. Over the last few decades, a number of diversified and promising "omics" technologies have emerged. These "omics"-based approaches have proved persuasive for elucidating how the modifications in the DNA, RNA, proteins and the metabolites within a plant are responsible for instigating response towards abiotic stresses.

This chapter will provide an insight towards various omics technologies and their applications for the development of varieties with enhanced stress tolerance.

S. Bagati · R. Mahajan (✉) · A. A. Dar
School of Biotechnology, Sher-e-Kashmir University of Agricultural Sciences & Technology of Jammu, Jammu, Jammu & Kashmir, India
e-mail: pisces418@gmail.com

M. Nazir · S. M. Zargar (✉)
Division of Plant Biotechnology, Sher-e-Kashmir University of Agricultural Sciences & Technology of Kashmir, Srinagar, Jammu & Kashmir, India
e-mail: smzargar@gmail.com

S. M. Zargar, M. Y. Zargar (eds.), *Abiotic Stress-Mediated Sensing and Signaling in Plants: An Omics Perspective*, https://doi.org/10.1007/978-981-10-7479-0_1

Although, it has been anticipated that significant potential prevails in the "omics" technologies towards the development of genetically improved crop plants, the integration of bioinformatics with the omics approaches will provide a foundation for attaining further in-depth information about how the plants response to environmental stresses.

Keywords
Abiotic stress · Omics · RNA · Protein · DNA · Metabolites

1.1 Introduction

Stress is any stimulus that surpasses the usual range of homeostatic regulation in any living being (Fraire-Velazquez 2011). Abiotic stresses (water deficit, high temperature, low temperature and high salinity) pose a serious threat to the food security worldwide. These major stresses have a negative influence on the plant's survival, thereby reducing the amount of biomass and yield by up to 50–70% (Bray et al. 2000; Kaur et al. 2008; Thakur et al. 2010). The exposure of a plant to a stress level above the threshold leads to the activation of a cascade of responses at physiological, biochemical, morphological and molecular levels which helps it to withstand the stress. In plants, stress tolerance is a quantitative trait regulated by assorted genes which makes it strenuous to unravel and comprehend the molecular mechanisms and various complex signalling pathways involved in activation and deactivation of stress responses (Collins et al. 2008; Chawla et al. 2011). Initially traditional breeding methods with marginal success were utilized to exploit the natural genetic variation within the germplasm to improve the tolerance to abiotic stresses (Flowers and Yeo 1995). Keeping in mind the need to develop genetically improved crop plants with abiotic stress tolerance, an array of "omics" approaches are emerging rapidly. These approaches, viz. genomics, proteomics, transcriptomics and metabolomics, are known to be the four axes of plant system biology which enable the scientists to decipher the complexity of plant stress responses (Yuan et al. 2008; Chawla et al. 2011). Genomics involves the study of the genome of an organism, transcriptomics explains the organization and functions of the sense and the nonsense RNA or transcriptome, proteomics deals with structural and functional analysis of protein as well as elucidates the regulatory pathway involved in post-translational protein modification and metabolomics acts as a potent contrivance for the analysis of various metabolites; further a unified analysis may prove to be a competent means for elucidating the intricate networks underlying abiotic stress tolerance. These omics approaches brought in a paradigm shift in research related to the behaviour of plants towards various environmental disturbances and paved a way for better understanding of the various aspects related to abiotic stress tolerance. Since plants alter their "omics" profiles to combat the changing environment, hence integration of phenotypic, genetic, transcriptomic, proteomic and metabolomic approaches that complement each other will help in identification of stress tolerance genes, their underlying mechanisms that will be utilized to select and develop stress tolerant and high-yielding crop plants (Takeda and Matsuoka 2008). The chapter aims at providing a detailed comprehension of the various available omics technologies that play a promising role in tailoring genotypes with improved performance under various abiotic stresses in near future for the sustainability of crop production.

1.2 Abiotic Stress: Causes, Physiological Responses and Effects on Plant Biochemistry

Abiotic stress in a broader sense comprises of multiple stresses such as heat, cold, excessive light, drought, waterlogging, UV-B radiation, osmotic shock and salinity (Figs. 1.1 and 1.2). All these stresses dramatically affect the plants' growth and metabolism leading to loss of yield. Plants being sessile organisms give rise to stress signals in response to the adverse changes in their surrounding environment which initiate a cascade of events/responses involving changes in plant's morphology, physiology and metabolism (Hasegawa et al. 2000; Hazen et al. 2003; Shao et al. 2007; Agrawal et al. 2010) in order to combat as well as adapt to the stress situation by maintaining homeostasis (Hazen et al. 2003). Based on their response to the stressful environment, plants have been classified into two groups: (a) glycophytes

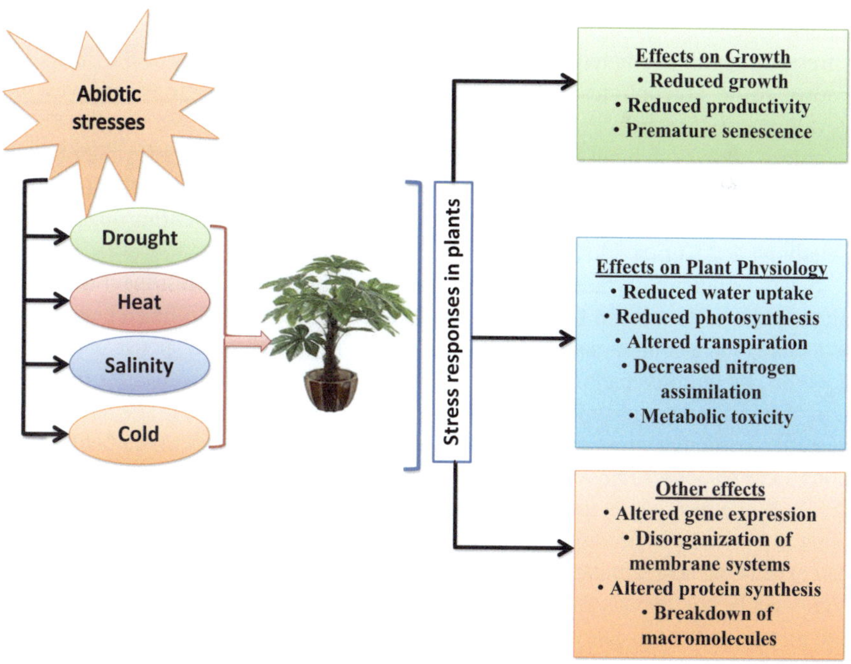

Fig. 1.1 Various abiotic stresses and stress responses in plants

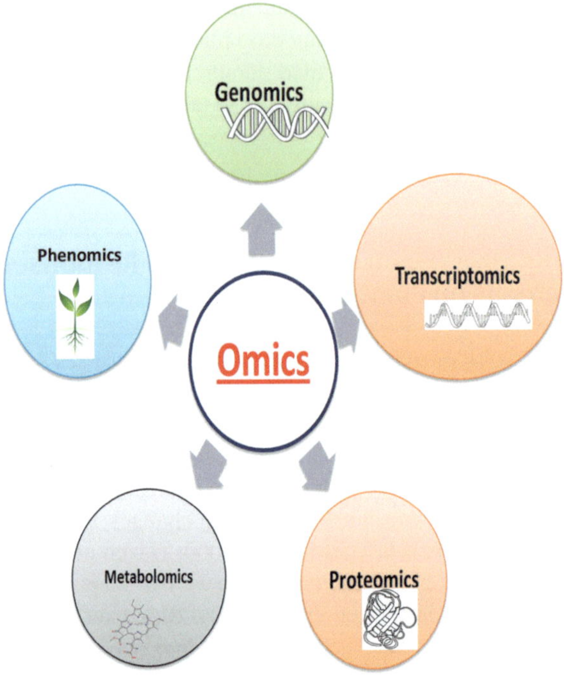

Fig. 1.2 Various omics approaches for studying abiotic stress responses in plants

(stress-susceptible) and halophytes (stress-tolerant). Since majority of the plants are known to be glycophytes, their survival under adverse conditions involves the development of tolerance, avoidance or resistance mechanisms. The mechanism of tolerance allows the plant to develop tolerance for any stress and helps the plant to overcome the stress without any injury, whereas the mechanism of avoidance prevents it from getting exposed to stressful conditions, thereby preventing any damage (Madlung and Comai 2004).

Several studies have been conducted to decipher how abiotic stresses influence the plant systems. It has been found that stresses are interrelated and affect a plant's system biology in a similar manner (Racz et al. 2008). Abiotic stress response in plants is governed by multiple factors like stage of growth at which the stress occurs, the duration of stress and the genotype, making it very intricate (Blum 1996). The response of plant to a particular stress may vary from lab to field conditions, as under field conditions these responses may be altered by the concurrence of other stresses (Misra et al. 2002). While responding to abiotic stress, a plant goes through a number of modifications in various physiological processes such as stomatal opening, photosynthesis, cell expansion, up-regulation of antioxidants and accumulation of organic solutes like amino acids, polyamines and carbohydrates. Exposure to any kind of stress limits the cell growth and expansion which is determined by the flexibility of the cell membrane and the maintenance of turgor pressure (Stepien and Klobus 2006). Many a times a plant is unable to accustom itself to the hostile conditions hence become susceptible and falls prey to them (Wang et al. 2003). The higher the amount of elasticity exhibited by a plant cell wall, the more competently it can deal with the adverse environment. Thus, maintenance of the cell wall elasticity becomes a prerequisite for proper growth and elongation of cells (Vincour and Altman 2005).

Relative water content, one of the most important physiological parameters for stress tolerance in plants, is defined as water content of a given amount of leaf relative to its fully turgid state. During dehydration, plants prevent water loss and control the turgor pressure by regulating the processes like transpiration and transportation (Buckley et al. 2003). Due to the decrease in the amount of CO_2 available, reduced stomatal conductance is observed during water stress (Flexas et al. 2004). Several studies have reported that the root/shoot ratio of plants increases during drought stress because roots in comparison to shoots are less sensitive to such stress (Wu and Cosgrove 2000). When the amount of water available for the plant is limited, roots instigate the shoots via xylem by generating a signal cascade so that the plant is able to adapt to the stress. Thus, a pervasive root system may prove helpful for stress tolerance in plants. The signal cascade is governed by several factors including the growth hormones like abscisic acid (ABA), cytokinins, ethylene and malate, etc. Increased amount of ABA levels results in the outflow of K^+ ions from the guard cells, leading to reduced turgor pressure and stomata closure (Guerrero and Mullet 1986). Photosynthesis which is an important process in plants is also extremely influenced by such stresses leading to reduction in plant growth and annual yield. Different components of photosynthetic apparatus are directly affected by abiotic stress, leading to the reduction in the photosynthetic efficiency of the plant which further affects the amount of CO_2 and stomatal conductance. The decreased amount of CO_2 leads to increased activity of RUBISCO enzyme during photorespiration leading to higher salt accumulation and the decreased stomatal conductance (Menconi et al. 1995). Environmental stresses hamper several important processes like the electron transport system, the carbon reduction cycle and the stomatal control of the CO_2 supply, resulting in enhanced amassing of carbohydrates, peroxidative disintegration of lipids and disrupted water balance (Allen and Ort 2001; Sudhir and Murthy 2004). The prolonged exposure of crop plants to water deficit conditions results in impaired gaseous exchange leading to reduced leaf size, early leaf death, oxidation of chloroplast lipids and alterations in structural conformation of pigments and proteins. During stress, the older leaves droop earlier as they are the most affected in comparison to the new and actively growing leaves (Parida and Das 2005). Several other photosynthetic mechanisms that are influenced by abiotic stress are the "non-stomatal" processes that include alterations in the synthesis of chlorophyll, changes in the structure and functions of chloroplasts and disrupted stacking, transit and circulation of assimilates. Magnesium ion forms an integral part of the structure of chlorophyll and acts as the enzyme cofactor. When exposed to stress, the plants' deficiency in magnesium ion has shown more degradation of chlorophyll due to increased oxidation of RUBISCO which contributes to decreased photosynthesis. In plants like *Arabidopsis*, salinity stress has been reported to have a negative effect on the amount of chlorophyll in the stress-susceptible species (Stepien and Johnson 2009).

Under abiotic stresses, for lowering the osmotic potential, amassment of various inorganic and organic solutes in the cytosol of plant cells has been reported (Rhodes and Samaras 1994). Among all the solutes, proline occupies a premier place with respect to stress tolerance. Proline accumulation marks the onset of the plants' defence mechanism against stress. Proline, a major signalling molecule known to alter the mitochondrial activity, affects the mechanisms of cell proliferation and cell death, instigates the expression of the stress tolerance genes, preserves the quaternary structure of complex proteins by checking the protein salvation, helps in keeping the cell membrane intact by preventing the oxidation of membrane lipid molecules, stabilizes the subcellular organelles and maintains the cellular potential by scavenging free radicals so that the plant can guard itself from any injury and is able to recover easily (Ashraf and Foolad 2007). A strong positive correlation exists between proline accumulation and stress tolerance wherein stress-tolerant plants have higher amounts of accumulated proline as compared to stress-sensitive plants.

Formation of reactive oxygen species (ROS) responsible for the burst of oxidative stress is the foremost biochemical response of plants towards all the abiotic stresses. ROS which comprise of free radicals, peroxides and oxygen ions exist as secondary messengers which activate the subsequent defence reactions in plants undergoing stress. Increase in the amount of ROS in the plants is detrimental as it leads to increased lipid peroxidation, protein disintegration, DNA fragmentation and ultimately cell death (Anjum et al. 2011). Therefore, it becomes important for the plant to maintain the balance between the production and the destruction of the ROS molecules during abiotic stresses for maintaining its proper growth and metabolism (Apel and Hirt 2004). For this purpose, plants have inbuilt antioxidant enzyme-catalysed defence machinery which consists of the enzymes like ascorbate peroxidase (APX), superoxide dismutase (SOD) and peroxidase and catalase (CAT). Apart from the ROS scavenging enzymes which help the plant in the protective clean up, there are certain non-enzymatic antioxidant molecules like glutathione, ascorbate and carotenoids which in cooperation with enzymatic components help in maintaining the intactness of the photosynthetic membranes under oxidative stress (Horváth et al. 2007). Destruction of ROS by the scavenging enzymes occurs step by step like the enzyme SOD leads to dismutation of O_2^- present in the chloroplast, mitochondria, cytoplasm and peroxisome of plant cells to H_2O_2, and then H_2O_2 generated is scavenged by the enzyme peroxidase, whereas the enzyme CAT plays an important role in eliminating H_2O_2 from mitochondria and micro-body and thus helps in mitigating the detrimental effects of oxidative stress (Shigeoka et al. 2002). Thus, in order to minimize the damage caused by ROS during abiotic stresses, a combined activity of all the antioxidant enzymes becomes obligatory.

1.3 Gene "Omics": Methods to Address Abiotic Stress

The shortcomings of the conventional breeding approaches such as the dependence on the available variation in germplasm, labour intensiveness and its time-consuming nature restrained its use for development of abiotic stress-tolerant varieties (Ashraf et al. 2008; Zhang et al. 2008). It has been found that the abiotic stress tolerance mechanisms in plants comprise of various elementary processes involving the water transporters, cell signalling modules (heat-shock proteins, specific transcription factors, molecular chaperones and late embryogenesis proteins), reactive oxygen species, osmolyte adjustment and ion accumulation. All these processes in plants are regulated by the interactions of a number of genes at the molecular level (Flowers 2004). The candidate genes having a significant role in organization of tolerance mechanisms provide an opportunity for carrying out research in important crop plants. For successful implication of breeding programmes for the generation of varieties with a wide range of adaptability, consolidated use of different aspects such as plant and cell physiology, molecular biology and genetics has been found to be persuasive (Rashid et al. 2014). For that reason, genome-based studies using genomics tools and techniques have been recommended as the promising approach to uphold the work of crop improvement and for having comprehensive knowledge of the changes in the regulatory gene expression, protein synthesis as well as the changes occurring in the production of the metabolites as soon as the plant encounters any abiotic stress. Hence, uncovering the candidate genes and the signalling pathways involved in stress tolerance is a must for the development of strategies for generation of genetically improved crops.

Functional genomics, an integral part of genomics approaches, has immensely contributed towards the recognition of candidate genes for abiotic stress imperviousness. Over the years, the advancements in the field of functional genomics have made it possible for the researchers to study the genome sequences of certain model crops. Recent advances in the field of genomics and bioinformatics have led to the development of efficacious, high-throughput methods as well as genetic database resources for genome-wide screening of stress-associated genes and identification of the stress-resistant gene families across species based on similarity (Gambino and Gribaudo 2012). Apart from the sequenced genomes, the development and availability of expressed sequence tags (ESTs) and cDNA sequences have made it possible to gather information regarding the stress-responsive genes from the plants which are yet to be sequenced (Marques et al. 2009). Using the genomics approaches, the modifications in the behaviour of candidate genes such as over- or under-expression have become possible resulting in effective response of crop plants towards stress. Such studies will make it easy to understand various molecular and metabolic pathways responsible for plants adaptation to abiotic stresses (Arpat et al. 2004; Micheletto et al. 2007).

Gene tagging using molecular marker technology, a promising genomics approach, has outperformed the conventional method of candidate gene recognition using mutagenic agents (both physical and chemical) for functional investigation on a wider range (Lukowitz et al. 2000). Stress response and tolerance in plants are polygenic and quantitative in nature. Recent developments in the field of genomics brought in a valuable tool known as "QTL mapping" which allowed the researchers to dissect the genetic makeup of crop plants as well as the pattern of inheritance of the traits controlled by numerous genes performing simultaneously, known as the quantitative trait loci (QTL). Further, improvements in marker technology and QTL mapping have led to the unfolding of the novel and effective breeding strategies such as marker-assisted selection (MAS) and breeding by design (Peleman and Voort 2003). Such improvements in the breeding methods created a scope for elucidating the complexity of abiotic plant stress responses and prompted the researchers to develop a new outlook towards development of crop plants with increased tolerance and improved productivity.

Preliminary approaches using the available genome sequences involved ample amount of human and financial resources which instigated the technologists to develop ultrahigh-throughput next-generation sequencing (NGS) technologies which remarkably raised the pace at which the sequencing of plant genomes was carried out. NGS technologies when combined with genome-wide association studies (GWAS) have brought in enormous opportunities in genome-wide identification of stress-related genes by making it easier to identify the promising molecular markers associated with stress response in plants and by aiding the development of comparative genomics which allows to unravel the existing diversity within or across species (Ma et al. 2012). NGS technology when united with high-throughput transcriptome profiling has proved to be a sturdy tool for studying the changes occurring in the transcripts at the genome-wide level during stress response in plants (Molina et al. 2011). In order to decipher the complex mechanisms involved in stress signalling and plant adaptation mechanisms, one major requirement is to analyse the mode of action of diverse genes responsible for stress response. Apart from the genes, the mitogen-activated protein kinase (MAPK) class plays a significant part in signalling mechanisms during various abiotic stresses (Pitzschke et al. 2009). The recent developments in the field of crop genomics have paved up new opportunities towards the development of improved stress tolerance and increased productivity in plants. These technologies have proved to be immensely beneficial for unravelling the role of genome sequences, stress-related transcript assemblies, potent changes at the proteome and metabolome levels, protein-protein interactions and mutant screening and selection in abiotic stress tolerance in plants (Perez-Clemente et al. 2012).

1.3.1 Structural Genomics Tools for Abiotic Stress Tolerance in Plants

Structural genomics marked the beginning of the genomics research and provided the ground for the transition into the era of functional genomics. Structural genomics comprise of genome sequencing, mapping and cloning of the traits of interest for the plant biologists.

1.3.1.1 Genome Sequencing

Deciphering the location of various genes in an organism marks the beginning towards explaining the mode of action of living organisms towards environmental stress. The presence of varied genome size, number of genes and huge amount of repetitive sequences and polyploidization events in plants makes it difficult to bring to light the precise sequence and location of the genes (Green 2001). The inability of the EST sequencing/databases to show low-abundance transcripts or transcripts during abiotic stresses forced the researchers to develop and adapt more powerful approaches like whole-genome sequencing. Sequencing of whole genome can be carried out via two approaches: whole-genome shotgun (WGS) and clone-by-clone sequencing. For retrieval of the necessary information regarding genome function and organization along with the complete collection of genes within an organism which is useful for positional cloning strategy, WGS strategy is expected to be theoretically superior than the clone-based strategy which is less complex in context to assembling the genome (Subudhi 2011). The most ponderous task during genome sequencing is the annotation of the genome sequences as it becomes difficult to differentiate between the coding and the non-coding sequences (introns). With the advent of NGS technologies, genome sequencing has turned out to be speedy, effective, less laborious and cost-effective by many folds. At present, a number of high-throughput sequencing platforms like Roche 454, Illumina, Pac-bio, ABI SOLiD, Helicos, Ion Torrent and Oxford Nanopore are available. A sequenced plant genome acts as a reference genome which is used for genome annotation as well as for pinpointing the genetic discrepancies for huge number of sequences within a definite period of time (Akpjnar et al. 2013). In India and across the globe, approximately more than 30 plants have been subjected to whole-genome sequencing in order to generate ample amounts of genomic resources for crop improvement and food security. Till date, more than 30 plants have already been sequenced using whole-genome de novo sequencing (http://genomevolution.org). The Beijing Genome Institute of China has undertaken "The Million Plant and Animal Genomes Project" in collaboration with scientists worldwide to sequence the genome of thousands of economically important plant/animal species which will result in the production of vast amount of genomic resources and information which will speed up the process of developing improved varieties to ensure food security and lead to improved ecological conservation as well as development of new energy sources (www.genomics.cn). In India, NGS platforms have been successfully used to sequence the pigeon pea and chickpea genomes recently (Mir et al. 2013).

Like NGS, whole-genome resequencing (WGS) is another approach that is practically suitable for the development of useful genomic resources and information even in the presence of rare alleles generated during evaluation of biparental mapping populations for linkage disequilibrium (LD) and genetic relationships between accessions (Cosart et al. 2011; Schuenemann et al. 2011). One best example of whole-genome resequencing efforts is 1001 Genomes Project (2008), the largest resequencing project which aimed at uncovering the sequence variations at genome-wide level in 1001 accessions of *Arabidopsis thaliana* (Lister and Ecker 2009; Cao et al. 2011). Apart from *Arabidopsis thaliana*, a number of whole-genome resequencing projects are being carried out in various crop varieties, like rice and maize (Lai et al. 2010; Huang et al. 2013). The whole-genome resequencing makes it possible to detect a large number of both small- and large-scale variations including insertions and deletions present within the genome and helps in determining the after effects of these on the gene functions and their linkage pattern. Table 1.1 enlists the abiotic stress-related genomic resources identified through high-throughput sequencing.

Table 1.1 Abiotic stress-related genomic resources identified through high-throughput genome sequencing platforms

Plant	Genome size (Mb)	Sequencing platform used	Information revealed	References
Chickpea (*Cicer arietinum* L.)	740	Roche 454	880 genes were up-regulated by drought stress	Molina et al. (2008)
Rice (*Oryza sativa* L.)	489	Illumina	213 (shoot) and 436 (root) transcript tags were differentially expressed under salinity stress	Mizuno et al. (2010)
Soybean (*Glycine max*)	975	Illumina	3231 genes related to nitrogen use efficiency were identified	Hao et al. (2004)
Chickpea (*Cicer. arietinum* L.)	9740	Roche 454	363 and 106 specific transcripts, respectively, were up- or downregulated under salinity stress	Molina et al. (2011)
Common bean (*Phaseolus vulgaris* L.)	587	Roche 454	611 up- and 728 downregulated genes in PEG-treated root tips were identified	Yang et al. (2011b)
Cucumber (*Cucumis sativus* L.)	880	Illumina	5787 genes were differentially expressed under waterlogged condition	Qi et al. (2012)
Rice (*Oryza. sativa* L.)	489	Illumina	18,833 unigenes were identified; of these, 40 were highly up-regulated under atrazine stress condition	Zhang et al. (2012)
Sugarcane (*Saccharum* spp.)	3961	Illumina	75,404 unigenes were identified. Of these, 213 were up-regulated under drought stress	Kido et al. (2012)

1.3.1.2 Molecular/Genetic/QTL Mapping

The pattern of inheritance of abiotic stress tolerance traits in plants is complex because of their multigenic nature. The recent developments in the field of marker technology and precision mapping methods have made it possible to access and utilize the naturally existing variation for complex quantitative traits like abiotic stress tolerance in plant species for crop improvement (Doerge 2002). The mapping populations like immortal recombinant inbred lines (RILs) and introgression lines along with advanced backcross approach tend to be a reliable resource for studying the effects of abiotic stresses on the phenotype in diversified environments with multiple replications and are used for successful transfer of abiotic stress tolerance genes from the wild varieties to the cultivated ones, respectively (Zamir 2001).

For alteration of plant's behaviour during stress, analysis of the interrelationship between molecular markers and phenotypic studies of plants provides access to the genetic variations at both qualitative and quantitative levels that nurture a plant's response under abiotic stress. QTL mapping is one important strategic approach which apart from allowing the determination of the location of the genes/QTLs for multigenic traits on a linkage map helps in evaluating the individual and cumulative effects of QTLs (Jansen and Nap 2001). Hence, QTL mapping is an exciting approach that makes use of available markers such as RFLP, AFLP, RAPD, SSR and SNP and the phenotypic information for determination of factors responsible for improved stress tolerance in plants through marker-assisted selection, map-based/positional cloning and gene pyramiding (Salvi and Tuberosa 2005; Ashraf et al. 2008). Genetic genomics, a new approach, reveals the relationship amid the genome and transcriptome using QTL mapping (Hansen et al. 2007). QTL mapping when combined with gene expression data allows the mapping of QTLs regulating the amount of transcript of each gene (eQTLs) which helps in elucidating the genetics underlying the gene expression during stress leading to a better insight of the transcriptional regulation which can be used for breeding and selection of high yielding and elite cultivars of crops (Jansen and Nap 2001; Langridge et al. 2006). Until recently, a number of QTLs associated with tolerance to various abiotic stresses, viz. high salinity, drought and high temperatures, have been recognized and used in crop improvement (http://www.gramene.org/qtl/). It is clear that detection and introduction of valuable QTLs is of immense importance to crop improvement as it remarkably contributes to the natural variation during abiotic stress response. It has been found that along with natural allelic variations, mutant populations with induced mutations are enormously promising to reveal the variability lurking the abiotic stress tolerance in plants by employing techniques like TILLING (targeting induced local lesions in genome) (Till et al. 2004, 2007; Cooper et al. 2008; Suzuki et al. 2008).

1.3.1.3 Map-Based Cloning

Another promising approach for cloning the QTLs for abiotic stress tolerance in plants is positional gene cloning. This strategy helps in developing a deep insight towards the regulation of abiotic stress tolerance at the molecular level and involves the use of fine mapping succeeded by physical mapping of the markers tightly linked to the mutant or natural allele of interest. This QTL cloning approach has profound relevance for MAS, genetic engineering and EcOTILLING and has been successfully applied for the traits with large additive effects and higher heritability (Salathia et al. 2007). Due to the development and availability of PCR-based markers and saturated molecular maps of various crops, the efficacy and the time intensiveness of the gene cloning approach have increased and decreased, respectively, making it popular among the researchers working on abiotic stress tolerance (Papdi et al. 2009). Apart from fine mapping, sequencing strategies are used for identifying the target genes/regions/mutations in both wild and mutant type (Jander 2006). For speeding up the research in the area of fine mapping and map-based cloning, several high-throughput marker technologies such as simple sequence repeats, indel polymorphism, single-nucleotide polymorphisms and array-based genotyping procedures are being developed (Salathia et al. 2007; Rios et al. 2008).

Along with whole-genome approach, candidate gene approach can serve as an alternate strategy for map-based QTL cloning wherein appropriate information is available about the role of genes/QTLs involved in a biochemical pathway or the

Table 1.2 List of QTLs associated with various abiotic stresses identified in different crops

Crop	QTL identified and relevant abiotic stress	References
Pearl millet	Osmotic potential, water-related attributes, cell Membrane stability	Serraj et al. (2004)
Arabidopsis	*ALMT1*/aluminium tolerance	Hoekenga et al. (2006)
Rice (*Oryza sativa* L.)	*Sub1A*, *Sub1B* and *Sub1C* on chromosome 9/ submergence tolerance	Xu et al. (2006)
Maize (*Zea mays* L.)	Five QTLs located on chromosomes *1, 2, 3, 8 and 10*/drought tolerance	Ribaut and Ragot (2006)
Wheat (*Triticum aestivum* L.)	*SKC1*/salt tolerance	Byrt et al. (2007)
Barley (*Hordeum vulgare* L.)	*Alp*/aluminium tolerance	Furukawa et al. (2007)
Rice (*Oryza sativa* L.)	*QTL9* (on chromosome 9)/higher biomass yield under drought stress	Steele et al. (2007)
Sorghum [*Sorghum bicolor* (L.) Moench]	*Stg1, Stg2, Stg3 or Stg4*/delayed leaf senescence and better grain yield at maturity under drought stress conditions	Harris et al. (2007)
Wheat (*Triticum aestivum* L.)	*C-repeat-binding factor (CBF) genes*/frost tolerance	Knox et al. (2008)
Maize (*Zea mays* L.)	Ionic balance, osmotic adjustment	Feng-Ling et al. (2008)
Cotton (*Gossypium* spp.)	*qtl12.1*(linkage group 12)/improved biomass production; panicle number under drought stress conditions	Levi et al. (2009)
Rice (*Oryza sativa* L.)	*QTL2* (on chromosome 2), *QTL9* (on chromosome 9), *QTL11*(on chromosome 11) *and QTL12.1* (on chromosome 12)/drought tolerance	Steele (2009)

genes which control the similar trait in other plant species such as the aluminium tolerance (*ALMT1*), or *HKT1* genes of *Arabidopsis* are similar to those controlling these traits in cereals (Hoekenga et al. 2006, Rus et al. 2006). Much precision of potential genes for abiotic stress tolerance can be achieved by carrying out expression profiling of the genes located in the QTL region of interest (Gorantla et al. 2005). Table 1.2 enlists the applications of QTL-based map-based cloning for abiotic stress tolerance in plants.

1.3.1.4 Association Mapping/Genome-Wide Association Studies (GWAS)

Until recently, classical linkage mapping developed using biparental population has been successfully utilized for identification of major stress-related genes and mapping of QTLs (Frary et al. 2000, Komatsuda et al. 2007). The several shortcomings of this so far useful method lead to the advent of association mapping (AM) or genome-wide association studies (GWAS) which served as a promising substitute to classical biparental-based mapping technique (Gupta et al. 2005; Hall and Ingvarsson 2010; Maccaferri et al. 2011). AM and GWAS efficiently unravel the genetics underlying intricate traits such as stress tolerance from many descendants (Abdurakhmonov and Abdukarimov 2008; Zhao et al. 2011). Association mapping works on the principle of linkage disequilibrium which exploits the association between the molecular and the phenotypic variation in a diverse population using the historical recombination events (Moragues et al. 2010; Cosart et al. 2011; Schuenemann et al. 2011). This approach has been used for mapping the abiotic stress tolerance traits in forest tree species (Gonzalez-Martinez et al. 2008; Eckert et al. 2009) as well as for obtaining the haplotypes at many genetic loci using sequencing or high-throughput SNP analysis (Thornsberry et al. 2001; Aranzana et al. 2005).

On the other hand, GWAS when carried out in diverse germplasm collections paves a way towards the discovery and recognition of new genes and alleles associated with abiotic stress tolerance in crop plants (Mackay et al. 2009; Hall and Ingvarsson 2010). The basic requirement for GWAS, i.e. scanning the genetic diversity within the population at genome-wide level and understanding the order of population structure and the decline of LD, can be achieved by using efficacious genotyping strategies including high-density maps, phenotyping resources and reference genome sequence of best quality (Rafalski 2010). The output obtained by GWAS needs to be validated using linkage analysis. In maize, the AM approach has been employed for identification of loci associated with carbohydrate and ABA accumulation during drought stress at flowering stage. Using this approach, it was concluded that among the 1229 SNPs in 540 candidate genes, one SNP in the maize homologue of the Arabidopsis MADS-box gene, PISTILLATA, was significantly related to phaseic acid in ears of well-watered plants, and one SNP of pyruvate dehydrogenase kinase which regulates carbon flux into respiration was found to be associated with silk sugar content in maize. Another SNP of aldehyde oxidase gene was significantly correlated with ABA contents in silks of the low water-stressed plants. Therefore, the three identified SNPs served as the most valuable genomics tools used for identification of drought-tolerant cultivars of maize (Setter et al. 2011). Various important QTLs involved in tolerance against abiotic stresses have been identified, mapped, cloned and incorporated into elite cultivars/genotypes, have been kept in records and are well maintained through web portals such as QlicRice: an online interface for abiotic stress-responsive QTLs in rice (Smita et al. 2011). Several modifications in the AM like nested association mapping, a combination of leverages of the biparental linkage analysis and association mapping, have also been successfully used in maize to carry out genome-wide scanning of QTLs for complex traits (Yu et al. 2008).

1.3.2 Functional Genomics Tools for Abiotic Stress Tolerance in Plants

Functional genomics involves the development of genome-wide experimental approaches which utilizes the information revealed by structural genomics for assessment of the gene function. Functional genomics deals with the approaches that are used to determine the function of genes through phenotypic evaluation of mutants. Several methods used for functional genomic studies include targeted gene deletions, insertional mutagenesis and RNA interference, overexpression, gene silencing, insertional mutagenesis and target induced local lesion in genome (TILLING). All these methods are of immense importance for studying the complex nature of regulatory networks related to abiotic stress response, acclimatization and stress tolerance mechanisms in crop species.

1.3.2.1 Targeted Gene Replacement by Homologous Recombination

This method is one robust and efficient tool for studying the gene function. Several success stories using this method have been reported in plants, but the shortcomings of this method like inefficiency due to low frequency of recombination events, existence of arbitrary integration events, non-homologous end joining and absence of specific gene selection system pose a need for the development of improved and efficient selection techniques to decipher the molecular mechanism underlying homologous recombination (Sutton et al. 2007; Heiter and Bogushi 1997; Kempin et al. 1997; Beetham et al. 1999; Terada et al. 2007).

1.3.2.2 Insertional Mutagenesis and Target Induced Local Lesion in Genome (TILLING)

Mutant population and mutation techniques are known as an important resource for breeding programmes and genomic studies for development and breeding of abiotic stress-tolerant lines (Henikoff and Comai 2003; Till et al. 2004). Usefulness and efficiency of a mutated population for genomics research depend upon the methods by which the mutation has been induced, i.e. irradiation (physical) or chemical methods (Comai and Henikoff 2006). Mutational genomics has emerged as a valuable strategy to understand the plant stress response at molecular level based on knowledge gathered from mutants of *Arabidopsis* and other studies carried on several model plants (Papdi et al. 2010). Emergence of high-throughput genomic platforms such as cDNA-amplified fragment length polymorphism (AFLP), single-strand conformational polymorphism (SSCP), serial analysis of gene expression (SAGE), microarray, differential display, targeting induced local lesions in genome (TILLING), high-resolution melt (HRM) analysis, etc. permits quick and comprehensive genome-wide analysis of mutational events.

The development of TILLING technology during the last decade has led to the revival of the utilization of chemically induced mutagenesis in genomics research. TILLING is a high-throughput, cost-effective, reverse genetics strategy used for identification of polymorphisms (specifically arising due to point mutations) in the genes of interest by heteroduplex analysis in a chemically mutagenized population (Colbert et al. 2001; Till et al. 2007). TILLING enables the researchers to identify the existing allelic variations prior to phenotyping and hence is predominantly used for identification and discovery of mutants and SNPs (Comai and Henikoff 2006; Cordeiro et al. 2006). The traditional approaches based on reverse genetics make use of transposons for knocking out a specific gene for decoding the accurate phenotype suffering from several drawbacks like requirement of time- and labour-intensive transgenic or sophisticated tissue culture techniques and inability to determine the consequences of partial loss of function of an active gene when a complete gene is striked out. The TILLING approach overcomes all these shortcomings by allowing the creation and detection of broad range of genetic modifications in not only loss-of-function alleles but also in hypomorphic, hypermorphic and neomorphic mutations and has been effectively used to develop gene-derived functional markers by targeting the genes of interest (Jain et al. 2010). Since the TILLING technique is independent of size of genome, reproductive system or propagation time and can be utilized in non-transformable species, it has been effectively used in *Arabidopsis* and other different plant species including rice (Sato et al. 2006; Suzuki et al. 2008), maize (Till et al. 2004), wheat (Slade et al. 2005), sorghum (Xin et al. 2008), barley (Caldwell et al. 2004), pea (Triques et al. 2007) and soybean (Cooper et al. 2008) and in animals as well. The TILLING mutants have also been used for determining the plant stress responses like for assessing salt stress response in legume species, and TILLING mutants for a specific kinase were used (Lorenzo et al. 2009). Like TILLING, EcoTILLING is a rapid, cost-effective, less time-consuming and efficient method to study the haplotypes by making use of a lesser number of sequenced accessions. The technique of EcoTILLING has been successfully used in domesticated rice to detect 15 and 23 representative SNPs of OsCPK17 and SalT gene, respectively, across 375 accessions (Negrao et al. 2013). In another study, the use of EcoTILLING has been carried out for detection of candidate genes and SNPs associated with drought stress tolerance in the lowland and upland rice cultivars and barley genotypes, respectively (Naredo Ma et al. 2009; Cseri et al. 2011).

Several different gene silencing techniques such as co-suppression (Tissier et al. 1999), virus-induced gene silencing (Bouche and Bouchez 2001), double-stranded RNA-mediated interference (Waterhouse et al. 1998; Baulcombe 1999; Meins 2000) and chimeric oligonucleotides (Chuang and Meyerowitz 2000) have also been used to study gene function used in plants.

1.3.2.3 RNA Interference

With the availability of the genome editing tools, introduction of modifications at targeted sites in the genome has become possible. These targeted modifications allow the researchers and the plant biologists to understand the functions of various genomic components in different plant species, thereby paving a way for the development of abiotic stress tolerance in crop plants. Genome editing tools such as zinc-finger nucleases (ZFNs), transcriptional activator-like effector nucleases (TALENs) and clustered regularly interspaced short palindromic repeat (CRISPR)-Cas9 (CRISPR-associated nuclease 9) allow the incorporation of targeted mutations, insertions/deletions and précised sequence modification incorporating customized nucleases in different organisms (Voytas 2013; Mahfouz et al. 2014; Kumar and Jain 2015).

The simplistic nature of CRISPR-Cas9 system obtained from a prokaryotic defence system guided by RNA paved a way for plenty of options in genome editing and has been recognized as the most preferred method for genome engineering

(Bhaya et al. 2011; Harrison et al. 2014; Hsu et al. 2014; Sander and Joung 2014). Although this system has not been used for engineering plants for the development of stress tolerance till date, it allows the researchers to manipulate/alter any sequence in the genome (where PAM site is available) to unravel its function. CRISPR-Cas9 system has proved to be a successful tool for efficient genome editing in bacteria, animals and plants (Feng et al. 2013; Jiang et al. 2013; Li et al. 2013a, b; Nekrasov et al. 2013; Shan et al. 2013). A web tool CRISPR-P for designing sgRNAs in more than 20 plant species has also been developed recently (Lei et al. 2014).

In order to carry out CRISPR-Cas9-mediated plant genome editing, a number of vectors and tools have been developed (Xing et al. 2014; Kumar and Jain 2015). The resources like Addgene (https://www.addgene.org/crispr/plant/), a non-profit plasmid data repository, permit the use of CRISPR-Cas9 system in various applications (editing, transcriptional modulation and genetic screens) to dissect the molecular basis of abiotic stress response and generate stress-tolerant crop plants. Carrying out the multiplex genome editing enables the scientists to unravel the role and function of various genes involved in the same regulatory processes like abiotic stress responses (Li et al. 2013a; Mao et al. 2013; Zhou et al. 2014). Another way out for development of abiotic stress tolerance in plants is to identify and target the genes involved in various stress-associated gene regulatory networks, signal transduction and metabolite production pathways which can be later pyramided or stacked via HDR-mediated gene targeting using CRISPR-Cas9 technologies.

1.3.3 Comparative Genomics Tools for Abiotic Stress Tolerance in Plants

The availability of sequenced plant genomes, expression data and stress-related cDNA libraries has made the discovery of stress-related genes and pathways via comparative genomics easy. It is now possible to transfer the genes of interest/gene annotations from model crop species to the newly sequenced crops which have not been studied properly. The basic requirement for comparative genomic studies is the availability of the orthologous data sets having a common ancestor (Tran and Mochida 2010a). The stress-associated transcription factors (TFs), from orthologs of different plant species, have similar sequences and expression patterns which make it possible to identify the orthologous genes having the same functions in the crop species whose functional analysis is at a rudimentary level. Comparative genomics has been successfully applied to predict the stress-responsive TFs in soybean, maize, sorghum, barley and wheat using the known stress-responsive TFs in *Arabidopsis* and rice (Mochida et al. 2009, 2011; Tran and Mochida 2010b). Apart from identification of TFs, comparative genomics approach allows the analysis of the expression profiles' stress-associated transcripts of newly identified crop species for detection of stress-related genes by comparing the gene expression profiles between various known and new species (Walia et al. 2009).

The similarities and the differences among the stress responses originating in various model and plant species identified by comparative studies of genomic databases aid the scientists and researchers to anticipate and comprehend the role and function of novel genes in newly sequenced species and provide a way out for detection of species-specific stress-responsive genes and regulatory mechanisms. Therefore, it has been concluded that the comparative genomic studies will widen the potential of development of stress-tolerant crop species by incorporating the necessary information from model plants.

1.4 Transcript "Omics": A Key to Understand Abiotic Stress Response in Plants

While undergoing any kind of stress, plants tend to activate various stress responses within them, and in order to understand how these stress responses provide tolerance against adverse conditions, deciphering how subsequent physiological, biochemical and molecular responses are activated through signal recognition and transduction becomes a prerequisite (Komatsu et al. 2009; Ge et al. 2010; Le et al. 2012). One promising approach towards understanding the abiotic stress responses in plants is to identify the candidate genes involved in various biological processes and stress regulatory networks via genome-wide expression profiling and to control the transcriptional activation or repression of stress-responsive genes via transcriptome profiling (Chen et al. 2002). Initially low-throughput technologies like Northern blotting (Dita et al. 2006) were used for the development of a comprehensive insight of a plant's transcriptome during stress. The inefficiency of these technologies to analyse the entire genes present within a plant and the advancements in the field of sequencing technology lead to the advent of several high-throughput techniques like expressed sequence tags (ESTs) sequencing (Clement et al. 2008), serial analysis of gene expression (SAGE) (Velculescu et al. 1995), massively parallel signature sequences (MPSS) (Brenner et al. 2000), differential display and cDNA-AFLP which utilize the nucleotide sequence information and fragment sizing for determination of the level of transcripts, respectively. Microarray technology allows the indirect assessment of gene expression using the principle of nucleic acid hybridization of mRNA or cDNA fragments (Schena et al. 1995; Lee et al. 2004; Koh et al. 2007).

Transcriptomic approaches have been extensively exploited for the upliftment of abiotic stress tolerance in plants. The next-generation sequencing (NGS) strategies like RNA-Seq for sRNAs have revolutionized the field of transcriptomics and have paved a way for the improvement of plant genomic resources.

1.4.1 Expressed Sequence Tags (ESTs)

ESTs have been recognized as the simplest method to reveal the sequence-related information of the plant species undergoing abiotic stresses (Rudd 2003). This technique makes use of the cDNA libraries having about 10,000 clones of the genes involved in plant stress tolerance mechanisms (Pariset et al. 2009). In certain cases where the complete information about the genome sequence is lacking, EST technology has been considered as the link to the genome.

In the recent years, EST technology has enabled the researchers to generate a huge amount of data that can be further used for studying the plant stress tolerance mechanisms. Approximately, 449,101 ESTs have been reported for drought stress. Three hundred twelve thousand three hundred fifty-three ESTs, 103,898 ESTs, 252,595 ESTs, 19,384 ESTs and 135,578 ESTs associated with salt, low temperature, high temperature, nutrient deficiency and light stresses, respectively, have also been identified and are available on the National Center for Biotechnology Information browser (http://www.ncbi.nlm.nih.gov/) (Reddy et al. 2012).

1.4.2 SAGE

Like EST sequencing, serial analysis of gene expression (SAGE) also makes use of the sequence information. SAGE is a high-throughput and cost-effective technique used for differential analysis of the expressed genes (Donson et al. 2002). The technique involves mRNA extraction, cloning and sequencing. Specific tags are used to identify the relevant genes within the database, and the pattern of expression of differential genes is determined by the relative amount of the individual tags.

1.4.3 MPSS

A technology developed by Lynx Therapeutics Inc., California. massively parallel signature sequencing (MPSS), is a genome-wide transcriptional profiling approach which makes use of the cloning technique. The cDNA molecules are cloned onto micro-beads which are then sequenced for the generation of short cDNA tags. The ability of MPSS to generate ample amount of good quality data with effective data management makes it superior than SAGE in terms of speed and informativeness (Meyers et al. 2004).

1.4.4 Differential Display

It is another important method for differential studies of the cDNA fragments. This technique has been categorized by scientists as an effective approach for gene discovery (Velculescu et al. 1995). Certain limitations like the lack of reproducibility, lack of sensitivity and occurrence of false positives are some of the major constraints that limit the use of this technology in transcriptional profiling (Lee et al. 2004).

1.4.5 cDNA-AFLP

This technique has been introduced to overcome the limitations of the differential display technique and involves the use of adapters ligated to restriction fragments, specific primer sets for PCR amplification (Brenner et al. 2000). Several modifications of the cDNA-AFLP technology to increase its throughput have also been introduced such as the use of single restriction enzyme and the use of fluorescent labelling, multiplexing and capillary-based electrophoresis (Liang and Pardee 1992; Bachem et al. 1996). A combination of automation and cDNA-AFLP technique has been used for gene expression analysis [READS, (Cho et al. 2001); Gene Calling, (Breyne and Zabeau 2001); TOGA, (Prashar and Weissman 1996)].

1.4.6 DNA Microarrays

Based on the principle of Northern hybridization, DNA microarray technology is an enormously robust platform for monitoring the gene expression and functions and for elucidating the regulatory and biochemical pathways during various developmental stages as well as different environmental conditions (Subudhi 2011). Two types of DNA microarrays, i.e. cDNA arrays and oligoarrays, are commonly used. The difference between them is in cDNA arrays; robotics is used to immobilize the spotted cDNA fragments onto the slides, whereas in the case of oligoarrays, photolithographic mask is used to directly synthesize the oligonucleotides on a solid matrix (Shimkets et al. 1999; Sutcliffe et al. 2000). Oligoarrays have evolved as the most preferred technique in plant science as they are easier to print than cDNA clones and can be effectively used for SNP detection, identification of slice variant and gene families and do not require large-scale maintenance, PCR reactions as well as clone validation like cDNA microarrays (Schena et al. 1995).

Although the microarray technology is both powerful and pragmatic, certain limitations like time, labour intensiveness as well as its expensive nature and several technical issues arising due to contamination of DNA, uneven hybridizations and spurious hybridizations limit its use. Moreover, for the generation of reliable data, it is mandatory to have multiple biological and technical replications (Lockhart et al. 1996). Since a huge amount of data is generated in the microarray experiments, the statistical analysis and data interpretation becomes a challenging task while using DNA microarrays.

1.4.7 RNA-Seq

It is an advanced approach used for transcriptome profiling. Unlike microarray technology, RNA-Seq is a cost-effective and high-throughput technology which makes use of sequencing for analysis of transcriptomes. RNA-Seq technique is independent of the gene information and uses available genomic information for designing probes as well as has the potential to identify novel transcripts which makes it possible to study non-coding RNAs. Proteomic approaches have been used to carry out the comparative analysis of the accuracy of microarrays and the RNA-Seq which lead to the conclusion that RNA-Seq is a preferred technique for estimation of total expression levels (Fu et al. 2009).

Until recently, the RNA-Seq approach has been used for mapping the start site of transcription, carrying out strand-specific measurements, detecting gene fusions alternative splicing events, characterizing small RNA and investigating seven tissues and seven stages during seed development in soybean for developing an atlas of the soybean expression genes which will be a potential resource for understanding the expression patterns and functions of tissue-specific genes (Ozsolak and Milos 2010; Severin et al. 2010).

1.4.8 Combining QTL Mapping, GWAS and Transcriptome Profiling

Genome-wide studies along with QTL mapping are the potential approaches for identification of chromosomal regions associated with a particular phenotype (Hyten et al. 2007). Using both the approaches, the identification of potential candidate genes becomes difficult as both the loci comprise of hundreds of genes (Sonah et al. 2012). Similar problems are encountered during transcriptome profiling where a massive amount of genes are expressed differentially. Hence, it has been elucidated that a complementation lies between GWAS, QTL mapping and transcriptome profiling (Deshmukh et al. 2010; Sharma et al. 2011; Kadam et al. 2012). The combined QTL and GWAS method has been used to study the candidate QTLs for grain number in rice (Deshmukh et al. 2010; Sharma et al. 2011; Kadam et al. 2012). While studying the relative transcript abundance in a pair of soybean near-isogenic lines (NILs) using the Affymetrix Soy GeneChip and high-throughput Illumina whole-transcriptome sequencing, 13 candidate genes have been identified in the QTL segment of ~8.4 Mb (Bolon et al. 2010). All the three approaches can be combined together based on the inference that differential expressions of candidate genes regulate the quantitative traits (Xu et al. 2013). As a consequence, the resequencing and the transcriptome profiling of the QTL locus will be a profitable means of complimenting mapping efforts.

The transcriptome technologies are a promising way of obtaining a better insight of the amount of transcript in plants for which the genome sequence is unavailable (Trujillo et al. 2008). However, the discrepancy between the amount of protein and the levels of gene transcripts makes it important to subject the proteome for further validation and complementary analysis. Various ways in which the transcriptomics approaches have been useful while studying abiotic stress tolerance mechanisms in various crops are enlisted in Table 1.3.

Table 1.3 Applications of transcriptomics approaches for understanding abiotic stress tolerance mechanisms

Crop	Technology used	Outcome	References
Rice	SAGE	24 differentially expressed genes were identified of which 18 genes were an aerobically induced and six genes were repressed	Matsumura et al. (1999)
Salt-tolerant (FL478) and salt-sensitive (IR29) rice varieties	Rice oligoarray	Response of IR 29 was strikingly different from FL478 with induction of a large number genes induced in the former. Salt stress activated a number of genes in flavonoid pathway in IR 29 but not in FL 478 during vegetative growth stage	Ueda et al. (2006)
Soybean	Custom array containing 9728 cDNAs	Genes involved in DNA repair and RNA stability were induced; 48 differentially expressed genes were identified	O'Rourke et al. (2007)
Chickpea (*Cicer arietinum* L.)	High-resolution power of super SAGE coupled to the Roche 454 life/APG GS FLX titanium NGS technology	Characterized the complete transcriptome of chickpea plant's roots and nodules under drought stress and control conditions	Molina et al. (2008)
Soybean	HiCEP (29,388) high-coverage expression profiling	97 genes and 34 proteins differentially expressed genes during flood stress were identified	Komatsu et al. (2009)
Soybean seven tissues and seven stages during seed development	RNA-Seq	Expression atlas for soybean genes has been generated	Severin et al. (2010)
Chickpea (*Cicer arietinum* L.)	Combined high-throughput next-generation sequencing and transcript profiling for GWAS	363 and 106 transcripts showed increased and decreased expression (over threefold) in roots and nodules, respectively, during salt stress	Molina et al. (2011)
Sweet potato	Illumina paired-end RNA-Seq	Temperature stress-responsive genes were identified from transcriptome sequence, such as abscisic acid-responsive element-binding factors (AREB) and CBF TFs	Tao et al. (2012)
Switchgrass cultivar Alamo	Affymetrix gene chips	5365 differentially expressed probe sets during heat stress	Li et al. (2013b)
Cotton seedlings	Comparative microarray analysis	The functional genes and abiotic stress-related pathways were identified	Zhu et al. (2013)

(continued)

Table 1.3 (continued)

Crop	Technology used	Outcome	References
Transgenic rice plants	RNA sequencing-mediated expression profiling	Provided valuable information about the ER stress response in rice plants and led to the discovery of new genes related to ER stress	Wakasa et al. (2014)
Chenopodium quinoa	RNA-Seq analysis	Drought stress-tolerant genes were identified	Raney et al. (2014)
EST collections of chickpea	NGS (next-generation sequencing) platforms (Illumina and FLX/454)	A more extensive chickpea transcriptome assembly (CaTA v2) was developed	Kudapa et al. (2014)

1.5 Prote"Omics": An Approach to Unravel Stress Tolerance Mechanisms in Plants

An organism's proteome forms an essential link between its transcriptome and metabolome (Charulata 2015). The disparity between the mRNA abundance and level of protein accumulation has made it mandatory to use proteomics for evaluation of plant stress responses (Gygi et al. 1999). Proteins, the vital players during the plant stress responses, are translated from the functional portion of the genome. Proteomic investigations have enabled the researchers to identify the stress-responsive proteins and decipher the metabolic pathways underlying various abiotic stress responses and tolerance mechanisms (Soda et al. 2015).

The aeon of proteomics research commenced with the debut of two-dimensional (2D) gel electrophoresis techniques to separate the crude protein mixtures (Subudhi 2011). In the recent years, several improvements have been made in extraction, separation, quantification and identification of plant proteins as well as several new technologies like mass spectrometry, fluorescent 2D differential in-gel electrophoresis, gel-free approaches such as multidimensional protein identification technology (MudPIT) (Herbert et al. 2001) and LC-MS-based tagging approaches such as isotope-coded affinity tags (ICAT) (Gygi et al. 1999), stable isotope labelling by amino acids in cell culture (SILAC) (Martinovic et al. 2002; Ong et al. 2002, 2003; Ibarrola et al. 2004), isobaric tags for relative and absolute quantitation (iTRAQ) (Ross et al. 2004; Choe et al. 2007; Ghosh et al. 2011, 2013) have been introduced to reduce the errors, to perform large-scale protein analysis in a single gel for the identification of post-translationally modified proteins and to improve the feasibility and reproducibility while assaying proteins at a global level (Salekdah and Komatsu 2007; Reddy et al. 2012). Easy access to various protein databases and the advancements in the MS techniques have made it easy to recognize the proteins that are expressed in plants during various developmental stages as well as during abiotic stresses (Subudhi 2011). Both transcriptomics and proteomic studies have contributed to the research on responses in plants towards abiotic stresses by providing widespread knowledge related to the changes in the expression of genes and protein abundance during control and stressful conditions (Hakeem et al. 2012; Mizoi et al. 2012). It has been concluded that the identification and functional analysis of the proteome adds on to the information available at the transcriptional level and hence provides an improved insight towards the abiotic stress response pathways in plants (Ghosh and Xu 2014).

Apart from the mentioned proteomic investigation platforms, several other techniques like yeast two-hybrid system for detection of low-abundance proteins with weak interactions (Unlu et al. 1997), protein microarrays for detection of relative abundance of proteins (Bayer et al. 2005), fluorescence imaging spectrometry for determining the protein-protein interactions and fluorescent resonance energy transfer (FRET) for studying the protein fusions using green fluorescent protein (GFP) (Wilkins et al. 1996) are also available for rapid investigation of protein activity.

Utilization of proteomics approaches for studying abiotic stress responses in plants has provided new insights on how plants adapt to abiotic stresses. Table 1.4 summarizes various proteomic studies undertaken to study the plants response to various abiotic stresses.

Table 1.4 Summary of comparative proteomic analyses performed to study plants' response to different abiotic stresses. Plant species, stress treatment conditions, proteomic approaches and protein classes are identified/a number of differentially expressed protein spots in these studies are described

Stress/plant material	Treatment/ duration	Proteomic approach/ technique used	Protein classes identified/ number of differentially expressed protein spots	References
Drought				
Populus euphratica – leaf	35, 24, 8% rel. soil water and recovery (10 d after re-irrigation)	2D-DIGE MALDI-TOF/TOF	375	Bogeat-Triboulot et al. (2007)
Maize (line FR697) – primary root elongation zone-cell wall proteome	ψw 1.6 MPa (48 h)	2 DE LC-ESI	152	Zhu et al. (2007)
Wild watermelon	Stop watering	2 DE	Osmolytes and transmembrane H_2O channels	Yoshimura et al. (2008)
Wheat	18% peg 6000	2 DE	Proteosomal factors and protease inhibitors	Demirevska et al. (2008)
Rapeseed	Stop watering (1–7 days)	2 DE	Molecular chaperones and proteosomal factors	Mohammadi et al. (2012a)
Soybean	10% PEG 6000 (4 days)	2 DE	Metabolic enzymes	Mohammadi et al. (2012b)
Pea (*P. sativum* L.)	–	2 DE/MS	139 proteins increased over twofold during germination	Wang et al. (2012)
Cold (low temperature)				
Arabidopsis thaliana Col	0 – leaf 6 or 10 °C (7 d and recovery) 2D	DIGE MALDI-TOF	22	Amme et al. (2006)

(continued)

Table 1.4 (continued)

Stress/plant material	Treatment/ duration	Proteomic approach/ technique used	Protein classes identified/ number of differentially expressed protein spots	References
Rice	5 °C, 48 h	2 DE	Antioxidants	Hashimoto and Komatsu (2007)
Rice cv Nipponbare – seedlings	5 °C (48 h)	2-DE MALDI-TOF ESI-MS/ MS	39	Hashimoto and Komatsu (2007)
Rice	10 °C(24 and 72 h)	2 DE	Primary metabolism-associated enzymes	Lee et al. (2009)
Rice	15 °C, 10 °C, and 5 °C, 28 h	2 DE	Molecular chaperones	Hashimoto et al. (2009)
Pea	6–8 °C, 11 days	2 DE	Defence-related proteins	Dumont et al. (2011)
Tomato	–	2 DE	Some differentially translated proteins were found, including defensive, embryogenesis and photosynthesis proteins	Sanchez-Bel et al. (2012)
Onion (*Allium cepa* L.)	Different freezing treatments	2 DE	Injury-related and recovery-related proteins were identified which showed different accumulation patterns	Chen et al. (2013)
Salinity				
Tomato seedlings	120 mM NaCl	–	23 salt stress response proteins were identified. Heat-shock proteins, detoxifying enzymes, carbohydrate metabolism-associated proteins, ATP synthase, transcription- and translation-related proteins and photosynthetic metabolism-related proteins were identified	Chen et al. (2009)
Rice	200 mM NaCl 1, 3, and 6 h	2 DE	Ca++ signalling protein, plasma membrane receptors	Zhang et al. (2009)
Barley	250 mM NaCl, 13 days	2DE- MS	Detoxification-related proteins	Witzel et al. (2009)
Wheat	200 mM NaCl, 24 h	2 DE	Metabolic enzymes, enzymes involved in ETC and ATP synthesis	Peng et al. (2009)
Maize	25 mM NaCl, 1 h	2 DE	–	Zorb et al. (2010)

(continued)

Table 1.4 (continued)

Stress/plant material	Treatment/ duration	Proteomic approach/ technique used	Protein classes identified/ number of differentially expressed protein spots	References
Tomato seedlings	NaCl stress	2 DE	Total of 40 and 36 proteins with significant quantitative variations were identified in the leaf and root libraries, respectively. The proteins appeared mainly related to photosynthesis, protein degradation, metabolism and protein folding	Manaa et al. (2011)
Sugar beet	–	–	Glycine decarboxylase, ferredoxin-NADP reductase, aminomethyltransferase	Wakeel et al. (2011)
Potato	150 mM NaCl	2 DE	Photosynthesis- related proteins were found to be repressed	Evers et al. (2012)
Sorghum (*Sorghum bicolor* L.)	–	–	Malate dehydrogenase, APX: ROS scavenging	Ngara et al. (2012)
Barley (*Hordeum vulgare* L.)	–	–	*HvNHX1:* improved salt tolerance due to better ion homeostasis and cell redox homeostasis	Wu et al. (2014)

1.6 Metabol"Omics": A Tool to Investigate Plant Response to Abiotic Stress

Having the knowledge of genes, transcripts and proteins isn't sufficient while studying the stress responses until a wide range of information is available regarding the metabolites involved in such responses. Metabolomics, the newest "omics" technology, has evolved with the aim of determining and quantifying "all" metabolites in a biological system. Metabolomics deals with the identification and quantification of primary and secondary metabolites involved in various life processes (Deshmukh et al. 2014). The metabolism of plants varies during the type of abiotic stress it encounters. Therefore, metabolomics is a comprehensive approach for unravelling the metabolic pathways and metabolites that regulate the response of crop plants towards various abiotic stresses (Hoefgen and Nikiforova 2008; Shao et al. 2008; Urano et al. 2010; Loiacono and De Tullio 2012).

Metabolic pathways in plants are highly complex, and depending upon the nature of query, several approaches, i.e. metabolic fingerprinting, metabolite profiling/ metabolomics and targeted analysis, are used in metabolomics research (Fiehn 2002; Halket et al. 2005; Shulaev 2006). Metabolic fingerprinting approach has been extensively used for generating specific metabolic signatures associated with a specific stress response from a mass of samples without precise quantification (Shulaev et al. 2008). A number of techniques like nuclear magnetic resonance (NMR) (Krishnan et al. 2005), MS (Goodacre et al. 2003), Fourier transform ion cyclotron resonance mass spectrometry or Fourier transform infrared (FT-IR) spectroscopy (Johnson et al. 2003) can be used for generating fingerprints. Another approach, i.e. metabolite profiling, deals with the quantification of total metabolome of a sample, i.e. it helps in generating a snapshot of all the metabolites within the sample (Subudhi 2011). Numerous analytical techniques can be used for identification and quantification of the metabolites in a sample (Sumner et al. 2003; Shulaev 2006). These techniques include NMR, GC-MS, liquid chromatography-mass spectrometry (LC-MS), capillary electrophoresis-mass spectrometry (CE-MS), gas chromatography-mass spectrometry (GC-MS), nuclear magnetic resonance (NMR) and FT-IR spectroscopy. Till date, GC-MS is the most advanced and highly chosen analytical technique for metabolite profiling in plants (Roessner and Beckles 2009). The last approach, i.e. the targeted analysis, is aimed at precise identification of a specific metabolite or a target using a particular analytical technique for best results. Targeted analysis is performed while studying the effects of a particular type of stress. Comparative metabolic profiling can also be done using targeted analysis (Bajad and Shulaev 2007).

Among all the approaches, the most commonly used analytical platforms in metabolomics research include (a) indirect measurement of samples involving separation by chromatography, i.e. gas chromatography (GC), liquid chromatography (LC) or capillary electrophoresis (CE) followed by MS of the separated molecules, and (b) direct measurement of the metabolites without chromatographic separation by using direct infusion MS (Fourier transform ion cyclotron resonance mass spectrometry (FT-ICR-MS)). The direct methods of metabolite identification allow accurate mass determination and precise peak identifications. NMR, on the other hand, is another important method that identifies the metabolites within a sample without prior separation and also determines the molecular structure of the identified metabolites (Arbona et al. 2013).

1.6.1 Techniques Used for Plant Metabolomic Research

1.6.1.1 Gas Chromatography-Mass Spectrometry (GC-MS)

Among the chromatographic techniques, GC is the foremost that had been linked to MS (Eneroth et al. 1964). The robust, reproducible, user-friendly, less expensive nature and short running time of the GC-MS has made it the most extensively used technique for metabolic profiling. GC-MS makes use of the time-of-flight (TOF)-MS detector for mass detection (Kopka et al. 2004). GC-MS is frequently used for detection of primary metabolites like amino acids and carbohydrates (Subudhi 2011). The availability of the metabolite databases like NIST (Kumari et al. 2011), FiehnLib (Kind et al. 2009) and Golm metabolic databases (GMD) (Kopka et al. 2005) has made the peak annotation in GC-MS easier. The only disadvantage associated with GC-MS is that it can only be used for thermally stable volatile compounds having low molecular weight (approximately 1KDa) making it difficult to analyse the high molecular weight compounds (Obata and Fernie 2012).

1.6.1.2 Liquid Chromatography-Mass Spectrometry (LC-MS)

Overcoming the limitation of GC-MS, LC-MS doesn't require the presence of volatile compounds, and the sample components are separated in a liquid phase without any prior treatment (Obata and Fernie 2012). A wide range of diverse secondary metabolites like alkaloids, flavonoids, phenylpropanoids and (poly)phenols can be detected using LC-MS. Apart from non-volatile samples, LC-MS allows the detection of thermally unstable compounds with high molecular weight and greater polarity (Bowne et al. 2011). Recent advancements in LC like development of high-performance liquid chromatography (HPLC) and ultra-performance liquid chromatography (UPLC) have made this technique more powerful. The mass spectra generated by LC-MS is typically dependent on the type of instrument (i.e. QqQ, qTOF, Ion Trap, etc.) which makes it mandatory for the researchers to create their own "in-house" LC-MS reference library, thereby limiting its use (Moco et al. 2006).

1.6.1.3 Nuclear Magnetic Resonance (NMR)

NMR, an entirely different technique for metabolite characterization, is based on atomic interactions (Obata and Fernie 2012). It is a non-destructive technique and doesn't involve volatilization of samples. Although NMR isn't as sensitive as the MS-based techniques, it is preferred as it allows the detection of various metabolites irrespective of their size, charge, volatility and stability (Schripsema 2010). This method also provided the structural information regarding the detected metabolites for their clear identification. One of the major limitations of the NMR technique is the detection threshold, i.e. it requires larger sample volumes. The metabolites whose concentration is less than 5 nmol will not be detected. Another limiting factor to this technique is the huge instrument cost involved (Krishnan et al. 2005). Some metabolic studies undertaken in context to stress responses in various crops are enlisted in Table 1.5.

1.7 Phen"Omics": An Integrated Approach for Dissection of Stress Tolerance Mechanisms

Like other "omics" approaches, phenomics is another important tool for understanding the biological system. Phenomics deals with the measurement of all the physical and biochemical parameters of an individual that often change with the changing environment or due to genetic mutation (Charulata 2015). In genomics and plant breeding, attempts are being made to unravel the relationship between a genetic marker and the phenotype and to improve the crop phenotype for development of improved varieties under various environmental conditions. Thus, phenomics when combined with other "omics" technologies can revolutionize the plant breeding research (Deshmukh et al. 2014).

Table 1.5 Metabolic studies of plant stress responses

Crop	Stress	Technique used	Outcome	References
Arabidopsis plants	Cold stress	GC-TOF-MS	C-repeat/dehydration-responsive element-binding factor (CBF) 3 plays an important role in cold response pathway	Cook et al. (2004)
Bean	Phosphorus stress	GS-TOF-MS	Metabolites, including amino acids, polyols and sugars, were found to increase in P-stressed plants	Hernandez et al. (2007)
Arabidopsis	Temperature stress	GC-MS	Mass spectral tags responding to temperature stress were identified	Kaplan and Guy (2004)
Arabidopsis	Salt stress (100 mM NaCl)	GC-MS and LC-MS	Salt stress included the induction of the methylation cycle for the supply of methyl groups, the phenylpropanoid pathway for lignin production, glycine betaine production, co-induction of glycolysis and sucrose metabolism and co-reduction of the methylation cycle	Kim et al. (2007)
Rice	Chilling, salt and osmotic stress	Quantitative HPLC assay	Chilling-tolerant genotype possesses a more effective ROS scavenging system	Morsy et al. (2007)
Brassica rapa	Metal stress	1H NMR and two-dimensional NMR spectra	The effect of Mn on plant metabolism was smaller than those of Cu and Fe	Jahangir et al. (2008)
Arabidopsis leaves	Drought stress	–	Accumulation of many metabolites was observed, including amino acids such as proline, raffinose family oligosaccharides, γ-amino butyric acid (GABA) and tricarboxylic acid (TCA) cycle metabolites was observed	Urano et al. (2009)
Transgenic potato overexpressing the *AtDREB1A* gene	Temperature and salt stress tolerance	GC-MS	The levels of glutathione metabolite, γ-amino butyric acid and β-cyanoalanine were elevated	Iwaki et al. (2013)
Proso millet	–	GC-TOFMS	Diversity among primary metabolites and phenolic acids was determined	Kim et al. (2013)

Plant phenomics is an integrated technology involving several technologies like photonics (Kelley 2009), biology (Poorter et al. 2012), computers and robotics that will allow the scientists to get a better insight towards all the aspects of crop plants (Yang et al. 2013). Recent advancements and developments in the fields of image processing and automation technology have encouraged the researchers towards the real-time analysis of plant growth and developmental stages (Mielewczik et al. 2013). Several technologies that play a key role in plant phenomics are discussed below.

1.7.1 Visible Light Imaging

Since the development of the first digital camera in the year 1975, visible light imaging has been intensively used for studying crop plants. The ease of maintenance and cost-effective nature of this technique has made it an important approach for analysing the shoot biomass, yield- and panicle-related traits and root morphology (Iyer-Pascuzzi et al. 2010). The drawbacks of the earlier 2D imaging like loss of spatial and volumetric information lead to the advent of new three-dimensional (3D) mesh algorithms like RootReader3D and 3D virtual rice for precise estimation of root and shoot morphology (Yang et al. 2013). This combination of low-cost visible light imaging with 2D or 3D image analysis has always been the first and foremost choice while studying crop phenomics.

1.7.2 Infrared and Hyperspectral Imaging

It is well known that the movement of the molecules within an object leads to the emission of characteristic infrared radiations (Kastberger and Stachl 2003). For imaging the infrared radiations, two most popular devices that screen them are a near-infrared (NIR, wavelength of approximately 0.9–1.7 mm) imaging device and a far-infrared (far-IR, wavelength of approximately 7.5–13.5 mm) imaging device (Shibayama et al. 2011). Another important device in this context is a crop phenology recording system (CPRS) which makes use of both visible light and infrared imaging to establish the relationship between camera-derived indices and agronomic traits which has been developed to monitor rice growth (Sakamoto et al. 2011). For visualization of the temperature differences and plant drought resistance, far-IR (also called IR-thermal) imaging is a highly preferred method.

Apart from the visible and the IR imaging technologies, one more imaging technique, i.e. hyperspectral imaging technique, is widely used for studying plant architecture, health conditions and growth characteristics (Wallays et al. 2009; Liu et al. 2010; Singh et al. 2010).

1.7.3 3D Structural Tomography and Functional Imaging

Recently, a number of improved imaging techniques like 3D structural tomography and functional imaging have been introduced for better visualization of living plants. The X-ray computed tomography (CT) scanners equipped with an acceleration algorithm using the adaptive minimum enclosing rectangle (AMER) and graphics processing unit (GPU) is effectively used for estimating the tiller number in rice (Yang et al. 2011a, b; Jiang et al. 2012). OCM also known as optical coherence tomography (OCT) is a new technology based on photonics, has an approximately 1 mm spatial resolution and is used for in vivo 3D imaging of plant structures (Reeves et al. 2002). Another tomography technique with greater penetration and capability of detecting nonfluorescent signals, optical projection tomography (OPT), can be applied to visualize plant developmental stages and gene expression (Lee et al. 2006).

Magnetic resonance imaging (MRI), one more approach towards crop phenomics, provides information regarding structural organization and the internal processes occurring in vivo by imaging the water protons (Borisjuk et al. 2012). The structural imaging and functional imaging technologies (such as fluorescence imaging and positron emission tomography, PET) reveal the alterations occurring in plants at physiological levels, i.e. chlorophyll fluorescence imaging is one such technique that determines the photosynthetic efficiency and stress encountered by plants, whereas, PET visualizes the metabolism-related activities using the radiolabelled tracers (Baker 2008; Jahnke et al. 2009). Among all the imaging techniques, chlorophyll fluorescence imaging is widely used in plant phenomics (Yang et al. 2013).

1.7.4 Applications of Phenomics in Abiotic Stress Tolerance

Due to the complexity of the plant responses to abiotic stress, phenotyping the abiotic stress tolerance is a huge challenge (Roy et al. 2011). For this, high-throughput phenotyping techniques that can screen multiple traits in a plant under stress need to be incorporated (Berger et al. 2010). The imaging technologies like Scanalyzer 3D which efficiently analyse all the tolerance mechanisms in plants during salinity stress like Na+ exclusion, osmotic tolerance and tissue tolerance (Rajendran et al. 2009) have been used in cereals. Another modification of the Scanalyzer 3D enables the researchers to more accurately estimate the cereal biomass under salt stress conditions (Yang et al. 2013).

Recently, a method has been introduced that makes use of the light sources in greenhouse and field conditions to assess the growth of leaves in soybean under changing environments (Mielewczik et al. 2013). Martrack Leaf, a marker tracking approach, has made it possible to perform the high resolution and accurate 2D analysis of leaf expansion in soybean (Mielewczik et al. 2013).

IR-thermal imaging technique has been successfully used for quantification of osmotic stress response in cereal crops (Munns et al. 2010). The visible and the NIR digital imaging techniques enable the high-throughput screening of crops during nitrogen stress (Shibayama et al. 2009). Furthermore, it has been concluded that the combination of precise phenomics/phenotyping approaches and high-resolution genetic dissection can explain the functional gene polymorphisms and abiotic stress tolerance mechanisms in crop plants to a greater extent.

1.8 Importance of Online Databases for Effective Incorporation of "Omics" Approaches

With the advancements in technology, "omics" technologies are being tremendously used in research activities which give rise to huge amount of data and information handling which is a tedious job. For the storage, organization and easy accessibility of the available information, several user-friendly computational resources that act as the repositories of the information have been developed. These available data repositories known as databases are a storehouse of the information related to molecular markers, genes, microRNAs, siRNAs, proteins, metabolites and phenomics of various crop plants. Access to this galaxy of information within the databases has made it easy for the researchers to compare, store and extract required useful information. Various genomics-, transcriptomics-, proteomics- and metabolomics-related databases are enlisted in Table 1.6.

1.9 Conclusion and Future Perspectives

"Omics" approaches have proved to be the most candid and promising biotechnological applications for improving abiotic stress tolerance in plants. For in-depth knowledge of the plants' response to abiotic stress, acquiring information about how plants perform during various abiotic stresses and identification of various strategies/techniques involved in improved stress tolerance is a preliminary requirement. Various "omics" approaches have been utilized by researchers to develop better insight towards the cascade of cellular events and signals resulting in rapid responses and tolerance to various abiotic stresses. With the recent technological advancements, the cost and the technical difficulties associated with the omics technologies have reduced to a greater extent, thereby enabling the researchers to unravel various physiological, molecular and metabolic aspects related to plant stress tolerance, in turn regulate the gene expression as well as the phenotype of a plant under stress. These multidimensional "omics" approaches provide new opportunities and avenues for future research in context to abiotic stress tolerance. Utilization of phenomics technologies will enable the biologists to get a clear picture of plants' biological system. The continuous advent of high-throughput sequencing platforms will help to overcome the problems encountered due to large genome sizes of certain crops like millets. In comparison to genomics, transcriptomics and proteomics, metabolomics and phenomics studies still lag behind. However, the introduction of the novel metabolic profiling and improved imaging

Table 1.6 Online databases associated with various omics research in crop plants

S. no.	Genomics databases	Transcriptomics databases	Proteomics databases	Metabolomics databases
1.	National Center for Biotechnology Information	Soybean knowledge base, University of Missouri	Proteome analysis at EBI	The soybean metabolome database
2.	Gramene	Soybean transcription factors database, Missouri	Soybean transcription factors database, Missouri	BRENDA
3.	The Arabidopsis Information Resource (TAIR)	TIGR Arabidopsis arrays	Soybean proteins database	Platform plant metabolomics
4.	The Oryza Tag Line mutant database	Gene expression omnibus	ExPASy *A. thaliana* 2D-proteome database	Metabolic modelling
5.	TIGR rice genome annotation	NSF rice oligonucleotide array project	Swissprot	Iowa gene expression toolkit
6.	Maize genome resources	Zeamage	PlantsP: functional genomics of plant	Solcyc Solanaceae metabolic pathway annotations
7.	Gene ontology	Tomato expression database	Functional genomics of plant	Plant metabolome database
8.	Maize genetics and genomics database	Soybean transposable elements database	ExPASy: SIB bioinformatics resource portal	AraCyc Arabidopsis metabolic pathway annotations
9.	An integrated soybean genome database including BAC-based physical maps	Virtual centre for cellular expression profiling in rice	Database of *A. thaliana* annotation	MetAlign tool for GC- or LC-MS data analysis
10.	SoyBase and the soybean breeder's toolbox	PLEXdb	PlantPReS	Plant metabolic network

techniques with higher precision and resolution will revolutionize the study of complex abiotic stress responses.

The integration of the omics approaches in the near future will pave a way towards identification of the genes, proteins and metabolites that play a pivotal role in the molecular and signalling pathways that enhance the tolerance and adaptability of plants towards various abiotic stresses. In order to prevent the losses occurring due to intricate abiotic stresses, it becomes necessary to carry out extensive studies in agronomically important crops for overcoming the harmful abiotic restraints. The combined effort of all the "omics" technologies will prove to be a remarkable step towards figuring out the regulatory networks underlying stress tolerance mechanisms which can be further subjected to MAS, gene pyramiding or conventional breeding for the development of stress-tolerant varieties. Integration of the "omics" technologies with bioinformatics will provide a foundation for improved management of abiotic stress tolerance and attaining further in-depth knowledge about the plant's response to abiotic stresses. A consolidated use of "omics" technologies, MAS and transgenic technology will prove pragmatic in improving the abiotic stress tolerance in crop plants.

References

Abdurakhmonov IY, Abdukarimov A (2008) Application of association mapping to understanding the genetic diversity of plant germplasm resources. Int J Plant Genomics 5:1–18

Agrawal GK, Bourguignon J, Rolland J et al (2010) Plant organelle proteomics: collaborating or optimal cell function. Mass Spectrom Rev 30:772–853

Akpjnar BA, Stuart JL, Budak H (2013) Genomics approaches for crop improvement against abiotic stress. Sci World J 1–9

Allen DJ, Ort DR (2001) Impact of chilling temperatures on photosynthesis in warm climate plants. Trends Plant Sci 6:36–42

Amme S, Matros A, Schlesier B et al (2006) Proteome analysis of cold stress response in Arabidopsis thaliana using DIGE-technology. J Exp Bot 57:1537–1546

Anjum SA, Xie XY, Wang LC et al (2011) Morphological, physiological and biochemical responses of plants to drought stress. Afr J Agric Res 6(9):2026–2032

Apel K, Hirt H (2004) Reactive oxygen species: metabolism, oxidative stress, and signal transduction. Ann Rev Plant Biol 55:373–399

Aranzana MJ et al (2005) Genome-wide association mapping in *Arabidopsis* identifies previously known flowering time and pathogen resistance genes. PLoS Genet 1:e60

Arbona V, Manzi M, Olas CD et al (2013) Metabolomics as a tool to investigate abiotic stress in plants. Int J Mol Sci 14:4885–4911

Arpat A, Waugh M, Sullivan JP et al (2004) Functional genomics of cell elongation in developing cotton fibers. Plant Mol Biol 54:911–929

Ashraf M, Foolad MR (2007) Roles of glycine betaine and proline in improving plant abiotic stress resistance. Environ Exp Bot 59:206–216

Ashraf M, Athar HR, Harris PJC et al (2008) Some prospective strategies for improving crop salt tolerance. Adv Agron 97:45–110

Bachem CW, Hoeven RSV, Bruijn SMD et al (1996) Visualization of differential gene expression using a novel method of RNA fingerprinting based on AFLP: analysis of gene expression during potato tuber development. Plant J 9:745–753

Bajad S, Shulaev V (2007) Highly-parallel metabolomics approaches using LC-MS2 for pharmaceutical and environmental analysis. Trends Anal Chem 26:625–636

Baker NR (2008) Chlorophyll fluorescence: a probe of photosynthesis in vivo. Ann Rev Plant Biol 59:89–113

Baulcombe DC (1999) Fast forward genetics based on virus induced gene silencing. Curr Opin Plant Biol 2:109–113

Bayer E, Bottrill AR, Walshaw J et al (2005) *Arabidopsis* cell wall proteome defined using multidimensional protein identification technology. Proteomics 6:301–311

Beetham PR, Kipp PB, Sawycky XL et al (1999) A tool for functional plant genomics: chimeric RNA/DNA oligonucleotides cause in vivo gene-specific mutations. Proc Natl Acad Sci U S A 96:8774–8778

Berger B, Parent B, Tester M (2010) High-throughput shoot imaging to study drought responses. J Exp Bot 61:3519–3528

Bhaya D, Davison M, Rodolphe B (2011) CRISPR-Cas systems in bacteria and archaea: versatile small RNAs for adaptive defense and regulation. Ann Rev Genet 45:273–297

Blum A (1996) Crop responses of drought and the interpretation of adaptation. Plant Growth Regul 20:135–148

Bogeat-Triboulot MB, Brosche M, Renaut J et al (2007) Gradual soil water depletion results in reversible changes of gene expression, protein profiles, eco physiology, and growth performance in Populus euphratica, a poplar growing in arid regions. Plant Physiol 143:876–892

Bolon YT, Joseph B, Cannon SB et al (2010) Complementary genetic and genomic approaches help characterize the linkage group I seed protein QTL in soybean. BMC Plant Biol 10:41

Borisjuk L, Rolletschek H, Neuberger T (2012) Surveying the plant's world by magnetic resonance imaging. Plant J 70:129–146

Bouche N, Bouchez D (2001) Arabidopsis gene knockout: phenotypes wanted. Curr Opin Plant Biol 4:111–117

Bowne J, Bacic A, Tester MRU (2011) Abiotic stress and metabolomics. Ann Plant Rev 43:61–85

Bray EA, Bailey-Serres J, Weretilnyk E (2000) Responses to abiotic stresses. In: Gruissem W, Buchannan B, Jones R (eds) Biochemistry and molecular biology of plants. American Society of Plant Physiologists, Rockville, pp 1158–1249

Brenner S et al (2000) Gene expression analysis by massively parallel signature sequencing (MPSS) on microbead arrays. Nat Biotechnol 18:630–634

Breyne P, Zabeau M (2001) Genome-wide expression analysis of plant cell cycle modulated genes. Curr Opin Plant Biol 4:136–142

Buckley TN, Mott KA, Farquhar GD (2003) A hydro mechanical and biochemical model of stomatal conductance. Plant Cell Environ 26:1767–1785

Byrt CS, Platten JD, Spielmeyer W et al (2007) HKT1;5-like cation transporters linked to Na+ exclusion loci in wheat, *Nax2* and *Kna1*. Plant Physiol 143:1918–1928

Caldwell DG, McCallum N, Shaw P et al (2004) A structured mutant population for forward and reverse genetics in barley (*Hordeum vulgare* L.) Plant J 40:143–150

Cao J, Schneeberger K, Ossowski S et al (2011) Whole-genome sequencing of multiple Arabidopsis thaliana populations. Nat Genet 43:956–963

Charulata (2015) Advances in omics for enhancing abiotic stress tolerance in millets. Proc Indian Natl Sci Acad 81(2):397–415

Chawla K, Barah P, Kuiper M, Bones AM (2011) Systems biology: a promising tool to study abiotic stress responses. Omics Plant Abiotic Stress Tolerance 163–172

Chen et al (2002) Expression profile matrix of Arabidopsis transcription factor genes suggests their putative functions in response to environmental stresses. Plant Cell 14:559–574

Chen S, Gollop N, Heuer B (2009) Proteomic analysis of salt-stressed tomato (*Solanum lycopersicum*) seedlings: effect of genotype and exogenous application of glycine betaine. J Exp Bot 60:2005–2019

Chen K, Renaut J, Sergeant K et al (2013) Proteomic changes associated with freeze-thaw injury and post-thaw recovery in onion (*Allium cepa* L.) scales. Plant Cell Environ 36:892–905

Cho Y, Meade JD, Walden JC et al (2001) Multicolor fluorescent differential display. BioTechniques 30:562–573

Choe L, D'ascenzo M, Relkin NR et al (2007) 8-Plex quantitation of changes in cerebrospinal fluid protein expression in subjects undergoing intravenous immunoglobulin treatment for Alzheimer's disease. Proteomics 7:3651–3660

Chuang CF, Meyerowitz EM (2000) Specific and heritable genetic interference by double-stranded RNA in *Arabidopsis thaliana*. Proc Natl Acad Sci U S A 97:4985–4990

Clement M, Lambert A, Herouart D et al (2008) Identification of new up-regulated genes under drought stress in soybean nodules. Gene 426:15–22

Colbert T, Till BJ, Tompa R et al (2001) High-throughput screening for induced point mutations. Plant Physiol 126:480–484

Collins NC, Tardieu F, Tuberosa R (2008) Quantitative trait loci and crop performance under abiotic stress: where do we stand? Plant Physiol 147:469–486

Comai L, Henikoff S (2006) TILLING: practical single-nucleotide mutation discovery. Plant J 45:684–694

Cook D, Fowler S, Fiehn O et al (2004) A prominent role for the CBF cold response pathway in configuring the low-temperature metabolome of Arabidopsis. Proc Natl Acad Sci U S A 101:15243–11524

Cooper JL, Till BJ, Laport RG et al (2008) TILLING to detect induced mutations in soybean. BMC Plant Biol 8:9

Cordeiro GM, Eliott F, McIntyre CL et al (2006) Characterization of single nucleotide polymorphisms in sugarcane ESTs. Theor Appl Genet 113:331–343

Cosart T, Beja-Pereira A, Chen S et al (2011) Exome-wide DNA capture and next generation sequencing in domestic and wild species. BMC Genomics 12:347–355

Cseri AM, Korff MV, Nagy BB et al (2011) Allele mining and haplotype discovery in barley candidate genes for drought tolerance. Euphytica 181:341–356

Demirevska K, Simova-Stoilova L, Vassileva V et al (2008) Drought-induced leaf protein alterations in sensitive and tolerant wheat varieties. Gen Appl Plant Physiol Spec Issue 34:79–102

Deshmukh R, Singh A, Jain N et al (2010) Identification of candidate genes for grain number in rice (*Oryza sativa* L.) Funct Integr Genomics 10(339):347

Deshmukh R, Sonah H, Patil G et al (2014) Integrating omic approaches for abiotic stress tolerance in soybean. Front Plant Sci 5:1–12

Dita MA, Rispail N, Prats E et al (2006) Biotechnology approaches to overcome biotic and abiotic stress constraints in legumes. Euphytica 147:1–24

Doerge RW (2002) Mapping and analysis of quantitative trait loci in experimental populations. Nat Rev Genet 3:43–52

Donson J, Fang Y, Espiritu-Santo G et al (2002) Comprehensive gene expression analysis by transcript profiling. Plant Mol Biol 48:75–97

Dumont E, Bahrman N, Goulas E et al (2011) A proteomic approach to decipher chilling response from cold acclimation in pea (*Pisum sativum* L.) Plant Sci 180:86–98

Eckert AJ, Bower AD, Wegrzyn JL et al (2009) Association genetics of coastal Douglas Fir (*Pseudotsuga menziesu* var. menziesii, Pinaceae). I. Cold-hardiness related traits. Genetics 182:1289–1302

Eneroth P, Hellstrom K, Ryhage R (1964) Identification and quantification of neutral fecal steroids by gas-liquid chromatography and mass spectrometry: studies of human excretion during two dietary regimens. J Lipid Res 5:245–262

Evers D, Legay S, Lamoureux D et al (2012) Towards a synthetic view of potato cold and salt stress response by transcriptomic and proteomic analyses. Plant Mol Biol 78:503–514

Feng Z, Zhang B, Ding W et al (2013) Efficient genome editing in plants using a CRISPR/Cas system. Cell Res 23:1229–1232

Feng-ling FU, Zhi-Lei F, Shi-bing G et al (2008) Evaluation and quantitative inheritance of several drought-relative traits in maize. Agric Sci China 7(3):280–290

Fiehn O (2002) Metabolomics – the link between genotypes and phenotypes. Plant Mol Biol 48:155–171

Flexas J, Bota J, Loreto F et al (2004) Diffusive and metabolic limitations to photosynthesis under drought and salinity in C3 plants. Plant Biol 6:1–11

Flowers TJ (2004) Improving crop salt tolerance. J Exp Bot 55:1–13

Flowers TJ, Yeo AR (1995) Breeding for salinity resistance in crop plants: where next? Aust J Plant Physiol 22:875–884

Fraire-Velazquez S, Rodríguez-Guerra R, Sánchez- Calderón L (2011) Abiotic and biotic stress response crosstalk in plants. In: Shankar A (ed) Abiotic stress response in plants – physiological, biochemical and genetic perspectives. InTech, Rijeka, pp 3–26

Frary A, Nesbitt TC, Grandillo S et al (2000) fw2.2: a quantitative trait locus key to the evolution of tomato fruit size. Science 289:85–88

Fu X, Fu N, Guo S et al (2009) Estimating accuracy of RNA-Seq and microarrays with proteomics. BMC Genomics 10:161

Furukawa J, Yamaji N, Wang H et al (2007) An aluminum-activated citrate transporter in barley. Plant Cell Physiol 48:1081–1091

Gambino G, Gribaudo I (2012) Genetic transformation of fruit trees: current status and remaining challenges. Transgenic Res 21:1163–1181
Ge Y, Li Y, Zhu YM et al (2010) Global transcriptome profiling of wild soybean (*Glycinesoja*) roots under NaHCO3 treatment. BMC Plant Biol 10:153
Ghosh D, Xu J (2014) Abiotic stress responses in plant roots: a proteomics perspective. Front Plant Sci 5:1–13
Ghosh D, Yu H, Tan XF et al (2011) Identification of key players for colorectal cancer metastasis by iTRAQ quantitative proteomics profiling of isogenic SW480 and SW620 cell lines. J Proteome Res 10:4373–4387
Ghosh D, Li Z, Tan XF et al (2013) ITRAQ based quantitative proteomics approach validated the role of calcyclin binding protein (CacyBP) in promoting colorectal cancer metastasis. Mol Cell Proteomics 12:1865–1880
Gonzalez-Martinez SC, Huber D et al (2008) Association genetics in *Pinus taeda* L. II. Carbon isotope discrimination. Heredity 101:19–26
Goodacre R, York EV, Heald JK, Scott IM (2003) Chemometric discrimination of unfractionated plant extracts analyzed by electrospray mass spectrometry. Phytochemistry 62:859–863
Gorantla M, Babu PR, Lachagari VBR et al (2005) Functional genomics of drought stress response in rice: transcript mapping of annotated unigenes of an *indica* rice (*Oryza sativa* L. cv. Nagina 22). Curr Sci 89:496–514
Green ED (2001) Strategies for the systematic sequencing of complex genomes. Nat Rev Genet 2:573–583
Guerrero F, Mullet JE (1986) Increased abscisic acid biosynthesis during plant dehydration requires transcription. Plant Physiol 80:588–591
Gupta PK, Sachin R, Pawan LK (2005) Linkage disequilibrium and association studies in plants: present status and future prospects. Plant Mol Biol 57:461–485
Gygi SP, Rist B, Gerber SA et al (1999) Quantitative analysis of complex protein mixtures using isotope-coded affinity tags. Nat Biotechnol 17:994–999
Hakeem KR, Chandna R, Ahmad P et al (2012) Relevance of proteomic investigations in plant abiotic stress physiology. OMICS 16:621–635
Halket JM, Waterman D, Przyborowska AM et al (2005) Chemical derivatization and mass spectral libraries in metabolic profiling by GC/MS and LC/MS/MS. J Exp Bot 56:219–243
Hall DCT, Ingvarsson PK (2010) Using association mapping to dissect the genetic basis of complex traits in plants. Brief Funct Genom 9:157–165
Hansen BG, Halkier BA, Kliebenstein DJ (2007) Identifying the molecular basis of QTLs: eQTLs add a new dimension. Trends Plant Sci 13:72–77
Hao GP, Wu ZY, Cao MQ et al (2004) Nucleotide polymorphism in the drought induced transcription factor CBF4 region of *Arabidopsis thaliana* and its molecular evolution analyses. Yi Chuan Xue Bao 31:1415–1425
Harris K, Klein R, Mullet J (2007) Sorghum stay-green QTL individually reduces post-flowering drought-induced leaf senescence. J Exp Bot 58:327–338
Harrison MM, Jenkins BV, O'Connor-Giles KM et al (2014) A CRISPR view of development. Genes Dev 28:1859–1872
Hasegawa PM, Bressan RA, Zhu JA et al (2000) Plant cellular and molecular responses to high salinity. Ann Rev Plant Physiol Plant Mol Biol 51:463–499
Hashimoto M, Komatsu S (2007) Proteomic analysis of rice seedlings during cold stress. Proteomics 7:1293–1302
Hashimoto M, Toorchi M, Matsushita K et al (2009) Proteome analysis of rice root plasma membrane and detection of cold stress responsive proteins. Protein Pept Lett 16:685–697
Hazen SP, Wu Y, Kreps JA (2003) Gene expression profiling of plant responses to abiotic stress. Funct Integr Genomics 3:105–111
Henikoff S, Comai L (2003) Single-nucleotide mutations for plant functional genomics. Ann Rev Plant Biol 54:375–401
Herbert BR, Harry JL, Packer NH et al (2001) What place for polyacrylamide in proteomics? Trends Biotechnol 19:S3–S9

Hernandez G, Ramirez M, Valdes-Lopez O (2007) Phosphorus stress in common bean: root transcript and metabolic responses. Plant Physiol 144:752–767

Hieter P, Bogushi M (1997) Functional genomics: it's all how you read it. Science 278:601–602

Hoefgen R, Nikiforova VJ (2008) Metabolomics integrated with transcriptomics: assessing systems response to sulfur-deficiency stress. Physiol Plant 132:190–198

Hoekenga OA et al (2006) AtALMT1, which encodes a malate transporter, is identified as one of several genes critical for aluminum tolerance in *Arabidopsis*. Proc Natl Acad Sci U S A 103:9738–9743

Horváth E, Pál M, Szalai G et al (2007) Exogenous 4-hydroxybenzoic acid and salicylic acid modulate the effect of short-term drought and freezing stress on wheat plants. Biol Plant 51:480–487

Hsu PD, Lander ES, Zhang F (2014) Development and applications of CRISPR-Cas9 for genome engineering. Cell 157:1262–1278

Huang X, Lu T, Han B (2013) Resequencing rice genomes: an emerging new era of rice genomics. Trends Genet 29:225–232

Hyten DL, Choi IY, Song Q et al (2007) Highly variable patterns of linkage disequilibrium in multiple soybean populations. Genetics 175:1937–1944

Ibarrola N, Molina H, Iwahori A et al (2004) A novel proteomic approach for specific identification of tyrosine kinase substrates using [13C] tyrosine. J Biol Chem 279:15805–15813

Iwaki T, Guo L, Ryals JA et al (2013) Metabolic profiling of transgenic potato tubers expressing Arabidopsis dehydration response element-binding protein 1A (DREB1A). J Agric Food Chem 61:893–900

Iyer-Pascuzzi AS, Symonova O, Mileyko Y et al (2010) Imaging and analysis platform for automatic phenotyping and trait ranking of plant root systems. Plant Physiol 152:1148–1157

Jahangir M, Abdel-Farid IB, Choi YH et al (2008) Metal ion inducing metabolite accumulation in *Brassica rapa*. J Plant Physiol 165:1429–1437

Jahnke S, Menzel MI, van Dusschoten D, Roeb GW et al (2009) Combined MRI–PET dissects dynamic changes in plant structures and functions. Plant J 59:634–644

Jain SM, Ochatt SJ, Kulkarni VM et al (2010) In vitro culture for mutant development. Acta Hortic 865:59–68

Jander G (2006) Gene identification and cloning by molecular marker mapping. Methods Mol Biol 323:115–126

Jansen RC, Nap JP (2001) Genetical genomics: the added value from segregation. Trends Genet 17:388–391

Jiang N, Yang WN, Duan LF et al (2012) Acceleration of CT reconstruction for wheat tiller inspection based on adaptive minimum enclosing rectangle. Comput Electron Agric 85:123–133

Jiang W, Zhou H, Bi H et al (2013) Demonstration of CRISPR/Cas9/sgRNA- mediated targeted gene modification in *Arabidopsis*, tobacco, sorghum and rice. Nucl Acids Res 41:e188

Johnson HE, Broadhurst D, Goodacre R, Smith AR (2003) Metabolic fingerprinting of salt-stressed tomatoes. Phytochemistry 62:919–928

Kadam S, Singh K, Shukla S et al (2012) Genomic associations for drought tolerance on the short arm of wheat chromosome 4B. Funct Integr Genomics 12:447–464

Kaplan F, Guy CL (2004) β-amylase induction and the protective role of maltose during temperature shock. Plant Physiol 135:1674–1684

Kastberger G, Stachl R (2003) Infrared imaging technology and biological applications. Behav Res Methods Instrum Comput 35:429–439

Kaur G, Kumar S, Nayyar H, Upadhyaya HD (2008) Cold stress injury during the pod-filling phase in chickpea (*Cicer arietinum* L.): effects on quantitative and qualitative components of seeds. J Agron Crop Sci 194(6):457–464

Kelley B (2009) Agri-photonics. SPIE Prof 7:14–17

Kempin SA, Liljegren SJ, Block LM et al (1997) Targeted disruption in *Arabidopsis*. Nature 389:802–803

Kido EA, Neto JRCF, de Oliveira Silva RL et al (2012) New insights in the sugarcane transcriptome responding to drought stress as revealed by super sage. Sci World J 2012: 821062. doi: 10.1100/2012/821062

Kim JK, Bamba T, Harada K et al (2007) Time-course metabolic profiling in *Arabidopsis thaliana* cell cultures after salt stress treatment. J Exp Bot 58:415–424

Kim JK, Park SY, Yeo Y et al (2013) Metabolic profiling of millet (*Panicum miliaceum*) using gas chromatography–time-offlight mass spectrometry (GC-TOFMS) for quality assessment. Plant Omics J 6:73–78

Kind T, Wohlgemuth G, Lee do Y et al (2009) FiehnLib: mass spectral and retention index libraries for metabolomics based on quadrupole and time-offlight gas chromatography/mass spectrometry. Anal Chem 81:10038–10048

Knox AK, Li C, Vagujfalvi A et al (2008) Identification of candidate CBF genes for the frost tolerance locus *Fr-Am2* in *Triticum monococcum*. Plant Mol Biol 67:257–270

Koh S, Lee SC, Kim MK et al (2007) T-DNA tagged knockout mutation of rice *OsGSK1*, an orthologue of *Arabidopsis BIN2*, with enhanced tolerance to various abiotic stresses. Plant Mol Biol 65:453–466

Komatsu S, Yamamoto R, Nanjo Y et al (2009) A comprehensive analysis of the soybean genes and proteins expressed under flooding stress using transcriptome and proteome techniques. J Proteome Res 8:4766–4778

Komatsuda TM, Pourkeirandish CH, Azhaguvel P et al (2007) Six-rowed barley originated from a mutation in a homeodomain leucine zipper I– class homeobox gene. Proc Natl Acad Sci USA 104:1424–1429

Kopka J, Fernie AR, Weckwerth W et al (2004) Metabolite profiling in plant biology: platforms and destinations. Genome Biol 5(6):101–109

Kopka J, Schauer N, Krueger S et al (2005) GMD@ CSB.DB: the Golm Metabolome Database. Bioinformatics 21:1635–1638

Krishnan P, Kruger NJ, Ratcliffe RG (2005) Metabolite fingerprinting and profiling in plants using NMR. J Exp Bot 56(410):255–265

Kudapa H, Azam S, Sharpe AG et al (2014) Comprehensive transcriptome assembly of chickpea (*Cicer arietinum* L.) using sanger and next generation sequencing platforms: development and applications. PLoS One 9(1):e86039

Kumar V, Jain M (2015) The CRISPR-Cas system for plant genome editing: advances and opportunities. J Exp Bot 66:47–57

Kumari S, Stevens D, Kind T et al (2011) Applying in silico retention index and mass spectra matching for identification of unknown metabolites in accurate mass GC-TOF mass spectrometry. Anal Chem 83:5895–5902

Lai J, Li R, Xu X et al (2010) Genome-wide patterns of genetic variation among elite maize inbred lines. Nat Genet 42:1027–1030

Langridge P, Paltridge N, Fincher G (2006) Functional genomics of abiotic stress tolerance in cereals. Brief Funct Genomics Proteomics 4:343–354

Le DT, Nishiyama R, Watanabe Y et al (2012) Differential gene expression in soybean leaf tissues at late developmental stages under drought stress revealed by genome-wide transcriptome analysis. PLoS ONE 7:e49522

Lee S, Kim SH, Kim SJ et al (2004) Trapping and characterization of cold-responsive genes from T-DNA tagging lines in rice. Plant Sci 166:69–79

Lee K, Avondo J, Morrison H, Blot L, Stark M et al (2006) Visualizing plant development and gene expression in three dimensions using optical projection tomography. Plant Cell 18:2145–2156

Lee DG, Ahsan N, Lee SH et al (2009) Chilling stress-induced proteomic changes in rice roots. J Plant Physiol 166:1–11

Lei Y, Lu L, Liu HY et al (2014) CRISPR-P: a web tool for synthetic single- guide RNA design of CRISPR-system in plants. Mol Plant 7:1494–1496

Levi A, Ovnat L, Paterson AH et al (2009) Photosynthesis of cotton near-isogenic lines introgressed with QTLs for productivity and drought related traits. Plant Sci 177:88–96

Li J, Norville JE, Aach J et al (2013a) Multiplex and homologous recombination-mediated genome editing in *Arabidopsis* and *Nicotiana benthamiana* using guide RNA and Cas9. Nat Biotechnol 31:688–691

Li YF, Wang Y, Tang Y et al (2013b) Transcriptome analysis of heat stress response in switch grass (*Panicum virgatum* L.) BMC Plant Biol 13:153
Liang P, Pardee AB (1992) Differential display of eukaryotic messenger RNA by means of the polymerase chain reaction. Science 257:967–971
Lister R, Ecker JR (2009) Finding the fifth base: genome-wide sequencing of cytosine methylation. Genome Res 19:959–966
Liu ZY, Shi JJ, Zhang LW, Huang JF (2010) Discrimination of rice panicles by hyperspectral reflectance data based on principal component analysis and support vector classification. Biomed Biotechnol 11:71–78
Lockhart DJ, Dong H, Byrne MC et al (1996) Expression monitoring by hybridization to high-density oligonucleotide arrays. Nat Biotechnol 14:1675–1680
Loiacono FV, De Tullio MC (2012) Why we should stop inferring simple correlations between antioxidants and plant stress resistance: towards the antioxidomic era. OMICS 16:160–167
Lorenzo L, Merchan F, Laporte P et al (2009) A novel plant leucine-rich repeat receptor kinase regulates the response of *Medicago truncatula* roots to salt stress. Plant Cell 21(2):668–680
Lukowitz W, Gillmor CS, Scheible WR (2000) Positional cloning in Arabidopsis. Why it feels good to have a genome initiative working for you. Plant Physiol 123:795–806
Ma Y, Qin F, Tran LSP (2012) Contribution of genomics to gene discovery in plant abiotic stress responses. Mol Plant 5(6):1176–1178
Maccaferri M, Sanguineti MC, Demontis A et al (2011) Association mapping in durum wheat grown across a broad range of water regimes. J Exp Bot 62:409–438
Mackay TFC, Stone EA, Ayroles JF (2009) The genetics of quantitative traits: challenges and prospects. Nat Rev Genet 10:565–577
Madlung A, Comai L (2004) The effect of stress on genome regulation and structure. Ann Bot 94:481–495
Mahfouz MM, Piatek A, Stewart CN Jr (2014) Genome engineering via TALENs and CRISPR/Cas9 systems: challenges and perspectives. Plant Biotechnol J 12:1006–1014
Manaa A, Ahmed HB, Smiti S et al (2011) Salt-stress induced physiological and proteomic changes in tomato *(Solanum lycopersicum)* seedlings. OMICS 15:801–809
Mao Y, Zhang H, Xu N et al (2013) Application of the CRISPR-Cas system for efficient genome engineering in plants. Mol Plant 6:2008–2011
Marques MC, Alonso-Cantabrana H, Forment J et al (2009) A new set of ESTs and cDNA clones from full-length and normalized libraries for gene discovery and functional characterization in citrus. BMC Genomics 10:428
Martinovic S, Veenstra TD, Anderson GA et al (2002) Selective incorporation of isotopically labeled amino acids for identification of intact proteins on a proteome-wide level. J Mass Spectrom 37:99–107
Matsumura H, Nirasawa S, Terauchi R (1999) Transcript profiling in rice (*Oryza sativa* L.) seedlings using serial analysis of gene expression (SAGE). Plant J 20:719–726
Meins F (2000) RNA degradation and models for post-transcriptional gene-silencing. Plant Mol Biol 43:261–273
Menconi M, Sgherri CLM, Pinzino C et al (1995) Activated oxygen production and detoxification in wheat plants subjected to a water deficit programme. J Exp Bot 46:1123–1130
Meyers BC, Vu TH, Tej SS et al (2004) Analysis of the transcriptional complexity of *Arabidopsis thaliana* by massively parallel signature sequencing. Nat Biotechnol 22:1006–1011
Micheletto S, Rodriguez-Uribe L, Hernandez R et al (2007) Comparative transcript profiling in roots of *Phaseolus acutifolius* and *P. vulgaris* under water deficit stress. Plant Sci 73:510–520
Mielewczik M, Friedli M, Kirchgessner N, Walter A (2013) Diel leaf growth of soybean: a novel method to analyze two-dimensional leaf expansion in high temporal resolution based on a marker tracking approach (Martrack Leaf). Plant Methods 9:30
Mir RR, Saxena RK, Saxena KB et al (2013) Whole-genome scanning for mapping determinancy in Pigeon pea (*Cajanus spp.*) Plant Breed 132:472–478
Misra AN, Biswal AK, Misra M (2002) Physiological, biochemical and molecular aspects of water stress in plants, and their biotechnological applications. Proc Natl Acad Sci USA 72:115–134

Mizoi J, Shinozaki K, Yamaguchi-Shinozaki K (2012) AP2/ERF family transcription factors in plant abiotic stress responses. Biochim Biophys Acta 1819:86–96

Mizuno HYK, Sakai H, Kanamori et al (2010) Massive parallel sequencing of mRNA in identification of unannotated salinity stress-inducible transcripts in rice (*Oryza sativa* L.) BMC Genomics 11:683–696

Mochida K, Yoshida T, Sakurai T et al (2009) In silico analysis of transcription factor repertoire and prediction of stress responsive transcription factors in soybean. DNA Res 16:353–369

Mochida K, Yoshida T, Sakurai T et al (2011) In silico analysis of transcription factor repertoires and prediction of stress-responsive transcription factors from six major gramineae plants. DNA Res 18:321–332

Moco S, Bino RJ, Vorst O et al (2006) A liquid chromatography-mass spectrometry based metabolome database for tomato. Plant Physiol 141:1205–1218

Mohammadi PP, Moieni A, Hiraga S et al (2012a) Organ-specific proteomic analysis of drought-stressed soybean seedlings. J Proteome 75:1906–1923

Mohammadi PP, Moieni A, Komatsu S (2012b) Comparative proteome analysis of drought-sensitive and drought-tolerant rapeseed roots and their hybrid F1 line under drought stress. Amino Acids 43:2137–2152

Molina CBR, Horres R, Udupa SM et al (2008) Super SAGE: the drought stress-responsive transcriptome of chickpea roots. BMC Genomics 9:553–581

Molina C, Zaman-Allah M, Khan F et al (2011) The salt-responsive transcriptome of chickpea roots and nodules via deep super- SAGE. BMC Plant Biol 11:31

Moragues M, Comadran J, Waugh R et al (2010) Effects of ascertainment bias and marker number on estimations of barley diversity from high-throughput SNP genotype data. Theor Appl Genet 120:1525–1534

Morsy MR, Jouve L, Hausman JF et al (2007) Alteration of oxidative and carbohydrate metabolism under abiotic stress in two rice (*Oryza sativa* L.) genotypes contrasting in chilling tolerance. J Plant Physiol 164:157–167

Munns R, James RA, Sirault XRR, Furbank RT, Jones HG (2010) New phenotyping methods for screening wheat and barley for beneficial responses to water deficit. J Exp Bot 61:3499–3507

Naredo M, Cairns EBJ, Wang H et al (2009) EcoTILLING as a SNP discovery tool for drought candidate genes in Oryza sativa germplasm. Philipp J Crop Sci 34:10–16

Negrao S, Almadanim MC, Pires IS, Abreu et al (2013) New allelic variants found in key rice salt tolerance genes: an association study. Plant Biotechnol J 11:87–100

Nekrasov V, Staskawicz B, Weigel D et al (2013) Targeted mutagenesis in the model plant *Nicotiana benthamiana* usingCas9- guide endonuclease. Nat Biotechnol 31(691):693

Ngara R, Ndimba R, Borch-Jensen J et al (2012) Identification and profiling of salinity stress – responsive proteins in *Sorghum bicolor* seedlings. J Proteome 75:4139–4150

O'Rourke J, Charlson D, Gonzalez D et al (2007) Microarray analysis of iron deficiency chlorosis in near-isogenic soybean lines. BMC Genomics 8:476

Obata T, Fernie AR (2012) The use of metabolomics to dissect plant responses to abiotic stresses. Cell Mol Life Sci 69:3225–3243

Ong SE, Blagoev B, Kratchmarova I et al (2002) Stable isotope labeling by amino acids in cell culture, SILAC, as a simple and accurate approach to expression proteomics. Mol Cell Proteomics 1:376–386

Ong SE, Kratchmarova I, Mann M (2003) Properties of 13C-substituted arginine in stable isotope labeling by amino acids in cell culture (SILAC). J Proteome Res 2:173–181

Ozsolak F, Milos PM (2010) RNA sequencing: advances, challenges and opportunities. Nat Rev Genet 12:87–98

Papdi C, Joseph MP, Salamo IP et al (2009) Genetic technologies for the identification of plant genes controlling environmental stress responses. Funct Plant Biol 36:696–720

Papdi C, Leung J, Joseph MP et al (2010) Genetic screens to identify plant stress genes. In: Sunkar R (ed) Plant stress tolerance: methods in molecular biology, vol 639. Springer, Berlin, pp 121–139

Parida AK, Das AB (2005) Salt tolerance and salinity effects on plants: a rev. Eco Toxicol Environ Saf 60:324–349

Pariset L, Chillemi G, Bongiorni S et al (2009) Microarrays and high-throughput transcriptomic analysis in species with incomplete availability of genomic sequences. New Biotechnol 25:272–279

Peleman JD, van der Voort JR (2003) Breeding by design. Trends Plant Sci 8:330–334

Peng Z, Wang M, Li F et al (2009) A proteomic study of the response to salinity and drought stress in an introgression strain of bread wheat. Mol Cell Proteomics 8(2676):2686

Perez-Clemente R, Vives V, Zandalinas SI et al (2012) Biotechnological approaches to study plant responses to stress. BioMed Res Int. doi.10.1155/2013/654120

Pitzschke A, Schikora A, Hirt H (2009) MAPK cascade signalling networks in plant defence. Curr Opin Plant Biol 12(4):421–426

Poorter H, Fiorani F, Stitt M et al (2012) The art of growing plants for experimental purposes: a practical guide for the plant biologist. Funct Plant Biol 39:821–838

Prashar Y, Weissman SM (1996) Analysis of differential gene expression by display of 3′ end restriction fragments of cDNAs. Proc Natl Acad Sci U S A 93:659–663

Qi XH, Xu XW, Lin XJ et al (2012) Identification of differentially expressed genes in cucumber (*Cucumis sativus* L.) root under waterlogging stress by digital gene expression profile. Genomics 99:160–168

Rácz I, Páldi E, Szalai G et al (2008) S-methylmethionine reduces cell membrane damage in higher plants exposed to low-temperature stress. J Plant Physiol 165:1483–1490

Rafalski JA (2010) Association genetics in crop improvement. Curr Opin Plant Biol 13:174–180

Rajendran K, Tester M, Roy SJ (2009) Quantifying the three main components of salinity tolerance in cereals. Plant Cell Environ 32:237–249

Raney J, Reynolds D, Elzinga D et al (2014) Transcriptome analysis of drought induced stress in *Chenopodium quinoa*. Am J Plant Sci 5(3):338–357

Rashid B, Husnain T, Riazuddin S (2014) Genomic approaches and abiotic stress tolerance in plants. In: Ahmad P (ed) Emerging technologies and management of crop stress tolerance: biological techniques, vol 1. Elsevier Inc, pp 1–37. https://doi.org/10.1016/B978-0-12-800876-8.00001-1

Reddy DS, Bhatnagar-Mathur P, Vadez V et al (2012) Grain legumes (Soybean, Chickpea, and Peanut): omics approaches to enhance abiotic stress tolerance. In: Tuteja N, Gill SS, Tiburcio AF, Tuteja R (eds) Improving crop resistance to abiotic stress. Wiley-VCH Verlag GmbH & Co. KGaA, New York

Reeves A, Parsons RL, Hettinger JW, Medford JI (2002) In vivo three dimensional imaging of plants with optical coherence microscopy. J Microsc 208:177–189

Rhodes D, Samaras Y (1994) Genetic control of osmoregulation in plants. In: Strange K, Raton B (eds) Cellular and molecular physiology of cell volume regulation. CRC Press, Boca Raton, pp 347–361

Ribaut JM, Ragot M (2006) Marker-assisted selection to improve drought adaptation in maize: the backcross approach, perspectives, limitations, and alternatives. J Exp Bot 58:351–360

Ríos G, Naranjo MA, Iglesias DJ et al (2008) Characterization of hemizygous deletions in citrus using array-comparative genomic hybridization and micro synteny comparisons with the poplar genome. BMC Genomics 9:381

Roessner U, Beckles DM (2009) Metabolite measurements. In: Schwender J (ed) Plant metabolic networks. Springer, New York, pp 39–69

Ross PL, Huang YN, Marchese JN et al (2004) Multiplexed protein quantitation in *Saccharomyces cerevisiae* using amine-reactive isobaric tagging reagents. Mol Cell Proteomics 3:1154–1169

Roy SJ, Tucker EJ, Tester M (2011) Genetic analysis of abiotic stress tolerance in crops. Curr Opin Plant Biol 14:232–239

Rudd S (2003) Expressed sequence tags: alternative or complement to whole genome sequences? Trends Plant Sci 8:321–329

Rus A, Baxter I, Muthukumar B, Gustin J et al (2006) Natural variants of *AtHKT1* enhance Na accumulation in two wild populations of *Arabidopsis*. PLoS Genet 2:1964–1973

Sakamoto T, Shibayama M, Kimura A et al (2011) Assessment of digital camera-derived vegetation indices in quantitative monitoring of seasonal rice growth. ISPRS J Photogramm 66:872–882
Salathia N, Lee HN, Sangster TA et al (2007) Indel arrays: an affordable alternative for genotyping. Plant J 51:727–737
Salekdah GH, Komatsu S (2007) Crop proteomics: aim at sustainable agriculture of tomorrow. Proteomics 7:2976–2996
Salvi S, Tuberosa R (2005) To clone or not to clone plant QTLs: present and future challenges. Trends Plant Sci 10:297–304
Sanchez-Bel P, Egea I, Sanchez-Ballesta MT et al (2012) Proteome changes in tomato fruits prior to visible symptoms of chilling injury are linked to defensive mechanisms, uncoupling of photosynthetic processes and protein degradation machinery. Plant Cell Physiol 53:470–484
Sander JD, Joung JK (2014) CRISPR-Cas systems for editing, regulating and targeting genomes. Nat Biotechnol 32:347–155
Sato Y, Shirasawa K, Takahashi Y et al (2006) Mutant selection from progeny of gamma-ray-irradiated rice by DNA heteroduplex cleavage using Brassica petiole extract. Breed Sci 56:179–183
Schena M, Shalon D, Davis RW et al (1995) Quantitative monitoring of gene expression patterns with a complementary DNA microarray. Science 270:467–470
Schripsema J (2010) Application of NMR in plant metabolomics: techniques, problems and prospects. Phytochem Anal 21:14–21
Schuenemann VJ, Bos K, DeWitte S et al (2011) Targeted enrichment of ancient pathogens yielding the pPCP1 plasmid of Yersinia pestis from victims of the black death. Proc Natl Acad Sci U S A 108:746–752
Serraj R, Krishnamurthy L, Kashiwagi J et al (2004) Variation in root traits of chickpea (Cicer arietinum L.) grown under terminal drought. Field Crop Res 88:115–127
Setter TL, Yan J, Warburton M et al (2011) Genetic association mapping identifies single nucleotide polymorphisms in genes that affect abscisic acid levels in maize floral tissues during drought. J Exp Bot 62:701–716
Severin AJ, Woody JL, Bolon YT et al (2010) RNA-Seq Atlas of *Glycine max*: a guide to the soybean transcriptome. BMC Plant Biol 10:160
Shan Q, WangY LJ et al (2013) Targeted genome modification of crop plants using a CRISPR-Cas system. Nat Biotechnol 31:686–688
Shao HB, Guo QJ, Chu LY et al (2007) Understanding molecular mechanism of higher plant plasticity under abiotic stress. Colloids Surf B Biointerfaces 54:37–45
Shao HB, Chu LY, Jaleel CA et al (2008) Water-deficit stress induced anatomical changes in higher plants. C R Biol 331:215–225
Sharma A, Deshmukh RK, Jain N et al (2011) Combining QTL mapping and transcriptome profiling for an insight into genes for grain number in rice (*Oryza sativa* L.) Indian J Genet Plant Breed 7:115–119
Shibayama M, Sakamoto T, Takada E, Inoue A, Morita K, Takahashi W, Kimura A (2009) Continuous monitoring of visible and near-infrared band reflectance from a rice paddy for determining nitrogen uptake using digital cameras. Plant Prod Sci 12:293–306
Shibayama M, Sakamoto T, Takada E et al (2011) Regression-based models to predict rice leaf area index using biennial fixed point continuous observations of near infrared digital images. Plant Prod Sci 14:365–376
Shigeoka S, Ishikawa T, Tamoi M et al (2002) Regulation and function of ascorbate peroxidase isoenzymes. J Exp Bot 53:1305–1319
Shimkets RA, Lowe DG, Tai JT et al (1999) Gene expression analysis by transcript profiling coupled to a gene database query. Nat Biotechnol 17:798–803
Shulaev V (2006) Metabolomics technology and bioinformatics. Brief Bioinform 7:128–139
Shulaev V, Cortes D, Miller G et al (2008) Metabolomics for plant stress response. Physiol Plant 132:199–208

Singh CB, Jayas DS, Paliwal J et al (2010) Identification of insect-damaged wheat kernels using short-wave near infrared hyperspectral and digital colour imaging. Comput Electron Agric 73:118–125

Slade AJ, Fuerstenberg SI, Loeffler D et al (2005) A reverse genetic, non transgenic approach to wheat crop improvement by TILLING. Nat Biotechnol 23:75–81

Smita S, Lenka SK, Katiyar A et al (2011) QlicRice: a web interface for abiotic stress responsive QTL and loci interaction channels in rice. Database 1:1–9

Soda N, Wallace S, Karan R (2015) Omics study for abiotic stress responses in plants. Adv Plants Agric Res 2(1):00037

Sonah H, Deshmukh RK, Chand S et al (2012) Molecular mapping of quantitative trait loci for flag leaf length and other agronomic traits in rice (*Oryza sativa*). Cereal Res Commun 40:362–372

Steele K (2009) Novel upland rice variety bred using marker-assisted selection and client oriented breeding released in Jharkhand. Bangor University, India

Steele KA, Virk DS, Kumar R et al (2007) Field evaluation of upland rice lines selected for QTLs controlling root traits. Field Crop Res 101:180–186

Stepien P, Johnson NG (2009) Contrasting responses of photosynthesis to salt stress in the glycophyte *arabidopsis* and the halophyte *thellungiella*: role of the plastid terminal oxidase as an alternative electron sink. Plant Physiol 149:1154–1165

Stepien P, Klobus G (2006) Water relations and photosynthesis in *Cucumis sativus* L. leaves under salt stress. Biol Plant 50:610–616

Subudhi PK (2011) Omics approaches for abiotic stress tolerance in plants. In: Tuteja N, Gill SS, Tuteja R (eds) Omics and plant abiotic stress tolerance. Bentham Science Publishers Ltd., pp 10–38

Sudhir P, Murthy SDS (2004) Effects of salt stress on basic processes of photosynthesis. Photosynthetica 42:481–486

Sumner LW, Mendes P, Dixon RA (2003) Plant metabolomics: large scale phytochemistry in the functional genomics era. Phytochemistry 62:817–836

Sutcliffe JG, Foye PE, Erlander MG et al (2000) TOGA: an automated parsing technology for analyzing expression of nearly all genes. Proc Natl Acad Sci U S A 97:1976–1981

Sutton T, Baumann U, Hayes J et al (2007) Boron-toxicity tolerance in barley arising from efflux transporter amplification. Science 318:1446–1449

Suzuki T, Eiguchi M, Kumamaru T et al (2008) MNU-induced mutant pools and high performance TILLING enable finding of any gene mutation in rice. Mol Gen Genomics 279:213–223

Takeda S, Matsuoka M (2008) Genetic approaches to crop improvement: responding to environmental and population changes. Nat Rev Genet 9:444–457

Tao X, Gu YH, Wang HY et al (2012) Digital gene expression analysis based on integrated de novo transcriptome assembly of sweet potato [*Ipomoea batatas* (L.) Lam]. PLoS One 7:e36234

Terada R, Johzuka- Hisatomi Y, Saitoh M et al (2007) Gene targeting by homologous recombination as a biotechnological tool for rice functional genomics. Plant Physiol 144:846–856

Thakur P, Kumar S, Malik JA, Berger JD, Nayyar H (2010) Cold stress effects on reproductive development in grain crops: an overview. Environ Exp Bot 67(3):429–443

Thornsberry JM, Goodman MM, Doebley J et al (2001) Dwarf8 polymorphisms associate with variation in flowering time. Nat Genet 28:286–289

Till BJ, Reynolds SH, Weil C et al (2004) Discovery of induced point mutations in maize genes by TILLING. BMC Plant Biol 4:12

Till BJ, Cooper J, Tai TH et al (2007) Discovery of chemically induced mutations in rice by TILLING. BMC Plant Biol 7:19

Tissier AF, Marillonnet S, Klimyuk V et al (1999) Multiple independent defective suppressor-mutator transposon insertions in Arabidopsis: a tool for functional genomics. Plant Cell 11:1841–1852

Tran LS, Mochida K (2010a) A platform for functional prediction and comparative analyses of transcription factors of legumes and beyond. Plant Signal Behav 5:550–552

Tran LS, Mochida K (2010b) Identification and prediction of abiotic stress responsive transcription factors involved in abiotic stress signaling in soybean. Plant Signal Behav 5:255–257

Triques K, Sturbois B, Gallais S et al (2007) Characterization of *Arabidopsis thaliana* mismatch specific endonucleases: application to mutation discovery by TILLING in pea. Plant J 51:1116–1125

Trujillo LE, Sotolongo M, Menendez C et al (2008) SodERF3, a novel sugarcane ethylene responsive factor (ERF), enhances salt and drought tolerance when over expressed in tobacco plants. Plant Cell Physiol 49(4):512–525

Ueda A, Kathiresan A, Bennett J et al (2006) Comparative transcriptome analyses of barley and rice under salt stress. Theor Appl Genet 112:1286–1294

Unlu M, Morgan ME, Minden JS (1997) Difference gel electrophoresis: a single gel method for detecting changes in protein extracts. Electrophoresis 18:2071–2077

Urano K, Maruyama K, Ogata Y et al (2009) Characterization of the ABA-regulated global responses to dehydration in *Arabidopsis* by metabolomics. Plant J 57(6):1065–1078

Urano K, Kurihara Y, Seki M et al (2010) 'Omics' analyses of regulatory networks in plant abiotic stress responses. Curr Opin Plant Biol 13:132–138

Velculescu VE, Zhang L, Vogelstein B et al (1995) Serial analysis of gene expression. Science 270:448–487

Vinocur B, Altman A (2005) Recent advances in engineering plant tolerance to abiotic stress: achievements and limitations. Curr Opin Biotechnol 16:123–132

Voytas DF (2013) Plant genome engineering with sequence- specific nucleases. Ann Rev Plant Biol 64:327–350

Wakasa Y, Oono Y, Yazawa T et al (2014) RNA sequencing-mediated transcriptome analysis of rice plants in endoplasmic reticulum stress conditions. BMC Plant Biol 14:101

Wakeel A, Asif AR, Pitann B et al (2011) Proteome analysis of sugar beet (*Beta vulgaris* L.) elucidates constitutive adaptation during the first phase of salt stress. J Plant Physiol 168:519–526

Walia H, Wilson C, Ismail AM et al (2009) Comparing genomic expression patterns across plant species reveals highly diverged transcriptional dynamics in response to salt stress. BMC Genomics 10:398

Wallays C, Missotten B, De Baerdemaeker J et al (2009) Hyperspectral waveband selection for on-line measurement of grain cleanness. Biosyst Eng 104:1–7

Wang W, Vinocur B, Altman A (2003) Plant responses to drought, salinity and extreme temperatures: towards genetic engineering for stress tolerance. Planta 218(1):1–14

Wang WQ, Møller IM, Song SQ (2012) Proteomic analysis of embryonic axis of *Pisum sativum* seeds during germination and identification of proteins associated with loss of desiccation tolerance. J Proteome 77:68–86

Waterhouse PM, Graham MW, Wang MB (1998) Virus resistance and gene silencing in plants can be induced by simultaneous expression of sense and antisense RNA. Proc Natl Acad Sci U S A 95:13959–13964

Wilkins MR, Pasquali C, Appel RD et al (1996) From proteins to proteomes: large scale protein identification by two-dimensional electrophoresis and amino acid analysis. Biotechnology 14:61–65

Witzel K, Weidner A, Surabhi GK et al (2009) Salt stress-induced alterations in the root proteome of barley genotypes with contrasting response towards salinity. J Exp Bot 60:3545–3557

Wu Y, Cosgrove DJ (2000) Adaptation of roots to low water potentials by changes in cell wall extensibility and cell wall proteins. J Exp Bot 51:1543–1155

Wu D, Shen Q, Qiu L et al (2014) Identification of proteins associated with ion homeostasis and salt tolerance in barley. Proteomics 14:1381–1392

Xin Z, Wang ML, Barkley NA et al (2008) Applying genotyping (TILLING) and phenotyping analyses to elucidate gene function in a chemically induced sorghum mutant population. BMC Plant Biol 8:103

Xing HL, Dong L, Wang ZP et al (2014) A CRISPR/Cas9 tool kit for multiplex genome editing in plants. BMC Plant Biol 14:327

Xu K, Xu X, Fukao T et al (2006) Sub1A is an ethylene-response-factor-like gene that confers submergence tolerance to rice. Nature 442:705–708

Xu Y, Gao S, Yang Y, Huang M et al (2013) Transcriptome sequencing and whole genome expression profiling of chrysanthemum under dehydration stress. BMC Genomics 14:662

Yang WN, Xu XC, Duan LF et al (2011a) High-throughput measurement of rice tillers using a conveyor equipped with x-ray computed tomography. Rev Sci Instrum 82:025102

Yang ZB, Rotter B, Rao IM et al (2011b) Physiological and molecular analysis of polyethylene glycol-induced reduction of aluminium accumulation in the root tips of common bean (*Phaseolus vulgaris*). New Phytol 192:99–113

Yang W, Duan L, Chen G et al (2013) Plant phenomics and high-throughput phenotyping: accelerating rice functional genomics using multidisciplinary technologies. Curr Opin Plant Biol 16:180–187

Yoshimura K, Masuda A, Kuwano M et al (2008) Programmed proteome response for drought avoidance/tolerance in the root of a C3 xerophyte (wild watermelon) under water deficits. Plant Cell Physiol 49:226–241

Yu JJBH, McMullen MD, Buckler ES (2008) Genetic design and statistical power of nested association mapping in maize. Genetics 178:539–551

Yuan JS, Galbraith DW, Dai SY, Griffin P, Stewart CN Jr (2008) Plant systems biology comes of age. Trends Plant Sci 13:165–171

Zamir D (2001) Improving plant breeding with exotic genetic libraries. Nat Rev Genet 2:983–989

Zhang H, Li Y, Wang B et al (2008) Recent advances in cotton genomics. Int J Plant Genom 742304:20

Zhang L, Tian LH, Zhao JF et al (2009) Identification of an apoplastic protein involved in the initial phase of salt stress response in rice root by two-dimensional electrophoresis. Plant Physiol 149:916–928

Zhang JJ, Zhou JS, Song JB et al (2012) Molecular dissection of atrazine-responsive transcriptome and gene networks in rice by high-throughput sequencing. J Hazard Mater 219–220:57–68

Zhao K, Tung CW, Eizenga GC et al (2011) Genome-wide association mapping reveals a rich genetic architecture of complex traits in Oryza sativa. Nat Commun 2:467–475

Zhou H, Liu B, Weeks DP et al (2014) Large chromosomal deletions and heritable small genetic changes induced by CRISPR/Cas9 in rice. Nucl Acids Res 42:10903–10914

Zhu JM, Alvarez S, Marsh EL et al (2007) Cell wall proteome in the maize primary root elongation zone II region-specific changes in water soluble and lightly ionically bound proteins under water deficit. Plant Physiol 145:1533–1548

Zhu YN, Shi DQ, Ruan MB et al (2013) Transcriptome analysis reveals crosstalk of responsive genes to multiple abiotic stresses in cotton (*Gossypium hirsutum* L.) PLoS One 8(11):e80218

Zorb C, Schmitt S, Muhling KH (2010) Proteomic changes in maize roots after short-term adjustment to saline growth conditions. Proteomics 10:4441–4449

Second Messengers: Central Regulators in Plant Abiotic Stress Response

2

Muskan Jain, Preeti Nagar, Parul Goel, Anil Kumar Singh, Sumita Kumari, and Ananda Mustafiz

Abstract

Plants differ from animals by lacking the ability to escape from their environmental conditions. Plants adapt to the seasonal as well as nonseasonal perturbations by means of stress-responsive genes. Manipulation of such genes has been shown to provide abiotic stress tolerance in plants. Since abiotic stress is a polygenic trait, overexpression of single stress-responsive gene would not serve the purpose of getting stress-tolerant plants. So, the focus needs to be shifted towards the "master regulators" which are critical for plant growth and development and play an important role in integrating various stress signals and controlling downstream stress responses by modulating gene expression machinery. In plants, there are various second messengers including calcium, ROS, phosphoinositides, cyclic nucleotides, etc., which are known to initiate the downstream signaling cascade leading to response against different, multiple, and simultaneous ambi-

Muskan Jain and Preeti Nagar contributed equally to this work.

M. Jain · P. Nagar · A. Mustafiz (✉)
Laboratory of Plant Molecular Biology, Faculty of Life Sciences and Biotechnology, South Asian University, New Delhi, India
e-mail: amustafiz@sau.ac.in

P. Goel
Department of Biotechnology, CSIR-Institute of Himalayan Bioresource Technology, Palampur, Himachal Pradesh, India

A. K. Singh
ICAR-Indian Institute of Agricultural Biotechnology, PDU Campus, IINRG, Ranchi, Jharkhand, India

S. Kumari
School of Biotechnology, Sher-e-Kashmir University of Agricultural Sciences and Technology, Jammu, India

S. M. Zargar, M. Y. Zargar (eds.), *Abiotic Stress-Mediated Sensing and Signaling in Plants: An Omics Perspective*, https://doi.org/10.1007/978-981-10-7479-0_2

ent cues. A better understanding of these elements will allow us to engineer a particular stress-responsive pathway, to achieve better stress-tolerant plants.

Keywords
Abiotic stress · Second messengers · Calcium · Calmodulin (CaM) · Calmodulin binding proteins (CaMBPs) · Calmodulin like sensors (CMLs) · Calcineurin B like sensors (CBLs) · Calcineurin B like interacting protein kinases (CIPKs) · Calcium dependent protein kinases (CDPKs) · Reactive oxygen species (ROS) · Cyclic nucleotides · Phosphoinositides · Phosphatidic acid (PA) · Cross talk

2.1 Introduction

Abiotic stress includes all the environmental factors that affect the growth and development of plants in a negative manner. Common forms of abiotic stresses include salinity, drought, flooding, extreme temperatures, extreme pH of soil, etc. (Apel and Hirt 2004; Duque et al. 2013). These unfavorable environmental conditions severely affect plant growth and development leading to reduced productivity; hence, they have to adapt themselves for such kind of adverse environmental conditions (Osakabe et al. 2013). Plants have developed a large array of physical and biochemical strategies for this. Under stress conditions, plants respond by specifically enhancing the expression of various stress-responsive genes (Hasanuzzaman et al. 2013; Nakashima et al. 2014; Sah et al. 2016). The fundamental strategy that might be employed for producing stress-tolerant varieties is the identification of crucial genes involved in stress tolerance. Although the target of generating stress tolerance response can be achieved by the overexpression of such genes in transgenic plants, it may have some negative effects on normal growth and developmental processes and hence compromising with quality and yield also (Thompson et al. 2000; Tung et al. 2008; Estrada-Melo et al. 2015; Li et al. 2016). Moreover, plants are constantly exposed to multitude of stresses at a given time rather than a single stress, and manipulation of an individual gene can only provide tolerance for that particular stress. So, the next strategy that could be followed to overcome the above mentioned bottlenecks is to identify and modulate the regulatory genes for signal perception and transduction, which are able to control a whole battery of genes that may be crucial for multiple stress tolerance (Atkinson and Urwin 2012). But to implicate this, an in-depth and systematic understanding of signal transduction pathways and different signaling molecules including the role of different second messengers in signal transduction and stress response generation is required along with their cross-talk information, so as to generate crop varieties that can withstand multiple stresses without any yield penalty. This chapter reviews the role of individual second messenger in establishing the stress tolerance response in plants to different stresses. It also explains the significance of cross talk between specific signaling pathways via second messengers in resisting the lethal abiotic stresses. This review will help to reveal the key regulatory and potential candidates of signaling pathway that can be exploited for development of transgenic crops which can withstand the adverse combination of abiotic stresses.

2.2 Second Messengers

Plants, being highly complicated immobile multicellular organism with multiple tissues and organs positioned above and below the soil layer, need short as well as long intra- and intercellular communications to develop a well-coordinated stress response. The perceived stress signal is amplified and then transmitted to trigger a response. A lot of molecules are involved in the process of stress perception and stress response. When there is any change in the environmental conditions, it is sensed by the sensors present in the plasma membrane of the cell. The sensors then, in turn, activate certain molecules that translate this external signal to intracellular signals (Knight et al. 1997). These molecules are known as second messengers. Second messengers are basically, a group of small intracellular signaling molecules, produced in response to an external stress signal received by the sensors and act as key regulatory components of signaling pathways that are meant to transduce the signal to the effectors by activating an array of downstream cascade. Most commonly studied second messengers include calcium, reactive oxygen species, lipid-derived signals, and cyclic nucleotides that work in a synergetic way as they cross talk with each other simultaneously. So, these master regulators can act as important candidates which may provide the opportunities to develop broad-spectrum stress-tolerant crops by connecting various stress response pathways with least negative pleiotropic effects on plant phenotype (Fig. 2.1).

2.2.1 Calcium

Calcium is an important macronutrient for plant system, and it regulates various physiological processes (Tuteja and Mahajan 2007; Tuteja 2007; Mahajan et al. 2006a; Pardo et al. 1998) as it has structural roles in the cell wall and membranes, and it regulates plant growth and development (Hepler 2005). Calcium is stored in a number of vesicular compartments in the cell from where it is released whenever required. To maintain calcium homeostasis in the cell, there are Ca^{2+} channels, pumps, and exchangers which work together to adapt to all kinds of stimulus received (Kudla et al. 2010). The Ca^{2+} concentrations in the apoplast is 0.1 mM, 1.0 mM in vacuoles, 1 mM in ER lumen and 200 nM is the free ion concentration in cytosol, and these concentrations are maintained at these levels to avoid toxicity (Sanders et al. 2002). The total Ca^{2+} concentration of cytosol is much higher due to its high affinity to various calcium-binding proteins (Bush 1995). However, the cytosolic calcium concentrations have been reported to change in response to various stresses such as cold, water, heat and salinity, and changing nitrate availability in soil (Mahajan et al. 2005; Tuteja 2007; Tuteja and Mahajan 2007; Takano et al. 1997; Gong et al. 1998; Scrase-Field and Knight 2003; Riveras et al. 2015). Also the calcium concentration in subcellular organelles like in mitochondria and nucleus is elevated following stress conditions. The calcium signals generated by each stimulus differ in terms of its spatial (cytosolic or organellar) and kinetic features (amplitude, duration, and frequency), thus tightly regulating the subsequent responses that

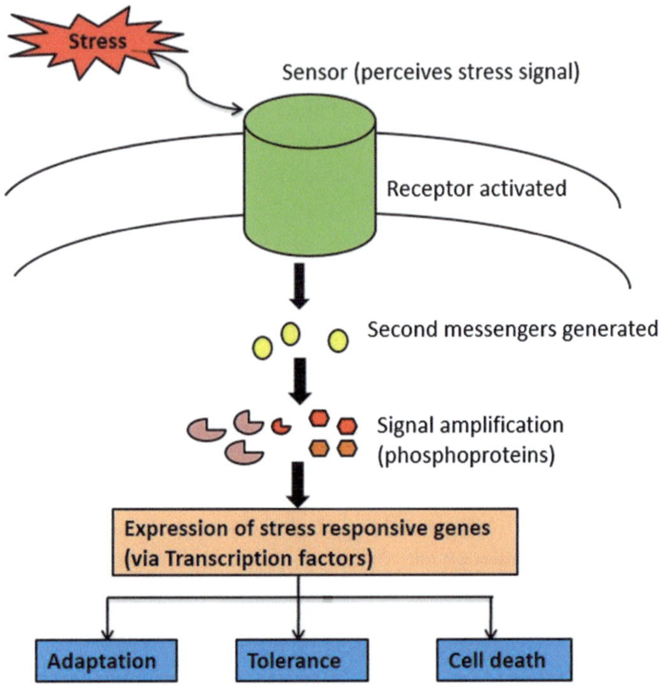

Fig. 2.1 General pathway of stress perception and response. The sensors present in the plasma membrane get activated on perceiving the stress in the environment, leading to the generation of second messengers. Second messengers initiate phosphoprotein cascade resulting in phosphorylation and dephosphorylation of various transcription factors which stimulate the expression of stress-responsive genes leading to stress response which may either be adaptation or tolerance to the stress present, and if the cell is not able to adapt, it dies

are generated (White and Broadley 2003; Pauly et al. 2001). Calcium release can occur either from extracellular space or from activation of PLC (phospholipase C) which hydrolyzes PIP_2 to IP_3 leading to the release of calcium from intracellular stores. Recently, it has been revealed that nitrate treatment can transiently increase cytoplasmic calcium concentration in *Arabidopsis* roots by activating phospholipase C (Riveras et al. 2015). Different calcium stores release calcium, depending on the type of stimulus received by the cell (White and Broadley 2003). Calcium signals carry specific information that distinguishes various abiotic stresses; so they are sometimes referred to as calcium signatures. For example, in tobacco seedlings, wind and cold induce the expression of NpCaM 1 (*Nicotiana plumbaginifolia* calmodulin 1 gene) in a Ca^{2+}-dependent manner. Although both stresses increase Ca^{2+} level in cytosol and nucleus, cytosolic calcium triggers NpCaM 1 induction by cold, whereas nuclear calcium is responsible for NpCaM 1 induction by wind (van Der Luit et al. 1999). During adverse stress effects, Ca^{2+} increases the content of proline, glycine, and betaine, thus improving the water status of plants (Geisler et al. 2000; Munns et al. 2006; Goldgur et al. 2007).

Calcium signaling has three phases: generation of a calcium signature, sensing the signal, and transduction of this calcium signal (Reddy and Reddy 2004).

Any modification in the concentration of calcium must be decoded in the target cells and organs to respond simultaneously to the stimulus. This is done by certain sensor proteins known as calcium-binding proteins (CBPs) that decode and relay the information encoded by Ca^{2+} signatures into specific protein-protein interactions, defined phosphorylation cascades, or transcriptional responses (Luan et al. 2002; Sanders et al. 2002; Finkler et al. 2007). Consequently, the dynamic interplay between Ca^{2+} signatures and Ca^{2+}-sensing proteins contributes to generating stimulus specificity of Ca^{2+} signaling.

Calcium sensors have been divided into two groups (Kudla et al. 2010; Sanders et al. 2002):

(A) *The sensor relays*, which, bind Ca^{2+} ions and undergo Ca^{2+}-induced conformational changes but lack other effector domains. To transmit the Ca^{2+} signal, sensor relay proteins interact with target proteins and regulate their activity. This class includes calmodulin (CaM) and calcineurin B-like (CBLs) proteins. Calcium-activated sensors after getting activated by the stress signal can:
   1. Directly bind to cis elements in promoters of stress-responsive genes
   2. Initiate a phosphorylation or dephosphorylation cascade of transcription factors
   3. Bind to DNA-binding proteins ultimately resulting in activation of genes involved in stress response

(B) *The sensor responders*, which, combine within one protein a sensing function (Ca^{2+}-binding and Ca^{2+}-induced conformational changes) with a response activity (e.g., protein kinase activity). It includes calcium-dependent protein kinases (CDPKs) as well as calmodulin-dependent protein kinases (CaMKs) and CIPKs (CBL-interacting protein kinases) (Fig. 2.2).

The major calcium sensors are discussed below.

2.2.1.1 Calmodulin (CaM) and Calmodulin-Like (CMLs) Sensors

Calmodulin is a small protein comprising of four Ca^{2+}-binding sites known as EF hands (Babu et al. 1988; Rhoads and Friedberg 1997; Luan et al. 2002). Calcium binding modifies CaM globular structure into an open conformation that allows interaction with target proteins known as CaM-binding proteins (CaMBPs) (Snedden and Fromm 2001; McCormack et al. 2005; Hoeflich and Ikura 2002; Yamniuk and Vogel 2005). This interaction translates the calcium signal into a biochemical response by either activating or inhibiting the CaM targets (Lee et al. 2000; Choi et al. 2005a, b; Yoo et al. 2005). CaMBPs are known to participate in regulating transcription, metabolism, ion transport, protein folding, cytoskeleton-associated functions, protein phosphorylation and dephosphorylation, and phospholipid metabolism as well (Snedden and Fromm 2001; Bouché et al. 2005; Kim et al. 2009; Du et al. 2011; Reddy et al. 2011; Yang and Poovaiah 2003; Reddy and Reddy 2004). Being quite similar to CaMs, CaM-like proteins (CMLs) also contain four

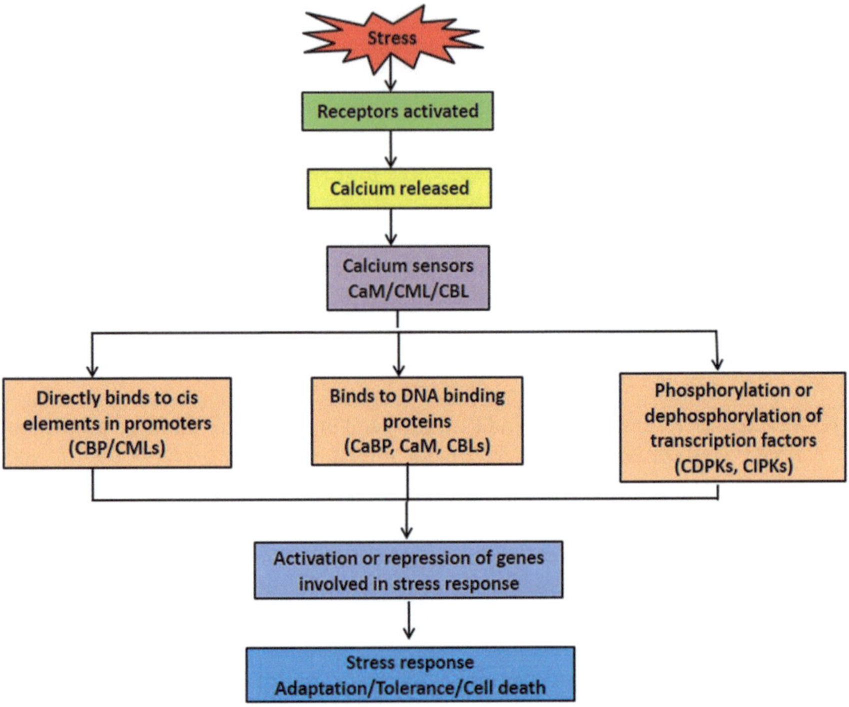

Fig. 2.2 General pathway involved in calcium signaling showing all the major participating components. Any stress signal in the environment is sensed by the receptors present on the plasma membrane which trigger the release of calcium from different sources, directly or indirectly. The changes in calcium concentration are sensed by the calcium sensor proteins (*CaM* calmodulin, *CMLs* calmodulin-like protein sensors, *CBLs* calcineurin B-like proteins). The sensor protein relays this signal to the effector proteins via different mechanisms, all of which lead to changes in gene expression resulting in stress response

EF hands and lack the other effector domains (McCormack and Braam 2003; Boonburapong and Buaboocha 2007; Perochon et al. 2011). CMLs have important roles in providing developmental, hormonal, and stress response (Bender and Snedden 2013). Multiple genes for CaM, i.e., four, six, and seven, have been reported in potato (*Solanum tuberosum*), tomato (*Lycopersicon esculentum*), and tobacco (*Nicotiana tabacum*), respectively (Zhao et al. 2013). The *Arabidopsis* genome contains 7 CaM genes encoding 4 isoforms and 50 genes encoding CMLs (McCormack and Braam 2003; McCormack et al. 2005). One hunred seventy-three protein targets for seven CaMs/CMLs in *Arabidopsis* have been identified (Popescu et al. 2007). The rice plant contains 5 CaM and 32 CML genes (Boonburapong and Buaboocha 2007). CaM exhibits high level of target specificities and also displays multiple subcellular localizations (Luan et al. 2002; Lee et al. 1999; Li et al. 2006; Yang and Poovaiah 2003). The reduced gene expression and reduced stress tolerance after treatment with CaM antagonists implied toward the involvement of CaMs in abiotic stress responses (Monroy et al. 1993; Liu et al. 2003). Various stresses such as cold, touch, heat shock, and salinity induce expression of CaMs and CMLs (Luan et al. 2002; Yang and Poovaiah 2003; Liu et al. 2003, 2005; Park et al. 2010; Gong et al. 1997a, b; Polisensky and Braam 1996; Delk et al. 2005; Xu et al. 2011). CaM3 provides tolerance against cold and heat stress, whereas CaM4 provides resistance to salinity (Townley and Knight 2002; Yang et al. 2010; Liu et al. 2005, 2007, 2008; Zhang et al. 2009a, b; Kutuzov et al. 1998; Jae et al. 2005). CaM7 binds to the promoter of light-responsive genes (Kushwaha et al. 2008). ABA regulates stomatal status, and the biosynthesis of ABA is dependent on cellular Ca^{2+} concentrations (Rabbani et al. 2003; NOCTOR 2006). Ca^{2+} via CaM/CMLs regulates ABA-induced signaling in drought stresses (Kaplan et al. 2006; Magnan et al. 2008). CaMs bind to MPK8 (MAP kinase 8) that are required for production of ROS after wounding (Takahashi et al. 2010). CaM also regulates ROS homeostasis and the entire antioxidant system including catalase, superoxide dismutase, and ascorbate peroxidase (Gong et al. 1997a, b; Yang and Poovaiah 2002; Medda et al. 2003; Mura et al. 2005; Gong and Li 1995; Larkindale and Knight 2002; Subbaiah and Sachs 2000; Ma et al. 2012; Shi et al. 2012; Takahashi et al. 2011; Du and Poovaiah 2004; Wang et al. 2015; Doherty et al. 2009; Kim et al. 2013a, b). CaM binds to and regulates various transcription factors generating the stress response, and the entire family of transcription factors is called as CAMTAs (calmodulin-binding transcriptional activators) (Reddy et al. 2011; Finkler et al. 2007). CAMTAs include representatives from bZIP, WRKY, NAC, and MYB families of transcription factors (Reddy et al. 2000; Popescu et al. 2007; Kim et al. 2007; Yoon et al. 2008).

2.2.1.2 Calmodulin-Binding Proteins (CaMBPs)

Since CaM does not have any enzymatic activity by itself, studying CaM-regulated proteins would provide insight into CaM functions in abiotic stresses. Many of the identified CaMBPs are induced by salinity, drought, or cold (Reddy and Reddy 2004). CaMBPs are divided into two classes: transduction proteins (includes protein kinases and transcription factors) and effector proteins (includes ion transporters and enzymes). CaM-regulated protein kinases (CaMKs) have not been well characterized in plants; only one CaMK has been identified in plants and it is from apple (Harper et al. 2004). Some CRKs (CDPK-related kinases) are stimulated by CaM in a calcium-dependent manner (Harper et al. 2004).

A CRK NtCBK2 is up-regulated by salt stress, thus indicating the role of CRKs in salinity stress tolerance (Hua et al. 2004). CaMs have been reported to show both positive and negative effects over transcription factors. GmCaM4 confers salt stress tolerance via activating transcription factor AtMYB2 (Yoo et al. 2005). CaM has been found to stimulate GLYI activity, thus providing salt tolerance (Bouché et al. 2005). During the heat shock response, CaM induces the expression of HSP genes, thus, increasing the DNA binding of heat shock transcription factors (Li et al. 2004; Liu et al. 2003). Contrarily, CaM also inhibits the transcriptional activation mediated by OsCBT (similar to CaM-binding transcription activator) induced in response to multiple stresses (Choi et al. 2005a, b; Yang and Poovaiah 2003). Ca^{2+}/CaM is also involved in GABA (γ-aminobutyric acid) regulation and also provides

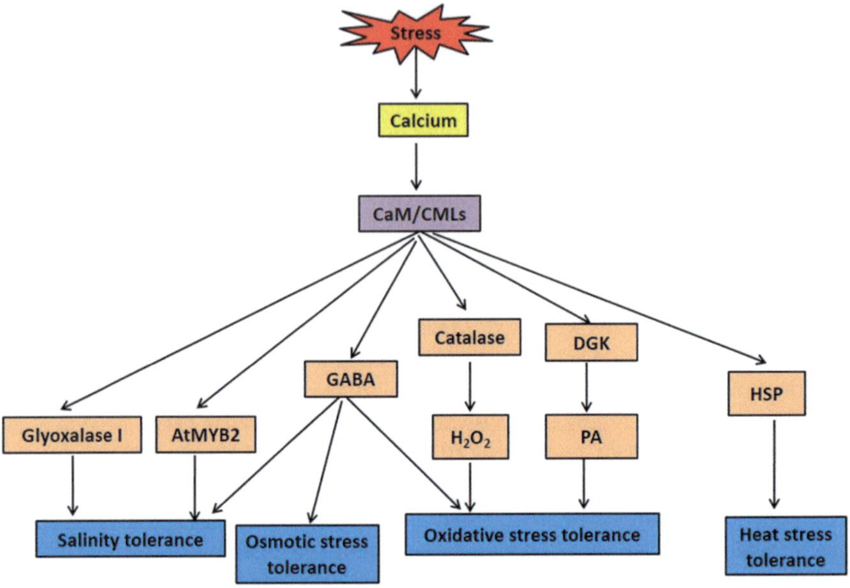

Fig. 2.3 A schematic diagram showing how calmodulin (*CaM*) functions to provide stress tolerance. After sensing stress signals, calcium release activates calmodulin which induces activation of various other molecules or genes or enzymes. CaM induces the expression of Glyoxalase I and another protein MYB2, both of which help in providing stress tolerance in case of salinity stress. CaM also induces the enzyme DGK (diacylglycerol kinase) which leads to accumulation of PA (phosphatidic acid) which is responsible for providing oxidative stress tolerance. CaM induces H_2O_2 detoxification by activating catalase. Also, HSP (heat shock proteins) are induced by CaM which provide tolerance to heat stress. By regulating GABA, CaM modulates responses in salinity, osmotic, and oxidative stress conditions. Thus, CaM acts as a second messenger in case of various stresses

tolerance to oxidative, heat, salt, and osmotic stress via regulation of effector proteins (Lee et al. 2010). In response to oxidative stress, CaM plays a dual role; CaM-activated NAD kinase induces H_2O_2 production, and on the other hand, it activates catalase enzymes inducing H_2O_2 detoxification (Yang and Poovaiah 2002). CaM also modifies cellular localization of target proteins. CaM recruits a tomato diacylglycerol kinase (LeDGK) to membranes where its substrate is located (Yang and Poovaiah 2003). As DGK produces phosphatidic acid (PA), involved in abiotic stress signaling, CaM may play a positive role in multiple stress responses by regulating PA signaling (Xiong et al. 2002; Bargmann and Munnik 2006) (Fig. 2.3).

2.2.1.3 Calcineurin B-Like Sensors

CBLs are small proteins composed of two globular domains connected by a short linker; each domain contains two EF-hand motifs, thus varying the response specificity by differing Ca^{2+} capacities and affinities (Nagae et al. 2003). CBLs are capable of sensing the calcium signal generated by stress conditions (Batistic and Kudla 2004; Luan et al. 2002; Batistič and Kudla 2012; Dodd et al. 2010). Upon calcium binding, CBLs undergo conformational changes allowing hydrophobic interactions with other proteins (Sánchez-Barrena et al. 2005). CBLs interact with certain protein kinases (known as CIPKs) which further transmit the signal of CBLs, generating a stress response (Shi et al. 1999). CBLs can also change their cellular localizations and help proteins to perform different functions (Luan et al. 2002). Several CBLs possess a myristoylation site which promotes membrane association (Kolukisaoglu et al. 2004; Ishitani et al. 2000; Cheong et al. 2007). Genes encoding CBLs and CIPKs have been identified from *Arabidopsis thaliana* (10 CBLs and 26 CIPKs), *Medicago truncatula* (9 CBLs and 11 CIPKs), *Triticum aestivum* (11 CBLs and 29 CIPKs), *Hordeum vulgare* (9 CBLs and 14 CIPKs), *Glycine max* (7 CBLs and 13 CIPKs), *Pinus* (2 CBLs and 7 CIPKs), and the moss *Physcomitrella patens* (4 CBLs and 3 CIPKs). Genes encoding CBLs or CIPKs have not been identified outside the plant kingdom by computer analysis indicating that the function of these genes is restricted to plants (Kolukisaoglu et al. 2004). Exposure to various stress conditions as cold, drought, salinity, and ABA leads to differential expression of CBL genes indicating their role in abiotic stress response (Kudla et al. 2010). Expression of CBL1 was strongly increased in response to wounding, drought, and cold treatments (Mahajan 2006; Kudla et al. 1999). The CBL1 also participates in decoding the Ca^{2+} signal under low nitrate concentration in *Arabidopsis* (Léran et al. 2015). AtCBL2 transcript increased on illumination of leaves with light (Nozawa et al. 2001). PsCBL was up-regulated in cold stress, salt stress, wounding, and salicylic acid and exogenously provided $CaCl_2$ (Mahajan 2006). CBL4 was highly up-regulated in saline stress (Xiong et al. 2002). In *Arabidopsis*, functional characterization of CBL7 has revealed its involvement in decoding Ca^{2+} signal during low nitrate response and regulating the expression of two genes encoding high-affinity nitrate transporters (NRT2.4, NRT2.5) (Ma et al. 2015). CBL9 and CBL10 have overlapping function with SOS3 in salt tolerance (Quan et al. 2007). CBL9 was induced by multiple stress signals such as cold, salt, drought, and ABA. CBL9 functions as a negative regulator of calcium-induced ABA signaling and ABA biosynthesis promoting seed germination (Pandey et al. 2004). OsCBL2 was up-regulated in response to GA (Hwang et al. 2005). CBL9 in combination with its target protein CIPK23 also participates in nitrate-mediated Ca^{2+} signaling under low nitrate condition (Ho et al. 2009).

2.2.1.4 Calcineurin B-Like-Interacting Protein Kinases (CIPKs)

A family of novel serine-threonine protein kinases was recognized as the main plant CBL partner (Shi et al. 1999; Halfter et al. 2000; Luan et al. 2002; Batistic and Kudla 2004). CIPKs or PKS (SOS2-like protein kinases) belong to SnRK3 family (Hrabak et al. 2003; Harper et al. 2004). Sequence analysis of CIPKs (CBL-interacting protein kinases) revealed a two-domain structure: the N-terminal catalytic and C-terminal regulatory domain (Guo et al. 2001; Shi et al. 1999; Batistič et al. 2010; Akaboshi et al. 2008; Marıa José Sánchez-Barrena et al. 2007a, b). Catalytic domain harbors an activation loop between the conserved amino acids DFG and APE. Regulatory domain is unique and harbors NAF motif which acts as an autoinhibitory domain and is also responsible for the interaction of CIPK with CBL. PPI domain is adjacent to

NAF motif and responsible for interaction with phosphatases (Ohta et al. 2003; Yunta et al. 2011; Lee et al. 2008). Catalytic and regulatory domains are connected by a junction domain which is also responsible for kinase activation (Batistič and Kudla 2009; Albrecht et al. 2001; Gong et al. 2002). Twenty-six CIPK genes were identified in *Arabidopsis* and 30 CIPK genes in rice (Kolukisaoglu et al. 2004; Batistič and Kudla 2009; Albrecht et al. 2001). CBL-CIPK interaction stimulates kinase activity and targets the complex to the plasma membrane, where CIPKs can phosphorylate specific substrates (Kolukisaoglu et al. 2004; D'Angelo et al. 2006). Differential stress induction of CIPK genes has been reported in distinct plant species, indicating their role in stress responses (Kolukisaoglu et al. 2004). CIPK1 is involved in osmotic stress signaling (D'Angelo et al. 2006). Expression of AtCIPK3 was induced strongly by cold, drought, salt, and wounding. The extracellular stress signal is transduced by the sensors inside the cell eliciting cytosolic calcium levels. The increased calcium is sensed by CBLs which interact with CIPKs, and this interaction mediates responses to various stresses such as cold, salt, drought, and ABA (D'Angelo et al. 2006; Cheong et al. 2007; Albrecht et al. 2001; Halfter et al. 2000; Kim et al. 2007; Pandey et al. 2004). The CIPKs include SOS genes (salt overly sensitive) which are known to participate in calcium-mediated pathway for salt stress tolerance (Mahajan et al. 2008; Liu and Zhu 1997). Various SOS genes and their interacting CBLs have been identified. The SOS1, SOS2/CIPK24, and SOS3/CBL4 function in salinity stress conditions (Xiong et al. 2002). Various genetic screening and mutant studies have been done on the SOS genes to know their functions and their roles in abiotic stresses (Halfter et al. 2000; Guo et al. 2001; Gong et al. 2002). In general pathway, the change in salt concentration is sensed by SOS3 protein which translates the signal down and activates SOS2, recruiting it to the plasma membrane where it activates SOS1 (Na^+ /H^+ anti-porter) (Quintero et al. 2002; Halfter et al. 2000). The excess Na^+ ions are removed out of the cell and cellular ion homeostasis is maintained. CBL1 is known to regulate the SOS pathway via interacting with SOS2/CIPK24 (Kolukisaoglu et al. 2004). In response to flooding stress, CBL-CIPKs are known to modulate metabolic switch (Lee et al. 2009). The CBL9-CIPK23 complex regulates the uptake of nitrate under low nitrate condition by regulating NRT1.1 (CHL1) transporter activity. The NRT1.1 is a dual-affinity nitrate transporter that can function as low- and high-affinity nitrate transporter depending on the phosphorylation or dephosphorylation at threonine 101 (Ho et al. 2009). During low nitrate stress, CBL9/CIPK23 complex phosphorylates NRT1.1 (CHL1) that acts as molecular switch to convert it into high-affinity nitrate transporter. The phosphorylation and dephosphorylation status of NRT1.1, thus, help plants to sense a wide range of nitrate concentration in soil. Recently, the ABI2-CBL1/CIPK23 complex acting downstream to NRT1.1 signaling has been found to be involved in regulating root nitrate uptake in *Arabidopsis* under nitrate-deficient conditions (Leran et al. 2015) (Fig. 2.4).

2.2.1.5 Calcium-Dependent Protein Kinases (CDPKs)

CDPKs comprise of a kinase domain linked to a CaM-like domain via a junction sequence that keeps the kinase inactive (Zhang and Choi 2001). Calcium binding to CaM-like domain induces conformational change displacing the autoinhibitory

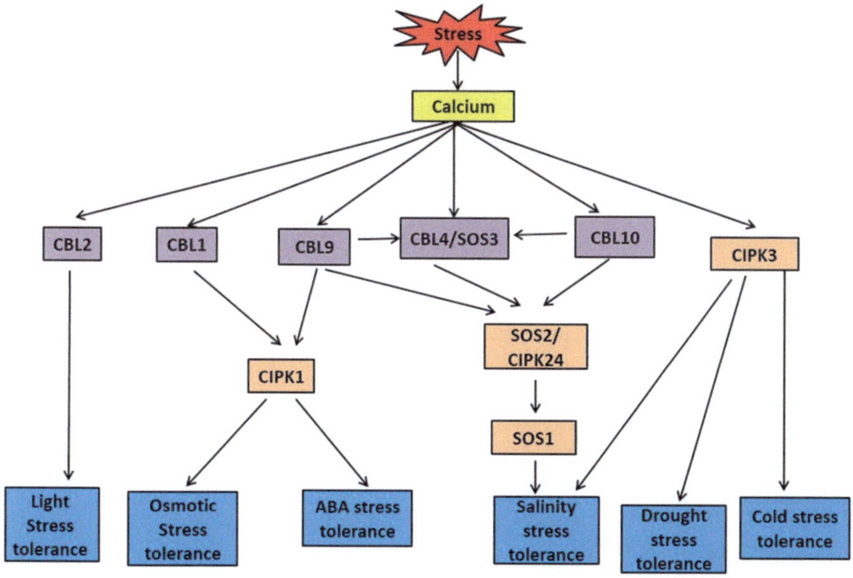

Fig. 2.4 CBLs acting as messengers to provide tolerance in multiple stresses. CBLs are activated by changes in calcium concentration. Different members of CBL family are induced following different calcium signatures. Different CBLs have different interacting partners (CIPKs), and this interaction leads to production of varied responses. CBL9 and CBL1 both interact with CIPK1. When CBL9 interacts with CIPK1, it provides ABA stress tolerance, and CIPK1-CBL1 provides osmotic stress tolerance. CBL10 and CBL4/SOS3 both induce same signaling pathway for salinity stress tolerance. CIPK3 is involved in cold stress tolerance, and CBL2 provides light stress tolerance

domain from kinase followed by intramolecular phosphorylation, activating the kinase (Weljie and Vogel 2004; Chandran et al. 2006). CDPKs belong to multigene families, 34 in *Arabidopsis* and 29 in rice (Cheng et al. 2002; Asano et al. 2005). CDPKs can regulate diverse targets as they can be localized in the cytoplasm and nucleoplasm, or they can be associated with the cytoskeleton, endoplasmic reticulum, or peroxisomes (Putnam-Evans et al. 1989; Dammann et al. 2003; Lu and Hrabak 2002). The role of CDPKs in different stresses as cold, osmotic, salinity, and ABA-mediated signaling in drought and salinity stress has been studied in different plants (Sheen 1996; Shao and Harmon 2003; Liu et al. 2006; Choi et al. 2005a, b; Zhu et al. 2007; Uno et al. 2000; Urao et al. 1994; Pei et al. 1996). Some other second messengers such as phospholipids may also regulate CDPK activity (Harper et al. 2004; Xiong et al. 2002). Expression of many CDPKs is increased by abiotic stresses; following which, there are changes in the intracellular localization of CDPKs, or there may be induction of the expression of other stress-responsive genes (Sheen 1996; Cheng et al. 2002). A CDPK from groundnut (AhCPK2) translocates to nucleus under hyperosmotic conditions; McCPK1 from ice plant moves to nucleus following exposure to low humidity and salt stress (Raichaudhuri et al. 2006; Patharkar and Cushman 2000; Chehab et al. 2004). CDPKs may also help in adaptive processes by being activated during long periods of stress exposure as was seen in rice where a CDPK was activated after 18–24 h of cold exposure (Martìn and Busconi 2001; Abbasi et al. 2004; Saijo et al. 2000). In drought conditions, CDPKs generate drought responses by regulating various ion channels and stomatal movements and also induce ABA-regulated responses in various stresses (Milla et al. 2006; Johansson et al. 1996; Li et al. 1998; Berkowitz et al. 2000; Pei et al. 1996; Choi et al. 2005a, b; Sheen 1996).

Certain other kinases known as calcium- and calmodulin-dependent kinases (CCaMKs) play a role in stress tolerance. Structurally, CCaMKs are quite similar to CDPKs, possessing a calcium-binding domain along with a calmodulin-binding domain (Patil et al. 1995). However their regulatory mechanism is complex pertaining to binding of both Ca^{2+} and calmodulin (Patil et al. 1995; Sathyanarayanan and Poovaiah 2004). They have been identified in many plant species and they have various functions, but their role in stress tolerance has not been explored much (Harper et al. 2004). The expression of a CCaMK from pea root was found to increase in cold and salt stresses (Pandey et al. 2002).

2.2.2 Reactive Oxygen Species (ROS)

Reactive oxygen species (ROS) are formed as a result of excitation or incomplete reduction of molecular oxygen. ROS contains both free radical (superoxide, $O_2^{\bullet-}$; peroxyl, $ROO^{\bullet}$; hydroxyl, $OH^{\bullet}$; etc.) and non-radical forms (hydrogen peroxide, H_2O_2; singlet oxygen, $^1O_2$; ozone; O_3; etc.) (Miller et al. 2008; Choudhury et al. 2013). In plants, ROS are usually formed by the inevitable leakage of electrons onto O_2 from the electron transport activities of chloroplasts, mitochondria, and plasma membranes (del Río et al. 2006; Heyno et al. 2011; Foyer et al. 2012; Roach and Krieger-Liszkay 2014). They are also produced enzymatically as a by-product of their catalytic reactions or metabolic pathways localized in various cellular compartments like peroxisomes. For example, H_2O_2 is produced by the peroxisomal flavin-containing enzymes glycolate oxidase and acyl-CoA oxidase, which are involved in the photorespiratory and fatty acid β-oxidation pathways, respectively, while cell wall-bounded peroxidases can also generate H_2O_2 (O'Brien et al. 2012), and $O_2^{\bullet-}$ by plasma membrane-localized NADPH oxidase (NOX) (Suzuki et al. 2011; Marino et al. 2012; Baxter et al. 2014). ROS is produced at normal physiological condition as well as under stress conditions with difference in their level at both conditions. Their production in plants under normal growth conditions is low; however, in response to various environmental stresses, ROS are drastically increased in plants disturbing the normal balance of $O_2^{\bullet-}$, $^{\bullet}OH$, and H_2O_2 in the intracellular environment. Environmental stresses such as drought, salinity, chilling, metal toxicity, and UV-B radiation as well as pathogens lead to enhanced generation of ROS in plants due to disruption of cellular homeostasis (Shah et al. 2001; Mittler 2002; Sharma and Dubey 2005; Miller et al. 2008; Han et al. 2009; You and Chan 2015). The accumulation of these toxic by-products of aerobic metabolism leads to oxidative stress which may cause damage to plant vital cellular organelles; destroy membranes by inducing the degradation of pigments, proteins, lipids, and nucleic acids; and inactivate enzymes, which ultimately results in cell death (Bailey-Serres and Mittler 2006; Sies 2014; Das and Roychoudhury 2014). It is in recent years that they have emerged as diffusible signal transducers in signal transduction pathway and also as second messenger in plant biology as well that activate the downstream signaling cascade to cater numerous biological processes including stomatal closure (Yan et al. 2007; Mittler and Blumwald 2015; Sierla et al. 2016), programmed cell death (Eric Lam 2004; Van Breusegem and Dat 2006; Petrov et al. 2015; Nath and Yan 2015), gravitropism (Joo et al. 2001; Sato et al. 2014; Krieger et al. 2016; Araniti et al. 2016), pollen-stigma interactions (Zafra et al. 2016), root hair growth (Foreman et al. 2003; Juárez et al. 2015), and acquisition of tolerance to both biotic and abiotic stresses (Bailey-Serres and Mittler 2006; Baxter et al. 2014; Sewelam et al. 2016; Shafi et al. 2015a, b). So, ROS has dual functionality in plants also, like animal system. But to utilize ROS as second messengers in intracellular signaling cascades and to ensure the plant survival, their low/moderate concentration has to be maintained in the cellular environment. Plant system has evolved the antioxidants and scavenging machinery to detoxify the ROS and maintain an equilibrium between the production and detoxification under favorable conditions. This delicate balance between ROS formation and scavenging is disturbed under stress conditions. But plant cells possess very efficient enzymatic (superoxide dismutase, SOD; catalase, CAT; ascorbate peroxidase, APX; glutathione reductase, GR; monodehydroascorbate reductase, MDHAR; dehydroascorbate reductase, DHAR; glutathione peroxidase, GPX; guaiacol peroxidase, GOPX; and glutathione S-transferase, GST) and nonenzymatic (ascorbic acid, ASH; proline; glutathione, GSH; phenolic compounds; alkaloids; nonprotein amino acids; and α-tocopherols) antioxidant defense systems that work to maintain ROS homeostasis under stress conditions also and thus help plants to escape deleterious effect of ROS-mediated oxidative injury (Gill and Tuteja 2010; Miller et al. 2010; Das and Roychoudhury 2014).

Plants can sense, transduce, and translate ROS signal into appropriate cellular responses. It has been suggested that downstream events modulated by ROS signaling include calcium mobilization, activation of G protein, protein phosphorylation [by calcium-dependent protein kinases (CDPKs) and mitogen-activated protein (MAP) kinases] and activation of phospholipid signaling and gene expression (Neill et al. 2002; Baxter-Burrell et al. 2002; Anthony et al. 2004).

The main target of H_2O_2 signaling is regulation of Ca^{2+} homeostasis (Petrov and Van Breusegem 2012) which in turn regulates antioxidative defenses in plants (Costa et al. 2010). This H_2O_2-mediated Ca^{2+} signaling is required for the regulation of guard cells during stomatal opening (Allen et al. 2000; Young et al. 2006). ROS signaling is also related to lipid signaling. Some studies suggested that phospholipase D (PLD) and phosphatidic acid (PA) play multiple roles in the signaling networks of plant response to abscisic acid and reactive oxygen species (Zhang et al. 2005). It has been shown that cold-induced activation of phospholipase C and phospholipase D (PLC and PLD), via G proteins, although not directly, leads to expression of different genes (Vergnolle et al. 2005; Zhang et al. 2011) (Fig. 2.5).

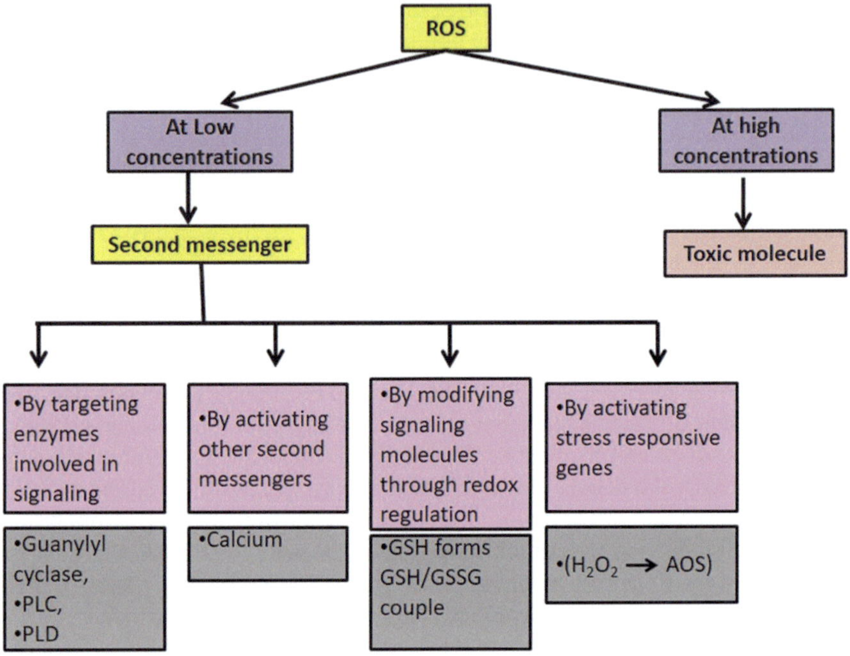

Fig. 2.5 ROS is known to be toxic for the cell, but it is at higher concentrations. At lower concentration, ROS is very useful for the cell, as it acts as second messenger. In response to various stresses, ROS levels increase inside the cell, and ROS tries to convey the message of stress by activating various signaling pathways inside the cell. ROS induces various enzymes such as guanylyl cyclase, PLC, and PLD which are important parts of signaling pathways. ROS also activates other second messengers so that the signal can be amplified and soon a stress response can be triggered. ROS modifies signaling molecules like GSH which indirectly serves as a signal for other molecules. ROS induces the expression of stress-responsive genes, such as H_2O_2 induces antioxidant scavenging systems (AOS)

MAPK cascades are important pathways in abiotic stress responses and enable extracellular stimuli to be transduced into intracellular changes (Zhou et al. 2014). In *Arabidopsis*, H_2O_2 activates the MAPKs, MPK3, and MPK6 via MAPKKK ANP1. Overexpression of ANP1 in transgenic plants resulted in increased tolerance to heat shock, freezing, and salt stress (Kovtun et al. 2000). H_2O_2 also increases expression of the *Arabidopsis* nucleotide diphosphate kinase 2 (NDPK 2) whose overexpression acts as negative feedback for H_2O_2 accumulation and thus helps in enhancing tolerance to multiple stresses including cold, salt, and oxidative stress (Moon et al. 2003). Similarly, in maize, a particular MAPK, MAP65-1a, was reported to enhance the antioxidant enzymes SOD and APX through the brassinosteroid signaling pathway positively to control H_2O_2 amplification (Zhu et al. 2013). Yuasa et al. (2001) has also reported that ROS mediated the activation of ATMPK6, which is one of the candidates for signal mediators in response to abiotic or biotic stresses in *Arabidopsis*. Similar to MAPKs, CDPKs also participate in development of abiotic stress tolerance response (Wei et al. 2014). For instance, the treatment of

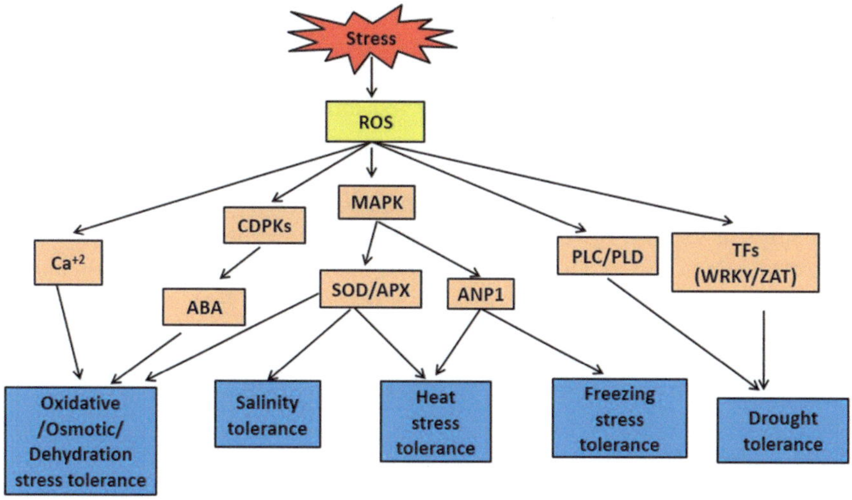

Fig. 2.6 On receiving a signal, ROS activates multiple pathways depending on the stress signal received to combat the stress. In case of oxidative, osmotic, and dehydration stresses, ROS activates Ca^{2+} and CDPKs which mediate ABA-dependent signaling pathways to provide tolerance against these conditions. ROS also activates phospholipases which function in drought signaling. Activation of various enzymes and MAPK pathway by ROS protects from oxidative, salinity, and heat stresses. ROS also induce various transcription factors protecting the plants from drought stress

tomato and wheat leaves with H_2O_2 increased the expression of CDPKs (Chico et al. 2002; Li et al. 2008). Certain CDPKs, CPK3, and CPK6 function in controlling of ABA-regulated stomatal signaling and guard cell ion channels also (Mori et al. 2006), which in turn regulate antioxidative defenses in plants (Fig. 2.6).

ROS-mediated regulation of certain transcription factors results in expression of a large number of stress-responsive genes (Smykowski et al. 2010; Shaikhali et al. 2012; Munné-Bosch et al. 2013). WRKY and zinc finger TFs are major classes of transcription factors that are involved in regulation of ROS-mediated stress response. For instance, zinc fingers like ZAT10 act as an inducer as well as a repressor of ROS-responsive genes under salt, drought, and osmotic stresses (Sakamoto et al. 2004; Mittler 2006), while ZAT6 positively regulates tolerance to drought, salt, and chilling stresses, by modulating ROS levels and SA-related gene expression (Shi et al. 2014).

Changes in environmental conditions, such as a decrease in water potential, light intensity, or humidity induce the accumulation of ABA and ROS levels in plants, sensed by some receptors present on leaf, which in turn would cause ABA-dependent stomatal closure by promoting the phosphorylation of SNF1 (sucrose non-fermenting kinase 1)-related protein kinase OPEN STOMATA1 (OST1)/SnRK2. OST1 is also involved in the production of ROS, particularly H_2O_2, by interacting with NOX. This OST1-dependent H_2O_2 formation could provide a signal for an increased release of further active OST1 through a positive feedback loop (Wang and Song 2008; Mittler and Blumwald 2015). This whole sensing and response mechanism occur within the leaf tissues, but a rapid, systemic, cell-to-cell communication is essential for whole plant acclimation to abiotic stresses. Here also ROS and Ca^{2+} waves mediate long-distance cell signaling in the form of SAA (systemic acquired acclimation) signals for the communication among leaves or different parts of whole plant, allowing a well-coordinated response generation by plant for different stresses (Gilroy et al. 2014) as seen in *Arabidopsis thaliana*, in response to heat stress in plants (Suzuki et al. 2013). It has been reported that H_2O_2 plays an important role in cold acclimation-induced chilling tolerance in tomato. By inducing a modest increase in H_2O_2, *RBOH1* gene expression and NADPH oxidase activity; cold acclimation modulates the expression and activity of ROS detoxifying enzymes and ensures stress cross-tolerance (Zhou et al. 2012, 2014).

2.2.3 Cyclic Nucleotides

The cyclic nucleotide monophosphates (cNMPs), particularly adenosine 3′,5′-cyclic monophosphate (cAMP) and guanosine 3′,5′-cyclic monophosphate (cGMP), are cyclic catalytic products of adenosine 5′-triphosphate (ATP) and guanosine 5′-triphosphate (GTP), synthesized by adenylate cyclase and guanylate cyclase, respectively, following their activation by Gα component of G protein after interacting with GPCRs (G protein-coupled receptors). As complex multimodule signaling pathways are responsible for generation of biotic and abiotic stress tolerance in plants which results in adaptive response, cyclic nucleotides also participate in these signal transduction pathways, restoring the cellular homeostasis and thus promoting survival. There has been an increase in evidences which suggest a list of physiological roles that involve these cNMPs (cAMP and cGMP) directly or indirectly, including several development processes other than photomorphogenesis (Moutinho et al. 2001; Pagnussat et al. 2003), chloroplast development (Barnes et al. 1995), cell cycle progression (Ehsan et al. 1998), etc., that helped to accept them as bona fide and important second messengers. In biotic and abiotic stress conditions, cNMPs have been found to be up-regulated and then mediate stress response (Choi and Xu 2010), for example, first, by activating defense gene induction and potentiation of ROS-induced cell death (Delledonne et al. 1998), and second, by protecting against salt stress (Maathuis and Sanders 2001). cNMPs are meant to regulate the activity of various proteins downstream the signaling pathway, and among them are protein kinases (PKA and PKG) which, once activated, phosphorylate an array of other cellular targets including other kinases, phosphatases, certain transcription factors, and several ion channels.

Ion homeostasis is basically disturbed under stress conditions affecting growth, development, and survival of all living organisms; hence, several transporters and channels become the crucial parameters for maintaining proper ion balance inside the cell required for its optimal growth and functioning. Many plant cation transporter and channel protein families such as cyclic nucleotide-gated ion channel (CNGC) have been implicated in providing biotic and abiotic stress tolerance (Jha et al. 2016). Plant CNGCs are ligand- and voltage-gated channels, usually present on plasma membranes, but some members are localized on intracellular organelles like tonoplast (Yuen and Christopher 2013). CNGCs function in sensory signal transduction and have competitive ligands like cyclic nucleotides either cAMP or cGMP and Ca^{2+}/calmodulin which compete for the overlapping binding sites at the C-terminus in the cytosolic part of the channel (Arazi et al. 2000; Li et al. 2005). The CNGC family members are known to be involved in the uptake of cations including essential macronutrients, Na^+, K^+, and Ca^{2+} as well as potentially toxic cations such as Na^+ or Pb^{2+}, and regulate plant growth and development (Yuen and Christopher 2010; Jha et al. 2016). Many CNGCs also participate in plant abiotic stress response by either regulating the sequestration and release of cations among intracellular stores or mediating Ca^{2+} signaling. In *Arabidopsis*, cAMP and/or cGMP regulate the expression of CNGC genes in all tissues by acting as signaling molecules and are believed to be involved in various physiological processes including tolerance to salt stress (Maathuis 2006). A study on *Arabidopsis* shows that phytosulfokine (PSK), a pentapeptide and cGMP, induces water influx for osmotic adjustment by CNGC activation (Ladwig et al. 2015). Tunc-Ozdemir et al. (2013) also showed that CNGC16, a pollen-expressing CNGC, is critical for heat or drought stress tolerance during reproductive development, which provides genetic evidence that helps in establishing a link between a stress-triggered cNMP signal and a downstream transcriptional heat shock response. It has also been reported by Yuen and Christopher (2013) that a group IV-A CNGCs mediate plant responses to salinity and pathogen infection. The reason behind this might be CNGCs' mediated redistribution of cations like Na^+ between the central vacuole and the cytosol or Ca^{2+} signaling.

In *Arabidopsis* seedlings, cGMP levels were found to increase in response to salt, osmotic and ozone stress (Donaldson et al. 2004; Ederli et al. 2008). The presence of higher concentration of salt in soil leads to combination of stresses like salinity stress, oxidative stress, as well as osmotic stress. cGMP downregulates sodium influx and provides tolerance in salinity and osmotic stress (Maathuis and Sanders 2001; Rubio et al. 2003). However, another pathway is also activated by cGMP via increasing Ca^{2+} influx, inducing SOS3-SOS2-SOS1 system which activates the Na^+/H^+ anti-porter in response to salinity stress (Pardo et al. 1998; Zhu 2002). These cyclic nucleotides, cGMPs, are induced via some phytohormones like auxins as well as natriuretic peptides mediating various processes including stomatal opening (Cousson and Vavasseur 1998; Pharmawati et al. 2001) and adventitious root formation (Nan et al. 2014). However, the mechanism of the participation of cGMP in auxin signaling to affect these growth and developmental processes is largely unknown. These studies have suggested the role of cGMP-mediated signaling in development of plant tolerance against abiotic stress like drought by modulating stomatal opening and root formation to cope up with reducing water level in the soil. Further study indicated that under salt stress in *Arabidopsis thaliana* roots, cGMP could regulate hydrogen peroxide accumulation in calcium-dependent pathway (Li et al. 2011). These evidences have suggested that these signaling molecules, cGMP, H_2O_2, and Ca^{2+}, may work in coordination with developing salt stress tolerance in

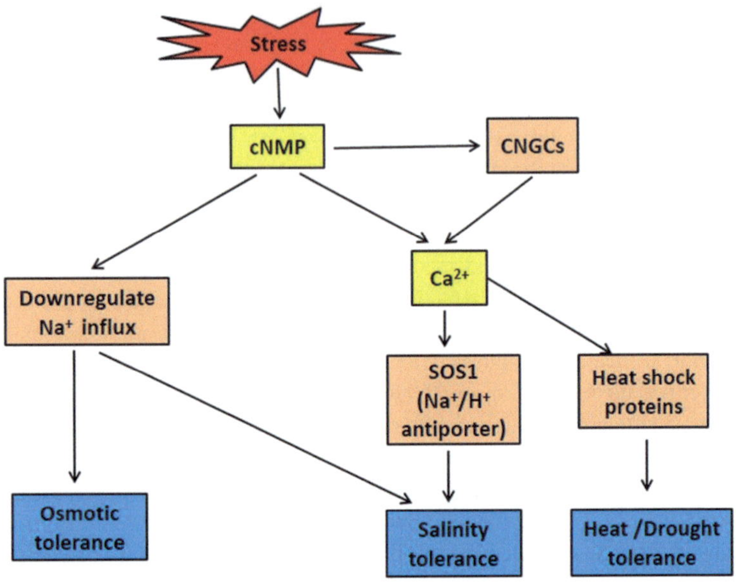

Fig. 2.7 Cyclic nucleotides act as second messengers. Abiotic stresses induce cGMP which lead to activation of CNGCs. cNMP and CNGCs trigger Ca^{2+} release which initiates the SOS pathway by SOS3-SOS2 complex to activate Na^{+}/H^{+} anti-porter (*SOS1*) which provides tolerance in case of salinity stress. By another pathway, cGMP downregulates Na^{+} influx providing osmotic stress tolerance. The SOS pathway is activated in case of salinity stress but not in osmotic stress. Ca^{2+} release activates heat shock proteins that participate in heat and drought tolerance

plants (Maathuis 2006; Donaldson et al. 2004). cGMP also promotes ethylene production and enhances ethylene perception in *Arabidopsis* under salt stress which stimulates the plasma membrane H(+)-ATPase activity, thus modulating the salt stress resistance pathway (Li et al. 2014).

The levels of cAMP have been shown to increase in response to biotic stress as well as abiotic stress and subsequently influence Ca^{2+} influx by targeting membrane CNGC, thereby increasing cytosolic free Ca^{2+} which, in turn, amplifies the signal as part of the cellular response. Köhler et al. 1999 had reported the activation of some CNGCs by cAMPs, produced by heat stress-activated adenylyl cyclase. Moreover, some voltage-independent channels (VICs) in the plasma membrane of *Arabidopsis* root cells have been reported to be highly sensitive to cAMP, and these channels are also implicated in salt tolerance in *Arabidopsis* seedlings by reducing sodium Na^{+} influx (Maathuis and Sanders 2001; Maathuis 2006) (Fig. 2.7).

2.2.4 Lipid Derivatives

Lipids are major components of biomembranes which serve as the platform for lipid signaling. Membrane lipids can respond to diverse stresses just by modulating the membrane fluidity or its physiochemical properties. However they have a significant role toward signaling in stress conditions. The signaling lipids are very less abundant and they have a rapid turnover. The modifying enzymes such as phospholipases, lipid kinases, or phosphatases generate signaling molecules from membrane lipids in response to various stimuli. The generated lipid signals can activate enzymes or recruit proteins to membranes leading to the activation of downstream signaling pathways resulting in specific cellular events and physiological responses. Lipid signaling forms a complex network responding to abiotic stress conditions. The major lipid signaling molecules include phosphatidylinositols, diacylglycerols, phosphatidic acid, certain sphingolipids, and lysophospholipids (Munnik and Testerink 2009; Xue et al. 2009; Munnik and Vermeer 2010).

2.2.4.1 Phosphoinositides

Phosphoinositides are inositol-containing phospholipids present in the plasma membrane derived from phosphatidylinositols (PI). They are rapidly formed in response to diverse stimuli through lipid kinases or phospholipase activation. Five isoforms of PI, phosphatidylinositol 3-phosphate (PI3P), phosphatidylinositol 4-phosphate (PI4P), phosphatidylinositol 5-phosphate (PI5P), phosphatidylinositol 3,5-bisphosphate (PI(3,5)P2), and phosphatidylinositol 4,5-bisphosphate PI(4,5)P2, have been identified in plants (Meijer and Munnik 2003; Balla 2013). These polyphosphoinositides are continuously being formed in the inner leaflet of the plasma membrane and are acted upon by a set of specific kinases and phosphatases and thereby are kept in a state of constant turnover (Testerink and Munnik 2005). The five isoforms are formed by the respective PI kinases present at the localization of the PIs (Van Leeuwen et al. 2007; Mishkind et al. 2009; Heilmann and Heilmann 2015). PIs and lipid kinases are present in nucleus as well, and their activities are elevated in stress conditions, suggesting their direct interaction with DNA-binding proteins in regulating gene expression (Bunney et al. 2000; Mishkind et al. 2009; Boss and Im 2012). PIs form part of various cellular signaling mechanisms as membrane trafficking, cytoskeleton organization, polar tip growth, and stress responses (Meijer and Munnik 2003; Di Paolo and De Camilli 2006; Thole and Nielsen 2008; Ischebeck et al. 2010). PIP2 levels increase in response to osmotic stress (Mikami et al. 1998; Pical et al. 1999; DeWald et al. 2001).

The PIs serve as the substrates of phosphoinositide-specific phospholipases (PI-PL) A2, C, and D leading to formation of signaling molecules. Phospholipase C (PI-PLC) is the most studied one; it gets activated in response to a stimulus. PLC acts upon PIP2 and releases DAG (diacylglycerol) and IP_3 (inositol

1,4,5,-triphosphate) (Tuteja and Sopory 2008; Tuteja and Mahajan 2007). PLC levels increased in salt and drought stress (Hirayama et al. 1995; Kopka et al. 1998). These increased PLC levels ultimately lead to increased IP_3 and DAG. The function of DAG in plants has not been well characterized. In plants, DAG can be further phosphorylated by DAG kinases to generate PA (Arisz et al. 2009). The decrease in DAG was found to compromise plant response to ABA and hyperosmotic stress and vice versa (Peters et al. 2010; Pical et al. 1999).

IP_3 levels increased in salt, osmotic, light, gravity, anoxia, hyperosmotic stress and in response to some plant hormones (Drøbak and Watkins 2000; Takahashi et al. 2001; Stevenson et al. 2000; DeWald et al. 2001). IP_3 induces Ca^{2+} release in the cytoplasm of guard cells triggering stomatal closure (Allen et al. 1995). IP_3 may also function as a precursor for formation of IP_6 (inositol hexakisphosphate) another signaling molecule in plants (Tsui and York 2010; Williams et al. 2015). IP_6 was reported to regulate Ca^{2+} stores and channel activities, and now its role in ABA-regulated Ca^{2+} release triggering stomatal closure has been identified (Yang et al. 2001; Larsson et al. 1997; Lemtiri-Chlieh et al. 2003).

2.2.4.2 Phosphatidic Acid (PA)

Phospholipase D (PLD) is also involved in signal transduction. PLD catalyzes the hydrolysis of the phosphodiester bond between the phosphate and the polar group of membrane phospholipids producing phosphatidic acid which acts as an important second messenger in plants (Wang 1999; Testerink and Munnik 2005; Tuteja and Mahajan 2007). PA is also produced via PLC and DAGK (Tuteja and Sopory 2008; Arisz et al. 2009). PA levels are increased due to increased activity of PLC or PLD enzymes via multiple abiotic stresses such as drought, salt, ethylene, ABA, osmotic, cold, frost, dehydration, wounding, temperature, and hyperosmolarity (Frank et al. 2000; Munnik et al. 2000; Takahashi et al. 2001; Testerink and Munnik 2005; Arisz et al. 2009, 2013; Li et al. 2009; Mishkind et al. 2009; Bargmann et al. 2009; Gasulla et al. 2013; Hong et al. 2008, 2010; Yu et al. 2010; Katagiri et al. 2005). PA usually does not act upon by itself rather it cross talks with various other signals. PA may recruit specific target proteins leading to translocation of the proteins, thus activating or repressing the protein function (Zhang et al. 2004; Testerink et al. 2007; Guo et al. 2011). PA binding translocates *Arabidopsis* MYB transcription factor into nucleus (Yao et al. 2013). PA interacts with H_2O_2, NO, and NADPH oxidase isoforms leading to ABA-induced ROS generation and stomatal closure (Distefano et al. 2008; Guo and Wang 2012; Zhang et al. 2009a, b). PA also targets other protein kinases like SnRK2 protein kinase, MAPK isoform MPK6, and sphingosine kinase (SPHK). PA was found to induce glyceraldehyde-3-phosphate dehydrogenase in response to salt stress in roots (Testerink et al. 2004; Yu et al. 2010; Guo and Wang 2012; Kim et al. 2013a, b; Guo et al. 2012; McLoughlin et al. 2013).

2.2.4.3 Other Lipid Derivatives

The role of some other lipid classes such as sphingolipids, lysophospholipids, and oxylipins in abiotic stress has recently been discovered. Sphingolipids are fatty acid derivatives of sphingosine. Plant sphingolipids mostly consist of glycosylated inositol phosphoryl ceramides or glucosylceramides (Markham et al. 2006; Berkey et al. 2012). These derivatives are involved in drought, cold, dehydration, heat, and ABA responses (Ng et al. 2001; Worrall et al. 2008; Guillas et al. 2013; Alden et al. 2011). PA binding activates sphingolipid formation stimulating stomatal closure in an ABA-dependent manner (Guo et al. 2011). Sphingolipids are involved in generating response to low-temperature conditions (Lynch 2012).

Lysophospholipids are phospholipids having only one acyl chain and are generally produced from glycerol- and sphingosine-based phospholipids in the membrane bilayer by the action of phospholipase A. Lysophospholipids accumulate in response to wounding, freezing, and various other stress conditions (Lee et al. 1997; Narváez-Vásquez et al. 1999; Welti et al. 2002, 2014). In response to a stimulus, the lysophospholipids are hydrolyzed from the membrane lipids and released into the extracellular space where they are recognized by extracellular receptors, thus, initiating diverse signaling pathways. In animals, the major target of this class of lipids is G protein-coupled receptors, whereas in plants there is a limited information about this (zu Heringdorf and Jakobs 2007; Taddese et al. 2014).

Oxylipins are oxidized fatty acids consisting of fatty acid hydroperoxides, divinyl ethers, and phytohormone jasmonic acid (JA) (Andreou et al. 2009). They can be synthesized enzymatically by lipoxygenases or dioxygenases or formed by autooxidation (Mosblech et al. 2009). Oxylipins have been implicated in a variety of environmental cues such as dehydration, salinity, etc. (Seltmann et al. 2010; Savchenko et al. 2014). Oxylipins were produced in response to water-deficient conditions in stress-tolerant chickpea varieties (De Domenico et al. 2012). Decreased oxylipin levels were correlated with increased sensitivity to drought (Grebner et al. 2013). Oxylipins activated oxidative stress-related MAPK-dependent signaling pathways and also induced the expression of stress genes in high-salinity conditions (Imbusch and Mueller 2000). Lysophospholipid level elevated in response to cold and freezing stress in *Arabidopsis* (Welti et al. 2002) (Fig. 2.8 and Table 2.1).

2.3 Conclusion

This is a well-known fact that plants are constantly confronted with multitude of abiotic stresses at a given point of time, rather than an individual stress that is most lethal for the crops. Plant responses to these stresses are complex as a consequence

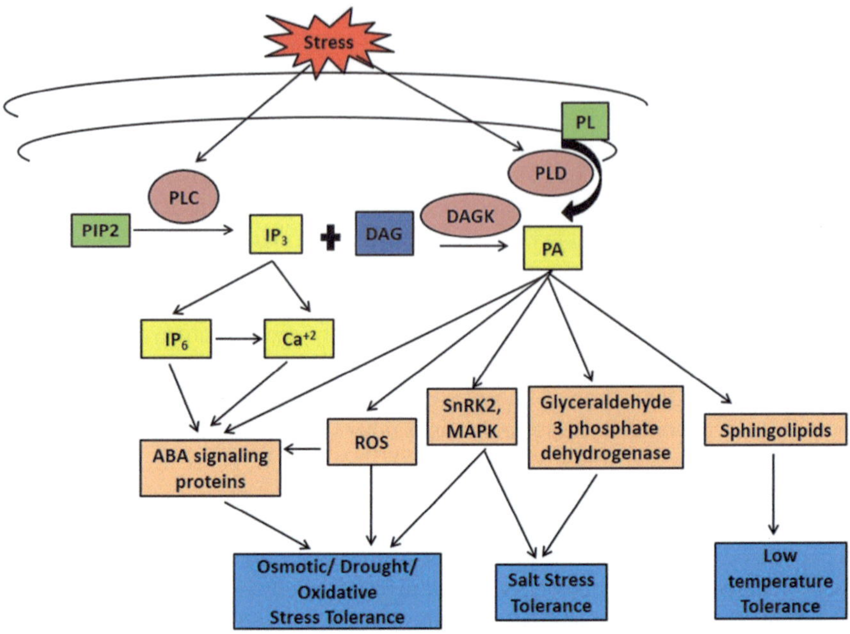

Fig. 2.8 Certain lipid derivatives also act as second messengers. Stress signals activate the enzyme, PLC, which catalyzes PIP2 to IP_3 and DAG. IP_3 induces calcium release which activates certain kinases to induce stress response. IP_3 acts as the substrate for generation of IP_6 which also helps in signal transduction. DAG is converted to PA by DAGK and PLD also forms PA from phospholipids. PA can activate ROS and ABA-signaling proteins and also kinases, both of which provide stress tolerance. PA induces sphingolipid activation which provides tolerance in low-temperature conditions

of the interplay of specific signaling pathways. Recent evidence shows that the interaction between several abiotic stresses leads to accumulation of a large number of signaling molecules, but only some of them are common, particularly some second messengers, phytohormones, transcription factors, and kinase cascades that point to a cross talk among the different signaling pathways. This cross talk may be synergistic and/or antagonistic, but in certain cases, such cross talk can help a plant to resist combinatorial stresses. Various novel approaches like OMICS-based studies have helped a lot to understand the complicated mechanism underlying the stress tolerance responses and also revealed various important and potential candidates of signaling pathway that can further be exploited for genetic manipulation to develop multi-stress-tolerant crops with least negative impacts on normal physiology (Fig. 2.9).

Table 2.1 Evidences revealing the role of different second messengers in exhibiting multiple stress tolerance:

Secondary messenger	Abiotic stress	Title	Species	References
Ca^{2+}/CaM/ CML	Cold	Distinct calcium-signaling pathways regulate calmodulin gene expression in tobacco	Tobacco	van Der Luit et al. (1999)
		Cold-induced changes in freezing tolerance, protein phosphorylation, and gene expression (evidence for a role of calcium)	Alfalfa	Monroy et al. (1993)
	Oxidative	Calcium-regulated phosphorylation of soybean serine acetyltransferase in response to oxidative stress	Soybean	Liu et al. (2006)
		Nitric oxide-activated calcium-/calmodulin-dependent protein kinase regulates the abscisic acid-induced antioxidant defense in maize	Maize	Ma et al. (2012)
		Calmodulin-dependent activation of MAP kinase for ROS homeostasis in *Arabidopsis*	*Arabidopsis*	Takahashi et al. (2011)
		Hydrogen peroxide homeostasis: activation of plant catalase by calcium/ calmodulin	*Arabidopsis*	Yang and Poovaiah (2002)
		A Ca^{2+}-/calmodulin-binding peroxidase from *Euphorbia latex:* novel aspects of calcium-hydrogen peroxide cross talk in the regulation of plant defenses	*Euphorbia*	Mura et al. (2005)
		Critical role of Ca^{2+} ions in the reaction mechanism of *Euphorbia characias* peroxidase	*Euphorbia*	Medda et al. (2003)
	Salinity	Direct interaction of a divergent CaM isoform and the transcription factor, MYB2, enhances salt tolerance in *Arabidopsis*	*Arabidopsis*	Jae et al. (2005)
		Proline accumulation and salt stress-induced gene expression in a salt-hypersensitive mutant of *Arabidopsis*	*Arabidopsis*	Liu and Zhu (1997)

(continued)

Table 2.1 (continued)

Secondary messenger	Abiotic stress	Title	Species	References
	Heat	Ca^{2+} and AtCaM3 are involved in the expression of heat shock protein gene in *Arabidopsis*	*Arabidopsis*	Liu et al. (2005)
		Effect of calcium and calmodulin on intrinsic heat tolerance in relation to antioxidant systems in maize seedlings	Maize	Gong et al. (1997a, b)
		Heat shock-induced changes in intracellular Ca^{2+} level in tobacco seedlings in relation to thermotolerance	Tobacco	Gong et al. (1998)
		Involvement of calcium and calmodulin in the acquisition of heat shock-induced thermotolerance in maize seedlings	Maize	Gong et al. (1997a, b)
		Ca^{2+} and calmodulin modulate DNA-binding activity of maize heat shock transcription factor in vitro	Maize	Li et al. (2004)
		The calmodulin-binding protein kinase 3 is part of heat shock signal transduction in *Arabidopsis thaliana*	*Arabidopsis*	Liu et al. (2008)
		Calmodulin-binding protein phosphatase PP7 is involved in thermotolerance in *Arabidopsis*	*Arabidopsis*	Liu et al. (2007)
		Calmodulin is involved in heat shock signal transduction in wheat	Wheat	Liu et al. (2003)
		Molecular and genetic evidence for the key role of AtCaM3 in heat shock signal transduction in *Arabidopsis*	*Arabidopsis*	Zhang et al. (2009a, b)
	Osmotic	The structure of *Arabidopsis thaliana* OST1 provides insights into the kinase regulation mechanism in response to osmotic stress	*Arabidopsis*	Yunta et al. (2011)
	Multiple	Rapid transcriptome changes induced by cytosolic Ca^{2+} transients reveal ABRE-related sequences as Ca^{2+}-responsive cis elements in *Arabidopsis*	*Arabidopsis*	Kaplan et al. (2006)

(continued)

Table 2.1 (continued)

Secondary messenger	Abiotic stress	Title	Species	References
		Mutations in AtCML9, a calmodulin-like protein from *Arabidopsis thaliana*, alter plant responses to abiotic stress and abscisic acid	*Arabidopsis*	Magnan et al. (2008)
		Calmodulin7 plays an important role as transcriptional regulator in *Arabidopsis* seedling development	*Arabidopsis*	Kushwaha et al. (2008)
		CML24, regulated in expression by diverse stimuli, encodes a potential Ca^{2+} sensor that functions in response to abscisic acid, day length, and ion stress	*Arabidopsis*	Delk et al. (2005)
		A Ca^{2+}-/CaM-dependent kinase from pea is stress regulated and in vitro phosphorylates a protein that binds to AtCaM5 promoter	Pea, *Arabidopsis*	Pandey et al. (2002)
		Cold-shock regulation of the *Arabidopsis* TCH genes and the effects of modulating intracellular calcium levels	*Arabidopsis*	Polisensky and Braam (1996)
		Calcium requirement for the induction of hydrotropism and enhancement of calcium-induced curvature by water stress in primary roots of pea, *Pisum sativum* L.	Pea	Takano et al. (1997)
		A membrane-associated NAC transcription factor regulates salt-responsive flowering via FLOWERING LOCUS T in *Arabidopsis*	*Arabidopsis*	Kim et al. (2007)
		Regulation of leaf senescence by NTL9-mediated osmotic stress signaling in *Arabidopsis*	*Arabidopsis*	Yoon et al. (2008)
		Roles of CAMTA transcription factors and salicylic acid in configuring the low-temperature transcriptome and freezing tolerance of *Arabidopsis*	*Arabidopsis*	Kim et al. (2013a, b)

(continued)

Table 2.1 (continued)

Secondary messenger	Abiotic stress	Title	Species	References
		A tobacco calmodulin-binding protein kinase (NtCBK2) induced by high-salt/GA treatment and its expression during floral development and embryogenesis	Tobacco	Hua et al. (2004)
Ca^{2+}/CBL	Light	An *Arabidopsis* SNF1-related protein kinase, AtSR1, interacts with a calcium-binding protein, AtCBL2, of which transcripts respond to light	*Arabidopsis*	Nozawa et al. (2001)
	Salt	SCABP8/CBL10, a putative calcium sensor, interacts with the protein kinase SOS2 to protect *Arabidopsis* shoots from salt stress	*Arabidopsis*	Quan et al. (2007)
		Stress signaling through Ca^{2+}-/calmodulin-dependent protein phosphatase calcineurin mediates salt adaptation in plants	Tobacco	Pardo et al. (1998)
		The structure of the *Arabidopsis thaliana* SOS3: molecular mechanism of sensing calcium for salt stress response	*Arabidopsis*	Sánchez-Barrena et al. (2005)
	Multiple stresses	CBL1, a calcium sensor that differentially regulates salt, drought, and cold responses in *Arabidopsis*	*Arabidopsis*	Cheong et al.
		Genes for calcineurin B-like proteins in *Arabidopsis* are differentially regulated by stress signals	*Arabidopsis*	Kudla et al. (1999)
		Protection against heat stress-induced oxidative damage in *Arabidopsis* involves calcium, abscisic acid, ethylene, and salicylic acid	*Arabidopsis*	Larkindale and Knight (2002)
Ca^{2+}/CDPK/CIPK	Temperature	A rice membrane-bound calcium-dependent protein kinase is activated in response to low temperature	Rice	Martìn and Busconi (2001)

(continued)

Table 2.1 (continued)

Secondary messenger	Abiotic stress	Title	Species	References
	Multiple	OsCDPK13, a calcium-dependent protein kinase gene from rice, is induced by cold and gibberellin in rice leaf sheath	Rice	Abbasi et al. (2004)
		Two calcineurin B-like calcium sensors, interacting with protein kinase CIPK23, regulate leaf transpiration and root potassium uptake in *Arabidopsis*	*Arabidopsis*	Cheong et al. (2007)
		Arabidopsis calcium-dependent protein kinase AtCPK32 interacts with ABF4, a transcriptional regulator of abscisic acid-responsive gene expression, and modulates its activity	*Arabidopsis*	Choi et al. (2005a, b)
		Novel rice OsSIPK is a multiple stress-responsive MAPK family member showing rhythmic expression at mRNA level	Rice	Lee et al. (2008)
		Alternative complex formation of the Ca^{2+}-regulated protein kinase CIPK1 controls abscisic acid-dependent and independent stress responses in *Arabidopsis*	*Arabidopsis*	D'Angelo et al. (2006)
		Autophosphorylation and subcellular localization dynamics of a salt and water deficit-induced calcium-dependent protein kinase from ice plant	Ice plant	Chehab et al. (2004)
		Two genes that encode Ca^{2+}-dependent protein kinases are induced by drought and high-salt stresses in *Arabidopsis thaliana*	*Arabidopsis*	Urao et al. (1994)
		In vivo phosphorylation of a recombinant peptide substrate of CDPK suggests involvement of CDPK in plant stress responses	Soybean, tobacco	Shao and Harmon (2003)

(continued)

Table 2.1 (continued)

Secondary messenger	Abiotic stress	Title	Species	References
		Overexpression of a single Ca^{2+}-dependent protein kinase confers both cold and salt/drought tolerance on rice plants	Rice	Saijo et al. (2000)
ROS	Oxidative stress	Functional analysis of oxidative stress-activated mitogen-activated protein kinase cascade in plants	*Arabidopsis thaliana*	Kovtun et al. (2000)
		MAP65-1a positively regulates H_2O_2 amplification and enhances brassinosteroid-induced antioxidant defense in maize	*Zea mays*	Zhu et al. (2013)
		Functional analysis of oxidative stress-activated mitogen-activated protein kinase cascade in plants	Tobacco	Kovtun et al. (2000)
		Zinc finger of *Arabidopsis Thaliana*12 (ZAT12) interacts with FER-like iron deficiency-induced transcription factor (FIT) linking iron deficiency and oxidative stress responses	*Arabidopsis thaliana*	Le et al. (2016)
		High redox responsive transcription factor1 levels result in accumulation of reactive oxygen species in *Arabidopsis thaliana* shoots and roots	*Arabidopsis thaliana*	Matsuo et al. (2015)
	Salt stress	An ROS-assisted calcium wave dependent on the AtRBOHD NADPH oxidase and TPC1 cation channel propagates the systemic response to salt stress	*Arabidopsis thaliana*	Evans et al. (2016)
		Sodium uptake in *Arabidopsis* roots is regulated by cyclic nucleotides	*Arabidopsis thaliana*	Maathuis and Sanders (2001)
	Cold stress	Hydrogen peroxide is involved in the cold acclimation-induced chilling tolerance of tomato plants	*Solanum lycopersicum*	Zhou et al. (2012)

(continued)

Table 2.1 (continued)

Secondary messenger	Abiotic stress	Title	Species	References
	Multiple stress	Biotic and abiotic stress responses through calcium-dependent protein kinase (CDPK) signaling in wheat (*Triticum aestivum* L.)	*Triticum aestivum* L.	Li et al. (2008)
		NDP kinase 2 interacts with two oxidative stress-activated MAPKs to regulate cellular redox state and enhances multiple stress tolerance in transgenic plants	*Arabidopsis thaliana*	Moon et al. (2003)
		A stress-responsive NAC transcription factor SNAC3 confers heat and drought tolerance through modulation of reactive oxygen species in rice	Rice	Fang et al. (2015)
		The cysteine2/histidine2-type transcription factor zinc finger of *Arabidopsis Thaliana*6 modulates biotic and abiotic stress responses by activating salicylic acid-related genes and C-repeat-binding factor genes in *Arabidopsis*	*Arabidopsis thaliana*	Shi et al. (2014)
cGMP	Salt stress	cGMP regulates hydrogen peroxide accumulation in calcium-dependent salt resistance pathway in *Arabidopsis thaliana* roots	*Arabidopsis thaliana*	Li et al. (2011)
		cGMP and ethylene are involved in maintaining ion homeostasis under salt stress in *Arabidopsis* roots	*Arabidopsis thaliana*	Li et al. (2014)
		The group IV-A cyclic nucleotide-gated channels, CNGC19 and CNGC20, localize to the vacuole membrane in *Arabidopsis thaliana*	*Arabidopsis thaliana*	Yuen and Christopher (2013)
	Heat or drought stress	A cyclic nucleotide-gated channel (CNGC16) in pollen is critical for stress tolerance in pollen reproductive development	*Arabidopsis thaliana*	Tunc-Ozdemir et al. (2013)

(continued)

Table 2.1 (continued)

Secondary messenger	Abiotic stress	Title	Species	References
PA/PI/IP_3/DAG	Low temperature	Rapid phosphatidic acid accumulation in response to low-temperature stress in *Arabidopsis* is generated through diacylglycerol kinase	*Arabidopsis*	Arisz et al. (2013)
	Heat	Heat stress activates phospholipase D and triggers PIP2 accumulation at the plasma membrane and nucleus	*Arabidopsis*, rice	Mishkind et al. (2009)
	Drought	Functional convergence of oxylipin and abscisic acid pathways controls stomatal closure in response to drought	*Arabidopsis*	Savchenko et al. (2014)
	Salinity	Identification of novel candidate phosphatidic acid-binding proteins involved in the salt stress response of *Arabidopsis thaliana* roots	*Arabidopsis*	McLoughlin et al. (2013)
		Phosphatidic acid mediates salt stress response by regulation of MPK6 in *Arabidopsis thaliana*	*Arabidopsis*	Yu et al. (2010)
	Osmotic	Hyperosmotic stress stimulates phospholipase D activity and elevates the levels of phosphatidic acid and diacylglycerol pyrophosphate	Tomato, alfalfa	Munnik et al. (2000)
		Hyperosmotic stress induces a rapid and transient increase in inositol 1,4,5-trisphosphate independent of abscisic acid in *Arabidopsis* cell culture	*Arabidopsis*	Takahashi et al. (2001)
	Multiple stress	The role of lipid metabolism in the acquisition of desiccation tolerance in *Craterostigma plantagineum*: a comparative approach	Blue gem	Gasulla et al. (2013)
		Multiple PLDs required for high salinity and water deficit tolerance in plants	Tomato, *Arabidopsis*	Bargmann et al. (2009)
		A gene encoding phosphatidylinositol-4-phosphate 5-kinase is induced by water stress and abscisic acid in *Arabidopsis thaliana*	*Arabidopsis*	Hirayama et al. (1995)

(continued)

Table 2.1 (continued)

Secondary messenger	Abiotic stress	Title	Species	References
		Non-specific phospholipase C NPC4 promotes responses to abscisic acid and tolerance to hyperosmotic stress in *Arabidopsis*	*Arabidopsis*	Peters et al. (2010)
		Nitric oxide-induced phosphatidic acid accumulation: a role for phospholipases C and D in stomatal closure	Fava bean	Distefano et al. (2008)
		Salinity and hyperosmotic stress induce rapid increases in phosphatidylinositol 4,5-bisphosphate, diacylglycerol pyrophosphate, and phosphatidylcholine in *Arabidopsis thaliana* cells	*Arabidopsis*	Pical et al. (1999)
		Water deficit triggers phospholipase D activity in the resurrection plant *Craterostigma plantagineum*	Blue gem	Frank et al. (2000)
		A gene encoding a phosphatidylinositol-specific phospholipase C is induced by dehydration and salt stress in *Arabidopsis thaliana*	*Arabidopsis*	Hirayama et al. (1995)
		Profiling membrane lipids in plant stress response role of phospholipase Dα in freezing-induced lipid changes in *Arabidopsis*	*Arabidopsis*	Welti et al. (2002)
		Phospholipase Dα1 and phosphatidic acid regulate NADPH oxidase activity and production of reactive oxygen species in ABA-mediated stomatal closure in *Arabidopsis*	*Arabidopsis*	Zhang et al. (2009a, b)
Sphingolipids	Low temperature	Long chain base changes triggered by a short exposure of *Arabidopsis* to low temperatures are altered by AHb1 nonsymbiotic hemoglobin overexpression	*Arabidopsis*	Guillas et al. (2013)

(continued)

Table 2.1 (continued)

Secondary messenger	Abiotic stress	Title	Species	References
	Drought	Drought-induced guard cell signal transduction involves sphingosine-1-phosphate	*C. communis*	Ng et al. (2001)
Oxylipins	Multiple	Analysis of oxidative stress and wound-inducible dinor isoprostanes F1 (phytoprostanes F1) in plants	Peppermint	Imbusch and Mueller (2000)

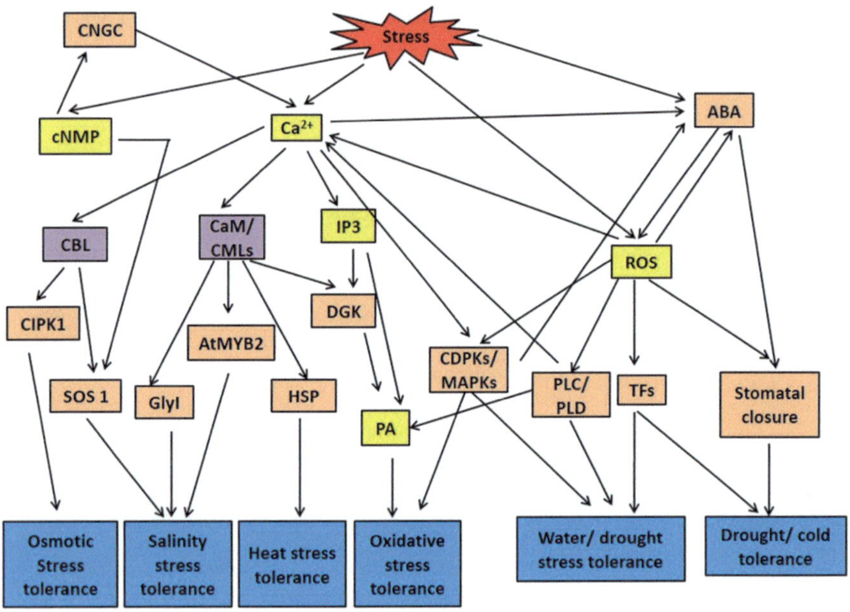

Fig. 2.9 Cross talk of the various second messengers and their signaling pathways. In response to multiple stresses, various signaling pathways are activated simultaneously which converge at one or the other point via some second messenger. For instance, stress increases Ca^{2+} levels which activate Ca^{2+} signaling, but this Ca^{2+} signal gets further amplified by the interaction of other signaling pathways like ROS, CNMP, and phospholipid-mediated pathways. This Ca^{2+} signaling is known to initiate an array of downstream cascade which controls the whole battery of stress-responsive candidates such as SOS1, GLYI, MYB2, HSP, DGK, CDPK, PLC/PLD, etc. Likewise, cyclic nucleotides also involved in accumulation of ROS through Ca^{2+} signaling can further activate phospholipid pathway which again regulates Ca^{2+}-mediated signal transduction. This type of synergistic cross talk among different signaling pathways shows how individual signaling pathways interact with each other to generate a combinatorial stress response in plants

Acknowledgements Funding and support from South Asian University are duly acknowledged.

References

Abbasi F, Onodera H, Toki S, Tanaka H, Komatsu S (2004) OsCDPK13, a calcium-dependent protein kinase gene from rice, is induced by cold and gibberellin in rice leaf sheath. Plant Mol Biol 55(4):541–552

Akaboshi M, Hashimoto H, Ishida H, Saijo S, Koizumi N, Sato M, Shimizu T (2008) The crystal structure of plant-specific calcium-binding protein AtCBL2 in complex with the regulatory domain of AtCIPK14. J Mol Biol 377(1):246–257

Albrecht V, Ritz O, Linder S, Harter K, Kudla J (2001) The NAF domain defines a novel protein–protein interaction module conserved in Ca2+-2+2+ 2+2+2+ 2+ regulated kinases. EMBO J 20(5):1051–1063

Alden KP, Dhondt-Cordelier S, McDonald KL, Reape TJ, Ng CK-Y, McCabe PF, Leaver CJ (2011) Sphingolipid long chain base phosphates can regulate apoptotic-like programmed cell death in plants. Biochem Biophys Res Commun 410(3):574–580

Allen GJ, Muir SR, Sanders D (1995) Release of Ca2+ from individual plant vacuoles by both InsP3 and cyclic ADP-ribose. Science (New York, NY) 268(5211):735–737

Allen GJ, Chu SP, Schumacher K, Shimazaki CT, Vafeados D, Kemper A, Hawke SD et al (2000) Alteration of stimulus-specific guard cell calcium oscillations and stomatal closing in Arabidopsis det3 mutant. Science (New York, NY) 289(5488):2338–2342

Andreou A, Brodhun F, Feussner I (2009) Biosynthesis of oxylipins in non-mammals. Prog Lipid Res 48(3):148–170

Anthony RG, Henriques R, Helfer A, Mészáros T, Rios G, Testerink C, Munnik T, Deák M, Koncz C, Bögre L (2004) A protein kinase target of a PDK1 signalling pathway is involved in root hair growth in Arabidopsis. EMBO J 23(3):572–581

Apel K, Hirt H (2004) Reactive oxygen species: metabolism, oxidative stress, and signal transduction. Annu Rev Plant Biol 55:373–399

Araniti F, Graña E, Krasuska U, Bogatek R, Reigosa MJ, Abenavoli MR, Sánchez-Moreiras AM (2016) Loss of gravitropism in Farnesene-treated Arabidopsis is due to microtubule malformations related to hormonal and ROS unbalance. PLoS One 11(8):e0160202

Arazi T, Kaplan B, Fromm H (2000) A high-affinity calmodulin-binding site in a tobacco plasma-membrane channel protein coincides with a characteristic element of cyclic nucleotide-binding domains. Plant Mol Biol 42(4):591–601

Arisz SA, Testerink C, Munnik T (2009) Plant PA signaling via diacylglycerol kinase. Biochim Biophys Acta Mol Cell Biol Lipids 1791:869

Arisz SA, van van Wijk R, Roels W, Zhu J-K, Haring MA, Munnik T (2013) Rapid phosphatidic acid accumulation in response to low temperature stress in Arabidopsis is generated through diacylglycerol kinase. Front Plant Sci 4:1

Asano T, Tanaka N, Yang G, Hayashi N, Komatsu S (2005) Genome-wide identification of the rice calcium-dependent protein kinase and its closely related kinase gene families: comprehensive analysis of the CDPKs gene family in rice. Plant Cell Physiol 46(2):356–366

Atkinson NJ, Urwin PE (2012) The interaction of plant biotic and abiotic stresses: from genes to the field. J Exp Bot 63:3523

Babu YS, Bugg CE, Cook WJ (1988) Structure of calmodulin refined at 2.2 Å resolution. J Mol Biol 204(1):191–204

Bailey-Serres J, Mittler R (2006) The roles of reactive oxygen species in plant cells. Plant Physiol 141(2):311

Balla T (2013) Phosphoinositides: tiny lipids with giant impact on cell regulation. Physiol Rev 93(3):1019–1137

Bargmann BO, Munnik T (2006) The role of phospholipase D in plant stress responses. Curr Opin Plant Biol 9(5):515–522

Bargmann BOR, Laxalt AM, Ter Riet B, Van Schooten B, Merquiol E, Testerink C, Haring MA, Bartels D, Munnik T (2009) Multiple PLDs required for high salinity and water deficit tolerance in plants. Plant Cell Physiol 50(1):78–89

Barnes SA, Quaggio RB, Chua N-H (1995) Phytochrome signal-transduction: characterization of pathways and isolation of mutants. Philos Trans R Soc Lond B: Biol Sci 350(1331):67–74

Batistic O, Kudla J (2004) Integration and channeling of calcium signaling through the CBL calcium sensor/CIPK protein kinase network. Planta 219(6):915–924

Batistič O, Kudla J (2009) Plant Calcineurin B-like proteins and their interacting protein kinases. Biochim Biophys Acta (BBA) Mol Cell Res 1793(6):985–992

Batistič O, Waadt R, Steinhorst L, Held K, Kudla J (2010) CBL-mediated targeting of CIPKs facilitates the decoding of calcium signals emanating from distinct cellular stores. Plant J 61(2):211–222

Batistič O, Kudla J (2012) Analysis of calcium signaling pathways in plants. BBA-Gen Subjects 1820(8):1283–1293

Baxter A, Mittler R, Suzuki N (2014) ROS as key players in plant stress signalling. J Exp Bot 65(5):1229–1240

Baxter-Burrell A, Yang Z, Springer PS, Bailey-Serres J (2002) RopGAP4-dependent Rop GTPase rheostat control of Arabidopsis oxygen deprivation tolerance. Science (New York, NY) 296(5575):2026–2028

Bender KW, Snedden WA (2013) Calmodulin-related proteins step out from the shadow of their namesake. Plant Physiol 163(2):486–495

Berkey R, Bendigeri D, Xiao S (2012) Sphingolipids and plant defense/disease: the 'death' connection and beyond. Front Plant Sci 3:68

Berkowitz G, Zhang X, Mercier R, Leng Q, Lawton M (2000) Co-expression of calcium-dependent protein kinase with the inward rectified guard cell K+ channel KAT1 alters current parameters in Xenopus Laevis Oocytes. Plant Cell Physiol 41(6):785–790

Boonburapong B, Buaboocha T (2007) Genome-wide identification and analyses of the rice calmodulin and related potential calcium sensor proteins. BMC Plant Biol 7(1):1

Boss WF, Im YJ (2012) Phosphoinositide signaling. Annu Rev Plant Biol 63:409–429

Bouché N, Yellin A, Snedden WA, Fromm H (2005) Plant-specific calmodulin-binding proteins. Annu Rev Plant Biol 56:435–466

Bunney TD, Watkins PAC, Beven AF, Shaw PJ, Hernandez LE, Lomonossoff GP, Shanks M, Peart J, Drøbak BK (2000) Association of phosphatidylinositol 3-kinase with nuclear transcription sites in higher plants. Plant Cell 12(9):1679–1687

Bush DS (1995) Calcium regulation in plant cells and its role in signaling. Annu Rev Plant Biol 46(1):95–122

Chandran V, Stollar EJ, Lindorff-Larsen K, Harper JF, Chazin WJ, Dobson CM, Luisi BF, Christodoulou J (2006) Structure of the regulatory apparatus of a calcium-dependent protein kinase (CDPK): a novel mode of calmodulin-target recognition. J Mol Biol 357(2):400–410

Chehab EW, Rahul Patharkar O, Hegeman AD, Taybi T, Cushman JC (2004) Autophosphorylation and subcellular localization dynamics of a salt-and water deficit-induced calcium-dependent protein kinase from ice plant. Plant Physiol 135(3):1430–1446

Cheng S-H, Willmann MR, Chen H-C, Sheen J (2002) Calcium signaling through protein kinases. The Arabidopsis calcium-dependent protein kinase gene family. Plant Physiol 129(2):469–485

Cheong YH, Pandey GK, Grant JJ, Batistic O, Li L, Kim B-G, Lee S-C, Kudla J, Luan S (2007) Two calcineurin B-like calcium sensors, interacting with protein kinase CIPK23, regulate leaf transpiration and root potassium uptake in Arabidopsis. Plant J 52(2):223–239

Chico JM, Raíces M, Téllez-Iñón MT, Ulloa RM (2002) A calcium-dependent protein kinase is systemically induced upon wounding in tomato plants. Plant Physiol 128(1):256–270

Choi YE, Xu JR (2010) The cAMP signaling pathway in fusarium Verticillioides is important for conidiation, plant infection, and stress responses but not fumonisin production. Mol Plant Microbe Interact 23(4):522–533

Choi H-i, Park H-J, Park JH, Kim S, Im M-Y, Seo H-H, Kim Y-W, Hwang I, Kim SY (2005a) Arabidopsis calcium-dependent protein kinase AtCPK32 interacts with ABF4, a transcriptional regulator of abscisic acid-responsive gene expression, and modulates its activity. Plant Physiol 139(4):1750–1761

Choi MS, Kim MC, Yoo JH, Moon BC, Koo SC, Park BO, Lee JH, Koo YD, Han HJ, Lee SY (2005b) Isolation of a calmodulin-binding transcription factor from rice (Oryza Sativa L.) J Biol Chem 280(49):40820–40831

Choudhury S, Panda P, Sahoo L, Panda SK (2013) Reactive oxygen species signaling in plants under abiotic stress. Plant Signal Behav 8(4):e23681

Costa A, Drago I, Behera S, Zottini M, Pizzo P, Schroeder JI, Pozzan T, Schiavo FL (2010) H2O2 in plant peroxisomes: an in vivo analysis uncovers a ca(2+)-dependent scavenging system. Plant J: For Cell Mol Biol 62(5):760–772

Cousson A, Vavasseur A (1998) Putative involvement of cytosolic Ca2+ and GTP-binding proteins in cyclic-GMP-mediated induction of stomatal opening by auxin in Commelina Communis L. Planta 206(2):308–314

D'Angelo C, Weinl S, Batistic O, Pandey GK, Cheong YH, Schültke S, Albrecht V et al (2006) Alternative complex formation of the Ca2+-regulated protein kinase CIPK1 controls abscisic acid-dependent and independent stress responses in Arabidopsis. Plant J 48(6):857–872

Dammann C, Ichida A, Hong B, Romanowsky SM, Hrabak EM, Harmon AC, Pickard BG, Harper JF (2003) Subcellular targeting of nine calciumdependent protein kinase isoforms from Arabidopsis. Plant Physiol 132(4):1840–1848

Das K, Roychoudhury A (2014) Reactive oxygen species (ROS) and response of antioxidants as ROS-scavengers during environmental stress in plants. Front Environ Sci 2:53

De Domenico S, Bonsegna S, Horres R, Pastor V, Taurino M, Poltronieri P, Imtiaz M, Kahl G, Flors V, Winter P (2012) Transcriptomic analysis of oxylipin biosynthesis genes and chemical profiling reveal an early induction of jasmonates in chickpea roots under drought stress. Plant Physiol Biochem 61:115–122

del Río LA, Sandalio LM, Corpas FJ, Palma JM, Barroso JB (2006) Reactive oxygen species and reactive nitrogen species in peroxisomes. Production, scavenging, and role in cell signaling. Plant Physiol 141(2):330–335

Delk NA, Johnson KA, Chowdhury NI, Braam J (2005) CML24, regulated in expression by diverse stimuli, encodes a potential Ca2+ sensor that functions in responses to abscisic acid, daylength, and ion stress. Plant Physiol 139(1):240–253

Delledonne M, Xia Y, Dixon RA, Lamb C (1998) Nitric oxide functions as a signal in plant disease resistance. Nature 394(6693):585–588

DeWald DB, Javad T, Jones CA, Shope JC, Cangelosi AR, Thompson JE, Prestwich GD, Hama H (2001) Rapid accumulation of phosphatidylinositol 4, 5-bisphosphate and inositol 1, 4, 5-trisphosphate correlates with calcium mobilization in salt-stressed Arabidopsis. Plant Physiol 126:759–769

Di Paolo G, De Camilli P (2006) Phosphoinositides in cell regulation and membrane dynamics. Nature 443(7112):651–657

Distefano AM, García-Mata C, Lamattina L, Laxalt AM (2008) Nitric oxide-induced phosphatidic acid accumulation: a role for phospholipases C and D in stomatal closure. Plant Cell Environ 31(2):187–194

Dodd AN, Kudla J, Sanders D (2010) The language of calcium signaling. Annu Rev Plant Biol 61:593–620

Doherty CJ, Van Buskirk HA, Myers SJ, Thomashow MF (2009) Roles for Arabidopsis CAMTA transcription factors in cold-regulated gene expression and freezing tolerance. Plant Cell 21(3):972–984

Donaldson L, Ludidi N, Knight MR, Gehring C, Denby K (2004) Salt and osmotic stress cause rapid increases in Arabidopsis Thaliana cGMP levels. FEBS Lett 569(1–3):317–320

Drøbak BK, Watkins PAC (2000) Inositol (1, 4, 5) trisphosphate production in plant cells: an early response to salinity and hyperosmotic stress. FEBS Lett 481(3):240–244

Du L, Poovaiah BW (2004) A novel family of Ca2+/calmodulin-binding proteins involved in transcriptional regulation: interaction with fsh/Ring3 class transcription activators. Plant Mol Biol 54(4):549–569

Du L, Yang T, Puthanveettil SV, Poovaiah BW (2011) Decoding of calcium signal through calmodulin: calmodulin-binding proteins in plants. Coding and decoding of calcium signals in plants. Springer, Berlin, pp 177–233

Duque AS, de Almeida AM, da Silva AB, da Silva JM, Farinha AP, Santos D, Fevereiro P, de Sousa Araújo S (2013) Abiotic stress responses in plants: unraveling the complexity of genes and networks to survive

Ederli L, Meier S, Borgogni A, Reale L, Ferranti F, Gehring C, Pasqualini S (2008) cGMP in ozone and NO dependent responses. Plant Signal Behav 3(1):36–37

Ehsan H, Reichheld J-P, Roef L, Witters E, Lardon F, Van Bockstaele D, Van Montagu M, Inzé D, Van Onckelen H (1998) Effect of indomethacin on cell cycle dependent cyclic AMP fluxes in tobacco BY-2 cells. FEBS Lett 422(2):165–169

Estrada-Melo AC, Chao, Reid MS, Jiang C-Z (2015) Overexpression of an ABA biosynthesis gene using a stress-inducible promoter enhances drought resistance in petunia. Hortic Res 2(April):15013

Lam E (2004) Plant cell biology: Controlled cell death, plant survival and development. Nat Rev Mol Cell Biol 5(4):305–315

Evans MJ, Choi WG, Gilroy S, Morris RJ (2016) A ROS-assisted calcium wave dependent on the AtRBOHD NADPH oxidase and TPC1 cation channel propagates the systemic response to salt stress. Plant Physiol 171(3):1771–1784

Fang Y, Liao K, Du H, Xu Y, Song H, Li X, Xiong L (2015) A stress-responsive NAC transcription factor SNAC3 confers heat and drought tolerance through modulation of reactive oxygen species in rice. J Exp Bot 66(21):6803–6817

Finkler A, Ashery-Padan R, Fromm H (2007) CAMTAs: calmodulin-binding transcription activators from plants to human. FEBS Lett 581:3893

Foreman J, Demidchik V, Bothwell JH, Mylona P, Miedema H, Torres MA, Linstead P, Costa S, Brownlee C, Jones JD, Davies JM (2003) Reactive oxygen species produced by NADPH oxidase regulate plant cell growth. Nature 422(6930):442–446

Foyer CH, Neukermans J, Queval G, Noctor G, Harbinson J (2012) Photosynthetic control of electron transport and the regulation of gene expression. J Exp Bot 63(4):1637–1661

Frank W, Munnik T, Kerkmann K, Salamini F, Bartels D (2000) Water deficit triggers phospholipase D activity in the resurrection plant craterostigma plantagineum. Plant Cell 12(1):111–123

Gasulla F, Dorp K, Dombrink I, Zähringer U, Gisch N, Dörmann P, Bartels D (2013) The role of lipid metabolism in the acquisition of desiccation tolerance in craterostigma plantagineum: a comparative approach. Plant J 75(5):726–741

Geisler M, Axelsen KB, Harper JF, Palmgren MG (2000) Molecular aspects of higher plant P-type ca 2+-ATPases. Biochim Biophys Acta (BBA)-Biomembr 1465(1):52–78

Gill SS, Tuteja N (2010) Reactive oxygen species and antioxidant machinery in abiotic stress tolerance in crop plants. Plant Physiol Biochem 48(12):909–930

Gilroy S, Suzuki N, Miller G, Choi W-G, Toyota M, Devireddy AR, Mittler R (2014) A tidal wave of signals: calcium and ROS at the forefront of rapid systemic signaling. Trends Plant Sci 19(10):623–630

Goldgur Y, Rom S, Ghirlando R, Shkolnik D, Shadrin N, Konrad Z, Bar-Zvi D (2007) Desiccation and zinc binding induce transition of tomato abscisic acid stress ripening 1, a water stress- and salt stress-regulated plant-specific protein, from unfolded to folded state. Plant Physiol 143(2):617–628

Gong M, Li Z-G (1995) Calmodulin-binding proteins from Zea Mays germs. Phytochemistry 40(5):1335–1339

Gong M, Chen S-N, Song Y-Q, Li Z-G (1997a) Effect of calcium and calmodulin on intrinsic heat tolerance in relation to antioxidant systems in maize seedlings. Funct Plant Biol 24(3):371–379

Gong M, Li Y-J, Dai X, Tian M, Li Z-G (1997b) Involvement of calcium and calmodulin in the acquisition of heat-shock induced thermotolerance in maize seedlings. J Plant Physiol 150(5):615–621

Gong M, van der Luit AH, Knight MR, Trewavas AJ (1998) Heat-shock-induced changes in intracellular Ca2+ level in tobacco seedlings in relation to thermotolerance. Plant Physiol 116(1):429–437

Gong D, Guo Y, Jagendorf AT, Zhu J-K (2002) Biochemical characterization of the Arabidopsis protein kinase SOS2 that functions in salt tolerance. Plant Physiol 130(1):256–264

Grebner W, Stingl NE, Oenel A, Mueller MJ, Berger S (2013) Lipoxygenase6-dependent oxylipin synthesis in roots is required for abiotic and biotic stress resistance of Arabidopsis. Plant Physiol 161(4):2159–2170

Guillas I, Guellim A, Rezé N, Baudouin E (2013) Long chain base changes triggered by a short exposure of Arabidopsis to low temperature are altered by AHb1 non-symbiotic haemoglobin overexpression. Plant Physiol Biochem 63:191–195

Guo L, Wang X (2012) Crosstalk between phospholipase D and sphingosine kinase in plant stress signaling. Front Plant Sci 3:51

Guo Y, Halfter U, Ishitani M, Zhu J-K (2001) Molecular characterization of functional domains in the protein kinase SOS2 that is required for plant salt tolerance. Plant Cell 13(6):1383–1400

Guo L, Mishra G, Taylor K, Wang X (2011) Phosphatidic acid binds and stimulates Arabidopsis sphingosine kinases. J Biol Chem 286(15):13336–13345

Guo L, Devaiah SP, Narasimhan R, Pan X, Zhang Y, Zhang W, Wang X (2012) Cytosolic Glyceraldehyde-3-phosphate dehydrogenases interact with phospholipase Dδ to transduce hydrogen peroxide signals in the Arabidopsis response to stress. Plant Cell 24(5):2200–2212

Halfter U, Ishitani M, Zhu J-K (2000) The Arabidopsis SOS2 protein kinase physically interacts with and is activated by the calcium-binding protein SOS3. Proc Natl Acad Sci 97(7):3735–3740

Han C, Liu Q, Yang Y (2009) Short-term effects of experimental warming and enhanced ultraviolet-B radiation on photosynthesis and antioxidant defense of Picea Asperata seedlings. Plant Growth Regul 58(2):153–162

Harper JF, Breton G, Harmon A (2004) Decoding Ca2+ signals through plant protein kinases. Annu Rev Plant Biol 55:263–288

Hasanuzzaman M, Nahar K, Alam MM, Roychowdhury R, Fujita M (2013) Physiological, biochemical, and molecular mechanisms of heat stress tolerance in plants. Int J Mol Sci 14(5):9643–9684

Heilmann M, Heilmann I (2015) Plant phosphoinositides—complex networks controlling growth and adaptation. Biochim Biophys Acta (BBA) Mol Cell Biol Lipids 1851(6):759–769

Hepler PK (2005) Calcium: a central regulator of plant growth and development. Plant Cell 17(8):2142–2155

Heyno E, Mary V, Schopfer P, Krieger-Liszkay A (2011) Oxygen activation at the plasma membrane: relation between superoxide and hydroxyl radical production by isolated membranes. Planta 234(1):35–45

Hirayama T, Ohto C, Mizoguchi T, Shinozaki K (1995) A gene encoding a phosphatidylinositol-specific phospholipase C is induced by dehydration and salt stress in Arabidopsis Thaliana. Proc Natl Acad Sci 92(9):3903–3907

Ho C-HL, Hu S-H, Tsay H-C, Yi-Fang (2009) CHL1 functions as a nitrate sensor in plants. Cell 138(6):1184–1194

Hoeflich KP, Ikura M (2002) Calmodulin in action: diversity in target recognition and activation mechanisms. Cell 108(6):739–742

Hong Y, Zheng S, Wang X (2008) Dual functions of phospholipase Dα1 in plant response to drought. Mol Plant 1(2):262–269

Hong Y, Zhang W, Wang X (2010) Phospholipase D and phosphatidic acid signalling in plant response to drought and salinity. Plant Cell Environ 33(4):627–635

Hrabak EM, Chan CWM, Gribskov M, Harper JF, Choi JH, Halford N, Kudla J, Luan S, Nimmo HG, Sussman MR (2003) The Arabidopsis CDPKSnRK superfamily of protein kinases. Plant Physiol 132(2):666–680

Hua W, Li R-J, Wang L, Ying-Tang L (2004) A tobacco calmodulin-binding protein kinase (NtCBK2) induced by high-salt/GA treatment and its expression during floral development and embryogenesis. Plant Sci 166(5):1253–1259

Hwang Y- s, Bethke PC, Cheong YH, Chang H-S, Zhu T, Jones RL (2005) A gibberellin-regulated Calcineurin B in Rice localizes to the tonoplast and is implicated in vacuole function. Plant Physiol 138(3):1347–1358

Imbusch R, Mueller MJ (2000) Analysis of oxidative stress and wound-inducible dinor isoprostanes F1 (phytoprostanes F1) in plants. Plant Physiol 124(3):1293–1304

Ischebeck T, Seiler S, Heilmann I (2010) At the poles across kingdoms: phosphoinositides and polar tip growth. Protoplasma 240(1–4):13–31

Ishitani M, Liu J, Halfter U, Kim C-S, Shi W, Zhu J-K (2000) SOS3 function in plant salt tolerance requires N-Myristoylation and calcium binding. Plant Cell 12(9):1667–1677

Jae HY, Chan YP, Jong CK, Won DH, Mi SC, Hyeong CP, Min CK et al (2005) Direct interaction of a divergent CaM isoform and the transcription factor, MYB2, enhances salt tolerance in Arabidopsis. J Biol Chem 280(5):3697–3706

Jha SK, Sharma M, Pandey GK (2016) Role of cyclic nucleotide gated channels in stress management in plants. Curr Genomics 17(4):315–329

Johansson I, Larsson C, Ek B, Kjellbom P (1996) The major integral proteins of spinach leaf plasma membranes are putative Aquaporins and are phosphorylated in response to Ca2+ and Apoplastic water potential. Plant Cell 8(7):1181–1191

Joo JH, Bae YS, Lee JS (2001) Role of auxin-induced reactive oxygen species in root gravitropism. Plant Physiol 126(3):1055–1060

Juárez SPD, Mangano S, Estevez JM (2015) Improved ROS measurement in root hair cells. Methods Mol Biol (Clifton, NJ) 1242:67–71

Kaplan B, Davydov O, Knight H, Galon Y, Knight MR, Fluhr R, Fromm H (2006) Rapid transcriptome changes induced by cytosolic Ca2+ transients reveal ABRE-related sequences as Ca2+-responsive cis elements in Arabidopsis. Plant Cell 18(10):2733–2748

Katagiri T, Ishiyama K, Kato T, Tabata S, Kobayashi M, Shinozaki K (2005) An important role of phosphatidic acid in ABA signaling during germination in Arabidopsis Thaliana. Plant J 43(1):107–117

Kim S-G, Kim S-Y, Park C-M (2007) A membrane-associated NAC transcription factor regulates salt-responsive flowering via FLOWERING LOCUS T in Arabidopsis. Planta 226(3):647–654

Kim MC, Chung WS, Yun D-J, Cho MJ (2009) Calcium and calmodulin-mediated regulation of gene expression in plants. Mol Plant 2(1):13–21

Kim S-C, Liang G, Wang X (2013a) Phosphatidic acid binds to cytosolic glyceraldehyde-3-phosphate dehydrogenase and promotes its cleavage in Arabidopsis. J Biol Chem 288(17):11834–11844

Kim YS, Park S, Gilmour SJ, Thomashow MF (2013b) Roles of CAMTA transcription factors and salicylic acid in configuring the low-temperature transcriptome and freezing tolerance of Arabidopsis. Plant J 75(3):364–376

Knight H, Trewavas AJ, Knight MR (1997) Calcium signalling in Arabidopsis Thaliana responding to drought and salinity. Plant J: Cell Mol Biol 12(5):1067–1078

Köhler C, Merkle T, Neuhaus G (1999) Characterisation of a novel gene family of putative cyclic nucleotide- and calmodulin-regulated ion channels in Arabidopsis Thaliana. Plant J: Cell Mol Biol 18(1):97–104

Kolukisaoglu U, Weinl S, Blazevic D, Batistic O, Kudla J (2004) Calcium sensors and their interacting protein kinases: genomics of the Arabidopsis and Rice CBL-CIPK signaling networks. Plant Physiol 134(1):43–58

Kopka J, Pical C, Gray JE, Müller-Röber B (1998) Molecular and enzymatic characterization of three phosphoinositide-specific phospholipase C isoforms from potato. Plant Physiol 116(1):239–250

Kovtun Y, Chiu WL, Tena G, Sheen J (2000) Functional analysis of oxidative stress-activated mitogen-activated protein kinase cascade in plants. Proc Natl Acad Sci U S A 97(6):2940–2945

Krieger G, Shkolnik D, Miller G, Fromm H (2016) Reactive oxygen species tune root tropic responses. Plant Physiol 172(2):1209–1220

Kudla J, Xu Q, Harter K, Gruissem W, Luan S (1999) Genes for Calcineurin B-like proteins in Arabidopsis are differentially regulated by stress signals. Proc Natl Acad Sci U S A 96(8):4718–4723
Kudla J, Batistic O, Hashimoto K (2010) Calcium signals: the lead currency of plant information processing. Plant Cell 22(3):541–563
Kushwaha R, Singh A, Chattopadhyay S (2008) Calmodulin7 plays an important role as transcriptional regulator in Arabidopsis seedling development. Plant Cell 20(7):1747–1759
Kutuzov MA, Evans DE, Andreeva AV (1998) Expression and characterization of PP7, a novel plant protein Ser/Thr phosphatase distantly related to RdgC/PPEF and PP5. FEBS Lett 440(1–2):147–152
Ladwig F, Dahlke RI, Stührwohldt N, Hartmann J, Harter K, Sauter M (2015) Phytosulfokine regulates growth in Arabidopsis through a response module at the plasma membrane that includes cyclic nucleotide-gated channel17, h+-atpase, and bak1[open]. Plant Cell 27(6):1718–1729
Larkindale J, Knight MR (2002) Protection against heat stress-induced oxidative damage in Arabidopsis involves calcium, abscisic acid, ethylene, and salicylic acid. Plant Physiol 128(2):682–695
Larsson O, Barker CJ, Sjöholm Å, Carlqvist H, Michell RH, Bertorello A, Nilsson T, Honkanen RE, Mayr GW, Zwiller J (1997) Inhibition of phosphatases and increased Ca2+ channel activity by inositol hexakisphosphate. Science 278(5337):471–474
Le CTT, Brumbarova T, Ivanov R, Stoof C, Weber E, Mohrbacher J, Fink-Straube C, Bauer P (2016) Zinc finger of Arabidopsis thaliana12 (zat12) interacts with fer-like iron deficiency-induced transcription factor (fit) Linking Iron Deficiency and Oxidative Stress Responses. Plant Physiol 170(1):540–557
Lee S, Suh S, Kim S, Crain RC, Kwak JM, Nam H-G, Lee Y (1997) Systemic elevation of phosphatidic acid and lysophospholipid levels in wounded plants. Plant J 12(3):547–556
Lee SH, Kim MC, Do Heo W, Kim JC, Chung WS, Park CY, Park HC, Cheong YH, Kim CY, Lee S-H (1999) Competitive binding of calmodulin isoforms to calmodulin-binding proteins: implication for the function of calmodulin isoforms in plants. Biochim Biophys Acta (BBA) Protein Struct Mol Enzymol 1433(1):56–67
Lee SH, David Johnson J, Walsh MP, Van Lierop JE, Sutherland C, Ande XU, Snedden WA, Danuta K-K, Fromm H, Narayanan N (2000) Differential regulation of Ca2+/calmodulin-dependent enzymes by plant calmodulin isoforms and free Ca2+ concentration. Biochem J 350(1):299–306
Lee M-O, Cho K, Kim S-H, Jeong S-H, Kim J-A, Jung Y-H, Shim J, Shibato J, Rakwal R, Tamogami S (2008) Novel Rice OsSIPK is a multiple stress responsive MAPK family member showing rhythmic expression at mRNA level. Planta 227(5):981–990
Lee K-W, Chen P-W, Lu C-A, Chen S, Ho T-HD YS-M (2009) Coordinated responses to oxygen and sugar deficiency allow rice seedlings to tolerate flooding. Sci Signal 2(91):61–64
Lee J-H, Kim Y-J, Jeong D-Y, Sathiyaraj G, Pulla RK, Shim J-S, In J-G, Yang D-C (2010) Isolation and characterization of a glutamate decarboxylase (GAD) gene and Their differential expression in response to abiotic stresses from Panax Ginseng CA Meyer. Mol Biol Rep 37(7):3455–3463
Lemtiri-Chlieh F, MacRobbie EAC, Webb AAR, Manison NF, Brownlee C, Skepper JN, Chen J, Prestwich GD, Brearley CA (2003) Inositol hexakisphosphate mobilizes an endomembrane store of calcium in guard cells. Proc Natl Acad Sci 100(17):10091–10095
Léran SE, Pervent KH, Hashimoto M, Corratgé-Faillie K, Offenborn C, Tillard JN, Gojon P, Kudla A, Benoît JL (2015) Nitrate sensing and uptake in Arabidopsis are enhanced by ABI2, a phosphatase inactivated by the stress hormone abscisic acid. Sci Signal 8(43):10–1126
Li J, Lee Y-RJ, Assmann SM (1998) Guard cells possess a calcium-dependent protein kinase that phosphorylates the KAT1 potassium channel. Plant Physiol 116(2):785–795
Li B, Liu H-T, Sun D-Y, Zhou R-G (2004) Ca2+ and calmodulin modulate DNA-binding activity of maize heat shock transcription factor in vitro. Plant Cell Physiol 45(5):627–634
Li X, Borsics T, Harrington HM, Christopher DA (2005) Arabidopsis AtCNGC10 rescues potassium channel mutants of E. coli, yeast and Arabidopsis and is regulated by calcium/calmodulin and cyclic GMP in E. coli. Funct Plant Biol 32(7):643–653

Li D-F, Li J, Ma L, Zhang L, Ying-Tang L (2006) Calmodulin isoform-specific activation of a rice calmodulin-binding kinase conferred by only three amino-acids of OsCaM61. FEBS Lett 580(18):4325–4331

Li A, Wang X, Leseberg CH, Jia J, Mao L (2008) Biotic and abiotic stress responses through calcium-dependent protein kinase (CDPK) signaling in wheat (Triticum Aestivum L.) Plant Signal Behav 3(9):654–656

Li M, Hong Y, Wang X (2009) Phospholipase D- and phosphatidic acid-mediated signaling in plants. Biochim Biophys Acta Mol Cell Biol Lipids 1791:927

Li J, Wang X, Zhang Y, Jia H, Bi Y (2011) cGMP regulates hydrogen peroxide accumulation in calcium-dependent salt resistance pathway in Arabidopsis Thaliana roots. Planta 234(4):709–722

Li J, Jia H, Wang J (2014) cGMP and ethylene are involved in maintaining ion homeostasis under salt stress in Arabidopsis roots. Plant Cell Rep 33(3):447–459

Liu J, Zhu J-K (1997) Proline accumulation and salt-stress-induced gene expression in a salt-hypersensitive mutant of Arabidopsis. Plant Physiol 114(2):591–596

Liu H-T, Li B, Shang Z-L, Li X-Z, Rui-Ling M, Sun D-Y, Zhou R-G (2003) Calmodulin is involved in heat shock signal transduction in wheat. Plant Physiol 132(3):1186–1195

Liu H-T, Sun D-Y, Zhou R-G (2005) Ca2+ and AtCaM3 are involved in the expression of heat shock protein gene in Arabidopsis. Plant Cell Environ 28(10):1276–1284

Liu F, Yoo B-C, Lee J-Y, Pan W, Harmon AC (2006) Calcium-regulated phosphorylation of soybean serine acetyltransferase in response to oxidative stress. J Biol Chem 281(37):27405–27415

Liu H-T, Li G-L, Chang HUI, Sun D-Y, Zhou R-G, Li B (2007) Calmodulin-binding protein phosphatase PP7 is involved in thermotolerance in Arabidopsis. Plant Cell Environ 30(2):156–164

Liu H-T, Gao F, Li G-L, Han J-L, Liu D-L, Sun D-Y, Zhou R-G (2008) The calmodulin-binding protein kinase 3 is part of heat-shock signal transduction in Arabidopsis Thaliana. Plant J 55(5):760–773

Luan S, Kudla J, Rodriguez-Concepcion M, Yalovsky S, Gruissem W (2002) Calmodulins and Calcineurin B-like proteins: calcium sensors for specific signal response coupling in plants. Plant Cell 14(Suppl):S389–S400

Lynch DV (2012) Evidence that sphingolipid signaling is involved in responding to low temperature. New Phytol 194(1):7–9

Ma F, Lu R, Liu H, Shi B, Zhang J, Tan M, Zhang A, Jiang M (2012) Nitric oxide-activated calcium/calmodulin-dependent protein kinase regulates the abscisic acid-induced antioxidant defence in maize. J Exp Bot 63(13):4835–4847

Ma QT, Zheng R-j, Wang X-j, Luan S-m, Sheng (2015) The calcium sensor CBL7 modulates plant responses to low nitrate in Arabidopsis. Biochem Biophys Res Commun 468(1–2):59–65

Maathuis FJM (2006) cGMP modulates gene transcription and cation transport in Arabidopsis roots. Plant J: Cell Mol Biol 45(5):700–711

Maathuis FJ, Sanders D (2001) Sodium uptake in Arabidopsis roots is regulated by cyclic nucleotides. Plant Physiol 127(4):1617–1625

Magnan F, Ranty B, Charpenteau M, Sotta B, Galaud J-P, Aldon D (2008) Mutations in AtCML9, a calmodulin-like protein from Arabidopsis Thaliana, alter plant responses to abiotic stress and abscisic acid. Plant J 56(4):575–589

Mahajan S (2006) Cloning and characterization of CBL-CIPK signalling components from a legume (Pisum Sativum). FEBS J 273(5):907–925. https://doi.org/10.1111/j.1742-4658.2006.05111.x

Mahajan S, Mahajan S, Tuteja N, Tuteja N (2005) Cold, salinity and drought stresses: an overview. Arch Biochem Biophys 444(2):139–158

Mahajan S, Sopory SK, Tuteja N (2006) CBL-CIPK paradigm: role in calcium and stress signaling in plants. Proc Indian Natl Sci Acad 72(2):63.

Mahajan S, Pandey GK, Tuteja N (2008) Calcium-and salt-stress signaling in plants: shedding light on SOS pathway. Arch Biochem Biophys 471(2):146–158

Marino D, Dunand C, Puppo A, Pauly N (2012) A burst of plant NADPH oxidases. Trends Plant Sci 17(1):9–15

Markham JE, Li J, Cahoon EB, Jaworski JG (2006) Separation and identification of major plant sphingolipid classes from leaves. J Biol Chem 281(32):22684–22694
Martìn ML, Busconi L (2001) A rice membrane-bound calcium-dependent protein kinase is activated in response to low temperature. Plant Physiol 125(3):1442–1449
Matsuo M, Johnson JM, Hieno A, Tokizawa M, Nomoto M, Tada Y, Godfrey R, Obokata J, Sherameti I, Yamamoto YY, Böhmer FD (2015) High redox responsive transcription factor1 levels result in accumulation of reactive oxygen species in Arabidopsis thaliana shoots and roots. Mol Plant 8(8):1253–1273
McCormack E, Braam J (2003) Calmodulins and related potential calcium sensors of Arabidopsis. New Phytol 159(3):585–598
McCormack E, Tsai Y-C, Braam J (2005) Handling calcium signaling: Arabidopsis CaMs and CMLs. Trends Plant Sci 10(8):383–389
McLoughlin F, Arisz SA, Dekker HL, Kramer G, de Koster CG, Haring MA, Munnik T, Testerink C (2013) Identification of novel candidate phosphatidic acid-binding proteins involved in the salt-stress response of Arabidopsis Thaliana roots. Biochem J 450(3):573–581
Medda R, Padiglia A, Longu S, Bellelli A, Arcovito A, Cavallo S, Pedersen JZ, Floris G (2003) Critical role of Ca2+ ions in the reaction mechanism of euphorbia characias peroxidase. Biochemistry 42(29):8909–8918
Meijer HJG, Munnik T (2003) Phospholipid-based signaling in plants. Annu Rev Plant Biol 54(1):265–306
Mikami K, Katagiri T, Iuchi S, Yamaguchi-Shinozaki K, Shinozaki K (1998) A gene encoding phosphatidylinositol-4-phosphate 5-kinase is induced by water stress and abscisic acid in Arabidopsis Thaliana. Plant J 15(4):563–568
Milla R, Miguel A, Uno Y, Chang I-F, Townsend J, Maher EA, Quilici D, Cushman JC (2006) A novel yeast two-hybrid approach to identify CDPK substrates: characterization of the interaction between AtCPK11 and AtDi19, a nuclear zinc finger protein1. FEBS Lett 580(3):904–911
Miller G, Shulaev V, Mittler R (2008) Reactive oxygen signaling and abiotic stress. Physiol Plant 133(3):481–489
Miller GAD, Suzuki N, Sultan C-Y, Mittler RON (2010) Reactive oxygen species homeostasis and signalling during drought and salinity stresses. Plant Cell Environ 4:453–467
Mishkind M, Vermeer JEM, Darwish E, Munnik T (2009) Heat stress activates phospholipase D and triggers PIP2 accumulation at the plasma membrane and nucleus. Plant J 60(1):10–21
Mittler R (2002) Oxidative stress, antioxidants and stress tolerance. Trends Plant Sci 7(9):405–410
Mittler R (2006) Abiotic stress, the field environment and stress combination. Trends Plant Sci 11(1):15–19
Mittler R, Blumwald E (2015) The roles of ROS and ABA in systemic acquired acclimation. Plant Cell 27(1):64–70
Monroy AF, Sarhan F, Dhindsa RS (1993) Cold-induced changes in freezing tolerance, protein phosphorylation, and gene expression (evidence for a role of calcium). Plant Physiol 102(4):1227–1235
Moon H, Lee B, Choi G, Dongjin S, Theertha Prasad D, Lee O, Kwak S-S et al (2003) NDP kinase 2 interacts with two oxidative stress-activated MAPKs to regulate cellular redox state and enhances multiple stress tolerance in transgenic plants. Proc Natl Acad Sci U S A 100(1):358–363
Mori IC, Murata Y, Yang Y, Munemasa S, Wang Y-F, Andreoli S, Tiriac H et al (2006) CDPKs CPK6 and CPK3 function in ABA regulation of guard cell S-type anion- and ca(2+)-permeable channels and stomatal closure. PLoS Biol 4(10):e327
Mosblech A, Feussner I, Heilmann I (2009) Oxylipins: structurally diverse metabolites from fatty acid oxidation. Plant Physiol Biochem 47(6):511–517
Moutinho A, Hussey PJ, Trewavas AJ, Malhó R (2001) cAMP acts as a second messenger in pollen tube growth and reorientation. Proc Natl Acad Sci U S A 98(18):10481–10486
Munnik T, Testerink C (2009) Plant phospholipid signaling: 'in a Nutshell'. J Lipid Res 50(Suppl):S260–S265

Munnik T, Vermeer JEM (2010) Osmotic stress-induced phosphoinositide and inositol phosphate signalling in plants. Plant Cell Environ 33(4):655–669

Munnik T, Meijer HJG, Ter Riet B, Hirt H, Frank W, Bartels D, Musgrave A (2000) Hyperosmotic stress stimulates phospholipase D activity and elevates the levels of phosphatidic acid and diacylglycerol pyrophosphate. Plant J 22(2):147–154

Munns R, James RA, Läuchli A (2006) Approaches to increasing the salt tolerance of wheat and other cereals. J Exp Bot 57(5):1025–1043

Mura A, Medda R, Longu S, Floris G, Rinaldi AC, Padiglia A (2005) A Ca2+/calmodulin-binding peroxidase from euphorbia latex: novel aspects of calcium-hydrogen peroxide cross-talk in the regulation of plant defenses. Biochemistry 44(43):14120–14130

Nagae M, Nozawa A, Koizumi N, Sano H, Hashimoto H, Sato M, Shimizu T (2003) The crystal structure of the novel calcium-binding protein AtCBL2 from Arabidopsis Thaliana. J Biol Chem 278(43):42240–42246

Nakashima K, Yamaguchi-Shinozaki K, Shinozaki K (2014) The transcriptional regulatory network in the drought response and its crosstalk in abiotic stress responses including drought, cold, and heat. Front Plant Sci 5:170

Nan W, Wang X, Yang L, Hu Y, Wei Y, Liang X, Mao L, Bi Y (2014) Cyclic GMP is involved in auxin signalling during Arabidopsis root growth and development. J Exp Bot 65(6):1571–1583

Narváez-Vásquez J, Florin-Christensen J, Ryan CA (1999) Positional specificity of a phospholipase a activity induced by wounding, systemin, and oligosaccharide elicitors in tomato leaves. Plant Cell 11(11):2249–2260

Nath K, Yan L (2015) A paradigm of reactive oxygen species and programed cell death in plants. J Cell Sci Ther 6:202

Neill S, Desikan R, Hancock J (2002) Hydrogen peroxide signalling. Curr Opin Plant Biol 5(5):388–395

Ng CK-Y, Carr K, McAinsh MR, Powell B, Hetherington AM (2001) Drought-induced guard cell signal transduction involves Sphingosine-1- phosphate. Nature 410(6828):596–599

Noctor G (2006) Metabolic signalling in defence and stress: the central roles of soluble redox couples. Plant Cell Environ 29(3):409–425

Nozawa A, Koizumi N, Sano H (2001) An Arabidopsis SNF1-related protein kinase, AtSR1, interacts with a calcium-binding protein, AtCBL2, ofwhich transcripts respond to light. Plant Cell Physiol 42(9):976–981

O'Brien JA, Daudi A, Butt VS, Paul Bolwell G (2012) Reactive oxygen species and their role in plant defence and cell wall metabolism. Planta 236(3):765–779

Ohta M, Guo Y, Halfter U, Zhu J-K (2003) A novel domain in the protein kinase SOS2 mediates interaction with the protein phosphatase 2C ABI2. Proc Natl Acad Sci 100(20):11771–11776

Osakabe Y, Yamaguchi-Shinozaki K, Shinozaki K, Tran L-SP (2013) Sensing the environment: key roles of membrane-localized kinases in plant perception and response to abiotic stress. J Exp Bot 64(2):445–458

Pagnussat GC, Lanteri ML, Lamattina L (2003) Nitric oxide and cyclic GMP are messengers in the indole acetic acid-induced adventitious rooting process. Plant Physiol 132(3):1241–1248

Pandey S, Tiwari SB, Tyagi W, Reddy MK, Upadhyaya KC, Sopory SK (2002) A Ca2+/CaM-dependent kinase from pea is stress regulated and in vitro phosphorylates a protein that binds to AtCaM5 promoter. Eur J Biochem 269(13):3193–3204

Pandey GK, Cheong YH, Kim K-N, Grant JJ, Li L, Hung W, D'Angelo C, Weinl S, Kudla J, Luan S (2004) The calcium sensor calcineurin B-like 9 modulates abscisic acid sensitivity and biosynthesis in Arabidopsis. Plant Cell 16(7):1912–1924

Pardo JM, Reddy MP, Yang S, Maggio A, Huh GH, Matsumoto T, Coca MA et al (1998) Stress signaling through Ca2+/calmodulin-dependent protein phosphatase calcineurin mediates salt adaptation in plants. Proc Natl Acad Sci U S A 95(16):9681–9686

Park HC, Park CY, Koo SC, Cheong MS, Kim KE, Kim MC, Lim CO, Lee SY, Yun D-J, Chung WS (2010) AtCML8, a calmodulin-like protein, differentially activating CaM-dependent enzymes in Arabidopsis Thaliana. Plant Cell Rep 29(11):1297–1304

Patharkar OR, Cushman JC (2000) A stress-induced calcium-dependent protein kinase from Mesembryanthemum crystallinum phosphorylates a two-component pseudo-response regulator. Plant J 24(5):679–691

Patil S, Takezawa D, Poovaiah BW (1995) Chimeric plant calcium/calmodulin-dependent protein kinase gene with a neural Visinin-like calciumbinding domain. Proc Natl Acad Sci 92(11):4897–4901

Pauly N, Knight MR, Thuleau P, Graziana A, Muto S, Ranjeva R, Mazars C (2001) The nucleus together with the cytosol generates patterns of specific cellular calcium signatures in tobacco suspension culture cells. Cell Calcium 30(6):413–421

Pei Z-M, Ward JM, Harper JF, Schroeder JI (1996) A novel chloride channel in Vicia Faba guard cell vacuoles activated by the serine/threonine kinase, CDPK. EMBO J 15(23):6564

Perochon A, Aldon D, Galaud J-P, Ranty B (2011) Calmodulin and calmodulin-like proteins in plant calcium signaling. Biochimie 93(12):2048–2053

Peters C, Li M, Narasimhan R, Roth M, Welti R, Wang X (2010) Nonspecific phospholipase C NPC4 promotes responses to abscisic acid and tolerance to hyperosmotic stress in Arabidopsis. Plant Cell 22(8):2642–2659

Petrov VD, Van Breusegem F (2012) Hydrogen peroxide–a central hub for information flow in plant cells. AoB PLANTS 2012:pls014

Petrov V, Hille J, Mueller-Roeber B, Gechev TS (2015) ROS-mediated abiotic stress-induced programmed cell death in plants. Front Plant Sci 6:69

Pharmawati M, Maryani MM, Nikolakopoulos T, Gehring CA, Irving HR (2001) Cyclic GMP modulates stomatal opening induced by natriuretic peptides and immunoreactive analogues. Plant Physiol Biochem 39(5):385–394

Pical C, Westergren T, Dove SK, Larsson C, Sommarin M (1999) Salinity and hyperosmotic stress induce rapid increases in phosphatidylinositol 4, 5-bisphosphate, diacylglycerol pyrophosphate, and phosphatidylcholine in Arabidopsis Thaliana cells. J Biol Chem 274(53):38232–38240

Polisensky DH, Braam J (1996) Cold-shock regulation of the Arabidopsis TCH genes and the effects of modulating intracellular calcium levels. Plant Physiol 111(4):1271–1279

Popescu SC, Popescu GV, Bachan S, Zhang Z, Seay M, Gerstein M, Snyder M, Dinesh-Kumar SP (2007) Differential binding of calmodulinrelated proteins to their targets revealed through high-density Arabidopsis protein microarrays. Proc Natl Acad Sci 104(11):4730–4735

Putnam-Evans C, Harmon AC, Palevitz BA, Fechheimer M, Cormier MJ (1989) Calcium-dependent protein kinase is localized with F-actin in plant cells. Cell Motil Cytoskeleton 12(1):12–22

Quan R, Lin H, Mendoza I, Zhang Y, Cao W, Yang Y, Shang M, Chen S, Pardo JM, Guo Y (2007) SCABP8/CBL10, a putative calcium sensor, interacts with the protein kinase SOS2 to protect Arabidopsis shoots from salt stress. Plant Cell 19(4):1415–1431

Quintero FJ, Ohta M, Shi H, Zhu J-K, Pardo JM (2002) Reconstitution in yeast of the Arabidopsis SOS signaling pathway for Na+ homeostasis. Proc Natl Acad Sci 99(13):9061–9066

Rabbani MA, Maruyama K, Hiroshi A, Ayub Khan M, Katsura K, Ito Y, Yoshiwara K, Seki M, Shinozaki K, Yamaguchi-Shinozaki K (2003) Monitoring expression profiles of rice genes under cold, drought, and high-salinity stresses and abscisic acid application using cDNA microarray and RNA gel-blot analyses. Plant Physiol 133(4):1755–1767

Raichaudhuri A, Bhattacharyya R, Chaudhuri S, Chakrabarti P, DasGupta M (2006) Domain analysis of a groundnut calcium-dependent protein kinase nuclear localization sequence in the junction domain is coupled with nonconsensus calcium binding domains. J Biol Chem 281(15):10399–10409

Reddy VS, Reddy ASN (2004) Proteomics of calcium-signaling components in plants. Phytochemistry 65(12):1745–1776

Reddy ASN, Reddy VS, Golovkin M (2000) A calmodulin binding protein from Arabidopsis is induced by ethylene and contains a DNA-binding motif. Biochem Biophys Res Commun 279(3):762–769

Reddy ASN, Ali GS, Celesnik H, Day IS (2011) Coping with stresses: roles of calcium-and calcium/calmodulin-regulated gene expression. Plant Cell 23(6):2010–2032

Rhoads AR, Friedberg F (1997) Sequence motifs for calmodulin recognition. FASEB J 11(5):331–340

Riveras EA, Vidal JM, Oses EA, Vega C, Gutiérrez A, Rodrigo A (2015) The calcium ion is a second messenger in the nitrate signaling pathway of Arabidopsis. Plant Physiol 169(2):1397–1404

Roach T, Krieger-Liszkay AK (2014) Regulation of photosynthetic electron transport and photoinhibition. Curr Protein Pept Sci 15(4):351–362

Rubio F, Flores P, Navarro JM, Martí V (2003) Effects of Ca 2+, K+ and cGMP on Na+ uptake in pepper plants. Plant Sci 165(5):1043–1049

Sah SK, Reddy KR, Li J (2016) Abscisic acid and abiotic stress tolerance in crop plants. Front Plant Sci 7

Saijo Y, Hata S, Kyozuka J, Shimamoto K, Izui K (2000) Over-expression of a single Ca2+-dependent protein kinase confers both cold and salt/drought tolerance on rice plants. Plant J 23(3):319–327

Sakamoto H, Maruyama K, Sakuma Y, Meshi T, Iwabuchi M, Shinozaki K, Yamaguchi-Shinozaki K (2004) Arabidopsis Cys2/His2-type zinc-finger proteins function as transcription repressors under drought, cold, and high-salinity stress conditions. Plant Physiol 136(1):2734–2746

Sánchez-Barrena MJ, Martínez-Ripoll M, Zhu JK, Albert A (2005) The structure of the Arabidopsis Thaliana SOS3: molecular mechanism of sensing calcium for salt stress response. J Mol Biol 345(5):1253–1264

Sánchez-Barrena MJ, Fujii H, Angulo I, Martínez-Ripoll M, Zhu J-K, Albert A (2007a) The structure of the C-terminal domain of the protein kinase AtSOS2 bound to the calcium sensor AtSOS3. Mol Cell 26(3):427–435

Sánchez-Barrena MJ, Moreno-Pérez S, Angulo I, Martínez-Ripoll M, Albert A (2007b) The complex between SOS3 and SOS2 regulatory domain from Arabidopsis Thaliana: cloning, expression, purification, crystallization and preliminary X-ray analysis. Acta Crystallogr Sect F Struct Biol Cryst Commun 63(7):568–570

Sanders D, Pelloux J, Brownlee C, Harper JF (2002) Calcium at the crossroads of signaling. Plant Cell 14(Suppl):S401–S417

Sathyanarayanan PV, Poovaiah BW (2004) Decoding Ca2+ signals in plants. Crit Rev Plant Sci 23(1):1–11

Sato EM, Hijazi H, Bennett MJ, Vissenberg K, Swarup R (2014) New insights into root gravitropic signalling. J Exp Bot 66(8):2155–2165

Savchenko T, Kolla VA, Wang C-Q, Nasafi Z, Hicks DR, Phadungchob B, Chehab WE, Brandizzi F, Froehlich J, Dehesh K (2014) Functional convergence of oxylipin and abscisic acid pathways controls stomatal closure in response to drought. Plant Physiol 164(3):1151–1160

Scrase-Field SAMG, Knight MR (2003) Calcium: just a chemical switch? Curr Opin Plant Biol 6(5):500–506

Seltmann MA, Stingl NE, Lautenschlaeger JK, Krischke M, Mueller MJ, Berger S (2010) Differential impact of lipoxygenase 2 and Jasmonates on natural and stress-induced senescence in Arabidopsis. Plant Physiol 152(4):1940–1950

Sewelam N, Kazan K, Schenk PM (2016) Global plant stress signaling: reactive oxygen species at the cross-road. Front Plant Sci 7:187

Shafi A, Chauhan R, Gill T, Swarnkar MK, Sreenivasulu Y, Kumar S, Kumar N, Shankar R, Ahuja PS, Singh AK (2015a) Expression of SOD and APX genes positively regulates secondary cell wall biosynthesis and promotes plant growth and yield in *Arabidopsis* under salt stress. Plant Mol Biol 87:615–631

Shafi A, Gill T, Sreenivasulu Y, Kumar S, Ahuja PS, Singh AK (2015b) Improved callus induction, shoot regeneration, and salt stress tolerance in Arabidopsis overexpressing superoxide dismutase from *Potentilla atrosanguinea*. Protoplasma 252:41–51

Shah K, Kumar RG, Verma S, Dubey RS (2001) Effect of cadmium on lipid peroxidation, superoxide anion generation and activities of antioxidant enzymes in growing rice seedlings. Plant Sci 161(6):1135–1144

Shaikhali J, Norén L, de Dios Barajas-López J, Srivastava V, König J, Sauer UH, Wingsle G, Dietz K-J, Strand Å (2012) Redox-mediated mechanisms regulate DNA binding activity of the

G-group of basic region leucine zipper (bZIP) transcription factors in Arabidopsis. J Biol Chem 287(33):27510–27525
Shao J, Harmon AC (2003) In vivo phosphorylation of a recombinant peptide substrate of CDPK suggests involvement of CDPK in plant stress responses. Plant Mol Biol 53(5):731–740
Sharma P, Dubey RS (2005) Drought induces oxidative stress and enhances the activities of antioxidant enzymes in growing Rice seedlings. Plant Growth Regul 46(3):209–221
Sheen J (1996) Ca2 plus-dependent protein kinases and stress signal transduction in plants. Science 274(5294):1900
Shi J, Kim K-N, Ritz O, Albrecht V, Gupta R, Harter K, Luan S, Kudla J (1999) Novel protein kinases associated with calcineurin B–like calcium sensors in Arabidopsis. Plant Cell 11(12):2393–2405
Shi B, Ni L, Zhang A, Cao J, Zhang H, Qin T, Tan M, Zhang J, Jiang M (2012) OsDMI3 is a novel component of abscisic acid signaling in the induction of antioxidant defense in leaves of Rice. Mol Plant 5(6):1359–1374
Shi H, Wang X, Ye T, Chen F, Deng J, Yang P, Zhang Y, Chan Z (2014) The Cysteine2/Histidine2-type transcription factor zinc finger of arabidopsis thaliana6 modulates biotic and abiotic stress responses by activating salicylic acid-related genes and c-repeat-binding factor genes in Arabidopsis. Plant Physiol 165(3):1367–1379
Sierla M, Waszczak C, Vahisalu T, Kangasjärvi J (2016) Reactive oxygen species in the regulation of stomatal movements. Plant Physiol 171(3):1569–1580
Sies H (2014) Role of metabolic H2O2 generation redox signaling and oxidative stress. J Biol Chem 289(13):8735–8741
Smykowski A, Zimmermann P, Zentgraf U (2010) G-box binding Factor1 reduces CATALASE2 expression and regulates the onset of leaf senescence in Arabidopsis. Plant Physiol 153(3):1321–1331
Snedden WA, Fromm H (2001) Calmodulin as a versatile calcium signal transducer in plants. New Phytol 151(1):35–66
Stevenson JM, Perera IY, Heilmann I, Persson S, Boss WF (2000) Inositol signaling and plant growth. Trends Plant Sci 5(6):252–258
Subbaiah CC, Sachs MM (2000) Maize cap1 encodes a novel SERCA-type calcium-ATPase with a calmodulin-binding domain. J Biol Chem 275(28):21678–21687
Suzuki N, Miller G, Morales J, Shulaev V, Torres MA, Mittler R (2011) Respiratory burst oxidases: the engines of ROS signaling. Curr Opin Plant Biol 14(6):691–699
Suzuki N, Miller G, Salazar C, Mondal HA, Shulaev E, Cortes DF, Shuman JL et al (2013) Temporal-spatial interaction between reactive oxygen species and abscisic acid regulates rapid systemic acclimation in plants. Plant Cell 25(9):3553–3569
SX L, Hrabak EM (2002) An Arabidopsis calcium-dependent protein kinase is associated with the endoplasmic reticulum. Plant Physiol 128(3):1008–1021
Taddese B, Upton GJG, Bailey GR, Jordan SRD, Abdulla NY, Reeves PJ, Reynolds CA (2014) Do plants contain G protein-coupled receptors? Plant Physiol 164(1):287–307
Takahashi S, Katagiri T, Hirayama T, Yamaguchi-Shinozaki K, Shinozaki K (2001) Hyperosmotic stress induces a rapid and transient increase in inositol 1,4,5-trisphosphate independent of abscisic acid in Arabidopsis cell culture. Plant Cell Physiol 42(2):214–222
Takahashi H, Kamakura H, Sato Y, Shiono K, Abiko T, Tsutsumi N, Nagamura Y, Nishizawa NK, Nakazono M (2010) A method for obtaining high quality RNA from paraffin sections of plant tissues by laser microdissection. J Plant Res 123(6):807–813
Takahashi F, Mizoguchi T, Yoshida R, Ichimura K, Shinozaki K (2011) Calmodulin-dependent activation of MAP kinase for ROS homeostasis in Arabidopsis. Mol Cell 41(6):649–660
Takano M, Takahashi H, Suge H (1997) Calcium requirement for the induction of hydrotropism and enhancement of calcium-induced curvature by water stress in primary roots of pea, Pisum Sativum L. Plant Cell Physiol 38(4):385–391
Testerink C (2011) Molecular, cellular, and physiological responses to phosphatidic acid formation in plants. J Exp Bot 62(7):2349–2361

Testerink C, Munnik T (2005) Phosphatidic acid: a multifunctional stress signaling lipid in plants. Trends Plant Sci 10:368

Testerink C, Dekker HL, Lim ZY, Johns MK, Holmes AB, De Koster CG, Ktistakis NT, Munnik T (2004) Isolation and identification of phosphatidic acid targets from plants. Plant J 39(4):527–536

Testerink C, Larsen PB, van der Does D, van Himbergen JAJ, Munnik T (2007) Phosphatidic acid binds to and inhibits the activity of Arabidopsis CTR1. J Exp Bot 58(14):3905–3914

Thole JM, Nielsen E (2008) Phosphoinositides in plants: novel functions in membrane trafficking. Curr Opin Plant Biol 11(6):620–631

Thompson AJ, Jackson AC, Symonds RC, Mulholland BJ, Dadswell AR, Blake PS, Burbidge A, Taylor IB (2000) Ectopic expression of a tomato 9-cis-epoxycarotenoid dioxygenase gene causes over-production of abscisic acid. Plant J: Cell Mol Biol 23(3):363–374

Townley HE, Knight MR (2002) Calmodulin as a potential negative regulator of Arabidopsis COR gene expression. Plant Physiol 128(4):1169–1172

Tsui MM, York JD (2010) Roles of inositol phosphates and inositol pyrophosphates in development, cell signaling and nuclear processes. Adv Enzyme Regul 50(1):324

Tunc-Ozdemir M, Tang C, Ishka MR, Brown E, Groves NR, Myers CT, Rato C et al (2013) A cyclic nucleotide-gated channel (CNGC16) in pollen is critical for stress tolerance in pollen reproductive development. Plant Physiol 161(2):1010–1020

Tung SA, Smeeton R, White CA, Black CR, Taylor IB, Hilton HW, Thompson AJ (2008) Overexpression of LeNCED1 in tomato (Solanum Lycopersicum L.) with the rbcS3C promoter allows recovery of lines that accumulate very high levels of abscisic acid and exhibit severe phenotypes. Plant Cell Environ 31(7):968–981

Tuteja N (2007) Mechanisms of high salinity tolerance in plants. Methods Enzymol 428:419

Tuteja N, Mahajan S (2007) Calcium signaling network in plants: an overview. Plant Signal Behav 2(2):79–85

Tuteja N, Sopory SK (2008) Plant signaling in stress: G-protein coupled receptors, heterotrimeric G-proteins and signal coupling via phospholipases. Plant Signal Behav 3(2):79–86

Uno Y, Furihata T, Abe H, Yoshida R, Shinozaki K, Yamaguchi-Shinozaki K (2000) Arabidopsis basic leucine zipper transcription factors involved in an abscisic acid-dependent signal transduction pathway under drought and high-salinity conditions. Proc Natl Acad Sci 97(21):11632–11637

Urao T, Katagiri T, Mizoguchi T, Yamaguchi-Shinozaki K, Hayashida N, Shinozaki K (1994) Two genes that encode Ca2+-dependent protein kinases are induced by drought and high-salt stresses in Arabidopsis Thaliana. Mol Gen Genet MGG 244(4):331–340

Van Breusegem F, Dat JF (2006) Reactive oxygen species in plant cell death. Plant Physiol 141(2):384–390

van Der Luit AH, Olivari C, Haley A, Knight MR, Trewavas AJ (1999) Distinct calcium signaling pathways regulate calmodulin gene expression in tobacco. Plant Physiol 121(3):705–714

Van Leeuwen W, Vermeer JEM, Gadella TWJ, Munnik T (2007) Visualization of phosphatidylinositol 4, 5-bisphosphate in the plasma membrane of suspension-cultured tobacco BY-2 cells and whole Arabidopsis seedlings. Plant J 52(6):1014–1026

Vergnolle C, Vaultier M-N, Taconnat L, Renou J-P, Kader J-C, Zachowski A, Ruelland E (2005) The cold-induced early activation of phospholipase C and D pathways determines the response of two distinct clusters of genes in Arabidopsis cell suspensions. Plant Physiol 139(3):1217–1233

Wang X (1999) The role of phospholipase D in signaling cascades. Plant Physiol 120(3):645–652

Wang P, Song C-P (2008) Guard-cell signalling for hydrogen peroxide and abscisic acid. New Phytol 178(4):703–718

Wang G, Zeng H, Hu X, Zhu Y, Yang C, Shen C, Wang H, Poovaiah BW, Du L (2015) Identification and expression analyses of calmodulinbinding transcription activator genes in soybean. Plant and Soil 386(1–2):205–221

Wei S, Hu W, Deng X, Zhang Y, Liu X, Zhao X, Luo Q et al (2014) A Rice calcium-dependent protein kinase OsCPK9 positively regulates drought stress tolerance and spikelet fertility. BMC Plant Biol 14:133

Weljie AM, Vogel HJ (2004) Unexpected structure of the Ca2+-regulatory region from soybean calcium-dependent protein kinase-α. J Biol Chem 279(34):35494–35502

Welti R, Li W, Li M, Sang Y, Biesiada H, Zhou H-E, Rajashekar CB, Williams TD, Wang X (2002) Profiling membrane lipids in plant stress responses role of phospholipase Dα in freezing-induced lipid changes in Arabidopsis. J Biol Chem 277(35):31994–31992

White PJ, Broadley MR (2003) Calcium in plants. Ann Bot 92:487

Wi SJ, Seo S, Cho K, Nam MH, Park KY (2014) Lysophosphatidylcholine enhances susceptibility in signaling pathway against pathogen infection through biphasic production of reactive oxygen species and ethylene in tobacco plants. Phytochemistry 104:48–59

Williams SP, Gillaspy GE, Perera IY (2015) Biosynthesis and possible functions of inositol pyrophosphates in plants. Front Plant Sci 6:67

Worrall D, Liang Y-K, Alvarez S, Holroyd GH, Spiegel S, Panagopulos M, Gray JE, Hetherington AM (2008) Involvement of sphingosine kinase in plant cell signalling. Plant J 56(1):64–72

Xiong L, Schumaker KS, Zhu J-k (2002) Cell signaling during cold, drought, and salt stress. Plant Cell Suppl:S165–S184

G-Y X, Rocha PSCF, Wang M-L, M-L X, Cui Y-C, Li L-Y, Zhu Y-X, Xia X (2011) A novel rice calmodulin-like gene, OsMSR2, enhances drought and salt tolerance and increases ABA sensitivity in Arabidopsis. Planta 234(1):47–59

Xue H-W, Xu C, Yu M (2009) Function and regulation of phospholipid signalling in plants. Biochem J 421(2):145–156

Yamniuk AP, Vogel HJ (2005) Structural investigation into the differential target enzyme regulation displayed by plant calmodulin isoforms. Biochemistry 44(8):3101–3111

Yan J, Tsuichihara N, Etoh T, Iwai S (2007) Reactive oxygen species and nitric oxide are involved in ABA inhibition of stomatal opening. Plant Cell Environ 30(10):1320–1325

Yang T, Poovaiah BW (2002) Hydrogen peroxide homeostasis: activation of plant catalase by calcium/calmodulin. Proc Natl Acad Sci 99(6):4097–4102

Yang T, Poovaiah BW (2003) Calcium/calmodulin-mediated signal network in plants. Trends Plant Sci 8(10):505–512

Yang S-N, Yu J, Mayr GW, Hofmann F, Larsson O, Berggren P-O (2001) Inositol hexakisphosphate increases L-type Ca2+ channel activity by stimulation of adenylyl cyclase. FASEB J 15(10):1753–1763

Yang T, Ali GS, Yang L, Du L, Reddy ASN, Poovaiah BW (2010) Calcium/calmodulin-regulated receptor-like kinase CRLK1 interacts with MEKK1 in plants. Plant Signal Behav 5(8):991–994

Yao H, Wang G, Liang G, Wang X (2013) Phosphatidic acid interacts with a MYB transcription factor and regulates its nuclear localization and function in Arabidopsis. Plant Cell 25(12):5030–5042

Yuen CY, Christopher DA (2010) The rtole of cyclic nucleotide-gated channels in cation nutrition and abiotic stress. In: Ion channels and plant stress responses. Springer, Berlin, pp 137–157

Yoo JH, Park CY, Kim JC, Do Heo W, Cheong MS, Park HC, Kim MC, Moon BC, Choi MS, Kang YH (2005) Direct interaction of a divergent CaMisoform and the transcription factor, MYB2, enhances salt tolerance in Arabidopsis. J Biol Chem 280(5):3697–3706

Yoon H-K, Kim S-G, Kim S-Y, Park C-M (2008) Regulation of leaf senescence by NTL9-mediated osmotic stress signaling in Arabidopsis. Mol Cells (Springer Sci Bus Media BV) 25(3)

You J, Chan Z (2015) ROS regulation during abiotic stress responses in crop plants. Front Plant Sci 6:1092

Young JJ, Mehta S, Israelsson M, Godoski J, Grill E, Schroeder JI (2006) CO(2) signaling in guard cells: calcium sensitivity response modulation, a ca(2+)-independent phase, and CO(2) insensitivity of the gca2 mutant. Proc Natl Acad Sci U S A 103(19):7506

Yu L, Nie J, Cao C, Jin Y, Yan M, Wang F, Liu J, Xiao Y, Liang Y, Zhang W (2010) Phosphatidic acid mediates salt stress response by regulation of MPK6 in Arabidopsis Thaliana. New Phytol 188(3):762–773

Yuasa T, Ichimura K, Mizoguchi T, Shinozaki K (2001) Oxidative stress activates ATMPK6, an Arabidopsis homologue of MAP kinase. Plant Cell Physiol 42(9):1012

Yuen CCY, Christopher DA (2013) The group IV-A cyclic nucleotide-gated channels, CNGC19 and CNGC20, localize to the vacuole membrane in Arabidopsis Thaliana. AoB Plants 5 (February)

Yunta C, Martínez-Ripoll M, Zhu J-K, Albert A (2011) The structure of Arabidopsis Thaliana OST1 provides insights into the kinase regulation mechanism in response to osmotic stress. J Mol Biol 414(1):135–144

Zafra A, Rejón JD, Hiscock SJ, de Dios Alché J (2016) Patterns of ROS accumulation in the stigmas of angiosperms and visions into their multifunctionality in plant reproduction. Front Plant Sci 7:1112

Zhang XS, Choi JH (2001) Molecular evolution of calmodulin-like domain protein kinases (CDPKs) in plants and protists. J Mol Evol 53(3):214–224

Zhang W, Qin C, Zhao J, Wang X (2004) Phospholipase Dα1-derived phosphatidic acid interacts with ABI1 phosphatase 2C and regulates abscisic acid signaling. Proc Natl Acad Sci U S A 101(25):9508–9513

Zhang W, Yu L, Zhang Y, Wang X (2005) Phospholipase D in the signaling networks of plant response to abscisic acid and reactive oxygen species. Biochim Biophys Acta 1736(1):1–9

Zhang W, Zhou R-G, Gao Y-J, Zheng S-Z, Xu P, Zhang S-Q, Sun D-Y (2009a) Molecular and genetic evidence for the key role of AtCaM3 in heat-shock signal transduction in Arabidopsis. Plant Physiol 149(4):1773–1784

Zhang Y, Zhu H, Zhang Q, Li M, Yan M, Wang R, Wang L, Welti R, Zhang W, Wang X (2009b) Phospholipase Dα1 and phosphatidic acid regulate NADPH oxidase activity and production of reactive oxygen species in ABA-mediated stomatal closure in Arabidopsis. Plant Cell 21(8):2357–2377

Zhang W, Jeon BW, Assmann SM (2011) Heterotrimeric G-protein regulation of ROS signalling and calcium currents in Arabidopsis guard cells. J Exp Bot 62(7):2371–2379

Zhao Y, Liu W, Xu Y-P, Cao J-Y, Braam J, Cai X-Z (2013) Genome-wid identification and functional analyses of calmodulin genes in Solanaceous species. BMC Plant Biol 13(1):1

Zhou J, Wang J, Shi K, Xia XJ, Zhou YH, Jing Quan Y (2012) Hydrogen peroxide is involved in the cold acclimation-induced chilling tolerance of tomato plants. Plant Physiol Biochem: PPB 60(November):141–149

Zhou J, Xia X-J, Zhou Y-H, Shi K, Chen Z, Jing-Quan Y (2014) RBOH1-dependent H2O2 production and subsequent activation of MPK1/2 play an important role in acclimation-induced cross-tolerance in tomato. J Exp Bot 65(2):595–607

Zhu J-K (2002) Salt and drought stress signal transduction in plants. Annu Rev Plant Biol 53:247–273

Zhu S-Y, X-C Y, Wang X-J, Zhao R, Li Y, Fan R-C, Shang Y, Shu-Yuan D, Wang X-F, Fu-Qing W (2007) Two calcium-dependent protein kinases, CPK4 and CPK11, regulate abscisic acid signal transduction in Arabidopsis. Plant Cell 19(10):3019–3036

Zhu Y, Zuo M, Liang Y, Jiang M, Zhang J, Scheller HV, Tan M, Zhang A (2013) MAP65-1a positively regulates H2O2 amplification and enhances brassinosteroid-induced antioxidant defence in maize. J Exp Bot 64(12):3787–3802

zu Heringdorf DM, Jakobs KH (2007) Lysophospholipid receptors: signalling, pharmacology and regulation by lysophospholipid metabolism. Biochim Biophys Acta (BBA) Biomembr 1768(4):923–940

Signaling Peptides: Hidden Molecular Messengers of Abiotic Stress Perception and Response in Plants

3

Jebi Sudan, Devyani Sharma, Ananda Mustafiz, and Sumita Kumari

Abstract

Abiotic stress constitutes a threat to plant growth and development owing to undesired morphological, physiological, biochemical and molecular changes leading to yield losses and also restricting the areas where the crops can be grown. However, the sessile nature of plants allows them to swiftly recognize and respond to various adverse climatic conditions. This rapid response is primarily due to effective cell-to-cell communication through various intricate defence machineries that enable the plants to sense stress and relay the signal to downstream response regulators. Till a few decades back, phytohormones were considered the major players of cell-to-cell communication. However, following the discovery of first signaling peptide, systemin in tomato, there has been a paradigm shift in our understanding of the role of these peptides in signaling in plants. Genome-wide approaches using the tools of bioinformatics, genetic screens and biochemical assays have led to the discovery of several novel plant peptides over the few years. Small signaling/peptide hormones/secreted peptides are now established in plants as molecular messengers because of their involvement in key developmental processes such as meristem maintenance, organ abscission, cell elongation, cell proliferation and differentiation, gravitropism and defence against abiotic and biotic aggressors. A better understanding of these signaling molecules might steer us towards manipulating them for engineering hitherto elusive stress tolerance trait in plants.

J. Sudan · S. Kumari (✉)
School of Biotechnology, Sher-e-Kashmir University of Agricultural Sciences and Technology, Jammu, India
e-mail: sumitaslsjnu@gmail.com

D. Sharma · A. Mustafiz
Laboratory of Plant Molecular Biology, Faculty of Life Sciences and Biotechnology, South Asian University, Akbar Bhawan, Chanakyapuri, New Delhi 110021, India

S. M. Zargar, M. Y. Zargar (eds.), *Abiotic Stress-Mediated Sensing and Signaling in Plants: An Omics Perspective*, https://doi.org/10.1007/978-981-10-7479-0_3

Keywords
Signaling peptides · Abiotic stress · Cysteine-rich peptides (CRPs)

3.1 Introduction

The immobile nature of plants renders them more sensitive to various abiotic stresses (drought, temperature, high salt, heavy metal toxicity, submergence, high UV exposure and mineral deficiency) prevailing in the environment. However, unlike biotic stressors, plants are rarely exposed to a single abiotic stress. Exposure of plants to multiple abiotic stresses frequently induces a series of morphological, biochemical and molecular changes, and their symptoms overlap with one another (Qin et al. 2011). Abiotic stress not only causes serious economic losses in terms of yield and quality of crop plants but also limits the geographical region where a particular crop can be grown. These stresses cause the overproduction of reactive oxygen species {(ROS) (O^-, O^{2-}, O^+, O^{2+})} that induces oxidative stress which is responsible for multiple detrimental changes in plant cells. However, in due course of evolution, plants have evolved a plethora of defence mechanisms to alleviate stress-related symptoms. These defence mechanisms include various morphological modifications and stress-inducible genes that start with the onset of stress and enable the plants to maintain homeostasis up to certain critical level of stress (Hirayama and Shinozaki 2010). Morphological modifications include various alterations in the shape, size, growing habits, growing season, etc. Stress-induced genes can be classified into two major types (Akpinar et al. 2012): (1) genes that code for functional proteins (whose expression provides abiotic stress tolerance) and (2) genes that code for regulatory proteins (mainly signaling peptides and transcription factors that are responsible from signal reception to induction of functional proteins).

Signaling peptides (from the Greek peptos meaning “digestive”) also called “peptide hormones” or “secreted peptides” are the smallest biological molecules of the plant proteome (usually <20 amino acids in the mature form and rarely more than ~120 amino acids as a full-length precursor) and often present in very low (nanomolar range) physiological concentrations. They have immense roles in plant growth and development, reproduction as well as biotic and abiotic stress responses (Albert 2013; Matsubayashi 2014). Peptides can interact with pathogens through their antimicrobial properties (Goyal and Mattoo 2014) and also with various signaling pathways through their cell-to-cell communication (Murphy et al. 2012; Araya et al. 2014; Tavormina et al. 2015).

Peptide signaling molecules in animals are well characterized, and a lot of progress has been done since the discovery of “nerve growth factor” by Rita Levi-Montalcini in 1954 and epidermal growth factor by Stanley Cohen in 1962 (Cowan 2001). However, the importance of signaling peptides in plants has been recognized since the last few decades. Plant signaling peptides have important roles in growth, development, defence, stress tolerance and in maintaining homeostasis (Lindsey

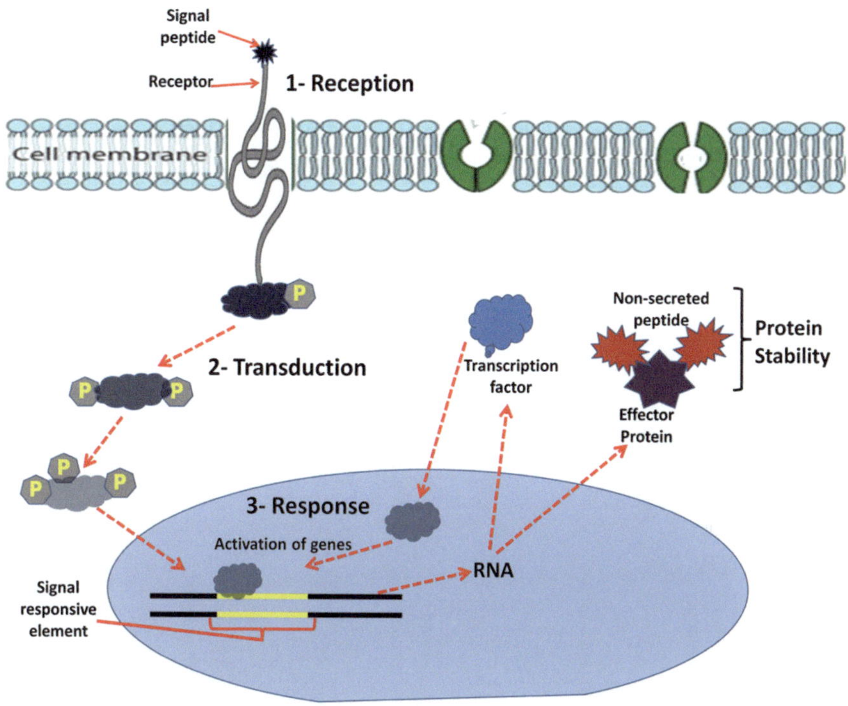

Fig. 3.1 Schematic diagram showing involvement of plant peptides in signaling

et al. 2002; Ghorbani et al. 2014). Their vital role in stress tolerance enables the plant to adapt for diverse climatic conditions. These signaling peptides act as the silent heroes in the process of providing stress tolerance to the plants. Their part starts from the perception of stress signal to relay of that signal information through a specified signal transduction pathway that leads to expression of some stress-responsive genes (Akpinar et al. 2012) (Fig. 3.1).

The discovery of these signal peptides began with the isolation of the first plant peptide (systemin) from tomato leaves by Clarence A. Ryan in 1991 followed by the discovery of hydroxyproline-rich glycopeptides in tobacco (Ryan and Pearce 1998; Ren and Lu 2006) and AtPEPs (*Arabidopsis thaliana* plant elicitor peptides) in *Arabidopsis thaliana* (Qi et al. 2010). Systemin was found to be involved in the defence response against insect herbivores and modification of root growth (Ryan and Pearce 2003; Narvaez and Orozco 2008). They are also shown to be involved in the plant responses against salt and UV stresses (Dombrowski 2003; Holley et al. 2003). With the discovery of these signal peptides, a plethora of other peptides were also discovered (Table 3.1). Until the discovery of the 18-amino-acid systemin peptides as regulators of signaling events in plants (Pearce et al. 1991), most intercellular communication involved in plant growth and development had been explained on the basis of signaling by the six non-peptide plant hormones: auxin, cytokinin, ethylene, gibberellin, abscisic acid and brassinolides. There is no doubt about the

Table 3.1 List of various signal peptides discovered in plants

Gene family	Plant species in which first discovered	Conserved motifs	Ligand/ receptor	Biological function	Synthesis process	Basic features	References
CLAVATA/ endosperm surrounding region (ESR)	*Arabidopsis*	12–14 amino acids (aa)	CLV1, CLV2, BAM1, RPK2	Meristem elongation, vascular development	Non-functional precursor derived (NFPD)	Post-translationally modified (PTM)	Kondo et al. (2006) and De Young and Clark (2008)
PROPEP	*Arabidopsis*	23 aa	AtPEP1 Pep1R	Related to innate defence responses	NFPD	Non-cys-rich/ non-PTM	Huffaker et al. (2006), Pearce et al. (2008) and Qi et al. (2010)
Systemin	*Solanum lycopersicum*	18 aa	SR160	Systemic defence responses	NFPD	Non-cys-rich/ non-PTM	Pearce et al. (1991)
POLARIS (PLS)	*Arabidopsis*	36 aa	POL	Root elongation and development	Non-precursor derived (NPD)	ORFs	Chilley et al. (2006)
ROT FOUR- LIKE/ DEVIL	*Arabidopsis*	29 aa	ROT4	Proliferation of cell and leaf	Non-precursor derived (NPD)	ORFs	Wen et al. (2004) and Ikeuchi et al. (2011)
Rapid alkalinization factor (RALF)	*Nicotiana attenuata*	16 aa	Feronia, RLK 1	Innate immune responses, respond to danger signals	NFPD	Cys rich/PTM	Silverstein et al. (2007) and Wu et al. (2007)
C-terminally encoded peptide (CEP)	*Arabidopsis*	14 aa	CEPR1, CEPR2	Negative role on root growth, stress responses and development	NFPD	PTM	Ohyama et al. (2008) and Sui et al. (2016)

ENOD40	*Glycine max*	–	ENOD40	Root nodule formation and legume development	NPD	ORF	Rohrig et al. (2002) and Gultyaev and Roussis (2007)
Phytosulphokine (PSK)	*Arabidopsis*	20 aa	PSKR1	Proliferation and differentiation of cell	NFPD	PTM	Yang et al. (2001) and Kutschmar et al. (2009)
Embryo surrounding factor (ESF1)	*Arabidopsis*	68 aa	SSP	Determine the embryo patterning in angiosperms	NFPD	Cys-rich/PTM	Costa et al. (2014)
Epidermal patterning factor (EPF)/ STOMAGEN	*Arabidopsis*	–	EPF1, EPF2	Helps in division of epidermal cell for formation of guard cell	NFPD	Cys-rich/PTM	Hara et al. (2007) Hara et al. (2009)
GOLVEN/root growth factor	*Arabidopsis*	18 aa	–	Determine root development	NFPD	PTM	Matsuzaki et al. (2010) and Whitford et al. (2012)
Inflorescence deficient in abscission (IDA)	*Arabidopsis*	–	HAE, HSL	Inhibits floral abscission	NFPD	PTM	Butenko et al. (2003) and Stenvik et al. (2008)
Plant natriuretic peptide	*Arabidopsis*	–	–	Inhibits stomatal closure through ABA, promotes ion movement	–	–	Rafudeen et al. (2003) and Wang et al. (2011)

(continued)

Table 3.1 (continued)

Gene family	Plant species in which first discovered	Conserved motifs	Ligand/ receptor	Biological function	Synthesis process	Basic features	References
Plant peptide containing sulphated tyrosine (PSY)	*Arabidopsis*	18 aa	PSYR1	Shows upregulation by wounding, promotes cell proliferation	NFPD	PTM	Amano et al. (2007)
S-locus cysteine-rich peptides (SCR/ SP11)	*Brassica spp.*	–	SRK	Inhibits self-fertilization	NFPD	Cys-rich/PTM	Schopfer et al. (1999)
Tapetum determinant	*Arabidopsis*	–	EMS1	Promotes development of anthers	–	–	Yang et al. (2003)
Cyclotides – cyclic peptides	*Viola hederacea*	28–37 aa	Cyclic cystine knot (CCK)	Protect plant from pests and pathogen	NFPD	Cys-rich/PTM	Craik et al. (1999) and Weidmann and Craik (2016)
Taximin	*Taxus baccata*	27 aa	–	Major role in plant metabolism	NFPD	Cys-rich/PTM	Onrubia et al. (2014)
Defensin-like polypeptide (LUREs)	*Torenia fournieri*	–	LURE 1, LURE 2	Attract the pollen tube to egg apparatus	NFPD	Cys-rich/PTM	Okuda et al. (2009)
HYPSYS	*Solanum lycopersicum*	15–20 aa	HYPSYS	Helps in biotic stress tolerance	NFPD	PTM	Chen et al. (2008) and Bhattacharya et al. (2013)
Ns-LTP	*Oryza sativa*	–	–	Reproduction embryogenesis, antimicrobial, somatic embryogenesis, nodulation and symbiosis, shoot cell expansion	NFPD	Cys-rich/PTM	van Loon et al. (2006) and Nawrot et al. (2014)

NCR	*Medicago truncatula*	–	–	Nodulation and symbiosis	NFPD	Cys-rich/PTM	Farkas et al. (2014)
							Durgo et al. (2015)
SNAKIN/GASA	*Solanum tuberosum*	15 aa	–	Antimicrobial, shoot cell expansion	NFPD	Cys-rich/PTM	Nahirñak et al. (2012) and Garcia et al. (2014)
KNOTTIN	*Mirabilis jalapa*	–	–	Antimicrobial	NFPD	Cys-rich/PTM	Chouabe et al. (2011) and De Souza Cândido et al. (2014)
Maize embryo sac4 (ES)	*Zea mays*	50 aa	–	Reproduction embryogenesis, male determinant	NFPD	Cys-rich/PTM	Takayama et al. (2000) and Higashiyama (2010)
THIONIN	*Arabidopsis*	–	–	Antimicrobial	NFPD	Cys-rich/PTM	Asano et al. (2013) and Ji et al. (2015)
HEVEIN	*Hevea brasiliensis*	–	–	Antimicrobial	NFPD	Cys-rich/PTM	De Souza Candido et al. (2014)
PDP	*Helianthus annuus*	–	–	Antimicrobial	NFPD	Cys-rich/PTM	
Ib-AMP	*Impatiens balsamina*	20 aa	–	Antimicrobial	NFPD	Cys-rich/PTM	Lee et al. (1999)

(continued)

Table 3.1 (continued)

Gene family	Plant species in which first discovered	Conserved motifs	Ligand/ receptor	Biological function	Synthesis process	Basic features	References
α-HAIRPININ	*Stellaria media*	–	–	Antimicrobial, trypsin inhibitory activity	NFPD	Cys-rich/PTM	De Souza Candido et al. (2014) and Slavokhotova et al. (2014)
ARACIN	*Arabidopsis*	40 aa	–	Antimicrobial	NFPD	Cys-rich/PTM	Neukermans et al. (2015
Maize egg apparatus (EA)	*Zea mays*	27–29 aa		Attracts maize pollen tube in vitro	NFPD	Cys-rich/PTM	Gray-Mitsumune and Matton (2006) and Marton et al. (2012)
Plant elicitor peptides (PEP)	*Arabidopsis*	23–36 aa	PEPR1, PEPR2	Plant immunity, senescence	NFPD	Non-cys-rich/ non-PTM	Ross et al. (2014) and Gully et al. (2015)
GRIM REAPER PEPTIDE (GRIp)	*Arabidopsis*	11 aa	PRK5	Cell death	NFPD	Non-cys-rich/ non-PTM	Wrzaczek et al. (2015)
CAPE	*Solanum lycopersicum*	11 aa	PRP	Shoot cell expansion, biotic stress tolerance, salt stress	Functional precursor derived (FPD)	–	Chen et al. (2014) and Chien et al. (2015)
INCEPTINS	*Vigna unguiculata*	11–13 aa	–	Biotic stress tolerance, anti-herbivore activity	Functional precursor derived (FPD)	–	Schmelz et al. (2006) and (2012)
KOD	*Arabidopsis*	25 aa	–	Programmed cell death	Non-precursor derived (NPD)	ORFs	Blanvillain et al. (2011)

Soybean subtilase peptide (gm-SUBPEP)	*Glycine max*	12 aa	–	Promote defence-related genes	Functional precursor derived	–	Pearce et al. (2010) and Yamaguchi and Huffaker (2011)
OSIP-108	*Arabidopsis*	–	–	Abiotic stress tolerance	Non-Precursor Derived (NPD)	ORFs	Spincemaille et al. (2014a, b)

CLV CLAVATA, *BAM* barely any meristem, *HYPSYS* hydroxyproline-rich glycopeptide systemins, *HAESA* HAE leucine-rich repeat receptor-like kinases, *RGF/CLEL/GLV* root growth factor/CLE-like/GOLVEN, *PSK* phytosulphokine, *PSKRs* phytosulphokine receptors, *PSY1* plant peptide containing sulphated tyrosine 1, *PDFs* plant defensins, *PR* pathogenesis-related protein family, *DEFLs* defensin-like genes, *NCR* nodule-specific cys-rich, *nsLTPs* nonspecific lipid transfer proteins, *SNAKIN/GASA* gibberellic acid stimulated-like, *PRK* pollen-specific receptor-like kinase, *CAPE* cysteine-rich secretory proteins, antigen5, and pathogenesis-related 1 protein significance of these hormones in plant growth, but discoveries over the past decade indicate that plant cell communication also makes use of small peptide signals and specific receptors (Matsubayashi 2011).

3.2 Peptide Classification System

The availability of the first plant genome in 2000, the generation of huge RNAseq data sets and improved biochemical isolation procedures and gene prediction tools have shown that plants possess thousands of genes encoding putative secreted extracellular peptide ligands. Based on structural characteristics as well as their biosynthetic pathway, these are classified into two major groups (Matsubayashi 2011; Murphy et al. 2012; Tabata and Sawa 2014).

3.2.1 Group I (Cysteine-Rich Peptides (CRPs))

These are characterized by the presence of an even number of cysteine residues with a length of about 40- to more than 100-amino-acid residues having intermolecular disulphide bonds that determine the three-dimensional structure of the mature protein (Pearce et al. 2001a, b). These peptides do not undergo post-translational modifications. Mature CRPs are generally longer than 20 amino acids and are particularly abundant during plant reproduction. They may or may not undergo proteolytic processing. So they can be further categorized into two subgroups:

1. CRPs with proteolytic cleavage including STOMAGEN, which belongs to the epidermal patterning factor (EPF) peptide family and is a positive regulator of stomatal density (Hara et al. 2007; Sugano et al. 2010); rapid alkalinization factor 1 (RALF1), which is essential for cell expansion and is recognized by the Feronia (FER) receptor (Haruta et al. 2014); and Taximin 1 and 2 (TAX1 and 2), a novel peptide isolated from *Arabidopsis*, which might indirectly affect organ separation by controlling the biosynthesis of hormone (Ingram and Gutierrez-Marcos 2015)
2. CRPs without proteolytic cleavage including the S-locus cysteine-rich protein/S-locus protein 11 (SCR/SP11) (Schopfer et al. 1999; Takayama et al. 2001) and LUREs (Okuda et al. 2009)

3.2.2 Group II: Small Post-translationally Modified Secreted Peptides (Non-CRPs)

These peptides are characterized by the small size (about ten amino acid residues) of the mature peptide. The propeptides corresponding to the mature peptides consist of approximately 70–120 amino acids and contain few or no cysteine residues and share a common tripartite structure: (i) a signal peptide at the N-terminal region, (ii) a C-terminal region that is usually conserved among different members of the family and corresponds to the mature peptide and (iii) a variable segment that links the two terminal domains (Matsubayashi 2014). They also undergo proteolytic cleavage. These peptides are involved in plant growth and development by regulating many intercellular communication processes during vegetative development and stress responses (Matsubayashi 2011). These include tracheary element differentiation (Matsubayashi et al. 1999), adventitious root formation factor (Amano et al. 2007), plant peptide containing sulphated tyrosine 1 (Amano et al. 2007), CLV3/embryo surrounding region-related (CLE) which modulate stem cell differentiation in various tissue types (Ohyama et al. 2009; Katsir et al. 2011), C-terminally encoded peptide (CEP) (Ohyama et al. 2008; Delay et al. 2013; Akker et al. 2016), and root growth factor (RGF)/GOLVEN (GLV)/CLE-like (CLEL) which promote root meristem growth by controlling PLT transcription factors through RLK-dependent signaling pathways (Matsuzaki et al. 2010; Meng et al. 2012; Fernandez et al. 2015).

3.3 Signaling Peptides Identified in Plants

Plants are immobile organisms, so they possess mechanisms to perceive and respond to the ever-changing environmental conditions like light, temperature, water availability, salinity, wind, etc. Signaling peptides among mobile ligands act as external signals which are perceived by receptors and elicit stress response as well as exchange this information by establishing cell-to-cell communication in response to abiotic and biotic stress conditions enabling the plant to survive. These signaling peptides may act as positive or negative regulators. Some of the signaling peptides which are well characterized in plants have been described below.

3.3.1 CLAVATA 3/Endosperm Surrounding Region (ESR)

"CLAVATA" originally gets its name from the *Arabidopsis CLAVATA* (*clv*) mutant that has a unique feature of having abnormal club-shaped (clava = club) fruit phenotype (Cock and McCormick 2001; Ni and Clark 2006). This phenotype is developed due to a defect in the stem cell of shoot meristem (Ni and Clark 2006). *CLV1* is a leucine-rich repeat receptor-like kinase (LRR-RLK) that is expressed in shoot apical meristem (SAM). These proteins are characterized by having a stretch of hydrophobic amino acids (signal peptide for secretion) at the amino terminal and a unique CLE amino acid sequence at the carboxyl terminal. The first member of CLE family that was specifically shown to be expressed in the endosperm region surrounding the embryo in *Zea mays* was termed as Endosperm Surrounding Region (ESR) gene as its mRNA was mostly present in ESR region (OpsahlFerstad et al. 1997; Bonello et al. 2000). All CLE peptides shared a common sequence elements having conserved proline residues at fourth, seventh and ninth positions. These proline residues are either hydroxylated (Pro4 or Pro7) or arabinosylated (Pro9) (Kondo et al. 2006; Huffaker et al. 2006).There are two types of CLEs:

1. A-type CLEs (CLV3, CLE1–27 and CLE40): These CLE peptides induce termination of the root or shoot apical meristem activity through terminal differentiation of stem cells (Ito et al. 2006; Hanada et al. 2007).
2. B-type CLEs (CLE41–44): These CLE peptides do not induce termination of either root or shoot meristem. However in *Arabidopsis*, these suppress the differentiation of xylem (Beers et al. 2004).

CLAVATA 3 (CLV3) constitutes the most studied member of the CLE family and acts as a ligand of CLV1/CLV2 receptor complex (Trotochaud et al. 1999). *CLV3* was reported to code for an extracellular protein, whose expression in root or shoot apical meristem controls cell proliferation and differentiation across the entire meristem (Fletcher et al. 1999). *CLV3* can shift from one region of meristem to interact with the receptor (CLV1 RLK) present in another meristem (Fletcher et al. 1999). CLV1 is known to exist in two (185 and 450 kD) protein forms. The inactive form of the receptor complex having CLV1 homodimer is of 185 kD which in the presence of functional CLV3 recruits an additional protein phosphate and a small GTP-binding protein and is then converted into large activated receptor protein complex of 450 kD, which later on associate with various signaling factors (Jeong et al. 1999 and Kohorn 1999).

3.3.2 POLARIS (PLS)

The POLARIS (PLS) peptide was first discovered in *Arabidopsis thaliana*, and it was shown to be involved in cell expansion and vascularization (Topping and Lindsey 1997). It was also observed that *pls* mutants that are characterized by short root phenotype are hyperresponsive to ethylene, have defective auxin transport and homeostasis and altered microtubule sensitivity to inhibitors (Chilley et al. 2006). POLARIS was also shown to play an important part in ethylene signaling and also regulate the root development (Chilley et al. 2006; Mudge 2016). The peptide is mostly expressed in seedling root, with low expression in aerial parts, and is required for proper auxin-cytokinin homeostasis (Casson et al. 2002).

3.3.3 Rapid Alkalinization Factor and RALF-Like

Rapid alkalinization factor (RALF) is a 5 kD signal peptide that was first identified from tobacco leaves while purifying systemin from them (Pearce et al. 2001a). RALF is a secreted peptide that is derived from the precursors of over 100 amino acids. Pre-RALF proteins consist of an N-terminal signal sequence followed by a variable region and at last a well-conserved C-terminal region that forms the active peptide (Pearce et al. 2001a). There are many cysteine-rich families of peptides that are homologous to RALF and are termed as RALF-like peptides (Olsen et al. 2002).

These peptides are very diverse and are expressed in most of the plant species. There are about 43, 39, 34 and 18 different RALF genes discovered in rice, *Arabidopsis*, maize and soybean, respectively (Convey et al. 2010; Cao and Shi 2012; Sharma et al. 2016). Out of these total RALF genes, about 50% are believed to be developed through tandem duplication. These peptides cause an elicitation of Ca^{2+} responses, altering the pH, and thus stimulate a mitogen-activated protein kinase (MAPK) that probably regulates various root developmental processes (Wu et al. 2007; Murphy et al. 2016). Recently, Murphy (2016) found that RALF34 also plays an important role in initiation and positioning of lateral roots. Apart from its profound role in the development of root, its overexpression also inhibits pollen germination, decreases tolerance capacity in plants and produces smaller fruits with fewer seeds (Covey et al. 2010; Atkinson et al. 2013; Chevalier et al. 2013; Moroto do Canto et al. 2014).

There are many studies related to the engineering of these peptide molecules that show the involvement of this peptide in different pathways. Matos et al. (2008) overexpressed the AtRALF1 (one of the isoforms) and found that the resultant plants have semidwarf phenotype. When the RALF isoform was silenced in tobacco, plants with longer roots were produced (Haruta and Constabel 2003).

3.3.4 DEVIL/ROT FOUR-LIKE (RTFL)

DEVIL (DVL) is a 40–145-amino-acid-long polypeptide that was first identified by Wen et al. (2004), and it is restricted to angiosperms (Wen et al. 2004). Homologous search for sequences similar to the DVL revealed the identification of 20 members of the DVL gene family (Wen and Walker 2006). DEVIL when overexpressed in *Arabidopsis* results in shortened stature, clustered inflorescences, rosette leaves, shortened pedicles and siliqua with pronged tips (Wen et al. 2004). Narita et al. (2004) proposed the characterization of an activation-tagged *Arabidopsis* line overexpressing another gene from this family. This gene was termed as ROTUNDIFOLIA4 (ROT4) and corresponds to the similar function as that of DVL1. About 22 putative homologs of DVL1 and ROT4 were later on identified in *Arabidopsis* genome and termed as ROT FOUR-LIKE/DEVIL (RTFL/DVL) family (Narita et al. 2004; Wen et al. 2004; Yamaguchi et al. 2013). These RTFL/DVL peptides have a common conserved domain of 30 amino acids near the C-terminus (RTF domain) (Narita et al. 2004). Various overexpression studies related to RTFL/DVL shed light in understanding the biological function of these gene families. Overexpression of various DVL genes resulted in the alteration of siliqua development giving rise to shorter and wider valves (Narita et al. 2004). ROT4 overexpression also retards the process of cell production in the meristematic zone within the leaf blade and hence results in shorter petioles (Ikeuchi et al. 2011). Combier et al. (2008) also overexpressed MtDVL1 in *Medicago truncatula* that results in the reduced nodule formation. When six RTFL gene members of *Arabidopsis* were overexpressed, a round leaf phenotype along with shorter primary roots and a lower rate of root elongation had been observed (Guo et al. 2015).

3.3.5 Phytosulphokines

Phytosulphokines (PSKs) are members of the growing family of sulphated peptides that have roles in localized paracrine signaling. PSK-a is a pentapeptide sulphated on the two tyrosine residues [Y(SO3H)IY(SO3H)TQ], which is a requirement for cellular activity as non-sulphated PSKs reduce the cellular activity by over 100-fold (Matsubayashi et al. 1997). PSKs were first discovered as a cell proliferation mediator needed to maintain low-density cell cultures (Matsubayashi and Sakagami 1996). Numerous precursor proteins for PSK (proPSK) have also been identified in different species that consist of an N-terminal secretory signal sequence and a common PSK sequence nearby C-terminus (Yang et al. 1999, 2001; Lorbiecke and Sauter 2002). PSK processing involves sulphating of proPSK by tyrosylprotein sulphotransferase. PSK acts in a paracrine or autocrine manner on nearby cells expressing the receptor and has been most clearly shown in roots where PSK enhances root elongation by controlling cell size (Matsubayashi et al. 2006; Kutschmar et al. 2009). PSKs influence gene expression and reduce expression of stress-responsive genes during differentiation of tracheary elements (Motose et al. 2009). PSK-a promotes tracheary element differentiation in *Zinnia elegans* mesophyll cells, and this process is hindered by CLE41–44 to regulate organogenesis (Matsubayashi et al. 1999; Ito et al. 2006; Motose et al. 2009). PSK-a, the natural ligand of the PSK receptor (PSKR1) (Matsubayashi et al. 2002), belongs to the large family of leucine-rich repeat receptor-like kinases (LRR-RLK). Lately, a guanylate cyclase (GC, enzymes that catalyse the conversion of GTP into cGMP) catalytic centre was discovered in PSKR1 embedded within the cytosolic kinase domain and dual kinase GC domain that has both serine and threonine kinase activity (Kwezi et al. 2011). Also, PSK-a itself, but not the non-sulphated peptide, rapidly raises cGMP levels in protoplasts. This elevation of the second messenger cGMP is vital for PSK downstream signaling (Ma et al. 2010), cyclic nucleotide-dependent phosphorylation and also direct or indirect activation of cGMP-dependent transcription (Maathuis 2006). Hence, it is now clear that cyclic nucleotides and particular cGMP play an important role in maintaining ion and water balance during biotic (Meier et al. 2009, 2010) and abiotic stresses (Donaldson et al. 2004). PSKs are classified as growth factors; on the other hand, their functions are growth regulation and include biotic and abiotic stress responses and reproductive processes. Therefore, it is reasonable to expect that PSK (and other plant signaling peptides) will also affect cellular homeostasis particularly during changes in growth rate when there is a need for quick adaptation to changes in ion and water balance.

3.3.6 GOLVEN (GLV)/Root Growth Factors (RGF)

GOLVEN (GLV)/root meristem growth factor (RGF)/CLE-like (CLEL) gene family has been described by three independent groups (Matsuzaki et al. 2010; Meng et al. 2012; Whitford et al. 2012). *GLV* genes were basically identified in *Arabidopsis thaliana*, and they are conserved throughout the plant kingdom (Whitford et al. 2012).

They code for small post-translationally modified peptides. The mature bioactive GLV peptides are produced by proteolytic cleavage of larger precursor proteins having the same conformation (Matsuzaki et al. 2010; Meng et al. 2012; Whitford et al. 2012) and similarity to other secreted signaling peptides. The GLV precursor proteins have a length of 86–163 amino acids. Two main domains can be differentiated in their primary sequence. The N-terminal domain contains a signal peptide (SP) which targets the protein to the secretory pathway, whereas the C-terminal domain contains a motif which is conserved among all GLV members that defines the family and thus codes for the functional secreted mature peptide. A third region with little sequence similarity between GLV pre-proteins connects the two above-mentioned domains (Matsuzaki et al. 2010; Meng et al. 2012; Whitford et al. 2012). Genes that code for the GLV/RGF/CLEL peptides were identified in three independent in silico studies (Matsuzakiet al. 2010; Meng et al. 2012; Whitford et al. 2012).

GLV peptide carries two types of post-translational modifications: tyrosine sulphation and hydroxylation of one of the proline residues. Most of the *Arabidopsis GLV* genes are active in specific portions of the root, suggesting they take part in different developmental processes. *GLV* gene expression is not only limited to roots but has been reported in shoot tissues also including hypocotyl, shoot apical meristem (SAM), cotyledon, leaf, stem and flower (Fernandez et al. 2013, 2015). GLV shoot patterns are precise to each gene. Although so far no clear functions for *GLV* genes have been imputed to distinct developmental processes in the shoot, their cell- and tissue-specific expression patterns suggest a potential involvement in the above-ground parts. Overexpression of *GLV* genes results in a prominent wavy and curly root phenotype which is further enhanced when the seedlings are germinated on the slanted plates. Referring to this phenotype, they were named GOLVEN (GLV) which means "waves" in Dutch (Whitford et al. 2012). Evidence shows that GLV peptides are non-cell autonomous molecules that probably transmit information to nearby cells. GLV overexpressed in plants indicate that they are secreted into the apoplast.

3.3.7 Systemin

Tomato plants respond to an injury by herbivorous insects with the induction of a systemic defence response which is characterized by the transcriptional activation of a large number of defence genes (Green and Ryan 1972; Bergey et al. 1996). The search for the signal molecule that permits tomato plants to act systemically to a stimulus led to the identification of the first plant peptide which has signaling function in 1991.This 18-amino-acid peptide was first isolated from the leaves of tomato plants on the basis of the ability of this peptide to induce the expression of defence genes. To underscore its key role and the systemic nature of the wound response, the peptide was named as systemin (Pearce et al. 1991). In succeeding years, it was found that systemin is both sufficient and necessary for systemic wound signal transduction. Whereas there is a clear indication for the additional signals to exist, a key role for systemin in wound signaling in tomato plants and other members of the Solanaceae family is now commonly accepted (Constabel et al. 1995). Sequence analysis of the defence gene isolated from the tomato leaves has the amino acid sequence AVQSKPPSKRDPPKMQTD as the basic structure of systemin. The most significant characteristic of this sequence is the palindromic structure around a central pair of basic residues with two proline doublets present at positions 6 and 7 and 12 and 13 (Pearce et al. 1991). Each of the 18 amino acids of systemin was separately replaced by the alanine to consider the involvement of single amino acid side chains to the biological activity of systemin. Systemin derivatives with Ala substitutions for any of the other proline residues, though, retained most (>10%) of the bioactivity (Pearce et al. 1993). Progressive deletions from the N- and C-termini indicated that the 18-amino-acid peptide is the nominal structure having full biological activity. Deletions from the N-terminus resulted in a continuous loss of activity, whereas the deletion of a single amino acid from the C-terminus inactivated the peptide. Interestingly, the C-terminally abridged peptide as well as the Thr17Ala derivative of systemin acted as competitive antagonists of the systemin activity (Pearce et al. 1993). Based on these observations, it was proposed that systemin probably binds to its receptor via the N-terminal part although the C-terminus is crucial to activate downstream responses (Pearce et al. 1991).

3.3.7.1 Inflorescence Deficient in Abscission and IDA-LIKE

Signaling peptides coordinate one of the developmental processes in plants called abscission. Abscission is useful for best possible plant growth, as the organs which are not necessary or functional anymore have to be removed by precise and programmed cell separation (Jinn et al. 2000; Stenvik et al. 2008). The IDA and IDA-LIKE (IDL) signaling peptides have been revealed to promote organ abscission by stimulating the cell separation or by inhibiting cellular repair mechanisms (Butenko et al. 2003; Stenvik et al. 2008). The IDA and IDL genes are expressed in the cell division zone that consists of the region where the organs are separated from the plant body (Stenvik et al. 2008). It is, therefore, believed that they signal through the LRR-RLKs HAESA (HAE) and HAESA-LIKE2 (HSL2) receptors (Stenvik et al. 2008). The IDA knockout mutants retain the floral organs, whereas the plants that overexpress the *IDA* or *IDL* genes display premature floral organ abscission, along with over-proliferation of the abscission zone and, furthermore, ectopic abscission of some organs that are usually not shed in *Arabidopsis* (Jinn et al. 2000; Stenvik et al. 2006; Kumpf et al. 2013). IDA signaling peptides are chiefly known for their role in floral organ abscission (Butenko et al. 2003), but, lately, a new function has also been assigned to the IDA-HAE/HSL2 signaling module, namely, a contribution to facilitate the passage of lateral root primordia (LRP) through the main root and to support in the lateral root (LR) emergence. In roots, the IDA is vigorously and constantly induced by auxin, whereas the hormone only briefly upregulates HAE and HSL2. Induction of IDA is dependent on the auxin influx carrier LIKE AUX1-3 (LAX3) and the AUXIN RESPONSIVE FACTOR7 (ARF7). Auxin, originating from the tip of the primordium, coordinates the cell separation in overlaying LRP tissues through signaling mediated by the IDA peptide. IDA and HAE have been made known to regulate LR emergence by promoting the degradation of the cell wall in the tissue layers overlaying the LRP as it grows outwards (Kumpf et al. 2013).

3.3.7.2 Epidermal Patterning Factor/EPF-Like STOMAGEN

Stomatal development relies on asymmetric cell divisions that are specifically coordinated in time and space by cell-to-cell communication networks. Stomata are generally separated from the neighbouring stomata by at least one cell, thus following the "one-cell-spacing rule". The signaling peptides have a major impact on the development of stomata (Shimada et al. 2011; Hunt and Gray 2009; Hunt et al. 2010; Rowe and Bergmann 2010). Multiple cysteine-rich signaling peptides that belong to the EPF family of signaling peptides, in particular EPF1, EPF2, EPF-LIKE6/CHALLAH (EPFL6/CHAL) and EPFL9/STOMAGEN, are linked with the regulation of stomatal density and positioning (Hara et al. 2007, 2009; Shimada et al. 2011; Hunt and Gray 2009; Lee et al. 2012). The overexpression of the EPF1 and EPF2 peptides reduces the stomatal density. Signaling which depends upon the EPF1 and EPF2 peptides needs the action of the LRR receptor-like proteins such as TOO MANY MOUTHS (TMM) and LRR-RLKs, ERECTA (ER) and ERECTA-LIKE1 (ERL1). EPF1 and EPF2 peptides bind to the ER and ERL1 receptors and EPF2 to TMM. As TMM lacks an intracellular domain (Shpak et al. 2005), an additional protein with an extracellular domain is used to intercede the signal transduction. As a result, TMM interacts straight with the ER receptor and forms a complex which starts the signaling pathway of EPF (Hara et al. 2007; Hara et al. 2009; Lee et al. 2012). EPFL9/STOMAGEN, another member of the EPF family, acts with EPF1 and EPF2 in stomatal development. EPFL9/STOMAGEN also requires TMM but, in this case, to positively control stomata formation, which depicts those peptide hormones from the identical family, can have contrast functions through the same receptor in plants (Kondo et al. 2006; Sugano et al. 2010). Also, EPFL6/CHAL negatively controls the stomatal development in stems and hypocotyls. The ER receptors were also to be involved in the signal transmission of the EPFL6/CHAL (Shimada et al. 2011; Ohki et al. 2011). Lately, a function different from stomatal development has also been assigned to the signaling module EPF4/EPF6-ER/ERL (Uchida et al. 2012). EPF4 and EPF6 are secreted from the endodermis and recognized by ER/ERL1 receptors in the phloem for maintaining the vascular development (Uchida and Tasaka 2013).

3.3.8 Taximin

Taximin is a plant-specific, small, cysteine-rich signaling peptide identified through a transcriptome survey of jasmonate-elicited suspension cells (Onrubiaet al. 2014). They are conserved across the entire higher plant kingdom, including dicots, monocots, gymnosperms, mosses and ferns. They are classified as CRP (cysteine-rich peptide) as the peptides of this family are rich in cysteine residues. They have six conserved cysteines and three conserved prolines, confirming that this peptide belongs to cysteine-enriched family of peptides. Taximin peptides have an N-terminal signal peptide as predicted by an in silico approach (Onrubia et al. 2014), and all homologues from diverse plants have an identical length with a high number of conserved cysteines and prolines and a conserved high concentration of hydrophobic AA residues (Onrubia et al. 2014). Expression of Taximin homologues was reported to be modulated by JA elicitation and/or biotic stresses in which JA signaling is involved suggesting a conserved role in JA signaling and/or plant defence processes for the Taximin peptides (Onrubia et al. 2014).

3.3.9 C-Terminally Encoded Peptides

C-terminally encoded peptides (CEP) constitute a multigene family of small post-translationally modified peptides (Ohyama et al. 2008).They are characterized by a conserved 15-amino-acid peptide domain which is present at or near C-terminal end. The mature peptide is represented by 14–15 amino acids which consist of one or two hydroxylated proline residues, and this 15-amino-acid peptide has been reported to be biologically active in roots (Ohyama et al. 2008). Their expression is prevalent in the shoot apical meristem and lateral root primordia during the development of plant, and overexpression of CEP(s) leads to reduced primary and lateral root elongation and also smaller shoot system. Some initial analysis showed that the overexpression of *AtCEP1* in roots resulted in reduced and comparatively less number of meristem cells (Ohyama et al. 2008). CEP expression is regulated by environmental factors like nitrogen limitation and increase in salt levels, CO_2 levels and osmotic pressure in both roots and shoots. *AtCEP3* knockout mutant showed increased expression of the different *CEP* genes under various environmental factors and increased shoot growth when grown in hydroponics.

3.3.10 PNPs (Plant Natriuretic Peptides)

The plant natriuretic peptides (PNPs) are a class of small peptides (14 kDa) which signal via guanosine 3′,5′-cyclic monophosphate (cGMP) and systemically affect plant salt and water balance and responses to biotrophic plant pathogens. They show homology to expansins as well as glucanases and cellulases but lack wall-binding domain (Ludidi et al. 2002). PNPs play role in ion and water homeostasis (Gehring and Irving 2003). The first evidence which confirmed the availability of PNP in plants was radioimmunoassays from the extracts of Florida Beauty (Vesely and Giordano 1991). They were later purified by immunoaffinity chromatography from several species including ivy and potato (Billington et al. 1997; Maryani et al. 2001; Rafudeen et al. 2003). PNPs consist of N-terminal signaling peptide secreted into extracellular space, which is a consistent location where PNPs can be identified in apoplast proteome (Boudart et al. 2004). Expression of PNP-A is induced by osmotic drought and saline conditions (Rafudeen et al. 2003; Wang et al. 2011). PNP-A acts on adjacent cells as it regulates its own expression and hence modulates the amount of PNP present in leaf and accordingly amplifies the levels when plant is under abiotic stress (Rafudeen et al. 2003; Wang et al. 2011). PNPs are secreted in the mesophyll cells of leaf and also expressed and act in paracrine or autocrine way on the cells present near to them for introducing cell water uptake (swelling) and changes in cellular cGMP levels (Wang et al. 2007, 2011). PNP expression is initiated by abiotic and biotic stimuli which lead to increase in PNP level in apoplast and also raises intracellular cGMP levels by receptors having guanylate cyclase activity. The photosynthetic capacity of a plant can be increased by PNP. They are reported to regulate various physiological processes, viz. regulation of stomatal aperture (Billington et al. 1997; Maryani et al. 2001; Wang et al. 2007), osmoticum-dependent volume changes in protoplasts (Suwastika and Gehring 1998; Maryani et al. 2001; Wang et al. 2007), enhancement of radial water movement out of xylem and modulation of developmental stage- and tissue-specific ion fluxes (Ludidi et al. 2002) and systemic acquired resistance immune response (Breitenbach et al. 2014).

3.4 Plant Peptides Implicated in Abiotic Stress Conditions

Plants employ a host of stress-mitigating mechanisms which help them to withstand changing environment. Although there are many reports regarding the involvement of signaling peptides in plant developmental processes, information regarding their involvement in plant abiotic and biotic stress is still in infancy stage. Nevertheless, there are an increasing number of studies over the past few years which have established these molecular messengers as important components of stress response. Peptides have been shown to directly interact with pathogens either through their antimicrobial activity or through mediating partners in signaling cascades/cell-to-cell communication underway during stress conditions (Costa et al. 2014; Haruta et al. 2014). Several defence-related peptides are involved in modulation of salt stress tolerance. Plant defence peptides (PDFs) are conserved across monocots and dicots and have antifungal activity as well as antibacterial activity. Their overexpression confers tolerance against fungi as well as bacterial pathogens in model as well as crop plants (Gaspar et al. 2014; van der Weerden and Anderson 2013). Overexpression of precursor of systemin peptides has been reported to confer better salt stress tolerance in tomato plants (Pearce et al. 1991). Moreover, systemins have also been implicated in deterrence of microbes and herbivores (Holley et al. 2003; Narvaez and Orozco (2008). Similarly, CAPE1 peptide elicits antibacterial activity, antiherbivory activity as well as involvement in salt stress responses (Chen et al. 2014; Chien et al. 2015). CAPE1 peptide in *Arabidopsis* is induced in response to salt stress and negatively regulates salt tolerance by suppressing several salt tolerance genes functioning in the production of osmolytes, detoxification, stomatal closure control and cell membrane protection (Tavormin et al. 2015). C-terminally encoded peptides (CEP) have also been postulated to negatively regulate salt stress, and CEP knockdown has been shown to improve salt tolerance in *Arabidopsis* (Ohyama et al. 2008). GRIM REAPER PEPTIDE (GRIp) identified in *Arabidopsis* serves as a trigger for oxidative stress and reactive oxygen species-dependent cell death in *Arabidopsis* (Wrzaczek et al. 2015). Overexpression and infiltration of specific oxidative stress-induced peptides (OSIP108) in *Arabidopsis* confers increased oxidative stress tolerance by suppressing apoptosis triggered by different oxidative stress inducers (Spincemaille et al. 2014a, b). Additionally, plant elicitor peptides (PEP) besides some specific stress conditions like starvation stress also contribute to immunity against attack by bacteria, fungi as well as herbivores. They are reported to interact with several plant hormone pathways involved in responses to abiotic stress, viz. SA (Huffaker et al. 2006), ethylene (Liu et al. 2013; Tintor et al. 2013) and JA (Flury et al. 2013). PIP-LIKE (PIPL) peptides, related to IDA/IDL and CEP peptides, have been implicated in both biotic and abiotic stresses (Vie and Najafi 2015). Sulphated peptides PSK and PSY1 display antagonistic effects against the fungi *Alternaria brassicicola* (Sauter 2015; Mosher and Kemmerling 2013). Peptides have also been shown to play a negative role in fungal growth by alkalinization of the local environment. *Arabidopsis* rapid alkalinization factor (RALF) peptides have been reported to have a role in regulating biotic and abiotic stress responses (Atkinson et al. 2013). Genetic manipulation of the EPF signaling pathway alters stomatal density and size, plant transpiration, rosette growth and tolerance to restricted water availability across a range of atmospheric CO_2 environments (Doheny-Adams et al. 2012). Plant natriuretic peptide (PNP), AtPNP-A and AtPNP-B also have a role in the regulation of homeostasis in abiotic and biotic stresses, and the former also controls the effects of ABA on stomata (Wang et al. 2011). These studies provide a necessary impetus for exploring the function of a range of peptides as important candidates for mediating plant stress response. Peptidomics and transcriptomics surveys with the help of in silico studies take the estimate of peptides much higher than the number reported so far. Functional genomics of myriad of these unreported peptides may help us to understand the complexities of plant stress response mechanisms and lead us towards the engineering of climate-resilient crop plants. A snapshot of various peptides involved in biotic as well as abiotic stress response has been provided in Table 3.2.

3.5 Methods to Isolate Plant Signaling Peptide Involved in Stress

Plants possess thousands of genes encoding putative secreted extracellular peptide ligands and their receptor components. However, only few peptides have been functionally characterized. A major hindrance in studying these signaling peptides is the small size of the genes encoding them. As most of the signal peptide precursors are shorter than 130 amino acids, genes that encode them are mostly ignored in genome annotations as their size is difficult to distinguish from open reading frame (Ghorbani et al. 2014). Moreover, the amount of mature signaling peptide isolated from crude plant tissue extracts is also very low (in nanograms), and the presence of other major peptides also makes it difficult to isolate the signaling peptides that are low in concentration. Consequently, identification of linear signaling pathways and prediction of peptide-receptor interaction become more challenging. Furthermore, functional analysis of these peptides through mutant analysis is difficult as loss of functional mutants is either not available for most of these peptides and if available there is no visible phenotype. The methods used for unravelling new peptides include various

Table 3.2 Plant signaling peptides as important candidates in stress response

S. no.	Peptide name	Plant species	Phenotypic effect	References
1.	EPF	*Arabidopsis thaliana*	Overexpression results in increased biomass in water limiting high CO_2 condition	Doheny-Adams et al. (2012)
2.	CEP	*Arabidopsis thaliana*	Overexpression leads to decrease in primary root length and increases shoot length and tolerance to salinity	Delay et al. (2013)
3.	CAPE	*Arabidopsis thaliana*	Reduced response in growth retardation and induced germination under high salt conditions	Chien et al. (2015)
4.	RALF	*Nicotiana tabacum*	Production of plants with longer roots, role in both abiotic and biotic stresses	Haruta and Constabel 2003 and Atkinson et al. (2013)
5.	SYSTEMIN	*Solanum lycopersicum*	Confer better salt stress tolerance	Pearce et al. (1991);
6.	GRIp	*Arabidopsis thaliana*	Induced ROS-dependent cell death	Wrzaczek et al. (2015)
7.	OSIP108	*Arabidopsis thaliana*	Confer increased oxidative stress tolerance	Spincemaille et al. (2014a, b)
8.	PEP	*Arabidopsis thaliana*	Increased plant immunity against bacteria, fungi and insect herbivores	Huffaker et al. (2006)
9.	RALF	*Arabidopsis thaliana*	Overexpression leads to semidwarf phenotype	Matos et al. (2008)
10.	PNP	*Arabidopsis thaliana*	Regulation of homeostasis in abiotic and biotic stresses	Wang and Irving (2011)

forward and reverse genetic approaches, bioinformatics tools and biochemical assays (Abrash et al. 2011). A schematic view of various techniques employed over the years to discern unknown plant peptides has been given in Fig. 3.2. EPF1 peptides were discovered through bioinformatics analysis for ligands of receptor-like kinases (Hara et al. 2007). Similarly, CEP family of peptides was also discovered through in silico analysis of small peptides in *Arabidopsis* (Ohyama et al. 2008). Genetic analysis of mutants has also been useful for the discovery of peptides conferring a specific phenotype. Club-shaped phenotypes in plants were associated with CLV peptides (Cock and McCormick 2001). Proteomics approaches, viz. LC-MS, mass spectrometry and immunoaffinity tagging, have been instrumental in the discovery of new peptides. Phenotype analysis and further biochemical purification through Edman degradation followed by mass spectrometry led to the identification of PSK peptides (Matasubayashi et al. 1997). PNP peptides in plants were discovered by capturing plant peptides using mammalian antibodies against PNP. Transcriptome surveys have also been used to identify new peptides, viz. Taximin peptides (Onrubia et al. 2014). Using the aforesaid techniques, several new

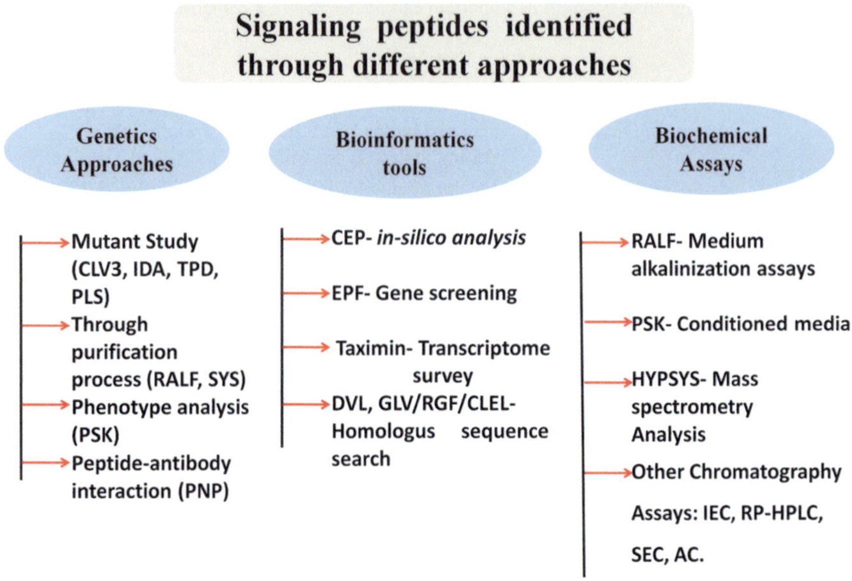

Fig. 3.2 Signaling peptide identification through various approaches

peptides have been discovered in the last few years, and some of them have been functionally characterized. They have been aptly recognized in plants as molecular messengers as they occupy centre stage in regulating developmental processes such as meristem maintenance, organ abscission, cell elongation, cell proliferation and differentiation, gravitropism and defence against biotic aggressors (Ghorbani et al. 2014). However, a vast majority of these peptides are yet uncharacterized. Techniques for a more comprehensive functional analysis of peptides are therefore urgently needed to decipher the precise role of the vast repertoire of these molecular messengers.

3.6 Future Prospects

We have come a long way in understanding the action of plant signaling peptides since the discovery of the first plant signaling peptide. They have been implicated in a host of plant growth and developmental processes with many of them exhibiting direct or indirect role in stress responses. Yet there is a need for comprehensive analysis of components of small peptide-triggered signaling cascades. Tools for tracking *in planta* movement and interacting partners of signaling peptides can give us better insight towards their wide roles in overall plant physiology. Additionally, deciphering gene networks directly regulated by them will aid in delineating the intricacies of molecular processes activated downstream of the peptide-receptor interaction.

References

Abrash EB, Davies KA, Bergmann DC (2011) Generation of signaling specificity in Arabidopsis by spatially restricted buffering of ligand receptor interactions. Plant Cell 23:2864–2879

Akker SEV, Lilley C, Yusup HB, Jones JT, Urwin PE (2016) Functional C-TERMINALLY ENCODED PEPTIDE (CEP) plant hormone domains evolved de novo in the plant parasite *Rotylenchulus reniformis*. Mol Plant Pathol 17(8):1265–1275

Akpinar B, Bihter A, Lucas SJ, Hikmet B (2012) Plant abiotic stress signaling. Plant Signal Behav 7:1450–1455

Albert M (2013) Peptides as triggers of plant defence. J Exp Bot 64:5269–5279

Amano Y, Tsubouchi H, Shinohara H, Ogawa M, Matsubayashi Y (2007) Tyrosine-sulfated glycopeptide involved in cellular proliferation and expansion in Arabidopsis. Proc Natl Acad Sci U S A 104:18333–18338

Araya T, Miyamoto M, Wibowo J, Suzuki A, Kojima S et al (2014) CLE-CLAVATA1 peptide-receptor signaling module regulates the expansion of plant root systems in a nitrogen dependent manner. Proc Natl Acad Sci U S A 111:2029–2034

Asano T, Miwa A, Maeda K, Kimura M, Nishiuchi T (2013) The secreted antifungal protein thionin 2.4 in Arabidopsis thaliana suppresses the toxicity of a fungal fruit body lectin from Fusarium graminearum. PLoS Pathog 9:e1003581

Atkinson NJ, Lilley CJ, Urwin PE (2013) Identification of genes involved in the response of Arabidopsis to simultaneous biotic and abiotic stresses. Plant Physiol 162:2028–2041

Beers EP, Jones AM, Dickeman AW (2004) The S8 serine, C1A cysteine and A1 aspartic protease families in Arabidopsis. Phytochemistry 65:43–58

Bergey DR, Howe GA, Ryan CA (1996) Polypeptide signaling for plant defensive genes exhibits analogies to defense signaling in animals. Proc Natl Acad Sci 93(22):12053–12058

Bhattacharya R, Koramutla MK, Negi M, Pearce G, Ryan CA (2013) Hydroxyproline-rich glycopeptide signals in potato elicit signalling associated with defense against insects and pathogens. Plant Sci 207:88–97

Billington T, Pharmawati M, Gehring CA (1997) Isolation and immunoaffinity purification of biologically active plant natriuretic peptide. Biochem Biophys Res Commun 235:722–772

Blanvillain R, Young B, Cai YM, Hecht V, Varoquaux F, Delorme V, Lancelin J-M, Delseny M, Gallois P (2011) The Arabidopsis peptide kiss of death is an inducer of programmed cell death. EMBO J 30:1173–1183

Bonello JF, Opsahl-Ferstad HG, Perez P, Dumas C, Rogowsky PM (2000) Esr genes show different levels of expression in the same region of maize endosperm. Gene 246:219–227

Boudart G, Jamet E, Rossignol M, Lafitte C, Borderies G, Jauneau A, Esquerre-Tugaye M, Pont-Lezica R (2004) Cell wall proteins in apoplastic fluids of Arabidopsis thaliana rosettes: identification by mass spectrometry and bioinformatics. Proteomics 5:212–221

Breitenbach HH, Wenig M, Wittek F, Jorda L, Maldonado-Alconada AM, Sarioglu H, Colby T, Knappe C, Bichlmeier M, Pabst E, Mackey D, Parker JE, Vlot AC (2014) Contrasting roles of the apoplastic aspartyl protease apoplastic, enhanced disease susceptibility1- dependent1 and legume lectin-like protein1 in arabidopsis systemic acquired resistance. Plant Physiol 165(2):791–809

Butenko MA, Patterson SE, Grini PE, Stenvik GE, Amundsen SS et al (2003) Inflorescence deficient in abscission controls floral organ abscission in Arabidopsis and identifies a novel family of putative ligands in plants. Plant Cell 15:2296–2307

Cao J, Shi F (2012) Evolution of the RALF gene family in plants: gene duplication and selection patterns. Evol Bioinforma 8:271–292

Casson SA, Paul MC, Jennifer FT, Marta E, Martin AS, Keith L (2002) The *POLARIS* gene of Arabidopsis encodes a predicted peptide required for correct root growth and leaf vascular patterning. Plant Cell 14:1705–1721

Chen YC, Siems WF, Pearce G, Ryan CA (2008) Six peptide wound signals derived from a single precursor protein in *Ipomoea batatas* leaves activate the expression of the defense gene sporamin. J Biol Chem 283:11469–11476

Chen Y-L, Lee C-Y, Cheng K-T, Chang W-H, Huang R-N, Nam HG, Chen Y-R (2014) Quantitative peptidomics study reveals that a wound-induced peptide from PR-1 regulates immune signaling in tomato. Plant Cell 26:4135–4148

Chevalier E, Hudon AL, Matton DP (2013) ScRALF3, a secreted RALF-like peptide involved in cell-cell communication between the sporophyte and the female gametophyte in a solanaceous species. Plant J 73:1019–1033

Chien P-S, Nam HG, Chen Y-R (2015) A salt-regulated peptide derived from the CAP superfamily protein negatively regulates salt-stress tolerance in Arabidopsis. J Exp Bot. pii: erv263 66:5301–5313

Chilley PM, Casson SA, Tarkowski P, Hawkins N, Wang KL-C, Hussey PJ, Beale M, Ecker JR, Sandberg GK, Lindsey K (2006) The POLARIS peptide of Arabidopsis regulates auxin transport and root growth via effects on ethylene signaling. Plant Cell 18:3058–3072

Chouabe C, Eyraud V, Da Silva P, Rahioui I, Royer C, Soulage C, Bonvallet R, Huss M, Gressent F (2011) New mode of action for a knottin protein bioinsecticide: pea albumin 1 subunit b (PA1b) is the first peptidic inhibitor of V-ATPase. J Biol Chem 286:36291–36296

Cock JM, McCormick S (2001) A large family of genes that share homology with CLAVATA3. Plant Physiol 126(3):939–942

Combier JP, Kuster H, Journet EP, Hohnjec N, Gamas P, Niebel A (2008) Evidence for the involvement in nodulation of the two small putative regulatory peptide-encoding genes MtRALFL1 and MtDVL1. Mol Plant-Microbe Interact 21:1118–1127

Constabel CP, Bergey DR, Ryan CA (1995) Systemin activates synthesis of wound-inducible tomato leaf polyphenol oxidase via the octadecanoid defense signaling pathway. Proc Natl Acad Sci 92(2):407–411

Costa LM, Marshall E, Tesfaye M, Silverstein KA, Mori M (2014) Central cell-derived peptides regulate early embryo patterning in flowering plants. Science 344:168–172

Covey PA, Subbaiah CC, Parsons RL, Pearce G, Lay FT, Anderson MA, Ryan CA, Bedinger PA (2010) A pollen-specific RALF from tomato that regulates pollen tube elongation. Plant Physiol 153:703–715

Cowan WM (2001) Viktor Hamburger and Rita Levi-Montalcini: the path to the discovery of nerve growth factor. Ann Rev Neurosci 24(1):551–600

Craik DJ, Daly NL, Bond T, Waine C (1999) Plant cyclotides: a unique family of cyclic and knotted proteins that defines the cyclic cystine knot structural motif. J Mol Biol 294:1327–1336

Delay C, Imin N, Djordjevic MA (2013) CEP genes regulate root and shoot development in response to environmental cues and are specific to seed plants. J Exp Bot 64(17):5383–5394

DeYoung BJ, Clark SE (2008) BAM receptors regulate stem cell specification and organ development through complex interactions with CLAVATA signaling. Genetics 180:895–904

Doheny-Adams T, Hunt L, Franks PJ, Beerling DJ, Gray JE (2012) Genetic manipulation of stomatal density influences stomatal size, plant growth and tolerance to restricted water supply across a growth carbon dioxide gradient. Philos Trans Royal Soc Lond B: Biol Sci 367(1588):547–555

Dombrowski JE (2003) Salt stress activation of wound-related genes in tomato plants. Plant Physiol 132(4):2098–2107

Donaldson L, Ludidi N, Knight MR, Gehring C, Denby K (2004) Salt and osmotic stress cause rapid increases in Arabidopsis thaliana cGMP levels. FEBS Lett 569:317–320. during stomatal development. Current Biology 19(10): 864–869

Durgo H, Klement E, Hunyadi-Gulyas E, Szucs A, Kereszt A, Medzihradszky KF, Kondorosi E (2015) Identification of nodule-specific cysteine-rich plant peptides in endosymbiotic bacteria. Proteomics 15:2291–2295

de Souza Cândido E, e Silva Cardoso MH, Sousa DA, Viana JC, de Oliveira-Júnior NG, Miranda V, Franco OL (2014) The use of versatile plant antimicrobial peptides in agribusiness and human health. Peptides 55:65–78

Farkas A, Maróti G, Durgő H, Györgypál Z, Lima RM, Medzihradszky KF, Kereszt A, Mergaert P, Kondorosi É (2014) Medicago truncatula symbiotic peptide NCR247 contributes to bacteroid differentiation through multiple mechanisms. Proc Natl Acad Sci U S A 111:5183–5188
Fernandez A, Drozdzecki A, Hoogewijs K, Nguyen A, Beeckman T, Madder A, Hilson P (2013) Transcriptional and functional classification of the GOLVEN/ROOT GROWTH FACTOR/CLE-like signaling peptides reveals their role in lateral root and hair formation. Plant Physiol 161:954–970
Fernandez A, Drozdzecki A, Hoogewijs K, Vassileva V, Madder A, Beeckman T, Hilson P (2015) The GLV6/RGF8/CLEL2 peptide regulates early pericycle divisions during lateral root initiation. J Exp Bot 66(17):5245–5256
Fletcher JC, Brand U, Running MP, Simon R, Meyerowitz EM (1999) Signaling of cell fate decisions by CLAVATA3 in Arabidopsis shoot meristems. Science 283:1119
Flury P, Klauser D, Schulze B, Boller T, Bartels S (2013) The anticipation of danger: microbe-associated molecular pattern perception enhances AtPep-triggered oxidative burst. Plant Physiol 161(4):2023–2035
García AN, Ayub ND, Fox AR, Gómez MC, Diéguez MJ, Pagano EM, Berini CA, Muschietti JP, Soto G (2014) Alfalfa snakin-1 prevents fungal colonization and probably coevolved with rhizobia. BMC Plant Biol 14:248
Gaspar YM, McKenna JA, McGinness BS, Hinch J, Poon S, Connelly AA, Anderson MA, Heath RL (2014) Field resistance to Fusarium oxysporum and Verticillium dahliae in transgenic cotton expressing the plant defensin NaD1. J Exp Bot 65:1541–1550
Gehring CA, Irving HR (2003) Natriuretic peptides – a class of heterologous molecules in plants. Int J Biochem Cell Biol 35:1318–1322
Ghorbani S, Fernandez A, Hilson P, Beeckman T (2014) Signaling peptides in plants. Cell Dev Biol 3:2
Goyal RK, Mattoo AK (2014) Multitasking antimicrobial peptides in plant development and host defense against biotic/abiotic stress. Plant Sci 228:135–149
Gray-Mitsumune M, Matton DP (2006) The egg apparatus 1 gene from maize is a member of a large gene family found in both monocots and dicots. Planta 223:618–625
Green TR, Ryan CA (1972) Wound-induced proteinase inhibitor in plant leaves: a possible defense mechanism against insects. Science 175(4023):776–777
Gully K, Hander T, Boller T, Bartels S (2015) Perception of Arabidopsis AtPep peptides, but not bacterial elicitors, accelerates starvation-induced senescence. Front Plant Sci 6:14
Gultyaev AP, Roussis A (2007) Identification of conserved secondary structures and expansion segments in enod40 RNAs reveals new enod40 homologues in plants. Nucleic Acids Res 35:3144–3152
Guo P, Yoshimura A, Ishikawa N et al (2015) Comparative analysis of the RTFL peptide family on the control of plant organogenesis. J Plant Res 128:497–510
Hanada K, Zhang X, Borevitz JO, Li WH, Shiu SH (2007) A large number of novel coding small open reading frames in the intergenic regions of the Arabidopsis thaliana genome are transcribed and/or under purifying selection. Genome Res 17:632–640
Hara K, Kajita R, Torii KU, Bergmann DC, Kakimoto T (2007) The secretory peptide gene EPF1 enforces the stomatal one-cell-spacing rule. Genes Dev 21:1720–1725
Hara K, Yokoo T, Kajita R, Onishi T, Yahata S (2009) Epidermal cell density is autoregulated via a secretory peptide, EPIDERMAL PATTERNING FACTOR 2 in Arabidopsis leaves. Plant Cell Physiol 50:1019–1031
Haruta M, Constabel CP (2003) Rapid alkalinization factors in poplar cell cultures. Peptide isolation, cDNA cloning, and differential expression in leaves and methyl jasmonate treated cells. Plant Physiol 131:814–823
Haruta M, Sabat G, Stecker K, Minkoff BB, Sussman MR (2014) A peptide hormone and its receptor protein kinase regulate plant cell expansion. Science 343(6169):408–411
Higashiyama T (2010) Peptide signaling in pollen-pistil interactions. Plant Cell Physiol 51:177–189
Hirayama T, Shinozaki K (2010) Research on plant abiotic stress responses in the post genome era: past, present and future. Plant J 61:1041–1052

Holley SR, Yalamanchili RD, Moura DS, Ryan CA, Stratmann JW (2003) Convergence of signaling pathways induced by systemin, oligosaccharide elicitors, and ultraviolet-B radiation at the level of mitogen-activated protein kinases in *Lycopersicon peruvianum* suspension-cultured cells. Plant Physiol 132:1728–1738

Huffaker A, Pearce G, Ryan CA (2006) An endogenous peptide signal in Arabidopsis activates components of the innate immune response. Proc Natl AcadSci USA 103:10098–10103

Hunt L, Gray JE (2009) The signaling peptide EPF2 controls asymmetric cell divisions during stomatal development. CurrBiol 19:864–869. 73

Hunt L, Bailey KJ, Gray JE (2010) The signalling peptide EPFL9 is a positive regulator of stomatal development. New Phytol 186:609–614. 74

Ikeuchi M, Yamaguchi Y, Kazama T, Ito T, Horiguchi G, Tsukaya H (2011) ROTUNDIFOLIA4 regulates cell proliferation along the body axis in Arabidopsis shoot. Plant Cell Physiol 52:59–69

Ingram G, Gutierrez-Marcos J (2015) Peptide signaling during angiosperm seed development. J Exp Bot 66(17):5151–5159

Ito Y, Nakanomyo I, Motose H, Iwamoto K, Sawa S, Dohmae N, Fukuda H (2006) Dodeca-CLE peptides as suppressors of plant stem cell differentiation. Science 313:842–845

Jeong S, Trotochaud AE, Clark SE (1999) The Arabidopsis *CLAVATA2* gene encodes a receptor-like protein required for the stability of the CLAVATA1 receptor-like kinase. Plant Cell 11:1925

Ji H, Gheysen G, Ullah C, Verbeek R, Shang C, De Vleesschauwer D, Höfte M, Kyndt T (2015) The role of thionins in rice defence against root pathogens. Mol Plant Pathol. in press

Jinn TL, Stone JM, Walker JC (2000) HAESA, an Arabidopsis leucine rich repeat receptor kinase, controls floral organ abscission. Genes Dev 14:108–117

Katsir L, Davies KA, Bergmann DC, Laux T (2011) Peptide signaling in plant development. Curr Biol 21(9):356–364

Kohorn BD (1999) Shuffling the deck: plant signalling plays a club. Trends Cell Biol 9:381

Kondo T, Sawa S, Kinoshita A, Mizuno S, Kakimoto T, Fukuda H, Sakagami Y (2006) A plant peptide encoded by CLV3 identified by in situ MALDI-TOF MS analysis. Science 313:845–848

Kumpf RP, Shi CL, Larrieu A, Stø IM, Butenko MA (2013) Floral organ abscission peptide IDA and its HAE/HSL2 receptors control cell separation during lateral root emergence. Proc Natl Acad Sci U S A 110:5235–5240

Kutschmar A, Rzewuski G, Stuhrwohldt N, GTS B, Inze D, Sauter M (2009) PSK-a promotes root growth in Arabidopsis. New Phytol 181:820–831

Kwezi L, Ruzvidzo O, Wheeler JI, Govender K, Iacuone S, Thopson PE, Gehring C, Irving HR (2011) The phytosulfokine (PSK) receptor is capable of guanylate cyclase activity and enabling cyclic GMP-dependent signalling in plants. J BiolChem 286:22580–22588

Lee DG, Shin SY, Kim DH, Seo MY, Kang JH, Lee Y, Kim KL, Hahm KS (1999) Antifungal mechanism of a cysteine-rich antimicrobial peptide, Ib-AMP1, from impatiens balsamina against *Candida albicans*. Biotechnol Lett 21:1047–1050

Lee JS, Kuroha T, Hnilova M, Khatayevich D, Kanaoka MM (2012) Direct interaction of ligand-receptor pairs specifying stomatal patterning. Genes Dev 26:126–136

Lindsey K, Casson S, Chilley P (2002) Peptides: new signalling molecules in plants. Trends Plant Sci 7:78–83

Liu Z, Wu Y, Yang F, Zhang Y, Chen S, Xie Q, Tian X, Zhou JM (2013) BIK1 interacts with PEPRs to mediate ethylene-induced immunity. PNAS USA 110:6205–6210

Lorbiecke R, Sauter M (2002) Comparative analysis of PSK peptide growth factor precursor homologs. Plant Sci 163:321–332

Ludidi NN, Heazlewood JL, Seoighe C, Irving HR, Gehring CA (2002) Expansin-like molecules: novel functions derived from common domains. J Mol Evol 54:587–594

Ma W, Yoshioka K, Gehring C, Berkowitz G (2010) The function of cyclic nucleotide-gated channels in biotic stress. In: Demidchik V, Maathuis F (eds) Ion channels and plant stress responses, signaling and communication in plants. Springer, Berlin, pp 159–174

Maathuis FJ (2006) cGMP modulates gene transcription and cation transport in Arabidopsis roots. Plant J 45:700–711

Márton ML, Fastner A, Uebler S, Dresselhaus T (2012) Overcoming hybridization barriers by the secretion of the maize pollen tube attractant ZmEA1 from Arabidopsis ovules. Curr Biol 22:1194–1198

Maryani MM, Bradley G, Cahill DM, Gehring CA (2001) Natriuretic peptides and immunoreactants modify osmoticum-dependent volume changes in Solanum tuberosum L. mesophyll cell protoplasts. Plant Sci 161:443–452

Matos JL, Fiori CS, Silva-Filho MC, Moura DS (2008) A conserved dibasic site is essential for correct processing of the peptide hormone AtRALF1 in *Arabidopsis thaliana*. FEBS 582:3343–3347

Matsubayashi Y (2011) Small post-translationally modified peptide signals in Arabidopsis. Arabidopsis Book/Am Soc Plant Biol 9:e0150

Matsubayashi Y (2014) Posttranslationally modified small-peptide signals in plants. Ann Rev Plant Biol 65:385–413

Matsubayashi Y, Sakagami Y (1996) Phytosulfokine, sulfated peptides that induce the proliferation of single mesophyll cells of Asparagus officinalis L. Proc Natl Acad Sci 93(15):7623–7627

Matsubayashi Y, Takagi L, Sakagami Y (1997) Phytosulfokine-alpha, a sulfated pentapeptide, stimulates the proliferation of rice cells by means of specific high- and low-affinity binding sites. Proc Natl Acad Sci U S A 94:13357–13362

Matsubayashi Y, Takagi L, Omura N, Morita A, Sakagami Y (1999) The endogenous sulfated pentapeptide phytosulfokine-α stimulates tracheary element differentiation of isolated mesophyll cells of Zinnia. Plant Physiol 120(4):1043–1048

Matsubayashi Y, Ogawa M, Morita A, Sakagami Y (2002) An LRR receptor kinase involved in perception of a peptide plant hormone, phytosulfokine. Science 296:1470–1472

Matsubayashi Y, Ogawa M, Kihara H, Niwa M, Sakagami Y (2006) Disruption and overexpression of Arabidopsis phytosulfokine receptor gene affects cellular longevity and potential for growth. Plant Physiol 142:45–53

Matsuzaki Y, Ogawa-Ohnishi M, Mori A, Matsubayashi Y (2010) Secreted peptide signals required for maintenance of root stem cell niche in Arabidopsis. Science 329:1065–1067

Meier S, Madeo L, Ederli L, Donaldson L, Pasqualini S, Gehring C (2009) Deciphering cGMP signatures and cGMP-dependent pathways in plant defence. Plant Signal Behav 4:307–309

Meier S, Ruzvidzo O, Morse M, Donaldson L, Kwezi L, Gehring C (2010) The Arabidopsis wall associated kinase-like 10 gene encodes a functional guanylyl cyclase and is co-expressed with pathogen defense related genes. PLoS One 5:e8904

Meng L, Buchanan BB, Feldman LJ, Luan S (2012) CLE-like (CLEL) peptides control the pattern of root growth and lateral root development in Arabidopsis. Proc Natl Acad Sci, USA 109:1760–1765

Morato do Canto A, Ceciliato PH, Ribeiro B, MFA O, GAA F, Silva-Filho MC, Moura DS (2014) Biological activity of nine recombinant AtRALF peptides: implications for their perception and function in Arabidopsis. Plant Physiol Biochem 75:45–54

Mosher S, Kemmerling B (2013) PSKR1 and PSY1R-mediated regulation of plant defense responses. Plant Signal Behav 8:e24119

Motose H, Iwamoto K, Endo S, Demura T, Sakagami Y, Matsubayashi Y, Moore KL, Fukuda H (2009) Involvement of phytosulfokine in the attenuation of stress response during the transdifferentiation of Zinnia mesophyll cells into tracheary elements. Plant Physiol 150:437–447

Mudge A (2016) The role of the POLARIS peptide in ethylene signalling and root development in Arabidopsis thaliana. Durham e-Theses (http://etheses.dur.ac.uk/11473)

Murphy E (2016) Characterising signalling components mediating root architecture in Arabidopsis thaliana. Nottingham e-Theses (http://eprints.nottingham.ac.uk/31976)

Murphy E, Smith S, De Smet I (2012) Small signaling peptides in Arabidopsis development: how cells communicate over a short distance. Plant Cell 24:3198–3217

Murphy E, Vu LD, Van den Broeck L (2016) RALFL34 regulates formative cell divisions in Arabidopsis pericycle during lateral root initiation. J Exp Bot 67:4863–4875

Nahirñak V, Almasia NI, Hopp HE, Vazquez-Rovere C (2012) Snakin/GASA proteins: involvement in hormone crosstalk and redox homeostasis. Plant Signal Behav 7:1004–1008

Narita NN, Moore S, Horiguchi G, Kubo M, Demura T, Fukuda H (2004) Overexpression of a novel small peptide ROTUNDIFOLIA4 decreases cell proliferation and alters leaf shape in *Arabidopsis thaliana*. Plant J 38:699–713

Narváez-Vásquez J, Orozco-Cárdenas ML (2008) Systemins and at peps: defense-related peptide signals. Induced plant resistance to herbivory. p 313

Nawrot R, Barylski J, Nowicki G, Broniarczyk J, Buchwald W, Gozdzicka-Józefiak A (2014) Plant antimicrobial peptides. Folia Microbiol (Praha) 59:181–196

Neukermans J, Inzé A, Mathys J, De Coninck B, van de Cotte B, Cammue BPA, Van Breusegem F (2015) ARACINs, Brassicaceae-specific peptides exhibiting antifungal activities against necrotrophic pathogens in Arabidopsis. Plant Physiol 167:1017–1029

Ni J, Clark SE (2006) Evidence for functional conservation, sufficiency and proteolytic processing of the CLAVATA3 CLE domain. Plant Physiol 140:726–733

Ohki S, Takeuchi M, Mori M (2011) The NMR structure of stomagen reveals the basis of stomatal density regulation by plant peptide. Nat Commun 12:373–386

Ohyama K, Ogawa M, Matasubayashi Y (2008) Identification of a biologically active, small, secreted peptide in Arabidopsis by in silico gene screening, followed by LC-MS-based structure analysis. Plant J 55:152–160

Ohyama K, Shinohara H, Ogawa-Ohnishi M, Matsubayashi Y (2009) A glycopeptide regulating stem cell fate in Arabidopsis thaliana. Nat Chem Biol 5(8):578–580

Okuda S, Tsutsui H, Shiina K, Sprunck S, Takeuchi H, Yui R, Higashiyama T et al (2009) Defensin-like polypeptide LUREs are pollen tube attractants secreted from synergid cells. Nature 458(7236):357–361

Olsen AN, Mundy J, Skriver K (2002) Peptomics, identification of novel cationic Arabidopsis peptides with conserved sequence motifs. In Silico Biol 2:441–451

Onrubia M, Pollier J, Van den Bossche R, Goethals M, Gevaert K, Moyano E, Vidal-Limon H, Cusidó RM, Palazón J, Goossens A (2014) Taximin, a conserved plant-specific peptide is involved in the modulation of plant-specialized metabolism. Plant Biotechnol J 12:971–983

Opsahl-Ferstad HG, LeDeunff E, Dumas C, Rogowsky PM (1997) ZmEsr, a novel endosperm-specific gene expressed in a restricted region around the maize embryo. Plant J 12(1):235–246

Pearce G, Strydom D, Johnson S, Ryan CA (1991) A polypeptide from tomato leaves induces wound-inducible proteinase inhibitor proteins. Science 253(5022):895–899

Pearce G, Johnson S, Ryan CA (1993) Structure-activity of deleted and substituted systemin, an 18-amino acid polypeptide inducer of plant defensive genes. J Biol Chem 268:212–216

Pearce G, Moura DS, Stratmann J, Ryan CA (2001a) RALF, a 5-kDa ubiquitous polypeptide in plants, arrests root growth and development. Proc Natl Acad Sci U S A 98:12843–12847

Pearce G, Moura DS, Stratmann J, Ryan CA (2001b) Production of multiple plant hormones from a single polyprotein precursor. Nature 411:817–820

Pearce G, Yamaguchi Y, Barona G, Ryan CA (2010) A subtilisin-like protein from soybean contains an embedded, cryptic signal that activates defense-related genes. Proc Natl Acad Sci 107(33):14921–14925

Pearce G, Yamaguchi Y, Munske G, Ryan CA (2008) Structure-activity studies of AtPep1, a plant peptide signal involved in the innate immune response. Peptides 29:2083–2089

Qi Z, Verma R, Gehring C, Yamaquchi Y, Zhao Y, Ryan CA, Berkowitz GA (2010) Ca2+ signaling by plant Arabidopsis thaliana pep peptides depends on AtPepR1, a receptor with guanylyl cyclase activity, and cGMP-activated Ca2+ channels. PNAS 107:21193–21198

Qin F, Shinozaki K, Yamaguchi-Shinozaki K (2011) Achievements and challenges in understanding plant abiotic stress responses and tolerance. Plant Cell Physiol 52:1569–1582

Rafudeen S, Gxaba G, Makgoke G, Bradley G, Pironcheva G, Raitt L, Irving H, Gehring C (2003) A role for plant natriuretic peptide immuno-analogues in NaCl- and drought-stress responses. Physiol Plant 119:554–562

Ren F, Lu YT (2006) Overexpression of tobacco hydroxyproline-rich glycopeptides systemin precursor A gene in transgenic tobacco enhances resistance against *Helicoverpa armigera* larvae. Plant Sci 171(2):286–292

Rohrig H, Schmidt J, Miklashevichs E, Schell J, John M (2002) Soybean ENOD40 encodes two peptides that bind to sucrose synthase. Proc Natl Acad Sci U S A 99:1915–1920

Ross A, Yamada K, Hiruma K, Yamashita-Yamada M, Lu X, Takano Y, Tsuda K, Saijo Y (2014) The Arabidopsis PEPR pathway couples local and systemic plant immunity. EMBO J 33:62–75

Rowe MH, Bergmann DC (2010) Complex signals for simple cells: the expanding ranks of signals and receptors guiding stomatal development. Curr Opin Plant Biol 13:548–555

Ryan CA, Pearce G (1998) Systemin: a polypeptide signal for plant defensive genes. Ann Rev Cell Dev Biol 14:1–17

Ryan CA, Pearce G (2003) Systemins: a functionally defined family of peptide signals that regulate defensive genes in Solanaceae species. Proc Natl Acad Sci U S A 2:14577–14580

Sauter M (2015) Phytosulfokine peptide signalling. J Exp Bot 66(17):5161–5169

Schmelz EA, Carroll MJ, LeClere S, Phipps SM, Meredith J, Chourey PS, Alborn HT, Teal PEA (2006) Fragments of ATP synthase mediate plant perception of insect attack. Proc Natl Acad Sci U S A 103:8894–8899

Schmelz EA, Huffaker A, Carroll MJ, Alborn HT, Ali JG, Teal PEA (2012) An amino acid substitution inhibits specialist herbivore production of an antagonist effector and recovers insect-induced plant defenses. Plant Physiol 160:1468–1478

Schopfer CR, Nasrallah ME, Nasrallah JB (1999) The male determinant of self-incompatibility in Brassica. Science 286:1697–1700

Sharma A, Hussain A et al (2016) Comprehensive analysis of plant rapid alkalization factor (*RALF*) genes. Plant Physiol Biochem 106:82–90

Shimada T, Sugano SS, Hara-Nishimura I (2011) Positive and negative peptide signals control stomatal density. Cell Mol Life Sci 68:2081–2088

Shpak ED, McAbee JM, Pillitteri LJ, Torii KU (2005) Stomatal patterning and differentiation by synergistic interactions of receptor kinases. Science 309:290–293

Silverstein KAT, Moskal WJ Jr, Wu HC, Underwood BA, Graham MA, Town CD, VandenBosch KA (2007) Small cysteine-rich peptides resembling antimicrobial peptides have been under-predicted in plants. Plant J 51:262–280

Slavokhotova AA, Rogozhin EA, Musolyamov AK, Andreev YA, Oparin PB, Berkut AA, Vassilevski AA, Egorov TA, Grishin EV, Odintsova TI (2014) Novel antifungal a-hairpinin peptide from Stellaria media seeds: structure, biosynthesis, gene structure and evolution. Plant Mol Biol 84:189–202

Spincemaille P, Chandhok G, Newcomb B, Verbeek J, Vriens K, Zibert A, Schmidt H, Hannun YA, van Pelt J, Cassiman D, Cammue BP, Thevissen K (2014a) The plant decapeptide OSIP108 prevents copper-induced apoptosis in yeast and human cells. Biochim Biophys Acta 1843:1207–1215

Spincemaille P, Pham DH, Chandhok G, Verbeek J, Zibert A, Libbrecht L, Schmidt H, Esguerra CV, de Witte PAM, Cammue BPA, Cassiman D, Thevissen K (2014b) The plant decapeptide OSIP108 prevents copper-induced toxicity in various models for Wilson disease. Toxicol Appl Pharmacol 280:345–351

Stenvik GE, Butenko MA, Urbanowicz BR, Rose JK, Aalen RB (2006) Overexpression of INFLORESCENCE DEFICIENT IN ABSCISSION activates cell separation in vestigial abscission zones in Arabidopsis. Plant Cell 18:1467–1476

Stenvik GE, Tandstad NM, Guo Y, Shi CL, Kristiansen W (2008) The EPIP peptide of INFLORESCENCE DEFICIENT IN ABSCISSION is sufficient to induce abscission in arabidopsis through the receptor-like kinases HAESA and HAESA-LIKE2. Plant Cell 20:1805–1817

Sugano SS, Shimada T, Imai Y, Okawa K, Tamai A et al (2010) Stomagen positively regulates stomatal density in Arabidopsis. Nature 463:241–244

Sui Z, Wang T, Li H, Zhang M et al (2016) Overexpression of peptide-encoding *OsCEP6.1* results in pleiotropic effects on growth in rice (*O. sativa*). Front Plant Sci 7:228

Suwastika IN, Gehring CA (1998) Natriuretic peptide hormones promote radial water movements from the xylem of Tradescantia shoots. Cell Mol Life Sci CMLS 54(10):1161–1167

Tabata R, Sawa S (2014) Maturation processes and structures of small secreted peptides in plants. Plant Prot 5:311

Takayama S, Shiba H, Iwano M, Shimosato H, Che FS, Kai N, Watanabe M, Suzuki G, Hinata K, Isogai A (2000) The pollen determinant of self-incompatibility in Brassica campestris. Proc Natl Acad Sci U S A 97:1920–1925

Takayama S, Shimosato H, Shiba H, Funato M, Che FS, Watanabe M, Isogai A (2001) Direct ligand-receptor complex interaction controls Brassica self-incompatibility. Nature 413(6855):534–538

Tavormina P, Coninck BD, Nikonorova N, Smet ID, Cammue BPA (2015) The plant peptidome: an expanding repertoire of structural features and biological functions. Plant Cell 13:1820–1845

Tintor N, Ross A, Kanehara K, Yamada K, Fan L, Kemmerling B, Nurnberger T, Tsuda K, Saijo Y (2013) Layered pattern receptor signaling via ethylene and endogenous elicitor peptides during *Arabidopsis* immunity to bacterial infection. PNAS USA 110(15):6211–6216

Topping JF, Lindsey K (1997) Promoter trap markers differentiate structural and positional components of polar development in *Arabidopsis*. Plant Cell 9(10):1713–1725

Trotochaud AE, Hao T, Wu G, Yang Z, Clark SE (1999) The CLAVATA1 receptor-like kinase requires CLAVATA3 for its assembly into a signaling complex that includes KAPP and a rho-related protein. Plant Cell 11:393

Uchida N, Tasaka M (2013) Regulation of plant vascular stem cells by endodermis-derived EPFL-family peptide hormones and phloem expressed ERECTA-family receptor kinases. J Exp Bot 64:5335–5343

Uchida N, Lee JS, Horst RJ, Lai HH, Kajita R et al (2012) Regulation of inflorescence architecture by inter tissue layer ligand-receptor communication between endodermis and phloem. Proc Natl Acad Sci USA 109:6337–6342

van der Weerden NL, Anderson MA (2013) Plant defensins: common fold, multiple functions. Fungal Biol Rev 26:121–131

van Loon LC, Rep M, Pieterse CMJ (2006) Significance of inducible defense-related proteins in infected plants. Ann Rev Phytopathol 44:135–162

Vesely DL, Giordano AT (1991) Atrial natriuretic peptide hormonal system in plants. Biochem Biophys Res Commun 179:695–700

Vie AK, Najafi J (2015) The *IDA/IDA-LIKE* and *PIP/PIP-LIKE* gene families in *Arabidopsis*: phylogenetic relationship, expression patterns, and transcriptional effect of the PIPL3 peptide. J Exp Bot 66(17):5351–5365

Wang YH, Irving HR (2011) Developing a model of plant hormone interactions. Plant Signal Behav 6:494–500

Wang YH, Gehring C, Cahill DM, Irving HR (2007) Plant natriuretic peptide active site determination and effects on cGMP and cell volume regulation. Funct Plant Biol 34:645–653

Weidmann J, Craik DJ (2016) Discovery, structure, function, and applications of cyclotides: circular proteins from plants. J Exp Bot 124:215–227

Wen J, Walker J (2006) DVL peptides are involved in plant development. In: Abba JK (ed) Handbook of biologically active peptides. Academic Press, Burlington, pp 17–22

Wen J, Lease KA, Walker JC (2004) DVL, a novel class of small polypeptides: overexpression alters Arabidopsis development. Plant J 37:668–677

Whitford R, Fernandez A, Tejos R, Perez AC, Kleine-Vehn J et al (2012) GOLVEN secretory peptides regulate auxin carrier turnover during plant gravitropic responses. Dev Cell 22:678–685

Wrzaczek M, Vainonen JP, Stael S, Tsiatsiani L, Gauthier A, Kaufholdt D, Bollhoner B, Lamminmaki A (2015) GRIM REAPER peptide binds to receptor kinase PRK5 to trigger cell death in Arabidopsis. EMBO J 34:55–66

Wu J, Kurten EL, Monshausen G, Hummel GM, Gilroy S, Baldwin IT (2007) NaRALF, a peptide signal essential for the regulation of root hair tip apoplastic pH in *Nicotiana attenuata*, is required for root hair development and plant growth in native soils. Plant J 52:877–890

Yamaguchi Y, Huffaker A (2011) Endogenous peptide elicitors in higher plants. Curr Opin Plant Biol 14:351–357

Yamaguchi T, Ikeuchi M, Tsukaya H (2013) ROTUNDIFOLIA4. In: Kastin AJ (ed) Handbook of biologically active peptides, 2nd edn. Elsevier, San Diego, pp 53–57

Yang H, Matsubayashi Y, Nakamura K, Sakagami Y (1999) Oryza sativa PSK gene encodes a precursor of phytosulfokine-alpha, a sulfated peptide growth factor found in plants. Proc Natl Acad Sci USA 96:13560–13565

Yang H, Matsubayashi Y, Nakamura K, Sakagami Y (2001) Diversity of Arabidopsis genes encoding precursors for phytosulfokine, a peptide growth factor. Plant Physiol 127:842–851

Yang SL, Xie LF, Mao HZ, Puah CS, Yang WC (2003) Tapetum determinant1 is required for cell specialization in the Arabidopsis anther. Plant Cell 15:2792–2804

Reactive Oxygen Species (ROS): A Way to Stress Survival in Plants

4

Pawan Saini, Mudasir Gani, Jashan Jot Kaur, Lal Chand Godara, Charan Singh, S. S. Chauhan, Rose Mary Francies, Ajay Bhardwaj, N. Bharat Kumar, and M. K. Ghosh

Abstract

Aerobic organisms which derive their energy from the reduction of oxygen are susceptible to the damaging actions of the small amounts of O_2^-, OH, and H_2O_2 that inevitably form during the metabolism of oxygen, especially in the reduction of oxygen by the electron transfer system of mitochondria. These three species together with unstable intermediates in the peroxidation of lipids are referred to as reactive oxygen species (ROS). They are formed as a natural by-product of the normal metabolism of oxygen and have important roles in cell signaling and homeostasis. ROS are thought to play a dual role in plant biology and are accumulated by many types of stresses. Some are highly toxic and rapidly detoxified by various cellular enzymatic and nonenzymatic mechanisms, whereas many are involved in various metabolic as well as physiological processes necessary for growth and development of plants. During environmental stress of plants (e.g., UV or heat exposure), ROS levels can increase dramatically. The ROS levels that

P. Saini · M. Gani (✉) · S. S. Chauhan · N. Bharat Kumar · M. K. Ghosh
Central Silk Board (CSB), Central Sericultural Research &Training Institute (CSR&TI), Pampore, Jammu & Kashmir, India
e-mail: mudasir32@gmail.com

J. J. Kaur
Punjab Agricultural University, Ludhiana, Punjab, India

L. C. Godara
Indian Council of Agricultural Research (ICAR), Central Agroforestry Research Institute (CAFRI), Jhansi, Uttar Pradesh, India

C. Singh
Indian Council of Agricultural Research (ICAR), Indian Institute of Wheat & Barley Research (IIWBR), Karnal, Haryana, India

R. M. Francies · A. Bhardwaj
Kerala Agricultural University, Thrissur, Kerala, India

S. M. Zargar, M. Y. Zargar (eds.), *Abiotic Stress-Mediated Sensing and Signaling in Plants: An Omics Perspective*, https://doi.org/10.1007/978-981-10-7479-0_4

are too low or too high impair plant growth and development, whereas maintaining ROS levels within the right range promotes plant health. Alterations in ROS levels that are part of the normal function of the plant should not exceed the threshold boundary between redox potentials and cytotoxic or cytostatic levels. Although recent studies have unraveled some of the key issues related to ROS like programmed cell death and cross talk with phytohormones during stress conditions, yet there are some unprecedented mechanisms which need to be expolred.

Keywords
Flowering · Gibberellic acid · Reactive Oxygen Species · Signalling pathways · Stress

4.1 Introduction

Plants are continuously affected by various complex types of interactions involving numerous environmental factors. The challenges of abiotic stress on plant growth and development are evident from the continuously changing climate (Bellard et al. 2012) and the constraints to crop production exacerbated with the increasing human population competing for environmental resources (Wallace et al. 2003). Climate change is predicted to affect the agricultural production with the adverse effects of several environmental factors including extreme level of light (high and low), radiation (UV-B and UV-A), temperature (high and low – chilling, freezing), water (drought, flooding, and submergence), minerals (toxicity/deficiency), etc. (Pereira 2016). Exposure of plants to biotic (symbionts, parasites, pathogens, herbivores, and competitors) and abiotic stress (drought, heat, salinity, alkalinity, freezing, chilling, mineral toxicity and deficiency, etc.) distorts plant growth, development, and metabolism implying physiological costs (Heil and Bostock 2002; Swarbrick et al. 2006) and thus leading to a reduction in fitness and ultimately low productivity. Hence, plants have evolved specific and complex gene regulatory mechanisms at transcriptional and posttranscriptional levels including genetic and epigenetic regulation to cope up with the abiotic stress. Abiotic stresses are the primary sources of yield losses as they may reduce the crop yields by 50% (Bray et al. 2000; Wang et al. 2003). Several stresses (biotic and abiotic) are known to affect the various phenological phases of plants in various manners which ultimately affect the crop yield which is directly correlated with the flowering of crop plants and have a great impact on flowering in a wide range of plants (Blazquez et al. 2003; Yang et al. 2004); as a result of that, both delayed and early flowering have been observed (Magome et al. 2004; Martinez et al. 2004). Many types of stresses lead to the accumulation of reactive oxygen species (ROS; e.g., O_2, H_2O_2, OH) which are partially reduced or excited forms of atmospheric oxygen (Mittler 2017).

The evolution of aerobic metabolic processes such as respiration and photosynthesis unavoidably led to the production of ROS in mitochondria, chloroplast, and peroxisomes since the appearance of oxygen-evolving photosynthetic organisms about 2.2–2.7 billion years ago (Apel and Hirt 2004; Mittler et al. 2011). The ROS arise in plant cell via a number of routes. In fact most cellular compartments have the potential to become a source of ROS. Most ROS in plant cells are formed via dismutation of superoxide, which arises as a result of single-electron transfer to molecular oxygen in electron transfer chains principally during the Mehler reactions in chloroplast (Asada 1994). The main characteristic of different ROS is their ability to cause oxidative damage to DNA or RNA, oxidations of polyunsaturated fatty acids in lipids (lipid peroxidation), oxidation of amino acids in proteins, and oxidative deactivation of specific enzymes by oxidation of cofactors during the stress conditions including superoxide (O^{-}_{2}), perhydroxyl radical (HO^{-}_{2}), hydrogen peroxide (H_2O_2), hydroxyl radical (OH), alkoxyl radical (RO), peroxy radical (ROO), organic hydroperoxide (ROOH), singlet oxygen (O_2), excited carbonyl (RO), etc. which leads to cell death and aging (Bhattacharjee 2005). These cytotoxic properties of a ROS molecule explain the evolution of complex arrays of nonenzymatic and enzymatic detoxification mechanisms in plants that maintain equilibrium between production and scavenging of ROS perturbed by a number of adverse environmental factors (Apel and Hirt 2004). ROS function in cells as signaling molecules in higher plants, and different ROS act as signaling molecules that can specify the response to each particular stress (drought, heat, salinity, etc.) (Mittler et al. 2004; Wagner et al. 2004). For example, ROS are known to trigger sexual induction in *Chlamydomonas reinhardtii* and *Volvox carteri* (Nedelcu 2005). In the recent years, a new role for ROS has been identified which control and regulate biological processes such as growth, cell cycle, programmed cell death, hormone signaling, and biotic (pathogen defense mechanism) and abiotic stress (drought, heat, salinity, and high light) response and development. These studies indicate a dual role for ROS in plant biology as both toxic by-products of aerobic metabolism and key regulators of growth, development, and defense pathways (Mittler et al. 2004).

In accordance with this, the new role of ROS as signaling molecule with emphasis on cross talk between ROS and phytohormones and role of ROS in gibberellin flowering pathways of plants to escape oxidative stress via reproduction is discussed in this chapter.

4.2 Production of ROS

In plants ROS are continuously produced as by-product of various metabolic pathways localized in different cellular organelles (Foyer and Harbinson 1994). The equilibrium between production and removal of ROS may be perturbed by a number of abiotic factors such as high light, drought, low temperature, high temperature, and mechanical stress (Elstner 1991; Malan et al. 1990; Prasad et al. 1994; Tsugane et al. 1999). Plants also produce ROS by activating various oxidases and peroxidases in response to environmental changes (Schopfer et al. 2001; Bolwell et al. 2002). The rapid increase in ROS concentration is called "oxidative burst." When pathogen attacks the defense reaction, it leads to the production of ROS, primarily as superoxide and H_2O_2, that finally leads to various defense mechanisms against the pathogens (Apostol et al. 1989). During abiotic stress primarily ROS is produced in chloroplast, mitochondria, peroxisomes, and apoplast (Dietz et al. 2016; Gilory et al. 2016; Huang et al. 2016; Kerchev et al. 2016; Rodriguez-Serrano et al. 2016; Takagi et al. 2016). Stomata closure enhances the production of ROS such as O^{-}_2, and due to abiotic stress, the chloroplast CO_2 becomes a limiting factor (Asada 2006; Sarvajeet and Narendra 2010; Baniulis et al. 2013; Kleine and Leister 2016; Mignolet-Spruyt et al. 2016) and thus initiates the signaling process and balances the energy distribution between PSI and PSII (Dietzel et al. 2008; Vainonen et al. 2008; Pesaresi et al. 2009). Also in chloroplast the major sources of ROS production are the Mehler reaction and the antenna pigments (Asada and Takahashi 1987).

Mitochondrial ROS accumulation during abiotic stress is typically mediated via electron leakage from complex I to III to produce O_2^-, which can be converted to H_2O_2 by MnSOD (Huang et al. 2016). Alteration in the levels of ROS during abiotic stress produced by mitochondria can induce retrograde signaling between mitochondria and nucleus and control plant acclimation (Woodson and Chory 2008). In peroxisomes, ROS production is triggered by enzymatic activity of glycolate oxidase due to enhanced photorespiration (Foyer and Noctor 2009; Sarvajeet and Narendra 2010; Baishnab and Ralf 2012). Other important sources of ROS in plants that have received little attention are detoxification reactions catalyzed by cytochrome P_{450} in cytoplasm and endoplasmic reticulum (ER). ROS are also generated in plants at plasma membrane level or extracellularly in apoplast. Plasma membrane NADPH-dependent oxidase (NADPH oxidase) has received a lot of attention as a source of ROS for oxidative burst (Dybing et al. 1976). During abiotic stress ROS production at apoplast is mediated by four different mechanisms. The most studied of the four are the plasma membrane (PM) NADPH oxidase-RBOH (respiratory burst oxidase homolog) proteins that link calcium and ROS signaling during the stress and produce superoxide in the apoplast (Gilory et al. 2014, 2016). Peroxidation of lipids is also a potential source of ROS production in phospholipids of cell membranes of plant cells. Lipid peroxidation in plant cells can also be initiated by enzyme lipoxygenase (Winston 1990). This enzyme is capable to initiate the formation of fatty acid hydroperoxides and ensuing peroxidation.

4.3 ROS Detoxification Mechanisms

The ROS detoxification mechanism consists of the antioxidant machinery which helps to mitigate the oxidative stress-induced damages. The antioxidant machinery has two arms with the enzymatic and nonenzymatic antioxidant components. The enzymes localized in the different subcellular compartments and comprising the antioxidant machinery include superoxide dismutase (SOD), catalase (CAT), ascorbate peroxidase (APX), monodehydroascorbate reductase (MDHAR), dehydroascorbate reductase (DHAR), glutathione reductase (GR), and guaiacol peroxidase (GPX) (Table 4.1).

Table 4.1 Gene annotation and expression of the reactive oxygen species scavenging network in *Arabidopsis*

Enzyme and reaction	Gene name	Localization
Superoxide dismutase (SOD) $O_{2^-} + O_{2^-} + 2H^+ \rightarrow H_2O_2 + O_2$	*FeSOD (FSD1)*	Chloroplast
	FeSOD (FSD2)	Chloroplast
	FeSOD (FSD3)	Chloroplast
	Cu/ZnSOD (CSD1)	Cytoplasm
	Cu/ZnSOD (CSD2)	Chloroplast
	Cu/ZnSOD (CSD3)	Peroxisome
	MnSOD (MSD1)	Mitochondria
Ascorbate peroxidase (APX) $2Asc + H_2O_2 \rightarrow 2MDA + 2H_2O$	*APX1*	Cytoplasm
	APX2	Cytoplasm
	APX3	Peroxisome, chloroplast
	APX4	Chloroplast
	APX5	Peroxisome
	APX6	Cytoplasm, chloroplast, mitochondria
	APX7	Mitochondria
	Stomatal – APX	Chloroplast, mitochondria
	Thylakoid – APX	Chloroplast
Monodehydroascorbate reductase (MDAR) $MDA + NAD(P)H + H^+ \rightarrow Asc + NAD(P)^-$	*MDAR1*	Chloroplast, mitochondria
	MDAR2	Cytoplasm
	MDAR3	Cytoplasm, mitochondria
	MDAR4	Cytoplasm
	MDAR5	Cytoplasm
Dehydroascorbate reductase (DHAR) $DHA + 2GSH \rightarrow Asc + GSSG$	*DHAR1*	Chloroplast, mitochondria
	DHAR2	Cytoplasm
	DHAR3	Cytoplasm, chloroplast
	DHAR4	Cytoplasm, chloroplast
	DHAR5	Cytoplasm, chloroplast
Glutathione reductase (GR) $GSSG + NAD(P)H \rightarrow 2GSH + NAD(P)^-$	*GR1*	Cytoplasm
	GR2	Chloroplast, mitochondria
Catalase (CAT) $2H_2O_2 \rightarrow 2H_2O + O_2$	*CAT1*	Peroxisome
	CAT2	Peroxisome
	CAT3	Peroxisome

(continued)

Table 4.1 (continued)

Enzyme and reaction	Gene name	Localization
Glutathione peroxidase (GPX) H_2O_2+2GSH→$2H_2O$+ GSSG	*GPX1*	Chloroplast
	GPX2	Cytoplasm, chloroplast
	GPX3	Mitochondria
	GPX4	Cytoplasm
	GPX5	Endoplasmic reticulum
	GPX7	Chloroplast
	GPX8	Cytoplasm, chloroplast
	Phospholipid GPX6	Chloroplast, mitochondria
Ferritin Fe+P→ P-Fe	*Ferritin1*	Chloroplast
	Ferritin2	Chloroplast, mitochondria
	Ferritin3	Chloroplast, mitochondria
	Ferritin4	Chloroplast
NADPH oxidase $NADPH + e^- + O_2 \rightarrow NADP^- + O_{2^-} + H^+$ ROS *producer*	*NADPH oxidase (RbohA)*	Membrane
	NADPH oxidase (RbohB)	Membrane
	NADPH oxidase (RbohC)	Membrane
	NADPH oxidase (RbohD)	Membrane
	NADPH oxidase (RbohE)	Membrane
	NADPH oxidase (RbohF)	Membrane
	NADPH oxidase (RbohG)	Membrane
	NADPH oxidase (RbohH)	Membrane
	NADPH oxidase (RbohI)	Membrane
	NADPH oxidase (RbohJ)	Membrane
NADPH oxidase – like alternative oxidase (AOX) $2e^-+2H^++O_2$→H_2O	*AOX putative*	Mitochondria
	AOX1A	Mitochondria
	AOX1B	Mitochondria
	AOX1C	Mitochondria
	Immutans	Chloroplast

(continued)

Table 4.1 (continued)

Enzyme and reaction	Gene name	Localization
Peroxiredoxin (Prx) 2P-SH + $H_2O_2 \rightarrow$ P-S-S-P + $2H_2O$	*1-Cys Prx*	Nucleus
	2-Cys Prx A	Chloroplast
	2-Cys Prx B	Chloroplast
	2-Cys Prx F	Mitochondria
	Prx Q	Chloroplast
	Type 2 Prx – related	Cytoplasm
	Type 2 Prx A	Membrane, chloroplast
	Type 2 Prx B	Cytoplasm
	Type 2 Prx C	Cytoplasm
	Type 2 Prx D	Cytoplasm
	Type 2 Prx E	Chloroplast, mitochondria

Source: Mittler et al. (2004)

4.3.1 Superoxide Dismutase (SOD)

SOD forms the first line of defense against ROS-induced damages. The SOD catalyzes the removal of $O^{\bullet -}_2$ by dismutating it into O_2 and H_2O_2. This removes the possibility of $OH^{\bullet}$ formation by the Haber-Weiss reaction. SODs are classified into three isozymes based on the metal ion it binds, MnSOD (localized in mitochondria), FeSOD (localized in chloroplasts), and Cu/ZnSOD (localized in cytosol, peroxisomes, and chloroplasts) (Mittler 2002). SOD has been found to be upregulated by abiotic stress conditions (Boguszewska et al. 2010).

4.3.2 Catalase

Catalase is a tetrameric heme-containing enzyme responsible for catalyzing the dismutation of H_2O_2 into H_2O and O_2. It has high affinity for H_2O_2 but lesser specificity for organic peroxides (ROOR). Peroxisomes are the hotspots of H_2O_2 production due to β-oxidation of fatty acids, photorespiration, purine catabolism, and oxidative stress (Mittler 2002). Angiosperms have been reported to have three *CAT* genes. *CAT1* is expressed in pollens and seeds (localized in peroxisomes and cytosol), *CAT2* is predominantly expressed in photosynthetic tissues but also in roots and seeds (localized in peroxisomes and cytosol), and finally *CAT3* is found to be expressed in leaves and vascular tissues (localized in the mitochondria). Stressful conditions demand greater energy generation and expenditure of the cell. This is fulfilled by increased catabolism which generates H_2O_2. CAT removes the H_2O_2 in an energy-efficient way (Das and Roychoudhary 2014).

4.3.3 Ascorbate Peroxidase (APX)

APX is an integral component of the ascorbate-glutathione (ASC-GSH) cycle. While CAT predominantly scavenges H_2O_2 in the peroxisomes, APX performs the same function in the cytosol and the chloroplast. The APX reduces H_2O_2 to H_2O and DHA, using ascorbic acid (AA) as a reducing agent. The APX family comprises of five isoforms based on different amino acids and locations, viz., cytosolic, mitochondrial, peroxisomal, and chloroplastid (stromal and thylakoidal) (Sharma and Dubey 2004).

4.3.4 Monodehydroascorbate Reductase (MDHAR)

MDHAR is responsible for regenerating ascorbic acid (AA) from the short-lived MDHA using NADPH as a reducing agent ultimately replenishing the cellular AA pool. Since it regenerates AA, it is co-localized with the APX in the peroxisomes and mitochondria where APX scavenges H_2O_2 and oxidizes AA in the process (Mittler 2002). MDHAR has several isozymes which are confined in chloroplast, mitochondria, peroxisomes, cytosol, and glyoxysomes.

4.3.5 Dehydroascorbate Reductase (DHAR)

DHAR reduces dehydroascorbate (DHA) to ascorbic acid (AA) using reduced glutathione (GSH) as an electron donor (Eltayeb et al. 2007). This makes it another agent apart from MDHAR which regenerates the cellular AA pool. It is critical in regulating the AA pool size in both symplast and apoplast thus maintaining the redox state of the plant cell (Chen and Gallie 2006). DHAR is found abundantly in seeds, roots, and both green and etiolated shoots.

4.3.6 Glutathione Reductase (GR)

GR is a flavoprotein oxidoreductase which uses NADPH as a reductant to reduce GSSG to GSH. Reduced glutathione (GSH) is used up to regenerate ascorbic acid (AA) from MDHA and DHA and as a result is converted to its oxidized form (GSSG). GR, a crucial enzyme of ASC-GSH cycle, catalyzes the formation of a disulfide bond in glutathione disulfide to maintain a high cellular GSH/GSSG ratio (Ghisla and Massey 1989). It is predominantly found in chloroplasts with small amounts occurring in the mitochondria and cytosol. GSH is a low molecular weight compound which plays the role of a reductant to prevent thiol groups from getting oxidized and react with detrimental ROS members like O_2 and OH (Edwards et al. 1990).

4.3.7 Glutathione Peroxidase (GPX)

GPX is a heme-containing enzyme composed of 40–50 kDa monomers which eliminates excess H_2O_2 both during normal metabolism and during stress. It plays a vital role in the biosynthesis of lignin as well as defends against biotic stress by degrading indole acetic acid (IAA) and utilizing H_2O_2 in the process. GPX prefers aromatic compounds like guaiacol and pyrogallol (Asada 1999) as electron donors. Since GPX is active intracellularly (cytosol, vacuole) in the cell wall and extracellularly, it is considered as the key enzyme in the removal of H_2O_2.

These enzymes all together with antioxidants detoxify O_2^- and H_2O_2. The balance between the SOD and the different peroxide-scavenging enzymes determines the level of O_2^- and H_2O_2 and prevents the formation of highly toxic HO· radical via Haber-Weiss reaction/Fenton reaction (Asada and Takahashi 1987; Bowler et al. 1991). Enzymes like monodehydroascorbate reductase and glutaredoxins are capable to oxidize the antioxidants like ascorbic acid and glutathione which are in reduced state by using NAD(P)H, and dehydroascorbate reductase and glutaredoxins are capable of reducing dehydroascorbic acid in plants with varying levels of efficiency (Chew et al. 2003).

4.4 ROS Signaling and Cross Talk Between ROS and Phytohormones

ROS are thought to be the unavoidable toxic by-products of aerobic metabolism but also function as signaling molecules. They are versatile in nature with different properties like specificity, reactivity and aid in transmission of signals across the membranes (Mittler 2017). The continual process of ROS production (metabolically and signaling) and detoxification is controlled by the ROS gene network in all the cellular compartments (Mittler et al. 2004).

As a signaling molecule, ROS offers several possible advantages which include the capacity of the cell to rapidly produce and scavenge different forms of ROS in a simultaneous manner, enabling rapid and dynamic changes in ROS levels and tight control over the subcellular localization of ROS signals in cells. The ROS signals are auto-propagating in nature and transferred throughout over a long distance in plants such as in *Arabidopsis thaliana* wherein signal propagates at a rate of 8.4 cm/min (Miller et al. 2009). Another key signaling advantage of ROS is their tight link to cellular homeostasis and metabolism. Almost any change in cellular homeostasis could lead to a change in the steady-state level of ROS in a particular compartment. Physiological conditions that favor photorespiration would enhance production of ROS in peroxisomes (Vanderauwera et al. 2011). It is easy to envision how a tight link between metabolism and ROS levels would make ROS good signals to monitor changes in cellular metabolism. Different biological processes among the different organisms generate ROS at different levels. ROS is very much helpful in assessing the toxic level of atmospheric oxygen and for the monitoring of various metabolic reactions as ROS signaling is highly integrated with the different signaling networks in plants which include kinases, calcium, and hormonal and cellular metabolic redox responses (Mittler et al. 2011).

The gibberellic acid (GA) increases or decreases the metabolism of plants in the presence of efflux favoring microbes which results in the production of antimicrobial compounds with nutritional value. A different GA concentration (mgl^{-1}) leads to flower bud formation and growth in Aprikola rose. In incompatible plant-pathogen interactions, H_2O_2 has been implicated in the elicitation of variety of defense responses. Plants undergo distinctive phase changes of juvenile to adult and adult to reproduction to ensure successful reproduction and better fitness to survive. Under normal conditions, environmental or external cues such as light (light intensity and duration of exposure) and low temperature are key factors in determining when plants flower (Srikanth and Schmidt 2011). The endogenous cues such as plant age, carbohydrate assimilates (mainly sucrose), and hormones (mainly gibberellic acid) coordinate with external cues to determine flowering time. Antimicrobial approaches primarily utilizing ROS comprise both bactericidal antibiotics and non-pharmacological methods such as photodynamic therapy, titanium dioxide photocatalysis, cold plasma, and medicinal honey.

Among kinase networks mitogen-activated protein kinase (MAPK) cascade is a very good example. MAPK cascade comprises more than 100 MAPK, MAPKK, and MAPKKK genes in rice and *Arabidopsis* (Hwang et al. 2002). MAPK cascades can be activated following ROS accumulation in the cell. Various abiotic stresses induce ROS production that in turn activates MAPK signaling cascades. NADPH oxidase is another example of ROS production that is accompanied by various signaling pathways and involved in calcium signaling and photophosphorylation (Kobayashi et al. 2007; Ogasawara et al. 2008; Demidchik et al. 2009).

4.5 Integration of ROS with Phytohormones

ROS signaling is integrated with hormonal signaling networks allowing plants to regulate growth and developmental process and adaptability against the environment.

The presence of an auxin biosynthesis, signaling, and transport apparatus in single-celled green algae is clear evidence of the evolutionary role played by auxin during the adaptation of plants to diverse land environments (De Smet et al. 2010). Interestingly, there is growing evidence that IAA (indole acetic acid) plays an integral part in plant adaptation to salinity stress (Fahad et al. 2015; Iqbal et al. 2014). It increases root and shoot growth of plants growing under salinity or heavy metal stresses (Sheng and Xia 2006; Egamberdieva 2009). Salinity reduced IAA levels in maize plants, but salicylic acid application effectively increased them (Fahad and Bano 2012), indicating that hormonal balance and cross talk are critical to signal perception, transduction, and mediation of stress response (Fahad et al. 2015). Auxin stimulates the transcription of a large number of genes called primary auxin response genes, and these genes have been identified and characterized in several plant species including rice, *Arabidopsis*, and soybean (Javid et al. 2011). The integration of ROS with auxin signaling networks triggered by environmental factors is known as the stress-induced morphogenetic response. In this response ROS and auxin metabolism interfere and lead to morphological changes that help to avoid the deleterious effects of environmental stress (Potters et al. 2009; Tognetti et al. 2010). Thus auxin is regarded as an influential constituent of defense responses via regulation of numerous genes and mediation of cross talk against abiotic stresses.

Abscisic acid (ABA) plays an important role during the seed development and dormancy, stomatal opening and closing, embryo morphogenesis, and synthesis of storage proteins and lipids in crop plants (Sreenivasulu et al. 2010). In response to environmental stresses, endogenous ABA levels increase rapidly, activating specific signaling pathways and modifying gene expression levels (O'Brien and Benkova 2013).Under water deficit condition, ABA concentration increased and water deficit is sensed by roots which begin to synthesize ABA within 1 h of the onset of water stress in maize (Giuliani et al. 2005). Abscisic acid (ABA) induced stomata closure as guard cell synthesizes ROS and accumulates in response to the water stress (Lee et al. 1999).

Polyamines accumulate in response to several abiotic stresses including drought, salinity, heat, chilling, heavy metal toxicity, etc. and play a protective role. Available evidence shows that ABA activates transcription of genes for polyamine biosynthesis, and in turn the polyamine putrescine activates ABA biosynthesis. Further polyamines promote the production of NO (nitrous oxide) and H_2O_2 a reactive oxygen species (ROS). The polyamines, NO, and ROS act together synergistically to promote ABA-induced stomatal closure (Alcazar et al. 2010).

Similar to ROS, salicylic acid (SA) involved in defense and cell death responses (Volt et al. 2009) and increased ROS concentration can cause SA accumulation in plants (Chamnongpol et al. 1998). Thus, enhanced level in cellular compartment (chloroplast) results in stomatal closure.

In seeds ROS play a key role in various events such as maturation, ripening, aging, and germination (Bailly et al. 1996, 2008; Lehner et al. 2006; Muller et al. 2009a, b). It had been reported that NADPH oxidases which produce superoxide anion (a type of ROS) regulate germination of barley seeds Ishibashi et al. 2010). H_2O_2 also accelerates the germination of pea (*Pisum sativum*) seeds and stimulates the early growth of seedlings (Barba-Espin et al. 2010). Production of H_2O_2 during the early imbibition period has been demonstrated in seeds of soybean (Puntarulo et al. 1988), maize (Hite et al. 1999), wheat (Caliskan and Cuming 1998), and zinnia (Ogawa and Iwabuchi 2001). ROS produced after imbibition are assumed to play a role in seed germination. There was a strong relationship between sunflower seed dormancy alleviation and accumulation of ROS and peroxidation products in cells of embryonic axes (Oracz et al. 2007, 2009). ROS regulates seed dormancy and germination of barley (Bahin et al. 2011) and *Arabidopsis* (Liu et al. 2010). In addition to this, H_2O_2 seems to function in GA and ABA signaling and regulates *α-amylase* production in aleurone cells. H_2O_2 repressed the *GAMYB* mRNA by PKABA and consequently promoted the production of *α-amylase* mRNA, thus suggesting that the H_2O_2 generated by GA in aleurone cells is a signal molecule that antagonizes ABA signaling (Ishibashi et al. 2012).

The gibberellins (GAs) are a large group of tetracyclic diterpenoid carboxylic acids, but only a few of them function as growth hormones in higher plants, predominant forms being GA_1 and GA_4 (Sponsel and Hedden 2004). The GAs show positive effects on seed germination, leaf expansion, stem elongation, flower and trichome initiation, and flower and fruit development (Yamaguchi 2008). They are essential for plants throughout their life cycle for growth-stimulatory functions. They also promote developmental phase transitions (Colebrook et al. 2014). Interestingly, there is increasing evidence for their vital roles in abiotic stress response and adaptation (Colebrook et al. 2014). Recently, experiments have been performed to investigate the role of GAs in osmotic stress response in *Arabidopsis thaliana* seedlings (Skirycz et al. 2011; Claeys et al. 2012). GAs are known to interact with all other phytohormones in numerous developmental and stimulus-response processes.

Gibberellin (GA) signaling is linked with the ROS production leading to destruction of DELLA proteins that regulate transcript levels of antioxidants which are directly related with growth and development (Achard et al. 2008b). Few GA receptor proteins interact and induce flowering by degrading certain F-box protein via ubiquitinase pathways in *Arabidopsis*. The presence of salt decreases the GA thus reducing flowering. The growth restrained by certain DELLA mutants is beneficial in certain long-day conditions for flowering and is also good in tolerating the long-time stressed environmental conditions that ultimately promote growth of the plants for further flowering. DELLAs must, therefore, also function as transcriptional activators, perhaps in partnership with transcription factors or by inactivating transcriptional repressors (Mutasa-Gottgenes 2009). Early responsive genes (also known as early DELLA responsive genes) encode the GA biosynthesis, GA receptors, and putative transcription factors (Eckardt 2007) (Fig. 4.1).

Commonly GAs function as growth hormones in higher plants, but forms of GA (GA1 and GA4) modulate the *FLC* and *SOC1* thus promoting flowering as reported in *Arabidopsis*. It controls flowering by modifying the time of day during which various floral integrators are expressed further initiating flowering under long-day conditions. But GA functions more critically under short-day conditions as *ga* mutants produced by various biosynthetic pathways and signaling pathways thus preventing the plant to flower. Sun (2008) suggested that there is a relationship between stress and delayed flowering in plants and reported that salt-induced flowering (delayed) can be restored with the exogenous application of GA.

Molecular genetic studies in addition to unique physiological approaches will be required to ascertain the position of ROS in the signal transduction pathways and also to understand how these short-lived endogenous signaling compounds are perceived and transduced to specific and nonspecific responses necessary for survival of plants. This will ultimately help us to screen better-performing plants under environmental stress for breeding program.

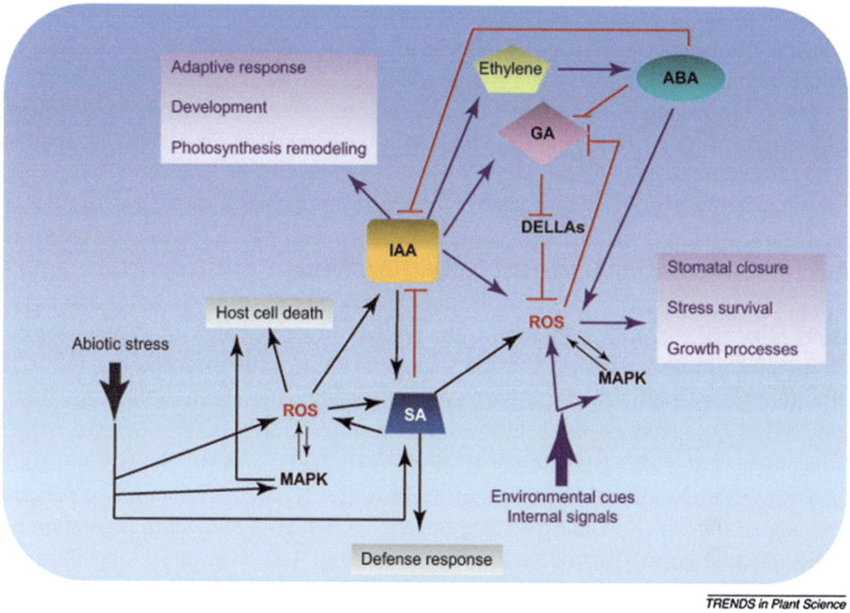

Fig. 4.1 ROS cross talk in relation to other hormone signals (Source: Mittler et al. 2011, Trends in Plant Science)

4.6 Role of Reactive Oxygen Species (ROS) in Stress Survival by Targeting Gibberellic Acid (GA) Promoting Flowering Pathway Mediated Through DELLA Proteins

Plant integrates multiple environmental signals during the growth regulation (Achard et al. 2006). Plant growth and development is regulated and coordinated through the action of several classes of plant hormones (phytohormones) which may act either close to or remote from their sites of synthesis to mediate genetically programmed developmental changes or responses to environmental stimuli (Davies 2010). Thus hormones have an important role in plant responses to abiotic stress such as drought, shading, flooding, or temperature from which the plant may attempt to escape by outgrowing the stress or more commonly may result in reduced growth in order that the plant can focus its resources on withstanding the stress (Skirycz and Inze 2010). When to flower is a key decision for the plant, affecting the adaptability of species to any given environment (Conti 2017). Early germinated clone plants of *Zantedeschia aethiopica* (Green Goddess) grew faster and flowered earlier at low temperatures followed by multiple shoot formation (Ngamau 2006). The role of HSP (heat shock proteins – required for thermo-tolerance) in flowering time is uncertain as its underlying mechanism in controlling flowering is unknown. The HSP's threshold level expression could be comparable with the strength of the stress prevailing in the environment. Salt-induced delayed flowering is due to suppressed

DELLA mutants because the transcript levels of few floral integrators are not reduced as the vegetative phase is lengthened and the transcription phase is reduced by the unknown mechanisms of DELLA (Park et al. 2016). Phytohormones regulate diverse plant processes including biotic and abiotic stress by modifying physiology and biochemistry of flowering time and plant hormones (Davis 2009; Diezel et al. 2011; Wasternack et al. 2013).

4.7 Gibberellic Acid and Flowering Time

Gibberellic acid (GA) is required for the normal growth of almost all plant organs through the promotion of cell division and or cell elongation. In addition to this, GA promotes transition from vegetative phase to floral induction both in short day (SD), long day (LD), and continuous light conditions in *Arabidopsis* (Wilson et al. 1992; Silverstone et al. 1997; Blazquez et al. 1998; Griffith et al. 2006). The GA signaling pathway comprises the biosynthesis of the bioactive forms of GA and their perception, signal transduction, and inactivation, each of which is subject to regulation by environmental signals including abiotic stress (Fig. 4.2). GAs are biosynthesized from trans-geranylgeranyl diphosphate (GGDP) formed in plastids (Kasahara et al. 2002) and catalyzed by six enzymes, of which *2-oxoglutarate-dependent dioxygenases*, *GA 20-oxidase (GA20ox)* and *GA3-oxidase (GA3ox)*, catalyze the final steps (reviewed by Yamaguschi 2008; Mutasa-Gottgens and Hedden 2009; Hedden and Thomas 2012). A third class of dioxygenase, *GA2-oxidase (GA2ox)*, produces inactive products and functions to enable GA turnover. Expression of certain paralogues within the *GA20ox*, *GA30x*, and *GA20x* gene families is regulated by GA action; the biosynthesis genes are being downregulated, while *GA20x* expression is upregulated (Thomas et al. 1999; Weston et al. 2008; O'Neill et al. 2010). Thus the homeostasis of GA depends on GA deactivation step.

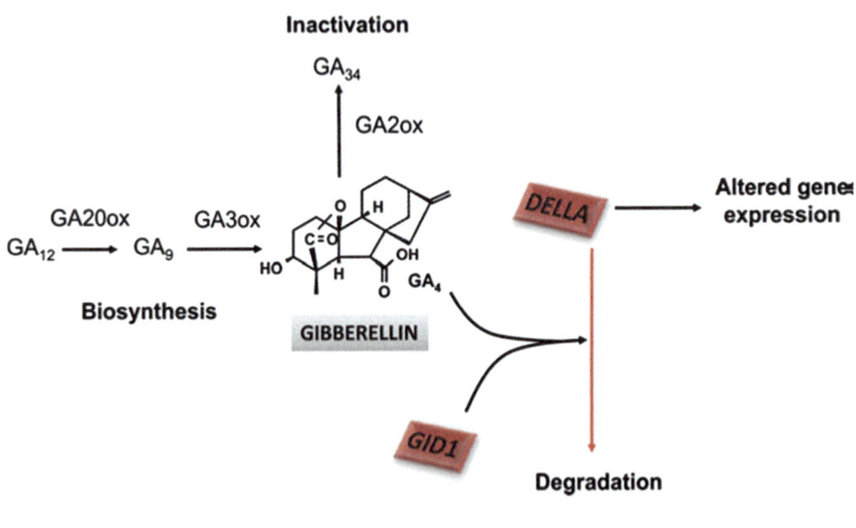

Fig. 4.2 Gibberellin signaling pathway (Source: Mutasa-Gottegens and Hedden 2009)

Flowering is a critical life-history trait that ensures seed production required for the survival of species (Kazan and Lyons 2016). The transition from vegetative to reproductive phase requires genetic and epigenetic reprogramming and reallocation of metabolic and biochemical resources throughout the plant (Andres and Coupland 2012; Blumel et al. 2014). Flowering is regulated by an elaborate network of genetic pathways responsive to endogenous and environmental stimuli. In *Arabidopsis thaliana*, the floral transition is controlled by complex genetic networks (Moon et al. 2005), and extensive molecular genetic analyses have elucidated five flowering pathways, including photoperiod, autonomous, vernalization, gibberellic acid (GA), and age pathways (Michaels 2009; Amasino 2010; Srikanth and Schmid 2011). The photoperiod and vernalization pathways respond to environmental signals such as light and temperature, whereas the autonomous and gibberellic acid (GA)-dependent pathways monitor the endogenous development state of the plant.

The autonomous pathways promote flowering independent of environmental conditions by repressing the *FLOWERING LOCUS (FLC)* gene that acts as a repressor of flowering (Michaels and Amasino 1999; Michaels and Amasino 2001; Sheldon et al. 1999). Vernalization, a long exposure to low temperature, also promotes flowering by repressing *FLC*. Therefore, *FLC* is a convergence point for autonomous and vernalization pathways. The genes *CONSTANS (CO)*, *GIGANTEA (GI)*, and *FT* are involved in the long-day (photoperiod) pathways, and mutations in these genes delay flowering under long day but not under short days (Koornneef et al. 1991, 1998).

Gibberellin (GA) pathways converge on a few floral integrator genes that promote flowering. *FLOWERING LOCUS (FT)*, a key floral integrator gene, encodes a component of the mobile signal "florigen" that activates floral meristem identity gene. During the long days (LD), FT is upregulated by the transcription factor *CONSTANS (CO)*. The FT protein travels from leaf to the meristem identity genes such as *LEAFY (LFY)* and *APETLLA1 (AP1)*, which then change the fate of the shoot apical meristem (SAM) from vegetative to floral. The *FLOWERING LOCUS (FLC)* gene antagonizes the GA and photoperiod pathways by repressing floral promoters such as *FT* and *SUPPRESSOR OF CONSTANS1 (SOC1)* (Kazan and Lyons 2016) (Fig. 4.3).

The GA signaling pathway has now been unraveled in *Arabidopsis* (Achard and Genschik 2009) and rice (Dai et al. 2007). The key regulators of the GA signaling pathways are the nuclear-localized growth-repressing DELLA proteins (Peng et al. 1997; Silverstone et al. 1998; Dill and Sun 2001; King et al. 2001). Rice has only one DELLA protein SLENDER RICE1 (SLR1), while *Arabidopsis* genome encodes five DELLAs, namely, GA-INSENSITIVE (GAI), REPRESSOR OF GA1-3 (RGA), RGA-LIKE1 (RGL1), RGL2, and RGL3 (Peng et al. 1997; Ikeda et al. 2001; Silverstone et al. 2001; Lee et al. 2002; Wen and Chang 2002). RGA and GAI repress stem elongation (Dill and Sun 2001; King et al. 2001), RGL2 inhibits seed germination (Lee et al. 2002); RGA, RGL1, and RGL2 together regulate floral development (Cheng et al. 2004; Tayler et al. 2004; Yu et al. 2004); and RGL3 improves the freezing tolerance (Achard et al. 2008a).

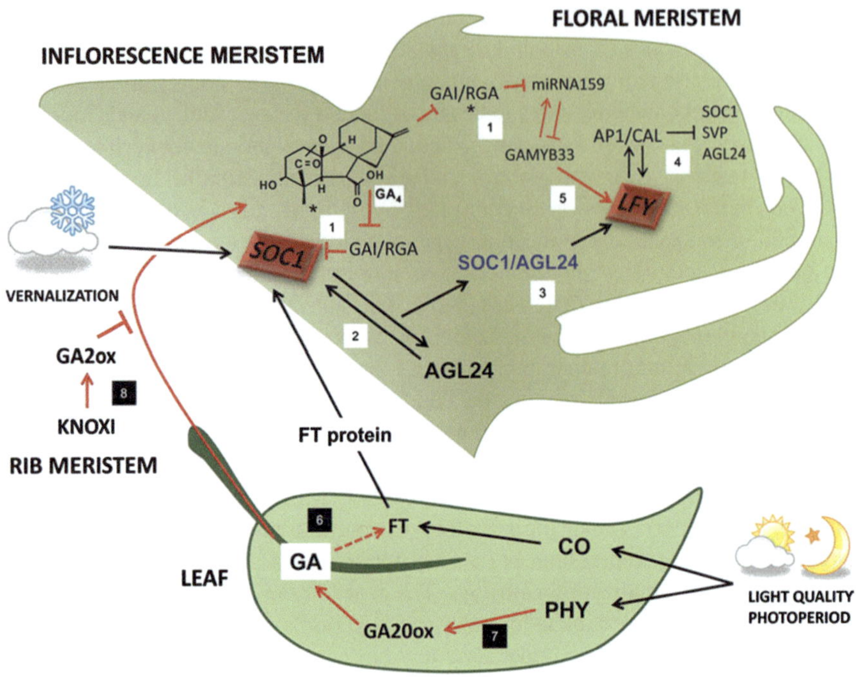

Fig. 4.3 GA-induced floral transition in *Arabidopsis* (Source: Mutasa-Gottegens and Hedden 2009)

DELLA restrains the plant growth, whereas GA promotes the growth by degradation of DELLA through binding of GA to soluble nuclear-localized receptors known as GID1 (GA-INSENSITIVE DWARF 1) (Ugeuchi-Tanaka et al. 2005). The binding of bioactive GA to GID1 promotes an interaction between GID1 and DELLA-domain DELLA protein and leads to rapid degradation of DELLA protein via the ubiquitin-proteasome pathway (Griffith et al. 2006; Ugeuchi-Tanaka et al. 2007; Willage et al. 2007).

The transition to reproductive development in rosette plant is frequently signified by bolting of the main axes, a process that involves GA-dependent cell division and elongation (Silverstone and Sun 2000) and has been directly correlated with bioactive GA in the shoot apex. GA also promotes flowering through the activation of genes encoding the floral integrator *SOC1*, *LEAFY (LFY)*, and *FT* (Mutasa-Gottgenes and Hedden 2009). Gibberellins promote expression of both *SOC1* (Bonhomme et al. 2000; Moon et al. 2003) and *LFY* (Blazquez et al. 1998) by independent DELLA (GAI/RGA)-mediated pathways directly or indirectly via *GAMYB* (Gocal et al. 1999, 2001) which modulate the expression of *SOC1* and *LFY*. Gibberellins therefore have an additional indirect route for upregulating *LFY* via *SOC1* and probably also control levels of *LFY* through the DELLA-dependent regulation of miRNA159 which in *Arabidopsis* negatively regulates MYB33 required for *LFY* transcription (Achard et al. 2004). The miRNA-regulated pathways and GA relate for the flowering time, and the initiation of stress-responsive floral integrators by binding to nuclear receptors further degrades the transcriptional regulator DELLAs (Han et al. 2013).

When bioactive GA levels are low, DELLA is relatively stable. By contrast, when bioactive GA level is high, DELLA proteins are ubiquitylated and rapidly degraded. Thus, DELLA protein abundance is inversely related to the amount in bioactive GA (Achard and Genschik 2009). GA function via the regulation of DELLA proteins negatively regulates GA signaling. GA signaling mediates stress tolerance through the control of cellular redox homeostasis. DELLA proteins appear to regulate ROS levels by controlling the expression of antioxidant genes. DELLA-mediated enhancement of ROS-scavenging capacity promotes survival under abiotic stress (Achard et al. 2008b).

Stress-regulated flowering is not formally recognized as floral transition pathway per se as the floral delay conditions with restrained growth provided an advantage over an escape for stress tolerance. In case of stress due to various biotic and abiotic factors, the florigens are the common threads for enactment of different flowering pathways (Robini et al. 2014).

4.8 Oxidative Stress Tolerance to Stress Survival in Plants

Environmental factors (temperature, humidity, light, water, CO_2, etc.) are responsible for the production of ROS in the cellular compartments. The balance between the production and the scavenging of the ROS is necessary for the maintenance of equilibrium and when the plants are exposed to environmental factors disturbs the equilibrium and produces the reactive oxygen species (ROS). The tolerance of plants to oxidative stress depends on the overall balance between production of ROS and antioxidative capability of the cell (Mittler 2002). A large number of enzymatic (SOD, CAT, APX, MDHAR, DHAR, GR AND GPX) as well as nonenzymatic (ascorbic acid, reduced glutathione (GSH), carotenoids, flavonoids, proline) scavenging/detoxification mechanisms are available in plants that efficiently scavenge ROS and avoid the oxidative stress (Dirk and Marc 1995). Thus, ROS plays an important and diverse role in the plant growth and development that are associated with the various physiological processes which are linked with the reproductive fitness.

Among all the physiological process, aging is most complex and least understood (Harman 1981). According to the free-radical theory, genetic as well as environmental factors are the basic cause of aging (Harman 1956). Extended longevity is related with stress tolerance, and it has been reported in *Caenorhabditis elegans* that mutants are resistant to oxidative stress and other environmental factors with the degree of tolerance being correlated to the extension of life span (Martin et al. 1996; Murakami and Johanson 1996). Thus, late flowering mutants can be considered as extended longevity mutants.

The phase transition from vegetative to reproductive phase is complex and regulated by the network of genes (Coupland 1995). The gigantea (gi-3) mutant which is responsible for delayed flowering in Arabidopsis has been found to be tolerant to paraquat-induced oxidative stress as the mutations reducing the growth rate can also delay flowering. A significant correlation between flowering time indices and paraquat tolerance has been observed in *gi* allele series. The *gi-3* mutant was more tolerant to H_2O_2. It is suggested that the link between longevity and the oxidative stress is not only limited to animal systems. Also, the sensitivity to oxidative stress together with the tolerance of flowering time/delayed leaf senescence mutants should lead to a better understanding of interactions between stress tolerance and aging (Kurepa et al. 1998).

The gi mutants have extremely pleiotropic effects on flowering time, phytochrome B signaling, circadian function, and starch metabolism (Koornneef et al. 1991; Emiert et al. 1995; Kurepa et al. 1998; Fowler et al. 1999; Park et al. 1999; Huq et al. 2000). In another study it shows that loss of function mutations in GI gene enhanced the oxidative stress tolerance. The oxidative tolerance to paraquat of *gi-3* is associated with the constitutive activation of SOD and APX genes (Cao et al. 2006).

When stressed, a sessile organism with a finite life cycle has few options which include ignoring the stress and maintaining vegetative phase, delaying reproductive phase, alleviating the stress through tolerance mechanisms, or escaping the stress via reproduction. *OXIDATIVE STRESS 2 (OXS2)* which is a zinc-finger transcription factor has been characterized which is cytoplasmic (for vegetative growth) until nuclear (stress tolerance) translocation is induced by stress. If the cytoplasmic form of *OXS2* accumulates more in concentration, then it leads to late flowering; on the other hand, if nuclear form accumulates in more concentration, then it favors early flowering. Null mutants of *OXS2* showed that the loss in function favors early flowering indicating that *OXS2* is needed to delay floral transition during non-stress conditions. The late flowering phenotype where cytoplasmic *OXS2* accumulates results in lower expression of *FT*, *FUL*, *SOC1*, and *LFY* floral integrator genes, whereas when *OXS2* accumulates in the nucleus, enhanced expression was observed for *SOC1*, *LFY*, *FUL*, and *AP1* which produces early flowering phenotype (Blanvillain et al. 2011) (Fig. 4.4).

O2L1 is similar to *OXS2*, cytoplasmic in the absence of stress but nuclear during stress. In contrast, *O2L2*, *O2L3*, and *O2L4* did not show stress-induced translocation. From mutant combination analysis, it is clear that *O2L1* is necessary for normal flowering. In this way, *OXS2* transcription factor targets floral integrator genes and bypass the oxidative stress in plants.

4.9 Conclusion

ROS primarily function in signal transduction and regulate various pathways during the plant acclimatization to stress. They initiate the cascade of kinase and further trigger the activation of transcription factors like zinc-finger-based *OXS2* which facilitate plants to maintain a basal level of ROS in the cellular compartments like chloroplast and mitochondria and bypass the stress which results in the survival of

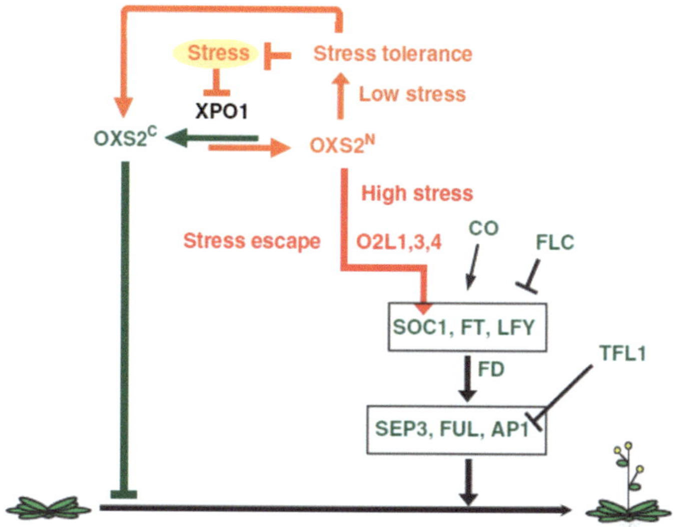

Fig. 4.4 Model of *OXS2* regulation of stress tolerance or stress escape (Source: Blanvillain et al. 2011)

plants under stress conditions. ROS supporting basic cellular processes, viability, and oxidative stress is only an outcome of a deliberate activation of a physiological cell death pathway. ROS are good but too much or too little can be detrimental as such; maintaining a basal level of ROS in cells is essential for life. The autoactivation of few stress-inducible flowering genes may be a commensurate response to form a pre-activation strategy to tolerate the stress effects. The stress-inducible changes due to few floral integrator genes (repressors/promoters) all along the life cycle of the plants are the key factors in the floral architecture of the plants. In higher plants different ROS act as signaling molecules that can specify the response to each particular stress (Mahalingam and Fedoroff 2003; Mittler et al. 2004; Wagner et al. 2004). For instance, H2O2 induces a mitogen-activated protein kinase cascade (Kovtun et al. 2000) and also serves as an intermediate in abscisic acid (ABA)-induced stomatal closure (Zhang et al. 2001) with varying degrees of temperatures and day lengths. Few nutrients like nitrates also promote flowering during normal and short days. Even the low nitrates can stimulate flowering. The GA biosynthesis gene promotes flowering in rice under the promoter of GA biosynthesis gene that induces flowering and also grain development. Given that biosynthetic pathways and convergence points for cross talk are still not clear, there is further scope to increase our understanding in this regard and to identify novel genes encoding phytohormone metabolisms to be targeted for engineering abiotic stress tolerance in crop plants. Recent findings have opened various avenues for targeting phytohormones for genetic engineering to confer abiotic stress tolerance in important crop species.

References

Achard P, Genschik P (2009) Releasing the brakes of plant growth: how GAs shutdown DELLA proteins. J Exp Bot 60(4):1085–1092

Achard P, Herr A, Baulcombe DC, Harberd NP (2004) Modulation of floral development by a gibberellin-regulated microRNA. Development 131:3357–3365

Achard P, Cheng H, De Grauwe L, Decat J, Schoutteten H, Moritz T, Van Der Straeten D, Peng J, Harberd NP (2006) Integration of plant responses to environmentally activated phytohormonal signals. Science 311:91–94

Achard P, Gong F, Cheminant S, Alioua M, Hedden P, Genschik P (2008a) The cold-inducible CBF1 factor-dependent signaling pathway modulates the accumulation of the growth-repressing DELLA proteins via its effect on gibberellin metabolism. Plant Cell 20:2117–2129

Achard P, Renou JP, Berthome R, Harberd NP, Genschik P (2008b) Plant DELLAs restrain growth and promote survival of adversity by reducing the levels of reactive oxygen species. Curr Biol 18:656–660

Alcazar R et al (2010) Polyamines: molecules with regulatory functions in plant abiotic stress tolerance. Planta 231:1237–1249

Amasino R (2010) Seasonal and developmental timing of flowering. Plant J 61:1001–1013

Andrés F, Coupland G (2012) The genetic basis of flowering responses to seasonal cues. Nat Rev Genet 13:627–639

Apel K, Hirt H (2004) Reactive oxygen species: metabolism, oxidative stress and signal transduction. Annu Rev Plant Biol 55:373–399

Apostol I, Heinstein PF, Low PS (1989) Rapid stimulation of an oxidative brust during elicidation of cultured plant cells. Role in defense and signal transduction. Plant Physiol 90:106–116

Asada K (1994) Production and action of active oxygen in photosynthetic tissue. In: Foyer CH, Mullineaux PM (eds) Causes of photooxidative stress and amelioration of defence systems in plants. CRC Press, Boca Raton, pp 77–104

Asada K (1999) The water-water cycle in chloroplast: scavenging of active oxygens and dissipation of excess photons. Annu Rev Plant Biol 50:601–639

Asada K (2006) Production and scavenging of reactive oxygen species in chloroplast and their functions. Plant Physiol 141:1621–1633

Asada K, Takahashi M (1987) Production and scavenging of active oxygen in photosynthesis. In: Kyle DJ et al (ed) Photoinhibition (Topics in photosynthesis). 9:227–287

Bahin E, Bailly C, Sotta B, Kranner I, Corbineau F, Leymarie J (2011) Crosstalk between reactive oxygen species and hormonal signaling pathways regulates grain dormancy in barley. Plant Cell Environ 34:980–993

Bailly C, Benamar A, Corbineau F, Co^me D (1996) Changes in superoxide dismutase, catalase and glutathione reductase activities as related to seed deterioration during accelerated aging of sunflower seeds. Physiol Plant 97:104–110

Bailly C, El-Maarouf-Bouteau H, Corbineau F (2008) From intracellular signaling networks to cell death: the dual role of reactive oxygen species in seed physiology. C R Biol 331:806–814

Baishnab CT, Ralf O (2012) Reactive oxygen species generation and signaling in plants. Plant Signal Behav 7:1621–1633

Baniulis D, Hasan SS, Stofleth JT, Cramer WA (2013) Mechanism of enhanced superoxide production in the cytochrome b(6)f complex of oxygenic photosynthesis. Biochemist 52:8975–8983

Barba-Espin G, Diaz-Vivancos P, Clemente-Moreno MJ, Albacete A, Faize L, Faize M, Perez-Alfocea F, Hernandez JA (2010) Interaction between hydrogen peroxide and plant hormones during germination and the early growth of pea seedlings. Plant Cell Environ 33:981–994

Bellard C, Bertelsmeier C, Leadley P, Thuiller W, Courchamp F (2012) Impacts of climate change on the future of biodiversity. Ecol Lett 15:365–377. https://doi.org/10.1111/j.1461-0248.2011.01736.x

Bhattacharjee S (2005) Reactive oxygen species and oxidative brust: roles in stress, senescence and signal transduction in plants. Curr Sci 89(7):1112–1121

Blanvillain R, Wei S, Wei P, Kim JH, Ow DW (2011) Stress tolerance to stress escape in plants: role of the *OXS2* zinc-finger transcription factor family. EMBO J 30:3812–3822
Blázquez MA, Green R, Nilsson O, Sussman MR, Weigel D (1998) Gibberellins promote flowering of *Arabidopsis* by activating the *LEAFY* promoter. Plant Cell 10:791–800
Blazquez MA, Ahn HH, Weigel D (2003) A thermosensory pathway controlling flowering time in *Arabidopsis thaliana*. Nat Genet 33:168–171
Blümel M, Dally N, Jung C (2014) Flowering time regulation in crops—what did we learn from Arabidopsis? Curr Opin Biotechnol 32C:121–129
Boguszewska D, Grudkowska M, Zagdanska B (2010) Drought-responsive antioxidant enzymes in potato (*solanum tuberosum* L.) Potato Res 53:373–382
Bolwell GP, Bindschedler LV, Blee KA, Butt VS, Davis DR et al (2002) The apoplastic oxidative brust in response to biotic stress in plants: a three-component system. J Exp Bot 53:1367–1376
Bonhomme F, Kurz B, Melzer S, Bernier G, Jacqmard A (2000) Cytokinin and gibberellin activate SaMADS A, a gene apparently involved in regulation of the floral transition in *Sinapis alba*. Plant J 24:103–111
Bowler C et al (1991) Manganese superoxide dismutase can reduce cellular damage mediated by oxygen radicals in transgenic plants. EMBO J 10:1723–1732
Bray EA, Bailey-Serres J, Weretilnyk E (2000) Responses to abiotic stresses. In: Buchanan BB, Gruissem W, Jones RL (eds) Biochemistry and molecular biology of plants. American Society of Plant Phytopathologists, Rockville, pp 1158–1203
Caliskan M, Cuming AC (1998) Spatial specificity of H2O2-generating oxalate oxidase gene expression during wheat embryo germination. Plant J 15:165–171
Cao S, Jiang S, Zhang R (2006) The role of *GIGANTEA* gene in mediating the oxidative stress response and in *Arabidopsis*. Plant Growth Regul 48:261–270
Chamnongpol S et al (1998) Defense activation and enhanced pathogen tolerance induced by H_2O_2 in transgenic tobacco. PNAS 95:5818–5823
Chen Z, Gallie DR (2006) Dehydroasorbate reductase affects leaf growth, development and function. Plant Physiol 142:775–787
Cheng H, Qin L, Lee S, Fu X, Richards DE, Cao D, Luo D, Harberd NP, Peng J (2004) Gibberellin regulates Arabidopsis floral development via suppression of DELLA protein function. Development 131:1055–1064
Chew O et al (2003) Characterization of the targeting signal of dual-targeted pea glutathione reductase. Plant Mol Biol 53:341–356
Claeys H, Skirycz A, Maleux K, Inze D (2012) DELLA signaling mediates stress-induced cell differentiation in *Arabidopsis* leaves through modulation of anaphase-promoting complex/cyclosome activity. Plant Physiol 159:739–747
Colebrook EH, Thomas SG, Phillips AL, Hedden P (2014) The role of gibberellin signaling in plant responses to abiotic stress. J Exp Biol 217:67–75
Conti H (2017) Hormonal control of the floral transition: can one catch them all? Develop Biol. https://doi.org/10.1016/j.ydbio.2017.03.024
Coupland G (1995) Genetic and environmental control of flowering time in Arabidopsis. Trends Genet 11:393–397
Dai M, Zhao Y, Ma Q, Hu Y, Hedden P, Zhang Q, Zhou DX (2007) The rice *YABBY1* gene is involved in the feedback regulation of gibberellin metabolism. Plant Physiol 144:121–133
Das K, Roychoudhary A (2014) Reactive oxygen species (ROS) and response of antioxidants as ROS- scavengers during environmental stress in plants. Front Plant Sci 2:1–13
Davies PJ (2010) The plant hormones: their nature, occurrence, and function. In: Davies PJ (ed) Plant hormones: biosynthesis, signal transduction, action! Springer, Dordrecht, pp 1–15
Davis SJ (2009) Integrating hormones into the floral-transition pathway of *Arabidopsis thaliana*. Plant Cell Environ 32:1201–1210
De Smet I, Voss U, Lau S, Wilson M, Shao N, Timme RE, Swarup R, Kerr I, Hodgman C, Bock R, Bennet M, Jurgens G, Beeckman T (2010) Unraveling the evolution of auxin signaling. Plant Physiol 155:209–221

Demidchik V et al (2009) Plant extracellular ATP signaling by plasma membrane NADPH oxidase and Ca2+ channels. Plant J 58:903–913
Dietz KJ, Turkan I, Krieger-Liszkay A (2016) Redox and reactive oxygen species-dependent signaling in and from the photosynthesizing chloroplast. Plant Physiol 171:1541–1550
Dietzel L, Brautigam K, Pfannschmidt T (2008) Photosynthetic acclimation: state transitions and adjustment of photosystem stoichiometry – functional relationship between short-term and long-term light quality acclimation in plants. FEBS J 275:1080–1088
Diezel C, Allmann S, Baldwin IT (2011) Mechanisms of optimal defense patterns in *Nicotiana attenuata*: flowering attenuates herbivory-elicited ethylene and jasmonate signaling. J Integr Plant Biol 53:971–983
Dill A, Sun T (2001) Synergistic derepression of gibberellin signaling by removing RGA and GAI function in *Arabidopsis thaliana*. Genetics 159:777–785
Dirk I, Marc VM (1995) Oxidative stress in plants. Curr Opin Biotechnol 6:153–158
Dybing E, Nelson JR, Mitchell JR, Sesame HA, Gillette JR (1976) Oxidation of a methyldopa and other catechols by cytochromes R450 generated superoxide anion: possible mechanism of methyldopa hepatitis. Mol Pharamacol 12:911–920
Eckardt NA (2007) GA signaling: direct targets of DELLA proteins. Plant Cell 19:2970
Edwards EA, Rawsthorne S, Mullineaux PM (1990) Subcellular distribution of multiple forms of glutathione reductase in leaves of pea (*Pisum sativum* L.) Planta 180(2):278–284
Egamberdieva D (2009) Alleviation of salt stress by plant growth regulators and IAA producing bacteria in wheat. Acta Physiol Plant 31:861–864
Eimert K, Wang SM, Lue WL, Chen J (1995) Monogenic recessive mutations causing both late floral initiation and excess starch accumulation in *Arabidopsis*. Plant Cell 7:1703–1712
Elstner EF (1991) Mechanism of oxygen activation in different compartments of plant cell. In: Pellend EJ, Steffen KL (eds) Active oxygen/oxidative stress in plant metabolism. Am. Soc. Plant Physiol., Rockville, pp 13–25
Eltayeb AE, Kawano N, Badawi GH, Kaminaka H, Saneklata T, Shibahara T et al (2007) Ovderexpression of monodehydroascorbate reductase in transgenic tobacco confers enhanced tolerance to ozone, salt and polyethylene glycol stresses. Planta 225:1255–1264
Fahad S, Bano A (2012) Effect of salicylic acid on physiological and biochemical characterization of maize grown in saline area. Pak J Bot 44:1433–1438
Fahad S, Hussain S, Bano A, Saud S, Hassan S, Shan D, Khan FA, Khan F, Chen YT, Wu C, Tabassum MA, Chun MX, Afzal M, Jan A, Jan MT, Huang JL (2015) Potential role of phytohormones and plant growth-promoting rhizobacteria in abiotic stresses: consequences for changing environment. Environ Sci Pollut Res 22:4907–4921
Fowler S, Lee K, Onouchi H, Samach A, Richardson K, Morris B, Coupland G, Putterill J (1999) *GIGANTEA*: a circadian clock-controlled gene that regulates photoperiodic flowering in Arabidopsis and encodes a protein with several possible membrane-spanning domains. EMBO J 18:4679–4688
Foyer CH, Harbinson JC (1994) Oxygen metabolism and the regulation of photosynthetic electron transport. In: Foyer CH, Mullineaux PM (eds) Causes of photooxidative stress and amelioration of defence systems in plants. CRC Press, Boca Raton, pp 1–42
Foyer CH, Noctor G (2009) Redox regulation in photosynthetic organisms: signaling, acclimation and practical implications. Antioxid Redox Signal 11:861–905
Ghisla S, Massey V (1989) Mechanism of flavoprotein catalyzed reactions. Euro J Biochem 181(1):1–17
Gilory S, Suzuki N, Miller G, Choi WG, Toyota M, Devireddy AR, Mittler R (2014) A tidal wave of signals: calcium and ROS at the forefront of rapid systemic signaling. Trends Plant Sci 19:623–630
Gilory S, Bialasek M, Suzuki N, Gorecka M, Devireddy A, Karpinski S, Mittler R (2016) ROS, calcium and electric signals: key mediators of rapid systemic signaling in plants. Plant Physiol 171:1606–1615
Giuliani S, Sanguineti MC, Tuberosa R, Bellotti M, Salvi S, Landi P (2005) Root-ABA1 a major constitutive QTL affects maize root architecture and leaf ABA concentration at different water regimes. J Exp Bot 56:3061–3070

Gocal GFW, Poole AT, Gubler F, Watts RJ, Blundell C, King RW (1999) Long-day up-regulation of a *GAMYB* gene during *Lolium temulentum* inflorescence formation. Plant Physiol 119:1271–1278

Gocal GFW, Sheldon CC, Gubler F et al (2001) *GAMYB*-like genes, flowering, and gibberellin signaling in *Arabidopsis*. Plant Physiol 127:1682–1693

Griffiths J, Murase K, Rieu I, Zentella R, Zhang ZL, Powers SJ, Gong F, Phillips AL, Hedden P, Sun TP et al (2006) Genetic characterization and functional analysis of the GID1 gibberellin receptors in *Arabidopsis*. Plant Cell 18:3399–3414

Han Y, Zhang X, Wang Y, Ming F (2013) The suppression of *WRKY44* by *GIGANTEA*-miR172 pathway is involved in drought response of *Arabidopsis thaliana*. PLoS One 8:e73541

Harman D (1956) Aging: a theory based on free radical and radiation chemistry. J Gerontol 11:298–300

Harman D (1981) The aging process. Proc Natl Acad Sci USA 78:7124–7128

Hedden P, Thomas SG (2012) Gibberellin biosynthesis and its regulation. Biochem J 444:11–25

Heil M, Bostock RM (2002) Induced systemic resistance (ISR) against pathogens in the context of induced plant defenses. Ann Bot 89:503–512

Hite DR, Auh C, Scandalios JG (1999) Catalase activity and hydrogen peroxide levels are inversely correlated in maize scutella during seed germination. Redox Rep 4:29–34

Huang S, Van Aken O, Schwarzlander M, Belt K, Millar A (2016) The roles of mitochondrial reactive oxygen species in cellular signaling and stress responses in plants. Plant Physiol 171:1551–1559

Huq E, Tepperman JM, Quail PH (2000) *GIGANTEA* is a nuclear protein involved in phytochrome signaling in *Arabidopsis*. Proc Natl Acad Sci U S A 97:9789–9794

Hwang I, Chen HC, Sheen J (2002) Two component signal transduction pathways in *Arabidopsis*. Plant Physiol 129:500–515

Ikeda A, Ueguchi-Tanaka M, Sonoda Y, Kitano H, Koshioka M, Futsuhara Y, Matsuoka M, Yamaguchi J (2001) Slender rice, a constitutive gibberellin response mutant, is caused by a null mutation of the *SLR1* gene, an ortholog of the height-regulating gene GAI/RGA/RHT/D8. Plant Cell 13:999–1010

Iqbal N, Umar S, Khan NA, Khan MIR (2014) A new perspective of phytohormones in salinity tolerance: regulation of proline metabolism. Environ Exp Bot 100:34–42

Ishibashi Y, Tawaratsumida T, Zheng SH, Yuasa T, Iwaya-Inoue M (2010) NADPH oxidases act as key enzyme on germination and seedling growth in barley (L.). Plant Prod Sci 13(1):45–52

Ishibashi Y, Tawaratsumida T, Kondo K, Kasa S, Sakamoto M, Aoki N, Zheng SH, Yuasa T, Mari II (2012) Reactive oxygen species are involved in gibberellin/abscisic acid signaling in Barley Aleurone cells. Plant Physiol 158:1705–1714

Javid MG, Sorooshzadeh A, Moradi F, Sanavy SAMM, Allahdadi I (2011) The role of phytohormones in alleviating salt stress in crop plants. Aust J Crop Sci 5:726–734

Kasahara H, Hanada A, Kuzuyama T, Takagi M, Kamiya Y, Yamaguchi S (2002) Contribution of the mevalonate and methylerythritol phosphate pathways to the biosynthesis of gibberellins in *Arabidopsis*. J Biol Chem 277:45188–45194

Kazan K, Lyons R (2016) The link between flowering time and stress tolerance. J Exp Bot 67(1):47–60

Kerchev PI, Waszczak C, Lewansowska A et al (2016) Lack of *GLYCOLATE OXIDASE* 1, but not *GLYCOLATE OXIDASE* 2, attenuates the photorespiratory phenotype of CATALASE2-deficient *Arabidopsis*. Plant Physiol 171:1704–1719

King RW, Moritz T, Evans LT, Junttila O, Herlt AJ (2001) Long day induction of flowering in *Lolium temulentum* involves sequential increases in specific gibberellins at the shoot apex. Plant Physiol 127:624–632

Kleine T, Leister D (2016) Rerograde signaling: organelles go net-working. Biochim Biophys Acta 1857:1313–1325

Kobayashi M et al (2007) Calcium-dependent protein kinases regulate the production of reactive oxygen species by potato NADPH oxidase. Plant Cell 19:1065–1080

Koornneef M, Hanhart CJ, Van der Veen JH (1991) A genetic and physiological analysis of late flowering mutants in *Arabidopsis thaliana*. Mol Gen Genet 299:57–66

Koornneef M, Alonso-Blanco C, Blankestijin-de Vries H, Hanhart CJ, Peeters AJ (1998) Genetic interactions among late-flowering mutants of *Arabidopsis*. Genetics 148:885–892

Kovtun Y, Chiu WL, Tena G, Sheen J (2000) Functional analysis of oxidative stress-activated mitogen-activated protein kinase cascade in plants. PNAS 97:2940–2945

Kurepa J, Smalle J, Montagu MV, Linz D (1998) Oxidative stress tolerance and longevity in *Arabidopsis*: the late flowering mutant gigantea is tolerant to paraquat. Plant J 14(6):759–764

Lee S, Choi H, Suh S, Doo IS, Oh KY et al (1999) Oligigalacturonic acid and Chitosan reduce stomatal aperture by inducing the evolution of reactive oxygen species from guard cells of tomato and *Commelina communis*. Plant Physiol 121:147–152

Lee S, Cheng H, King KE, Wang W, Husssain A, Lo J, Harberd NP, Peng J (2002) Gibberellin regulates Arabidopsis seed germination via RGL2, a GAI/RGA-like gene whose expression is upregulated following imbibition. Genes Develop 16:646–658

Lehner A, Bailly C, Flechel B, Poels P, Co^me D, Corbineau F (2006) Changes in wheat seed germination ability, soluble carbohydrate and antioxidant enzyme activities in the embryo during the desiccation phase of maturation. J Cereal Sci 43:175–182

Liu Y, Ye N, Liu R, Chen M, Zhang J (2010) H_2O_2 mediates the regulation of ABA catabolism and GA biosynthesis in Arabidopsis seed dormancy and germination. J Exp Bot 61:2979–2990

Magome H, Yamaguchi S, Hanada A, Kamiya Y, Oda K (2004) *Dwarf and delayed-flowering 1,* a novel Arabidopsis mutant deficient in gibberellin biosynthesis because of overexpression of a putative *AP2* transcription factor. Plant J 37:720–729

Mahalingam R, Fedoroff N (2003) Stress response, cell death and signalling: the many faces of reactive oxygen species. Physiol Plant 119:56–68

Malan C, Gregling MM, Gressel J (1990) Correlation between CuZn superoxide dismutase and glutathione reductase and environmental and xenobiotic stress tolerance in maize inbreds. Plant Sci 69:157–166

Martin GM, Austad SN, Johnson TE (1996) Genetic analysis of aging: role of oxidative damage and environmental stresses. Nat Genet 13:25–34

Martinez C, Pons E, Prates G, Leon J (2004) Salicylic acid regulates flowering time and links defense response and reproductive development. Plant J 37:209–217

Michaels SD (2009) Flowering time regulation produces much fruit. Curr Opin Plant Biol 12:75–80

Michaels SD, Amasino RM (1999) *FLOWERING LOCUS C* encodes a novel MADS domain protein that acts as a repressor of flowering. Plant Cell 11:949–956

Michaels SD, Amasino RM (2001) Loss of *FLOWERING LOCUS C* activity eliminates the late-flowering phenotype of *FRIGIDA* and autonomous pathway mutations but not responsiveness to vernalization. Plant Cell 13:935–941

Mignolet-Spruyt L, Xu E, Idanheimo N, Hoeberichts FA, Muhlenbock P, Brosche M, Van Breusegem F, Kangasjarvi J (2016) Spreading the news: subcellular and orgnaller reactrive oxygen species production and signaling. J Exp Bot 67:3831–3844

Miller G, Schlauch K, Tam R, Cortes D, Shulaev V, Dangl JL, Mittler R (2009) The plant NADPH oxidase RBOHD mediates rapid systemic signaling in response to diverse stimuli. Sci Signal 2(84):ra45. https://doi.org/10.1126/scisignal.2000448

Mittler R (2002) Oxidative stress, anti-oxidants and stress tolerance. Trends Plant Sci 7:405–410

Mittler M (2017) ROS are good. Trends Plant Sci 22(1):11–19

Mittler R, Vanderauwera S, Gollery M, Van Breusegem F (2004) Reactive oxygen species gene network of plants. Trends Plant Sci 9:490–498

Mittler R, Vanderauwera S, Suzuki N, Miller G, Tognetti VB, Vanserpoele K, Gollery M, Shulaev V, Van Breusegem F (2011) ROS signaling: the new wave? Trends Plant Sci 16:300–309

Moon J, Suh SS, Lee H, Choi KR, Hong CB, Paek NC, Kim SG, Lee I (2003) The *SOC1* MADS-box gene integrates vernalization and gibberellin signals for flowering in *Arabidopsis*. Plant J 35:613–623

Moon J, Lee H, Kim M, Lee I (2005) Analysis of flowering pathway integrators in Arabidopsis. Plant Cell Physiol 46:292–299

Muller K, Carstens AC, Linkies A, Torres MA, Leubner-Metzger G (2009a) The NADPH-oxidase AtrbohB plays a role in Arabidopsis seed after-ripening. New Phytol 184:885–897

Muller K, Linkies A, Vreeburg RAM, Fry SC, Krieger-Liszkay A, Leubner-Metzger G (2009b) In vivo cell wall loosening by hydroxyl radicals during cress seed germination and elongation growth. Plant Physiol 150:1855–1865

Murakami S, Johnson TE (1996) A genetic pathways conferring life extension and resistance to UV stress in Caenorhabditis elegans. Genetics 143:1207–1218

Mutasa-Gottgens E, Hedden P (2009) Gibberellin as a factor in floral regulatory networks. J Exp Bot 60(7):1979–1989

Nedelcu AM (2005) Sex as a response to oxidative stress: stress genes co-opted for sex. Proc Biol Sci 272:1935–1940

Ngamau K (2006) Selection for early flowering, temperature and salt tolerance of *Zantedeschia aethiopica* 'Green Goddess'. https://doi.org/10.17660/ActaHortic.2008.766.19

O'Brien JA, Benkova E (2013) Cytokinin cross-talking during biotic and abiotic stress responses. Front Plant Sci 4:451. https://doi.org/10.3389/fpls.2013.00451

O'Neill DP, Davidson SE, Clarke VC, Yamauchi Y, Yamaguchi S, Kamiya Y, Reid JB, Ross JJ (2010) Regulation of the gibberellin pathway by auxin and DELLA proteins. Planta 232:1141–1149

Ogasawara Y et al (2008) Synergistic activation of the Arabidopsis NADPH oxidase AtrbohD by Ca2+ and phosphorylation. J Biol Chem 283:8885–8892

Ogawa K, Iwabuchi M (2001) A mechanism for promoting the germination of *Zinnia elegans* seeds by hydrogen peroxide. Plant Cell Physiol 42:286–291

Oracz K, El-Maarouf Bouteau H, Farrant JM, Cooper K, Belghazi M, Job C, Job D, Corbineau F, Bailly C (2007) ROS production and protein oxidation as a novel mechanism for seed dormancy alleviation. Plant J 50:452–465

Oracz K, El-Maarouf-Bouteau H, Kranner I, Bogatek R, Corbineau F, Bailly C (2009) The mechanisms involved in seed dormancy alleviation by hydrogen cyanide unravel the role of reactive oxygen species as key factors of cellular signaling during germination. Plant Physiol 150:494–505

Park DH, Somers DE, Kim YS, Choy YH, Lim HK, Soh MS, Kim HJ, Kay SA, Nam HG (1999) Control of circadian rhythms and photoperiodic flowering by the Arabidopsis *GIGANTEA* gene. Science 185:1579–1582

Park HJ, Kim W-Y, Yun D-J (2016) A new insight of salt stress signaling in plant. Mol Cells 39(6):447–459

Peng J, Carol P, Richards DE, King KE, Cowling RJ, Murphy GP, Harberd NP (1997) The Arabidopsis *GAI* gene defines a signaling pathway that negatively regulates gibberellin responses. Genes Dev 11:3194–3205

Pereira A (2016) Plant abiotic stress challenges from the changing environment. Front Plant Sci 7:1123–1125

Pesaresi P, Hertle A, Pribil M et al (2009) Arabidopsis *STN7* kinase provides a link between short- and long-term photosynthetic acclimation. Plant Cell 21:2402–2423

Potters G et al (2009) Different stresses, similar morphogenic responses: integrating a plethora of pathways. Plant Cell Environ 32:158–169

Prasad TK, Anderson MD, Martin BA, Stewart CR (1994) Evidence for chilling-induced oxidative stress in maize seedlings and a regulatory role for hydrogen peroxide. Plant Cell 6:65–74

Puntarulo S, Sanchez RA, Boveris A (1988) Hydrogen peroxide metabolism in soybean embryonic axes at the onset of germination. Plant Physiol 86:626–630

Riboni M, Test AR, Galbiati M, Tonelli C, Conti L (2014) Environmental stress and flowering time – the photoperiodic connection. Plant Sign Behav 9:e29036

Rodriguez-Serrano M, Romero-Puerteas MC, Sanz-Fernandez M, Hu J, Sandalio LM (2016) Peroxisomes extend peroxules in a fast response to stress via a reactive oxygen species-mediated induction of the peroxin *PEX11a*. Plant Physiol 171:1665–1674

Sarvajeet SG, Narendra T (2010) Reactive oxygen species and antioxidant machinery in abiotic stress tolerance in crop plants. Plant Physiol Biochem 48:909–930

Schopfer P, Plachy C, Frahry G (2001) Release of reactive oxygen intermediates (speroxide radicals, hydrogen peroxide, and hydrogen radicals) and peroxidase in germinating radish seeds controlled by light, gibberellin and abscisic acid. Plant Physiol 125:1591–1602

Sharma P, Dubey RS (2004) Ascorbate peroxidase from rice seedlings: properties of enzyme isoforms, effects of stresses and protective roles of osmolytes. Plant Sci 167:541–550

Sheldon CC, Burn JE, Perez PP, Metzger J, Edwards JA, Peacock WJ, Dennis ES (1999) The *FLF* MADS box gene: a repressor of flowering in Arabidopsis regulated by vernalization and methylation. Plant Cell 11:445–458

Sheng XF, Xia JJ (2006) Improvement of rape (*Brassica napus*) plant growth and cadmium uptake by cadmium-resistant bacteria. Chemosphere 64:1036–1042

Silverstone AL, Sun T (2000) Gibberellins and the green revolution. Trends Plant Sci 5:1–2

Silverstone AL, Chang CW, Krol E, Sun TP (1997) Developmental regulation of the gibberellin biosynthetic gene GA1 in *Arabidopsis thaliana*. Plant J 12:9–19

Silverstone AL, Ciampaglio CN, Sun TP (1998) The *Arabidopsis RGA* gene encodes a transcriptional regulator repressing the gibberellin signal transduction pathway. Plant Cell 10:155–169

Silverstone AL, Jung HS, Dill A, Kawaide H, Kamiya Y, Sun TP (2001) Repressing a repressor: gibberellin-induced rapid reduction of the RGA protein in Arabidopsis. Plant Cell 13:1555–1566

Skirycz A, Inzé D (2010) More from less: plant growth under limited water. Curr Opin Biotechnol 21:197–203

Skirycz A, Claeys H, Bodt SD, Oikawa A, Shinoda S, Andriankaja M, Maleux K, Eloy NB, Coppens F, Yoo SD, Saito K, Inze D (2011) Pause-and-stop: the effects of osmotic stress on cell proliferation during early leaf development in Arabidopsis and a role for ethylene signaling in cell cycle arrest. Plant Cell 23:1876–1888

Sponsel VM, Hedden P (2004) Gibberellin, biosynthesis and inactivation. In: Davies PJ (ed) Plant hormones biosynthesis, signal transduction, action! Springer, Dordrecht, pp 63–94

Sreenivasulu N, Radchuk V, Alawady A, Borisjuk L, Weier D, Staroske N, Fuchs J, Miersch O, Strickert M, Usadel B, Wobus U, Grimm B, Weber H, Weschke W (2010) De-regulation of abscisic acid contents causes abnormal endosperm development in the barley mutant *seg8*. Plant J 64:589–603

Srikanth A, Schmid M (2011) Regulation of flowering time: all roads lead to Rome. Cell Mol Life Sci 68:2013–2037

Sun T (2008) Gibberellin metabolism, perception and signaling pathways in Arabidopsis. Tha Arabidopsis Book. e0103. doi: https://doi.org/10.1199/tab.0103

Swarbrick PJ, Schulze-Lefert P, Scholes JD (2006) Metabolic consequences of susceptibility and resistance in barley leaves challenged with powdery mildew. Plant Cell Environ 29:1061–1076

Takagi D, Takumi S, Hashiguchi M, Sejima T, Miyake C (2016) Superoxifde and singlet oxygen produced within the thylakoid membranes both causes photosystem I photoinhibition. Plant Physiol 171:1626–1634

Thomas SG, Phillips AL, Hedden P (1999) Molecular cloning and functional expression of *gibberellin 2-oxidases*, multifunctional enzymes involved in gibberellin deactivation. PNAS, USA 96:4638–4703

Tognetti VB et al (2010) Perturbation of indole-3-butyric acid homeostasis by the UDP-glucosyltransferase UGT74E2 modulates Arabidopsis architecture and water stress tolerance. Plant Cell 22:2660–2679

Tsugane K, Kobayashi K, Niwa Y, Ohba Y, Wada K, Kobayashi H (1999) A recessive Arabidopsis mutant that grows enhanced active oxygen detoxification. Plant Cell 11:1195–1206

Tyler L, Thomas SG, Hu J, Dill A, Alonso JM, Ecker JR, Sun TP (2004) DELLA proteins and gibberellin-regulated seed germination and floral development in *Arabidopsis*. Plant Physiol 135:1008–1019

Ueguchi-Tanaka M, Ashikari M, Nakajima M et al (2005) GIBBERELLIN INSENSITIVE DWARF1 encodes a soluble receptor for gibberellin. Nature 437:693–698

Ueguchi-Tanaka M, Nakajima M, Motoyuki A, Matsuoka M (2007) Gibberellin receptor and its role in gibberellin signaling in plants. Annu Rev Plant Biol 58:183–198

Vainonen JP, Sakuragi Y, Stael S et al (2008) Light regulation of CaS, a novel phosphoprotein in the thylakoid membrane of *Arabidopsis thaliana*. FEBS J 275:1767–1777

Vanderauwera S, Suzuki N, Miller G, van de Cotte B, Morsa S, Ravanat L, Hegie A, Triantaphylides C, Shulaev V, Montagu MCEV, Breusegem FV, Mittler R (2011) Extranuclear protection of chromosomal DNA from oxidative stress. Proc Natl Acad Sci U S A 108:1711–1716

Volt AC et al (2009) Salicylic acid, a multifaceted hormone to combat disease. Annu Rev Phytopathol 47:177–206

Wagner D, Przybyla D, op den Camp R et al (2004) The genetic basis of singlet oxygen-induced stress responses of *Arabidopsis thaliana*. Science 306:1183–1185

Wallace JS, Acreman MC, Sullivan CA (2003) The sharing of water between society and ecosystems: from conflict to catchment–based co– management. Philos Trans R Soc Lond Ser B Biol Sci 358:2011–2026. https://doi.org/10.1098/rstb.2003.1383

Wang HY, Huang YC, Chen SF, Yeh KW (2003) Molecular cloning, characterization and gene expression of a water deficiency and chilling induced proteinase inhibitor I gene family from sweet potato (*Ipomoea batatas* Lam.) leaves. Plant Sci 165:191–203

Wasternack C, Forner S, Strnad M, Hause B (2013) Jasmonates in flower and seed development. Biochimie 95:79–85

Wen CK, Chang C (2002) *Arabidopsis RGL1* encodes a negative regulator of gibberellin responses. Plant Cell 14:87–100

Weston DE, Elliott RC, Lester DR, Rameau C, Reid JB, Murfet IC, Ross JJ (2008) The Pea DELLA proteins LA and CRY are important regulators of gibberellin synthesis and root growth. Plant Physiol 147:199–205

Willige BC, Ghosh S, Nill C, Zourelidou M, Dohmann EMN, Maier A, Schwechheimer C (2007) The DELLA domain of GA INSENSITIVE mediates the interaction with the GA INSENSITIVE DWARF1A gibberellin receptor of Arabidopsis. Plant Cell 19:1209–1220

Wilson RN, Heckman JW, Somerville CR (1992) Gibberellin is required for flowering in *Arabidopsis thaliana* under short days. Plant Physiol 100:403–408

Winston GW (1990) Free radicals in cells. In: Stress responses in plants: adaptations and acclimation mechanisms. Willy-Liss, p 57–86

Woodson JD, Chory J (2008) Coordination of gene expression between organeller and nuclear genomes. Nat Rev Genet 9:383–395

Yamaguchi S (2008) Gibberellin metabolism and its regulation. Annu Rev Plant Biol 59:225–251

Yang Y, Ma J, Chen Y, Wu M (2004) Nucleocytoplasmic shuttling of *receptor-interacting protein 3 (RIP3)*: identification of novel nuclear export and import signals in *RIP3*. J Biol Chem 279:38820–38829

Yu H, Ito T, Zhao Y, Peng J, Kumar P, Meyerowitz EM (2004) Floral homeotic genes are targets of gibberellin signaling in flower development. PNAS, USA 101:7827–7832

Zhang X, Zhang L, Dong F, Gao J, Galbraith DW, Song CP (2001) Hydrogen peroxide is involved in abscisic acid induced stomatal closure in *Vicia faba*. Plant Physiol 126:1438–1448

Role of Cuticular Wax in Adaptation to Abiotic Stress: A Molecular Perspective

5

Swati Singh, Sandip Das, and R. Geeta

Abstract

Environmental stresses have a major influence on growth and development of plant. In order to counteract these stresses, plants have evolved various physiological and morphological adaptations. Evolution of the cuticle was one of the major adaptive morphological novelties in the transition of plants from aquatic to terrestrial habitat. One adaptive strategy was the acquisition of an external protective covering, in the form of a hydrophobic layer of waxes covering the cuticle, through the recruitment and expansion of existing *KCS* gene (β-ketoacyl synthase) from aquatic algal ancestors. Along the plant lineage, *KCS* existed as single gene in algae as seen from analysis of the genome sequence of *Chlamydomonas reinhardtii* and *Volvox carteri* (Merchant et al., Science (New York, N.Y.) 318(5848):245–250, 2007; Prochnik et al., Science 329(5988):223–226, 2010). This single *KCS* gene expanded into a three-member gene family in the liverwort, *Marchantia polymorpha*, which represents the first group of plants that colonized the terrestrial environment (Qiu et al., Nature 394(6694):671–674, 1998; Kajikawa et al., Biosci Biotechnol Biochem 67(3):605–612, 2003a; Biosci Biotechnol Biochem 67(8):1667–1674, 2003b); and in *Arabidopsis thaliana*, KCS exists as a 21-member gene family (Costaglioli et al., Biochim Biophys Acta 1734(3):247–258, 2005; Song et al., Mol Genet Genomics: MGG 291(2):739–752, 2016). Thus, changes in the biosynthesis and accumulation pattern of cuticular wax, in response to changing environmental conditions, played a major role in plant protection against stress. Understanding the genetic machinery involved in biosynthesis of this continuous hydrophobic layer over the aerial plant parts is a major challenge. In the present chapter, we attempt to summarize

S. Singh · S. Das (✉) · R. Geeta
Department of Botany, University of Delhi, New Delhi, India
e-mail: sdas@botany.du.ac.in; sandipdas04@gmail.com; rgeeta@botany.du.ac.in

S. M. Zargar, M. Y. Zargar (eds.), *Abiotic Stress-Mediated Sensing and Signaling in Plants: An Omics Perspective*, https://doi.org/10.1007/978-981-10-7479-0_5

the knowledge on regulation and function of the myriad of genes involved in the biosynthesis and accumulation of cuticular waxes and to elucidate the role of cuticle and cuticular waxes in plant tolerance toward abiotic stresses.

Keywords
Abiotic stress · Cuticular wax · KCS

5.1 Introduction

The evolution of early land plants with a thallus-like basic plant body (e.g., bryophytes) to that with a complex assembly of organs and tissues occurred about 400–500 million years ago (Kenrick and Crane 1997; Magallon et al. 2013). The evolution of complex organs and tissues (e.g., protected reproductive organs, vascular tissue, stomata, and cuticle) was morphological adaptations to various abiotic and biotic stresses faced by early land plants in their newly acquired terrestrial habitat. Of several adaptations that evolved in plant in its earthbound territory, the cuticle stands out as an important trait for establishment and survival. Due to its resistant nature, fossil records of the cuticle date back to the boundary between late Silurian and early Devonian period. Simultaneous occurrence of stomata and cuticle fossil in early land plants suggests their role as paleo-ecophysiological adaptations toward a relatively dry and terrestrial habitat (Raven 1997, 2002). Since its appearance in land plants, the cuticle has been of significant importance as it represents the interface between biosphere and atmosphere, thereby acting as plant's first line of defense against stress (Riederer and Schreiber 1995).

In spite of the fact that the cuticle is generally considered free from the basic polysaccharide cell wall of the epidermis, the two structures are physically related and bear some overlapping features. The cuticle is a composite structure containing two noteworthy hydrophobic components: the insoluble polymer, cutin, and dissolvable lipids collectively termed as waxes (Kolattukudy 1980). Ultrastructural studies have proposed a generalized view of a bilayered cuticle with a thin outer "cuticle proper" of lamellate appearance, with a variable quantity of alternating layers supposed to contain wax and cutin, and an internal "cuticle layer" with a reticulate appearance due to impregnation of the epidermal cell wall during maturation (Holloway 1994; Jeffree 1996). Both intracuticular and epicuticular waxes are indispensable segments of the plant cuticle. The waxes together are composed of long-chain aliphatic lipids, triterpenoids, and minor secondary metabolites, for example, sterols and flavonoids. As cuticular wax is made up of varied compounds with diverse chemistry, it gives an impression of being extremely heterogeneous in structure. The cuticular waxes also demonstrate compositional and quantitative variability among various plant species and organs. This waxy layer performs important functions, for example, limiting non-stomatal water loss, protection against ultraviolet (UV) radiation (Reicosky and Hanover 1978; Barnes and Cardoso Vilhena 1996), and lowering of water retention on plant surface, thus minimizing dust, pollen, and air pollutant deposition (Kerstiens 1996; Barthlott and Neinhuis 1997). Surface wax also plays a key role in plant defense against microorganism (Jenks et al. 1994) and varied plant-insect interactions (Eigenbrode and Espelie 1995). Cutin and waxes together bear important function in cell-cell interactions, such as mediating pollen-stigma contact and stopping post-genital organ fusions (Tanaka and Machida 2008).

5.2 Wax Biosynthesis, Deposition, and Morphology

5.2.1 Biosynthesis and Transport

Wax biosynthesis and transport has been studied in several species such as tomato (*Lycopersicon esculentum*), soybean (*Glycine max*), rice (*Oryza sativa*), and barrel medic (*Medicago truncatula*) (Girard et al. 2012; Islam et al. 2009; Uppalapati et al. 2012; Luo et al. 2013; Rowland et al. 2006) aside from model organism *Arabidopsis*. Cuticular cutin and wax deposition has been reported to start from the late globular stage of embryo development (Tanaka et al. 2002; Suh 2005) in a tightly coordinated manner that continues in the adult, on vegetative as well as reproductive organs such as flowers, fruits, and seeds in *A. thaliana* (Jessen et al. 2011).

Biosynthesis of cuticle wax is broadly classified into two stages:

1. Synthesis of C16/C18 fatty acids (FA) in plastids
2. Formation of very-long-chain fatty acid (VLCFA) in the endoplasmic reticulum

The first stage of de novo fatty acid synthesis is marked by synthesis of 16-carbon and 18-carbon acyl chains by soluble enzyme complex (fatty acid synthase (FAS) complex) in plastidial stroma (Ohlrogge and Browse 1995). Once formed, these 16-carbon and 18-carbon acyl chains are transported to the endoplasmic reticulum and subjected to the second stage of elongation, to form VLCFAs. The fatty acid elongase (FAE) complex acts as the primary enzyme complex catalyzing this pathway by iterative addition of two carbon units from malonyl-CoA to either C16 or C18 acyl-CoA. Similar to C16/C18 FA synthesis, the wax biosynthetic pathway has four consecutive reactions:

(i) Condensation of two carbon moieties to acyl-CoA by 3-ketoacyl-coenzyme A synthase (KCS)
(ii) Reduction of KCS intermediate by 3-ketoacyl-coenzyme A reductase (KCR)
(iii) Dehydration of 3-hydroxyacyl-CoA by 3-hydroxyacyl-CoA dehydratase (HCD)
(iv) Final reduction of trans-2,3-enoyl-CoA by trans-2-enoyl-CoA reductase (ECR) (Bernard and Joubès 2013; Haslam and Kunst 2013)

Of all the enzymes in the pathway, the elongating/condensing enzyme, KCS, determines the substrate and tissue specificity of fatty acid elongation (Millar and

Kunst 1997). Many KCSs have so far been associated with cuticle formation; of these only *KCS6/CER6/CUT1* has been shown to be specifically involved in cuticular wax biosynthesis. Genes encoding KCR such as *KCR1*, *HCD* by *PAS2*, and ECR by *CER10* have been identified in *Arabidopsis*, but mutations in these genes are pleiotropic, thereby affecting not only cuticular wax biosynthesis but also amounts of sphingolipids and seed triacylglycerols (Zheng et al. 2005). Once formed, most of the epidermal VLCFA-CoAs enter either of the two ER-localized pathways: an acyl reduction pathway responsible for production of primary alcohols and wax esters or an alkane-forming/decarbonylation pathway that produces alkanes, aldehyde, ketones, and secondary alcohols (Li-Beisson et al. 2013).

5.2.1.1 Acyl Reduction Pathway

The fatty acyl reduction of VLCFA-acyl-CoA esters resulting in formation of primary alcohols has been studied in a variety of plants (Kolattukudy 1970, 1971; Vioque and Kolattukudy 1997; Metz et al. 2000). In *Brassica oleracea*, primary alcohol production is carried out in two steps by two separate enzymes. First, an NADH-dependent acyl-CoA reductase reduces VLCFAs to aldehydes, which are acted upon by an NADPH-dependent aldehyde reductase resulting in formation of primary alcohols (Kolattukudy 1971). However, in pea (*Pisum sativum*) and jojoba (*Simmondsia chinensis*), alcohol formation from VLCFA precursors proceeds via an aldehyde intermediate (without the release of free aldehyde) catalyzed by a single fatty acyl-CoA reductase (FAR) (Vioque and Kolattukudy 1997). FAR-related sequences have also been reported in public databases of corn (*Zea mays*), rice (*Oryza sativa*), cotton (*Gossypium hirsutum*), *Brassica napus*, and *Arabidopsis* (family of eight FAR-like proteins), thereby showing a nearly ubiquitous presence of an alcohol-generating reductase in plants (Kunst and Samuels 2003). Synthesis of wax esters, catalyzed by a fatty acyl-CoA/fatty alcohol acyltransferase (wax synthase, WS), acts as the final step for the acyl reduction pathway (Wu et al. 1981).

5.2.1.2 Decarbonylation Pathway

Of the total wax load on the stem of *Arabidopsis thaliana*, 90% is accounted for by the decarbonylation pathway (Millar et al. 1999). This pathway is responsible for production of aldehydes from VLCFA precursors by the action of a membrane-bound fatty acyl-CoA reductase. The aldehydes thus produced are decarbonylated to odd-chain alkanes, catalyzed by an aldehyde decarbonylase. Subsequently mid-chain alkane hydroxylase 1 (MAH1) causes hydroxylation of these odd-chain alkanes to produce secondary alcohols, which are eventually oxidized to yield ketones (Kolattukudy and Liu 1970; Kolattukudy et al. 1973; Greer et al. 2007).

Upon synthesis through different pathways, cuticular lipids are trafficked from ER to plasma membrane via three possible routes: vesicular transport (lipid rafts of sphingolipids and sterol micro-domains), trafficking via carrier proteins (acyl-CoA-binding proteins (ACBPs)), or direct transfer through membrane junctions (ER–PM junctions) (Leung et al. 2006; Xiao and Chye 2009). Lipid transfer proteins (LTPs) have also been suggested to be involved in the transport of cuticular lipids from the plasma membrane via the cell wall (Somerville et al. 2000).

5.2.2 Composition and Morphology

Cuticular lipids are a complicated mixture of aliphatic and aromatic components, with major derivatives of n-acyl alkanes with chain length of C20–C40 (von Wettstein-Knowles 1995; Jetter et al. 2006). The complexity of cuticular lipids is further widened by its substitution with different functional groups such as -hydroxyl, -carboxyl, or -ketoyl groups. Addition of these functional groups results in formation of compounds such as fatty acids, primary alcohols, aldehydes, β-diketones, and secondary alcohols (Walton 1990; Bianchi 1995; Kunst and Samuels 2003). Chemical composition and shape of the cuticle and cuticular waxes is extraordinarily variable among plant species, between organs or even individual cells of same species during organ ontogeny. A few examples of such variations include the presence of aromatic compounds consisting of flavonoids (Wollenweber 1984; Wollenweber and Schneider 2000) and compounds such as polymeric aldehydes in waxes of sugarcane (*Saccharum officinarum*), rice (*Oryza sativa*), and leaves and traps of the pitcher plant (*Nepenthes alata*) (Haas et al. 2001; Riedel et al. 2003). These waxes are either embedded as filler in the cutin matrix to provide hydrophobicity to the membrane or deposited as epicuticular waxes upon the cell surface, where they often form complicated three-dimensional crystalline microstructures. These are useful characters in plant systematics, considering the fact that precise surface wax forms are restricted to particular taxa (Barthlott 1981; Gülz 1994; Barthlott et al. 1998, 2003). It has been shown that epicuticular wax is chemically distinct from the intracuticular form (Jetter et al. 2000; Guhling et al. 2005). However, patterning of epicuticular waxes confers distinct macroscopic surface properties such as the glossy appearance common to many leaves and fruits (due to epicuticular films) or the dull, glaucous look of broccoli (*Brassica oleracea*) leaves and *A. thaliana* stems (due to epicuticular wax crystals).

5.3 Cuticular Wax and Abiotic Stress Tolerance: Molecular Machinery

5.3.1 Major Abiotic Stresses Influencing Cuticular Wax

Plants have evolved to exist in situations that might be suboptimal for upkeep of normal plant functions. Abiotic stress arises from exposure to climatic extremes such as drought, heat, cold, ultraviolet radiation, altitude, soil nutrient, and pollution. In response, plants evolve to deal with stress conditions through various mechanisms that fall comprehensively into two categories – evasion and resilience – which may also happen together. Evasion/avoidance includes a range of internal situations in which the plant cells are not stressed although external situations may be stressful. An instance of avoidance mechanism is the control of high leaf temperatures by transpiration and prevention of drought with the aid of water conservation. Tolerance toward stress helps the plant to resist the extremes of both internal and external stresses. Examples of tolerance include endurance during desiccation due to drought and revival on hydration. The evolution of specialized physiological mechanisms is one aspect of tolerance, while utilization of existing physiological mechanisms to confer mechanistic and morphological devices to shield plants from the consequences of stress situations is a frequent feature of avoidance. Thus, plants can acclimate to avoid or overcome stress via numerous physiological and biochemical mechanisms, together with modification of genes, to produce ecologically tailored tolerant phenotypes. Evolution of a waterproof cuticle over 400 million years ago was essential to the successful colonization of land by vegetation (Edwards et al. 1996). Thus, the primary interface between the plant and its surroundings, the cuticle, performs a key role in keeping the plant's integrity in inherently antagonistic surroundings (Fig. 5.1).

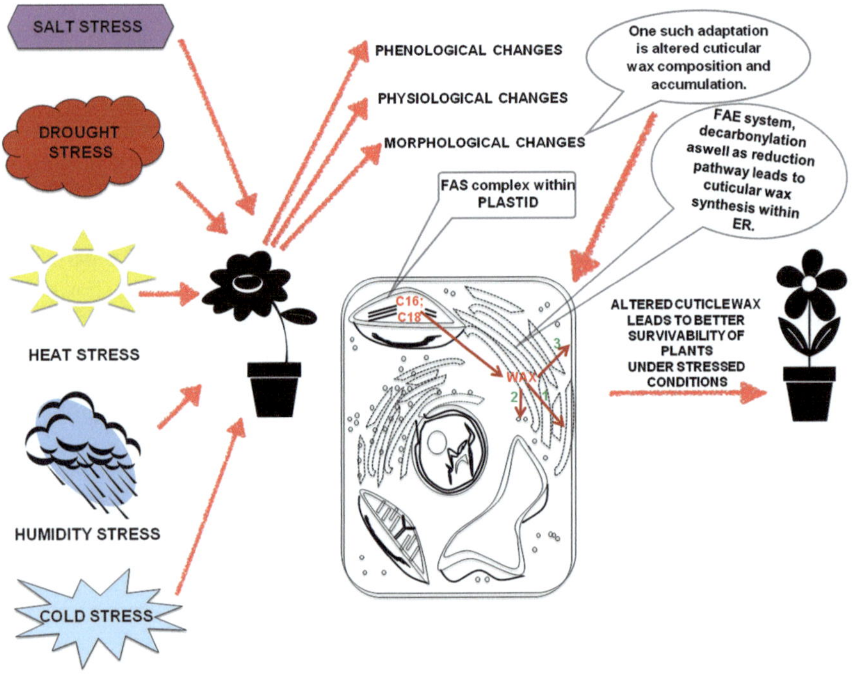

Fig. 5.1 Schematic diagram showing plant response toward abiotic stresses (salt, drought, temperature (heat/cold), and humidity), results in various phenological, physiological, and morphological adaptation. One of these morphological adaptations, to help plant cope with stress results, is altered cuticular wax composition and synthesis. This alteration is caused by modulating the normal wax biosynthetic pathway which essentially is the formation of C16/C18 compounds through FAS (fatty acid synthesis) system in plastid. The compounds are further transported to endoplasmic reticulum (ER), where they are processed through FAE (fatty acid elongase) complex (resulting in elongation of C16/C18 fatty acids to very-long-chain fatty acids) and further modified through decarbonylation and reduction pathway to yield cuticular wax. The cuticular wax once formed is transported to the plasma membrane and thereafter to cell wall. The transport is carried out through different pathways, (1) direct transfer through ER–PM junctions, (2) transport through vesicles (vesicular transport), and (3) transport mediated through lipid transfer proteins (LTPs). The altered waxy layer thus formed above the cell surface of aerial plant part helps the plant to better survive the extreme environmental conditions

5.3.1.1 Drought Stress

Drought stress prompts a wide range of physiological, morphological, and developmental changes in plants. Various regulatory proteins, transcription factors, and genes encoding proteins and enzymes involved in drought resistance reactions are induced under a dry spell (Shinozaki et al. 2003; Seki et al. 2007; Tran et al. 2007). Molecular and biochemical studies of knockout and overexpression transgenic plants suggest that essential plant responses to drought include stomatal closure, optimization of root development and water uptake, biosynthesis and accumulation of osmo-protectants, and metabolic modification of hormones and secondary metabolites (Schroeder et al. 2001; Seki et al. 2007; Mouillon et al. 2008; Seo et al. 2009). In addition, it has been proposed that the cuticle, being the outermost thin layer of leaves and stems, contributes to drought resistance. Cuticular wax can counteract non-stomatal water loss, allowing plants to acclimate to dry situations (Kerstiens 1996; Buda et al. 2013). Studies on *Arabidopsis*, soybean, and mosses have shown significant increase in leaf wax content upon subjection to drought stress, confirming the role of epicuticular wax in mediating drought stress (Kosma et al. 2009; Aharoni et al. 2004; Buda et al. 2013; Luo et al. 2013). A few transcription factors have been implicated in regulating cuticular wax biosynthesis in plants (Aharoni et al. 2004; Zhang et al. 2005, 2007; Cominelli et al. 2008). However, it is not yet clear as to how drought stress alerts are incorporated into gene regulating pathways modulating cuticular wax biosynthesis (Seo and Park 2011).

5.3.1.2 Temperature/Irradiation Stress

Temperature influences wax biosynthesis in several plants including *Brassica* species, where large amounts of waxes are produced at lower temperatures. Light and temperature have a prominent impact upon composition of cuticular waxes. The responses of several plants such as various *Brassica* species (Baker 1974; Reed and Tukey 1982; Shepherd et al. 1995), carnation (Reed and Tukey 1982), and barley (Nodskov Giese 1975) to high irradiation levels resulted in increases in cuticular wax thickness. As the biosynthesis of waxes is a streamlined process, it may be possible to relate changes in the composition of waxes to the channeling of acyl precursors into free fatty acids or to changes in the products of the reductive and decarbonylation pathways. Baker (1974) observed that by and large, normal (glaucous) plants of *B. oleracea* produced more unsaturated fatty acid and products of the reductive pathways (fatty acid, aldehyde, primary alcohol, ester) at lower light and higher temperature levels and more decarbonylation products (alkanes, secondary alcohol, ketones) at higher light and lower temperature levels. However, the effect of environmental factors such as light and temperatures differs in *glossy* (*gl*) mutants of *B. oleracea*: *gl2* and *gl3* mutants have more products of the reductive pathway at lower temperature, while *gl1*, *gl2*, and *gl3* mutants showed intermediate product distributions under high temperature and low light levels, respectively (Baker 1974). Subsequent studies on barley showed increased deposition rates and high wax density on leaf surfaces in dark-grown 4-day-old plants subjected to light (Nodskov Giese 1975). Exposure of the 4-day-old dark-grown plants to light resulted in significant increased rates of wax deposition in the first 24 h after exposure, then normalizing to the same rate of wax deposition as for light-grown plants. These findings indicated that light bears a stimulatory effect on rate of wax deposition. Normalization of wax deposition rate suggested that wax deposition is a density-limited phenomenon. In another study, glasshouse-grown *Brassica* species (kale, swede, etc.), when subjected to lower irradiation and higher temperature than outdoor-grown plants under higher irradiation and lower temperature, showed more reductive pathway products, whereas abundant decarbonylation products were detected in outdoor-grown plants. The results obtained for wax compositional changes were in accordance with the observations of Baker (1974) on influence of irradiation and temperature on wax composition. The distribution of various compound classes within the cuticular wax biosynthetic pathway can likewise change, suggesting that individual steps and transformations might be differentially influenced by environment. Baker's observations on the role of temperature were further supported by studies on *Citrus aurantium* demonstrating that day and night temperatures can also impact wax composition (Riederer and Schneider 1990). Leaves developing under high daytime temperatures bear reduced quantities per unit area of alkanes, fatty acids, primary alcohol, and alkyl esters (but not esters) compared to leaves developing under higher nighttime temperature (Shepherd and Griffiths 2006).

5.3.1.3 Cold/Frost Stress

Cold or frost is another important environmental stress impacting the composition and amount of epicuticular wax formation. Cold stress widely varies in its severity and effect on plants. Maize plants at four-leaf stage exposed to cold spell of 7 days showed reduction (29%) in cuticular waxes (Gauvrit and Gaillardon 1991). Increases in wax load and n-alkane of willow, *Salix* species, were correlated with their poor frost tolerance capabilities (Shepherd and Griffiths 2006). The effects of cold stress on epicuticular wax morphology and constituents and expression of wax-related genes have been examined in model plant system in order to understand the interactive mechanism between plant epicuticular wax and cold stress and to establish effects of abiotic stresses on crop and their respective resistance mechanisms to counteract these effects (Yu and Song Chao 2014; described below).

5.3.1.4 UV Irradiation Stress

Forms of electromagnetic radiation on plant cuticle cover a wide variety of energy levels and wavelengths: from high-energy, short-wavelength bright (UV) through visible and near infrared (IR), to lower-energy, longer-wavelength middle–far IR. Lately, amounts of UV-B radiation reaching the earth's surface have increased because of stratospheric ozone (O_3) depletion created by atmospheric pollutants such as chlorofluorocarbons (Farman et al. 1985; Kerr and McElroy 1993; Webb 1997; Holmes and Keiller 2002). UV radiation has a negative influence on plant development, diminishing leaf size, restricting the range accessible for solar energy capture, and a range of other detrimental effects including changes in cuticular wax biosynthesis (Zuk-Golaszewska et al. 2003; Lydon et al. 1986; Gwynn-Jones 2001; Flint and Caldwell 1984; Barnes et al. 1990; Tevini and Steinmüller 1987; Tevini et al. 1991; Beggs and Wellmann 1985). An expansion in wax thickness in response to higher light levels is the reaction of many plants including different *Brassica* species (Reed and Tukey 1982; Shepherd et al. 1995), barley (Nodskov Giese 1975), and carnation (Reed and Tukey 1982).

5.3.1.5 Humidity Stress

Soil and air humidity plays a crucial role in plant morphology and development patterns. Any alteration in environmental humidity index has a significant impact on cuticular wax morphology and deposition patterns. Baker (1974) studied *Brassica oleracea* var. *gemmifera* (brussels sprout) for studying micromorphology of epicuticular waxes, under different ranges of air humidity. Plants of *Brassica* grown at low air humidity showed changes in wax micromorphology, while such effects were not detected in other species – for β-diketone tubules of *Eucalyptus gunnii l.* or nonacosanol wax tubules of *Tropaeolum majus l.* (Koch et al. 2006).

Effects of these abiotic stresses upon cuticular wax biosynthesis and accumulation are mediated in two possible ways:

1. By modulating regulators of cuticular wax biosynthetic and transport genes
2. By directly affecting cuticular wax biosynthetic and transport genes

5.3.2 Modulation of Regulators of Cuticular Wax Biosynthesis and Transport Genes

The following section summarizes the information on regulation and role of various genes in mediating cuticular wax biosynthesis under abiotic stress (Fig. 5.2).

5.3.2.1 WIN1/SHN1

Members of the ethylene-responsive transcription factor (ERF) family are involved in a number of processes ranging from various developmental processes, signal transduction via hormones, to responses to environmental constraints (Feng et al. 2005; Sharma et al. 2010). Owing to its versatile roles, this gene family is considered as an important regulator (Mizoi et al. 2012). One of the crucial features of this gene family is the presence of a conserved AP2/ERF DNA-binding domain of approximately 60–70 amino acids (Sakuma et al. 2002; Wessler et al. 2005) first identified in tobacco and shown to bind specifically to the GCC box, present in the promoters of genes modulated via ethylene, pathogens, and wounding (Ohme-Takagi and Shinshi 1995). Evidence on members of the AP2/ERF family shows their active contribution in accumulation of cuticular wax. Cuticular wax biosynthesis involves a complex metabolic pathway regulated by various transcription elements (Nawrath 2006). *WAX INDUCER1/SHINE 1 (WIN1/SHN1)*, a *AP2/ERF* class of transcription factor from *Arabidopsis thaliana* with three members *AtSHN1*, *AtSHN2*, and *AtSHN3*, was one of the first TFs characterized for its involvement in cuticle metabolism by comparing the response of transgenic overexpression and inducible plants to drought stress (Aharoni et al. 2004; Broun et al. 2004). Comparisons were made of growth performance and wax biosynthesis/

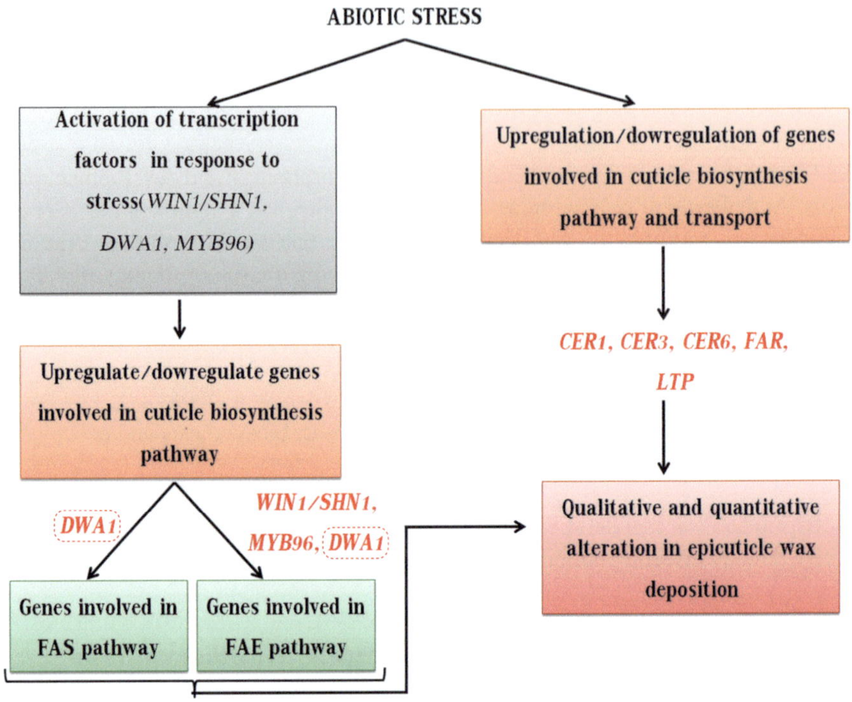

Fig. 5.2 Summary of various regulatory elements in genetic regulation of cuticular wax biosynthesis and transport under abiotic stress conditions. The role of DWA1, marked by *dashed boxes*, is based on indirect evidence based on Zhu et al. (2013)

accumulation in plants that had been exposed to dehydration stress for 12 days; these plants were mature non-transgenic plants (WT), a transgenic plant that constitutively overexpresses *WIN1/SHN1* (OX-*WIN1/SHN1* under CaMV 35S promoter), and inducible transgenic plants *GS-WIN1/SHN1* (*WIN1/SHN1* with an inducible promoter) grown with or without an inducer (15 days after induction). After 10 days, a prominent waxy phenotype was clearly visible in overexpression lines as well as lines that were induced for *WIN1/SHN1* expression, thereby differentiating the two from the untreated and wild type. Transgenic plants expressing *WIN1/SHN1* under inducible plant promoter displayed shriveling following 12 days of lack of hydration in the absence of inducer, while plants constitutively overexpressing *WIN1/SHN1* were not affected. To find out whether or not the plants recuperate from lack of hydration, the dehydrated plants were rewatered after 10 days and monitored for a week. While non-transgenic plants and non-induced GS-WIN1/SHN1 plants did not recuperate, all plants derived from lines that constitutively overexpress *SHN1/WIN1* gene recuperated. The study underlines the fact that in *A. thaliana*, overexpression of *WIN1/SHN1* leads to an increase in cuticular wax deposition and enhances drought tolerance, providing better chances of survival (Yang et al. 2011).

Due to the critical role of this transcription factor in wax biosynthesis, the role of *WIN1/SHN1* has been studied extensively not only in *Arabidopsis* but also in several other species such as *Medicago sativa l.* (Zhang et al. 2005) and crop species such as *Triticum* (Djemal and Khoudi 2015) and *Oryza* (Wang et al. 2012).

In *Medicago sativa* overexpression of SHINE transcription factor resulted in increased cuticular wax deposition with enhanced drought tolerance (Zhang et al. 2005). Altered expression of rice *SHN1/WIN1* homolog *wax synthesis regulator 1 (OsWR1)* resulted in change in wax biosynthesis and drought tolerance response (Wang et al. 2012).

Studies conducted in wheat identified the *TdSHN1* gene (wheat homolog of *WIN1/SHIN1*) (Djemal and Khoudi 2015). Expression analyses of *TdSHN1* under diverse stress regimes have demonstrated its role in abiotic stress tolerance. *TdSHN1* could perceive and interact with both the GCC and DRE boxes fused to 35S minimal promoter and setting off the transcription of the GUS reporter gene. Ectopic expression of *TdSHN1* in transgenic tobacco plants also revealed that *NtCER1* is the most induced gene when compared with expression levels of other stress-related genes in constitutively overexpressed lines (nearly 4.7-fold in transgenic, compared to WT plants). Constitutive overexpression also resulted in reduced chlorophyll leaching, water loss, and increased tolerance toward salt and drought stress. These results were in concordance with *CER1* overexpression in cucumber (Djemal and Khoudi 2016), demonstrating that the role of *WIN/SHN* transcription factor is similar to that of wax biosynthetic genes under abiotic stress conditions. In a number of crop species, prominent cuticular wax deposition is often related to advantageous drought resilience or potentially diminished transpiration and is thought to be a dry spell adaptation trait (Islam et al. 2009; Jordan et al. 1984; Jefferson et al. 1989). The high level of wax accumulation at the surface plant that overexpresses *WIN1/SHN1* is certainly associated with improved drought tolerance (Yang et al. 2011).

5.3.2.2 DWA1 (Drought-Induced Wax Accumulation 1)

Rice is one of the major staple crops and a model plant. The growing scarcity of water assets causing drought and unpredictable climate change has become a significant restricting element for rice growth and yield (Zhang 2007; Fukao and Xiong 2013). In recent years, *DWA1* gene, responsible for drought tolerance in rice, has been identified. The knockout mutant *dwa1* shows limited drought-induced wax accumulation and marked changes in cuticular wax composition, resulting in accelerated drought sensitivity (Zhu et al. 2013). *DWA1* has been characterized to encode a multidomain protein that carries a prokaryotic non-ribosomal peptide synthetase (NRPS)-like module that is conserved across vascular plants. The A domain of DWA1 protein has been shown to exhibit strong activity in activating long-chain fatty acids (LCFAs; C16:0/C18:0) but weak or no activity on very-long-chain fatty acids (VLCFAs; C20:0–C34:0). These results suggest that DWA1 may additionally act as an essential enzyme in the initiation of VLCFA biosynthesis for epicuticular wax creation under drought stress conditions in rice (Zhu et al. 2013).

The onset of drought stress is accompanied by an increase in cuticular wax accumulation as one of the major physiological responses. Drought-induced wax accumulation is principally a result of expansion in the level of alkanes, which are the most abundant component of the cuticular wax in *Arabidopsis*. In spite of FAs being the predominant component of cuticular wax in rice (Mao et al. 2012; Wang et al. 2012), it has been suggested that cuticular alkanes confer more imperviousness to water dispersion than VLCFAs (Kosma et al. 2009). In a complementary approach, Wang et al. (2012) showed that expression levels of genes such as *OSWR1*, *GL1-2*, and *Wda1*, which regulate wax production, were significantly suppressed in drought-stressed *dwa1* mutant plants, and showed a significant enhancement in their expression in the *DWA1* overexpression lines, under normal as well as drought stress conditions (Wang et al. 2012). Complementation of *dwa1* mutant with *DWA1* successfully recovered the drought-induced expression of these genes. The *dwa1* mutant had decreased levels of very-long-chain fatty acids, while plants overexpressing *DWA1* gene showed elevated level of very-long-chain fatty acids relative to its wild-type (WT) counterpart. Expression of many wax-associated genes such as *GL1-2*, *GL1-4*, *WSL1*, and *Wda1* was also greatly suppressed in *dwa1* mutant under drought conditions. In summary, *DWA1*, a recently reported mega-enzyme conserved across vascular plants, performs an important role in drought-initiated epicuticular wax deposition and contributes to drought resistance in rice. The results suggest that *DWA1* controls drought resistance by regulating dry spell instigated cuticular wax deposition in rice. This finding may have substantial implications for improving the drought resistance of crop varieties (Zhu et al. 2013).

5.3.2.3 MYB96

The transcription factor MYB96 has been characterized as an ABA-inducible factor with a pivotal role in drought resistance response. It regulates cuticular wax biosynthesis by binding to the promoters of genes encoding rate-limiting fatty acid elongation enzymes, such as *KCS1*, *KCS2*, *KCS6*, and *KCR1*. Evidence for the activity of MYB96 in regulating cuticular wax biosynthesis has been observed through overexpression and mutant analysis in *A. thaliana*. In constitutively overexpression lines, many of the genes required in cuticular wax biosynthesis were upregulated, thereby resulting in increased cuticular wax deposition; in contrast, diminished cuticular wax accumulation was observed in the knockout mutant (*myb96*), suggesting a role for MYB96 in drought response and cuticular wax biosynthesis. *MYB96* is a drought-responsive transcription factor gene, and its presence or absence in plant systems results in altered cuticular wax content, implying that MYB96 transcription factor integrates various drought stress response signals into gene regulatory framework that regulate cuticular wax biosynthesis under drought stress. MYB96 eventually provides a molecular mechanism through which regulation of cuticular wax biosynthesis enables responses to drought as well as pathogenic and mechanical stresses (Seo et al. 2011).

5.3.3 Direct Effect on Cuticular Wax Biosynthetic and Transport Genes

Wax accumulation in higher plants is known to be impacted by an assortment of natural variables, primarily light (von Wettstein-Knowles et al. 1980), drought, or freezing temperatures due to limited soil water around roots (Thomas and Barber 1974; Hadley 1989; Bengtson et al. 1979). One likely common mechanism through which all these environmental signals enhance wax accumulation and response of plant toward extreme stress conditions is by directly regulating the key wax biosynthetic enzymes involved in wax biosynthetic pathway (Hooker et al. 2002).

5.3.3.1 ECERIFERUM1 (CER1)

The total wax load in *Arabidopsis thaliana* accounts for nearly 50–80% of VLC alkanes (Bourdenx et al. 2011), which play important roles in drought resistance, plant–pathogen interplay, and pollen hydration (Aarts et al. 1995; Kosma et al. 2009). Identification of most of the genes involved in alkane biosynthesis has been done in recent years through analysis of various *ECERIFERUM* (*CER*) mutants, such as *eceriferum1* (*cer1*) and *eceriferum3* (*cer3/wax2*). Analysis of knockout mutant of *A. thaliana cer1* revealed reduced quantities of VLC alkanes and their derivatives and an increased accumulation of aldehydes (McNevin et al. 1993; Aarts et al. 1995; Jenks et al. 1995; Bourdenx et al. 2011). Upon comparison with wild type, the total wax load of *cer3* mutant was reduced by nearly 80% primarily due to decreased components of alkane-forming pathway. A hypothetical two-step alkane synthesis pathway has been proposed wherein acetyl-CoA is acted upon by *CER3* resulting in formation of VLC aldehyde as an intermediate compound. This is followed by action of *CER1* catalyzing the decarbonylation step of fatty aldehydes to VLC alkanes. However, this two-stage alkane synthesis pathway has been exhibited only in cyanobacteria, in which fatty acyl-ACPs are reduced to intermediate aldehydes catalyzed by acyl-ACP reductase; the aldehydes are further catalyzed by aldehyde-deformylating oxygenase (ADO) to produce alkanes right away (Schirmer et al. 2010; Das et al. 2011; Warui et al. 2011; Andre et al. 2013). Another model for VLC alkane biosynthesis on the lines of iterative approach was proposed (Bernard et al. 2012). In this model, *CER1* interacts with both *CER3* and cytochrome b5 (CYTB5) isoforms to catalyze the synthesis of VLC alkanes (Bernard et al. 2012). The CER1/CER3 complex may improve the efficiency of CYTB5 cofactors and hence act as the core component of the proposed VLC alkane biosynthesis pathway (Bernard et al. 2012). In *Arabidopsis*, *AtCER1* controls VLC alkane formation and plays a key position within the alkane-forming pathway. The gene product has been primarily located in the endoplasmic reticulum of epidermal cells of aerial plant organs. The expression of *AtCER1* could be triggered by abiotic stresses. Mutant and overexpression analysis of *AtCER1* showed bright green phenotype (indicating low/reduced wax load) and male sterility under low humidity stress in the loss-of-function mutant concomitant with increased odd-carbon-numbered and iso-branched alkanes. The expression of *AtCER1* impacted cuticle permeability (assessed extent of chlorophyll leaching and water loss) and plant imperviousness to various abiotic stresses (Aarts et al. 1995; Bourdenx et al. 2011). The direct effect of low temperature, drought, salt and ABA stress on induction of cucumber *CsCER1* has also been reported (Wang et al. 2014). Temperature plays a major role in cuticular wax morphology and composition. Low temperature stress leads to alteration of crystal structure and composition of epicuticular wax. Under cold stress, loss-of-function mutants of *cer1* in *Pinus* showed significantly reduced crystals on leaves (needles), while in mutant *Arabidopsis* plants, small dendrites were observed on the stem. Expression analysis of *CER1* in *A. thaliana* wild type (WT) showed strong induction of *CER1* by cold stress, suggesting that the plant can upregulate the expression of *CER1* to promote alkane synthesis under cold stress (Yu and Song Chao 2014).

The advent of global warming is likely to cause far more severe drought stress (in duration as well as intensity) on forest trees (Le Provost et al. 2013). However, studies on the role of the cuticle in drought stress adaptation in such long-lived organisms have been limited so far. One report on drought stress adaptation in relation to epicuticular wax in conifers has been reported for *Pinus pinaster* (maritime pine) (Le Provost et al. 2013). In this study two pine families (genetic units) with different growth rates were subjected to drought stress by rainfall exclusion in a greenhouse, and samples were collected at the time when drought stress was most severe. The results showed that the family with improved growth rates was more tolerant to drought stress, and many wax biosynthetic pathway genes were upregulated in this family. The upregulated genes included *CER6/KCS6*, *KCS4*, *KCR1*, *ECR/CER10*, *LACS3*, and decarbonylation pathway genes *CER1*, *CER2*, and *CER3*, which may therefore result in higher alkane production in the cuticle of plants subjected to drought stress. The study acts as a stepping stone to further work on drought stress in these long-lived tree species (Le Provost et al. 2013).

5.3.3.2 ECERIFERUM3 (CER3)/WAX2

Analysis of *eceriferum* (*cer*) mutants has been one major approach in identification of genes involved in cuticular wax biosynthetic process. The involvement of *CER3* in biosynthesis of alkane component of epicuticular wax in *Arabidopsis thaliana* was done through identification and analysis of *eceriferum3 (cer3/wax2)* mutant (McNevin et al. 1993; Jenks et al. 1995). Apart from *A. thaliana*, studies have been done in several other plant species. Characterization of cucumber *CsWAX2*, a *AtWAX2* homolog, has been reported by Wang et al. (2015a). *CsWAX2* was found to be highly expressed in the epidermis, where waxes are incorporated, and subcellular localization demonstrated that CsWAX2 protein is limited to the endoplasmic reticulum (ER). Abiotic stresses such as low temperature, drought, salt, and ABA induced transcriptional activation of *CsWAX2*, while the ectopic expression of *CsWAX2* in an *A. thaliana* wax2 mutant could only partially complement the smooth stem phenotype. Transgenic cucumber plants overexpressing *CsWAX2* gene caused changes in pollen viability, resistance of fruit to pathogen attack, and water loss compared to its wild-type counterpart. In order to further dissect the role of the gene under specific abiotic stress conditions, experiments were conducted on cucumber seedlings at three- to four-true-leaf stage under cold, drought, different concentrations of salt, and ABA stress regimes. The results indicated that *CsWAX2* transcripts were induced at high level under drought stress (nearly 80-fold in 3 h after treatment) and at lower levels under different concentrations and durations of salt stress (nearly sevenfold increase in 150 mM salt-treated plants upon comparison with mock-treated plants). Strong induction of *CsWAX2* was also observed with the application of ABA on cucumber seedlings. On the whole, different abiotic stresses were able to induce *CsWAX2* gene independently, thereby indicating that transcription of this gene is positively regulated by abiotic stresses (Wang et al. 2015a). Comparative analysis of overexpression and loss-of-function mutants of *CsWAX2*, thus, demonstrated increased wax load and resistance toward drought stress compared to its wild-type counterpart in the overexpression lines. On the other hand, *CsWAX2i* knockout mutants showed decreased wax load and sensitivity toward drought stress (Wang et al. 2015a).

A homolog of *Arabidopsis CER3/WAX2/YRE/FLP* is present in maize and annotated as *Glossy 1* (*GL1*) locus. Analysis of *gl1* shows significant reduction in the aldehyde and alcohol components of epicuticular wax, suggesting its essential role in the wax biosynthetic pathway. A scan for *GL1-like* homologs in a rice database (http://Rice.plantbiology.msu.edu/) identified 11 putative *GL1-like* genes in rice (*OsGL1-1* to *OsGL1-11*; Islam et al. 2009). On the basis of sequence similarity between all the 11 homologs, *OsGL1-1*, *OsGL1-2*, and *OsGL1-3* are most similar to maize *GL1* and *Arabidopsis WAX2/YRE/CER3/FLP* (Chen et al. 2003; Kurata et al. 2003; Sturaro et al. 2005; Bourdenx et al. 2011).Of these 11 genes, only 4 genes, *OsGL1-1/WSL2*, *OsGL1-2*, *OsGL1-5/Wda1*, and *OsGL1-6*, have been characterized (Jung et al. 2006; Islam et al. 2009; Qin et al. 2011; Mao et al. 2012; Zhou et al. 2013). In silico analysis of OsGL1-3 protein for domain structure, signaling peptide prediction, and subcellular localization revealed the presence of a conserved fatty acid hydroxylase domain (FAH domain) and a WAX2 C-terminal domain, and this protein is inferred to be membrane-bound. Majority of the characterized members of FAH/WAX2 families are membrane-bound proteins of the fatty acid elongase (FAE) complex and are wax biosynthetic enzymes responsible for production of VLCFA wax precursors (Samuels et al. 2008; Haslam et al. 2012; Mao et al. 2012). Comparative analysis of wild-type (WT) RNAi-silenced lines and overexpression lines of *OsGL1-3* showed higher wax crystal accumulation with greater wax load in overexpressing lines than the other two lines, indicating its role in cuticular wax biosynthesis. Further analysis of wax composition of overexpression lines showed alterations in both acyl reduction and alkane-forming pathways, with altered concentrations of C30-32 aldehyde and C30 primary alkanes in transgenic leaves. These results for *OsGL1-3* were in accordance with those obtained for *WSL/OsGL1-1* (Mao et al. 2012) and maize *GL1* (Sturaro et al. 2005), but not the same as that of rice *OsGL1-2* (Islam et al. 2009). These distinctions in cuticular wax compounds suggest unique roles of *GL*-related genes in cuticular wax biosynthesis. Altered expression, in *OsGL1-3* transgenic plants, of rice homologs of genes associated with wax biosynthesis, such as *LACS1* (a synthetase for VLCFAs C20–C30, particularly with most elevated action for C30 acids) (Lü et al. 2009), rice homologs of *KCR1* (*FAE* complex protein that performs the initial reduction) (Beaudoin et al. 2009), *PAS2* (*FAE* complex protein performs subsequent dehydration) (Bach et al.

2008), and *CER10* (FAE complex protein performs second reduction) (Zheng et al. 2005), additionally supports the contribution of *OsGL1-3* in wax synthesis. Study of the effects of abiotic stress on *OsGL1-3* showed that the gene was upregulated under ABA and PEG treatment, while no significant change was observed under heat and salt stress conditions. It can therefore be inferred that, in all likelihood, the expression of *OsGL1-3* is regulated via ABA-dependent drought signaling pathway to stimulate wax biosynthesis. As the overexpression lines of *OsGL1-3* lead to enhanced cuticular wax aggregation, reduced cuticular penetrability, decreased water loss, and improved drought tolerance, the expression of this gene under the influence of abiotic stress would thus be useful for plant against stress management (Zhou et al. 2015).

Freezing/low temperature stress had a major impact on *cer3* and *cer10* loss-of-function mutants in *A. thaliana*. Application of cold stress resulted in significant reduction of rod crystals of epicuticular wax in *cer3* mutant lines. Cold stress also caused considerable effect on wax composition of *cer3* with significant decrease in primary alcohol, while a significant increase in alkane and secondary alcohol composition. Expression of *CER3* gene in wild-type *Arabidopsis* plants is considerably downregulated under the influence of cold stress, indicating that the increase in wax load in response to cold stress may be the result of upregulation of the alkane biosynthetic branch of wax biosynthetic pathway (Zhou et al. 2015).

5.3.3.3 β-Ketoacyl-CoA Synthase (CER6/KCS6)

CER6/KCS6 is one of the condensing enzymes involved in fatty acid elongation process of cuticular wax biosynthesis (Hooker et al. 2002). Among the 21 members of the *KCS* family in *Arabidopsis*, the only wax-specific KCS is represented by *CER6*, characterized to be involved in the elongation of chain length longer than C24 (Millar et al. 1999; Vogg et al. 2004). Any change in the expression pattern of *CER6* is related to the total wax load of plant, as the gene is directly involved in fatty acid chain length elongation, which is a prime constituent of the epicuticular wax. In response to various abiotic stresses, the transcript level of *CER6* was monitored in comparison to untreated samples of *Arabidopsis*. Increased accumulation of *CER6* transcript was observed in response to osmotic stress caused by drought, ABA, PEG (polyethylene glycol), or salt upon comparison with levels of *rd29A* transcript (an extensively studied drought-inducible gene; Yamaguchi-Shinozaki and Shinozaki 1993). Under water stress, the mRNA level of *CER6* gene was nearly twofold higher than that of well-watered plants. Accumulation of increased transcript level of *CER6* was also obtained with increased application of PEG and salt stress, respectively. However, osmotic stress did not have much significant effect on *CER6* mRNA level upon comparison with *rd29A*, probably because the endogenous level of *CER6* is relatively high even under normal unstressed conditions.

The phytohormone ABA plays an important part in regulating the transcription of genes that respond to drought and salt stress. Action of these stresses triggers production of ABA, which eventually results in activation of several downstream genes, usually characterized by the presence of an ABRE (ABA-responsive element) motif in their promoter (Guiltinan et al. 1990). Exogenous application of

ABA also induces expression of *CER6*, another such gene responsible for activity under stress conditions that contains an ABRE motif in its promoter region (Hooker et al. 2002). Increased chain lengths longer than C26 obtained by induction of *CER6* under the influence of ABA have also been reported in *Lepidium* (Macková et al. 2013). In order to further examine the role of ABA in *CER6* expression, ABA-deficient mutants (*ab1*) under well-watered conditions were analyzed for *CER6* transcript levels. It was observed that CER6 transcript levels were nearly 50% reduced in mutant as compared to wild type, suggesting that ABA is involved in regulation of *CER6* levels even under well-watered conditions. In conclusion, increased *CER6* transcript level in ABA-treated seedlings and reduced *CER6* levels in ABA-deficient mutants show the role of ABA in induction of *CER6* transcription in response to drought stress (Hooker et al. 2002).

The presence, in the promoter, of I-box-like and GT1-like sequences, which are reported to be involved in light-responsive actions (Terzaghi and Cashmore 1995), suggests that *CER6* transcription may also be influenced by light. The effect of light on *CER6* was assessed in 8-day-old seedling and bolting shoots of *Arabidopsis*. The results showed undetectable *CER6* transcript level in etiolated seedlings; transcript was enhanced to significant levels 2 days after the seedlings were transferred to light conditions. *CER6* transcript level in the presence or absence of light was also assessed in developing bolting stems, which showed that the absence of light for 24 h significantly reduced *CER6* mRNA levels, while transcript levels were nearly undetectable in stems which were subjected to dark for 96 h. These observations indicate that light acts as a critical factor in transcription of the *CER6* gene and that its absence causes a quick decline in CER6 transcript levels (Hooker et al. 2002).

5.3.3.4 Fatty Acyl-CoA Reductase (FAR)

Primary alcohol is one of the major components of wax. Biosynthesis of primary alcohol is done by action of fatty acyl-CoA reductase (*FAR*) through the acyl reduction pathway wherein fatty acyl-CoA is converted to primary alcohol. In *A. thaliana*, *CER4* locus has been identified to encode a *FAR*. In wheat, three genes, *TaFar1*, *TaFar5*, and *TaTAA1a*, have been identified as homologs of *AtFAR*. Of the three homologs, *TaFaR1* and *TaFaR5* were found to be associated with biosynthesis of primary alcohol in anther and leaf cuticle (Wang et al. 2015c), while *TaTAA1a* was identified as an anther-specific gene involved in determining pollen fertility. Subsequently, Wang et al. (2016) were able to identify around 32 FAR-like genes in the wheat genome. Out of these 32 FAR-like genes, three, *TaFaR2*, *TaFaR3*, and *TaFaR4*, have been characterized to encode alcohol-forming *FAR* in leaf cuticle.

Cuticular wax accumulation is influenced by a varied set of environmental stresses, effects of which further influence expression of the underlying genes involved in its biosynthesis and transport (Jenks et al. 2002; Wang et al. 2015b; Domergue et al. 2010; Kosma et al. 2009). Induction of *TaFaR2*, *TaFaR3*, and *TaFaR4* in response to cold, PEG-mediated drought, salt, and ABA stresses is one example of such downstream activation of genes. The endogenous transcript level of all the three *TaFaRs* is differentially altered with the advent of environmental stresses. Transcript levels of *TaFaR4*s were highest at 12 h in response to cold stress, while those of *TaFaR2* and *TaFaR3* peaked at 4 and 2 h, respectively. Such differences in expression of the genes were also encountered when plants were treated with PEG 6000; expression level of *TaFaR4* was highest after 12 h, whereas those of *TaFaR2* and *TaFaR3* peaked after 6 h. Similarly, in the case of salt stress treatment mediated through NaCl, the three genes showed quick induction, peaking at 2 h posttreatment for *TaFaR2* and *TaFaR3* and after 4 h for *TaFaR4*. Exogenous application of ABA (response to drought stress) on wheat seedlings also increased the transcript levels of TaFaRs with earliest peaking observed for *TaFaR3* in 2 h, followed by TaFaR2 in 4 h, and of *TaFaR4* in 48 h after application of ABA. Drought stress in relation to air humidity was also verified in wheat seedlings; results showed significant upregulation of the three *TaFaRs* under low humid conditions (10–20% relative humidity). The upregulation in plants exposed to low relative humidity was concomitant with increased leaf wax accumulation (~28%), with an almost 26% increase in primary alcohol content. Humidity also had severe impact on wax morphology in wheat seedlings, where a clear-cut difference was obtained in the patterning of wax under low and high humidity. Under low humidity a dense array of platelet-shaped wax crystals were found covering the leaf surfaces, while under high humidity (~90–95% relative humidity), scattered crystals covered the leaf surface. The results presented conclusive evidence for upregulation of these *TaFaR* genes in cuticular wax biosynthesis in response to diverse abiotic stresses (Wang et al. 2016).

5.3.3.5 Lipid Transfer Protein (LTPs)

Transport of waxes from the site of synthesis to plant cells' exteriors most likely involves the participation of endoplasmic reticulum (ER), substrate ligands, transport vesicles, vesicle receptors, nonspecific lipid transfer proteins (*nsLTPs*), and many other secretory factors (Jenks et al. 2002; Kunst and Samuels 2003). Evidence for the role of transport vesicles (McFarlane et al. 2014), adenosine triphosphate-binding cassette (ABC) transporters (Pighin et al. 2004; Bird et al. 2007; Bessire et al. 2011; Panikashvili et al. 2011; Buda et al. 2013; Li et al. 2013), and glycosylphosphatidylinositol (GPI)-anchored lipid transfer protein (LTPG) (DeBono et al. 2009; Kim et al. 2012) in transporting of wax constituents from inside the cell to outside of the plasma layer have been provided in recent years. Hints for a role of nsLTPs in transporting wax to the epidermal cell surface from the extracellular space have also been seen (Kader 1996; Kunst and Samuels 2003).

nsLTPs are a group of small proteins of approximately 70–90 amino acid residues (Edstam et al. 2011), which can be classified as nsLTP1 (~9 kDa) and nsLTP2 (~7 kDa) as per the molecular mass of mature proteins (Douliez et al. 2000). nsLTPs were initially hypothesized to aid in the intracellular trafficking of lipids within organelles based on in vitro analysis showing that nsLTPs can bind and transfer all kinds of lipids between membranes (Kader et al. 1984). Subsequently, quite a few observations have revealed the role of nsLTPs as secretory proteins, which could probably be located extracellular, thereby falsifying the hypothesis of intracellular lipid transfer proposed earlier (Sterk et al. 1991; Thoma and Kaneko 1991). In vivo function of LTPs remains unclear; however, evidence suggests that nsLTPs might be involved in different aspects of plant physiology and cell biology, such as formation of cutin by transporting hydrophobic cutin monomers to the apoplast space (Sterk et al. 1991; Cameron 2005), plant defense against pathogens by acting as antimicrobial agents (Choi et al. 2012; Zhu et al. 2012; Pii et al. 2013), plant signaling (Pii et al. 2010), countering abiotic stresses (Guo et al. 2013), and plant organ development (Chae and Lord 2011; Zhang et al. 2010).

High sequence similarity to plant nsLTPs was shown in *TsnsLTP4* from *Thellungiella salsuginea* (Brassicaceae). In order to ascertain whether change of *TsnsLTP4* gene expression is linked to wax accumulation, *TsnsLTP4* was constitutively overexpressed in *A. thaliana* and silenced in *T. salsuginea*. Scanning electron microscope (SEM) analysis of cuticular wax patterning in transgenic *Arabidopsis* and *T. salsuginea* revealed that enormous, highly dense wax crystals were deposited on stems of transgenic *Arabidopsis* compared with the control, while smaller and more sparsely distributed wax crystals were observed on leaves and stems of silenced *T. salsuginea* compared to the wild type. The results illustrated that the transcript level of *TsnsLTP4* was directly proportional to wax load accumulation in plants, observed upon comparison of both the transgenic plants with their controls (Sun et al. 2015). Furthermore, induction of this gene under the influence of different environmental stresses through Northern blot analysis has been demonstrated. The analysis was performed on aerial tissue extracts of normal and stressed plants and showed a significant increase in transcript level of *TsnsLTP4* (~9.9-fold) under high vs low (37°/4°) temperatures, exogenous ABA application (~25.9-fold), and exposure to PEG (~9-fold) in comparison to normal control plants. However, salt stress resulted in only minor change in an increased accumulation of *TsnsLTP4* (~twofold in comparison to control) (Sun et al. 2015).

5.4 Conclusion

The environment has always been a major factor affecting plant growth and development. Owing to their sessile nature, plants are the most severely affected life forms facing the extremes of various environmental stresses. In order to cope with the extremes of environment, plants either avoid or adapt to these conditions. Various physiological and anatomical changes help the plant to adapt to these stress situations. Cuticular waxes have shielded plants from the vagaries of their surroundings since the advent of land plants. However, the underlying mechanism of wax accumulation and their role in abiotic and biotic stress tolerance is still unclear. Some of the major crucial issues that need answers to this enigmatic compound include regulation of genetic, molecular, and morphological entities involved in deposition and accumulation of wax in response to environmental stress. Plant stress tolerance in relation to wax in crop plants needs greater attention. Studies on non-model species would help in obtaining a greater representative image of the biology of the plant cuticle and its regulatory cues. Such studies could provide tools for manipulations of wax levels and to help raise transgenic plants with capacity to tolerate stresses.

Acknowledgments Research on cuticular waxes in the laboratory of RG and SD is supported by financial assistance from R&D grants of Delhi University and PURSE grant (DST, through Delhi University). SS would like to acknowledge UGC for award of JRF/SRF.

References

Aarts MGM et al (1995) Molecular characterization of the CER1 gene of Arabidopsis involved in epicuticular wax biosynthesis and pollen fertility. Plant Cell. American Society of Plant Biologists 7(December):2115–2127

Aharoni A et al (2004) The SHINE clade of AP2 domain transcription factors activates wax biosynthesis, alters cuticle properties, and confers drought tolerance when overexpressed in Arabidopsis. Plant Cell 16(9):2463–2480

Andre C et al (2013) Fusing catalase to an alkane-producing enzyme maintains enzymatic activity by converting the inhibitory byproduct H2O2 to the cosubstrate O2. Proc Natl Acad Sci U S A. National Academy of Sciences 110(8):3191–3196

Bach L et al (2008) The very-long-chain hydroxy fatty acyl-CoA dehydratase PASTICCINO2 is essential and limiting for plant development. Proc Natl Acad Sci U S A. National Academy of Sciences 105(38):14727–14731

Baker EA (1974) The influence of environment on leaf wax development in Brassica oleracea var. gemmifera. New Phytol. Blackwell Publishing Ltd 73(5):955–966

Barnes JD, Cardoso Vilhena J (1996) Interactions between electromagnetic radiation and the plant cuticle. In: Plant cuticles: an integrated functional approach. BIOS Scientific Publishers Ltd, Oxford, pp 157–174

Barnes PW, Flint SD, Caldwell MM (1990) Morphological responses of crop and weed species of different growth forms to ultraviolet-B radiation. Am J Bot 77:1354–1360

Barthlott W (1981) Epidermal and seed surface characters of plants: systematic applicability and some evolutionary aspects. Nord J Bot. Blackwell Publishing Ltd 1(3):345–355

Barthlott W, Neinhuis C (1997) Purity of the sacred lotus, or escape from contamination in biological surfaces. Planta. Springer-Verlag 202(1):1–8

Barthlott W et al (1998) Classification and terminology of plant epicuticular waxes. Bot J Linn Soc 126:237–260

Barthlott W et al (2003) Epicuticular waxes and vascular plant systematics: integrating micromorphological and chemical data. In: Deep morphology: toward a renaissance of morphology in plant systematics. (2). pp 189–206

Beaudoin F, Wu X, Li F, Haslam RP, Markham JE, Zheng H, Napier JA, Kunst L et al (2009) Functional characterization of the Arabidopsis beta-ketoacyl-coenzyme A reductase candidates of the fatty acid elongase. Plant Physiol 150(3):1174–1191

Beggs CJ, Wellmann E (1985) Analysis of light-controlled anthocyanin formation in coleoptiles of Zea mays L.: the role of UV-B, Blue, Red and Far-Red light. Photochem Photobiol. Blackwell Publishing Ltd 41(4):481–486

Bengtson C, Larsson S, Liljenberg C (1979) Water stress, epicuticular wax and cuticular transpiration rate. In: Advances in the biochemistry and physiology of plant lipids. Elsevier/North Holland Biomedical Press, New York, pp 269–274

Bernard A, Joubès J (2013) Arabidopsis cuticular waxes: advances in synthesis, export and regulation. Prog Lipid Res 52(1):110–129

Bernard A et al (2012) Reconstitution of plant alkane biosynthesis in yeast demonstrates that Arabidopsis ECERIFERUM1 and ECERIFERUM3 are core components of a very-long-chain alkane synthesis complex. Plant Cell. American Society of Plant Biologists 24(July):3106–3118

Bessire M et al (2011) A member of the pleiotropic drug resistance family of ATP binding cassette transporters is required for the formation of a functional cuticle in Arabidopsis. Plant Cell. American Society of Plant Biologists 23(5):1958–1970

Bianchi G (1995) Plant waxes. In: Hamilton RJ (ed) Waxes: chemistry, molecular biology and functions. The Oily Press, Scotland, pp 177–222
Bird D et al (2007) Characterization of Arabidopsis ABCG11/WBC11, an ATP binding cassette (ABC) transporter that is required for cuticular lipid secretion. Plant J. Blackwell Publishing Ltd 52(3):485–498
Bourdenx B et al (2011) Overexpression of Arabidopsis ECERIFERUM1 promotes wax very-long-chain alkane biosynthesis and influences plant response to biotic and abiotic stresses. Plant Physiol. Am Soc Plant Biol 156(1):29–45
Broun P et al (2004) WIN1, a transcriptional activator of epidermal wax accumulation in Arabidopsis. Proc Natl Acad Sci U S A 101(13):4706–4711
Buda GJ et al (2013) An ATP binding cassette transporter is required for cuticular wax deposition and desiccation tolerance in the moss Physcomitrella patens. Plant Cell. Am Soc Plant Biol 25(10):4000–4013
Cameron KD (2005) Increased accumulation of cuticular wax and expression of lipid transfer protein in response to periodic drying events in leaves of tree tobacco. Plant Physiol. American Society of Plant Biologists 140(1):176–183
Chae K, Lord EM (2011) Pollen tube growth and guidance: roles of small, secreted proteins. Ann Bot (Oxford University Press) 108:627–636
Chen X et al (2003) Cloning and characterization of the WAX2 gene of Arabidopsis involved in cuticle membrane and wax production. Plant Cell. American Society of Plant Biologists 15(5):1170–1185
Choi YE et al (2012) Tobacco NtLTP1, a glandular-specific lipid transfer protein, is required for lipid secretion from glandular trichomes. Plant J. Blackwell Publishing Ltd 70(3):480–491
Cominelli E et al (2008) Overexpression of the Arabidopsis AtMYB41 gene alters cell expansion and leaf surface permeability. Plant J. Blackwell Publishing Ltd 53(1):53–64
Costaglioli P et al (2005) Profiling candidate genes involved in wax biosynthesis in Arabidopsis thaliana by microarray analysis. Biochim Biophys Acta 1734(3):247–258
Das D et al (2011) Oxygen-independent decarbonylation of aldehydes by cyanobacterial aldehyde decarbonylase: a new reaction of diiron enzymes. Angew Chem Int Ed. WILEY-VCH Verlag 50(31):7148–7152
DeBono A et al (2009) Arabidopsis LTPG is a glycosylphosphatidylinositol-anchored lipid transfer protein required for export of lipids to the plant surface. Plant Cell 21(4):1230–1238
Djemal R, Khoudi H (2015) Isolation and molecular characterization of a novel WIN1/SHN1 ethylene-responsive transcription factor TdSHN1 from durum wheat (Triticum turgidum. L. subsp. durum). Protoplasma. Springer Vienna 252(6):1461–1473
Djemal R, Khoudi H (2016) TdSHN1, a WIN1/SHN1-type transcription factor, imparts multiple abiotic stress tolerance in transgenic tobacco. Environ Exp Bot 131:89–100
Domergue F et al (2010) Three Arabidopsis fatty acyl-coenzyme A reductases, FAR1, FAR4, and FAR5, generate primary fatty alcohols associated with suberin deposition. Plant Physiol. Am Soc Plant Biol 153(4):1539–1554
Douliez JP et al (2000) Structure, biological and technological functions of lipid transfer proteins and indolines, the major lipid binding proteins from cereal kernels. J Cereal Sci 32(1):1–20
Edstam MM et al (2011) Evolutionary history of the non-specific lipid transfer proteins. Mol Plant 4:947–964
Edwards D, Abbott GD, Raven J (1996) Cuticles of early land plants: a palaeoecophysiological evaluation. In: Kerstiens G (ed) Plant cuticles an integrated functional approach. BIOS Scientific Publishers Ltd., Oxford, pp 1–31
Eigenbrode SD, Espelie K (1995) Effects of plant epicuticular lipids on insect herbivores. AtIllu Rev Entomol 40(1):171–194
Farman JC, Gardiner BG, Shanklin JD (1985) Large losses of total ozone in Antarctica reveal seasonal ClOx/NOx interaction. Nature 315(6016):207–210
Feng JX et al (2005) An annotation update via cDNA sequence analysis and comprehensive profiling of developmental, hormonal or environmental responsiveness of the Arabidopsis

AP2/EREBP transcription factor gene family. Plant Mol Biol. Kluwer Academic Publishers 59(6):853–868
Flint SD, Caldwell MM (1984) Partial inhibition of in vitro pollen germination by simulated solar ultraviolet-B radiation. Ecology. Ecological Society of America 65(3):792–795. doi: https://doi.org/10.2307/1938051
Fukao T, Xiong L (2013) Genetic mechanisms conferring adaptation to submergence and drought in rice: Simple or complex? Curr Opin Plant Biol 16:196–204
Gauvrit C, Gaillardon P (1991) Effect of low-temperatures on 2,4-D behaviour in maize plants. Weed Res. Blackwell Publishing Ltd 31(3):135–142
Girard A-L et al (2012) Tomato GDSL1 is required for cutin deposition in the fruit cuticle. Plant Cell. American Society of Plant Biologists 24(7):3119–3134
Greer S et al (2007) The cytochrome P450 enzyme CYP96A15 is the midchain alkane hydroxylase responsible for formation of secondary alcohols and ketones in stem cuticular wax of Arabidopsis. Plant Physiol 145(3):653–667
Guhling O et al (2005) Surface composition of myrmecophilic plants: cuticular wax and glandular trichomes on leaves of Macaranga tanarius. J Chem Ecol. Springer-Verlag 31(10):2323–2341
Guiltinan MJ, Marcotte W, Quatrano RS (1990) A plant leucine zipper protein that recognizes an abscisic acid response element. Science 250(4978):267–271
Gülz P-G (1994) Epicuticular leaf waxes in the evolution of the plant kingdom. J Plant Physiol 143(4–5):453–464
Guo L et al (2013) Lipid transfer protein 3 as a target of MYB96 mediates freezing and drought stress in Arabidopsis. J Exp Bot. Oxford University Press 64(6):1755–1767
Gwynn-Jones, D (2001) Short-term impacts of enhanced UV-B radiation on photo-assimilate allocation and metabolism: a possible interpretation for time-dependent inhibition of growth. Plant Ecol. Kluwer Academic Publishers 154(1/2):67–73
Haas K, Brune T, Rücker E (2001) Epicuticular wax crystalloids in rice and sugar cane leaves are reinforced by polymeric aldehydes. J Appl Bot Angew Bot 75:178–187
Hadley NF (1989) Lipid water barriers in biological systems. Prog Lipid Res. Pergamon 28:1–33
Haslam TM, Kunst L (2013) Extending the story of very-long-chain fatty acid elongation. Plant Sci 210:93–107
Haslam TM et al (2012) Arabidopsis ECERIFERUM2 is a component of the fatty acid elongation machinery required for fatty acid extension to exceptional lengths. Plant Physiol. American Society of Plant Biologists 160(3):1164–1174
Holloway PJ (1994) SECTION I – REVIEWS plant cuticles: physicochemical characteristics and biosynthesis. NATO ASI Series Vol. G 36. In: Air pollutants and the leaf cuticle. Springer, Berlin, pp 1–13
Holmes MG, Keiller DR (2002) Effects of pubescence and waxes on the reflectance of leaves in the ultraviolet and photosynthetic wavebands: a comparison of a range of species. Plant Cell Environ. Blackwell Science, Ltd 25(1):85–93
Hooker TS, Millar AA, Kunst L (2002) Significance of the expression of the CER6 condensing enzyme for cuticular wax production in Arabidopsis. Plant Physiol 129(4):1568–1580
Islam MA et al (2009) Characterization of Glossy1-homologous genes in rice involved in leaf wax accumulation and drought resistance. Plant Mol Biol. Springer Netherlands 70(4):443–456
Jefferson PG et al (1989) Water stress and genotypic effects on epicuticular wax production of *Alfalfa* and crested wheatgrass in relation to yield and excised leaf water loss rate. Can J Plant Sci. NRC Research Press Ottawa, Canada 69(2):481–490
Jeffree CE (1996) Structure and ontogeny of plant cuticles. In: Plant cuticles. pp 33–82
Jenks MA et al (1994) Chemically induced cuticle mutation affecting epidermal conductance to water vapor and disease susceptibility in Sorghum bicolor (L.) Moench. Plant Physiol 105(4):1239–1245
Jenks MA et al (1995) Leaf epicuticular waxes of the eceriferum mutants in Arabidopsis. Plant Physiol. American Society of Plant Biologists 108(1):369–377
Jenks MA, Eigenbrode SD, Lemieux B (2002) Cuticular waxes of Arabidopsis. Arabidopsis Book 29(1):1

Jessen D et al (2011) Combined activity of LACS1 and LACS4 is required for proper pollen coat formation in Arabidopsis. Plant J. Blackwell Publishing Ltd 68(4):715–726
Jetter R, Schaffer S, Riederer M (2000) Leaf cuticular waxes are arranged in chemically and mechanically distinct layers: evidence from Prunus laurocerasus L. Plant Cell Environ 23(6):619–628
Jetter R, Kunst L, Samuels A (2006) Composition of plant cuticular waxes. In: Riederer M, Müller C (eds) Biology of the plant cuticle. Blackwell Publishing, Oxford, pp 145–181
Jordan WR et al (1984) Environmental physiology of sorghum. II. Epicuticular wax load and cuticular transpiration1. Crop Sci. Crop Science Society of America 24(6):1168
Jung K-H et al (2006) Wax-deficient anther1 is involved in cuticle and wax production in rice anther walls and is required for pollen development. Plant Cell. American Society of Plant Biologists 18(11):3015–3032
Kader J-C (1996) Lipid-transfer proteins in plants. Annu Rev Plant Physiol Plant Mol Biol. Annual Reviews 4139 El Camino Way, P.O. Box 10139, Palo Alto, CA 94303-0139, USA 47(1):627–654
Kader JC, Julienne M, Vergnolle C (1984) Purification and characterization of a spinach leaf protein capable of transferring phospholipids from liposomes to mitochondria or chloroplasts. Eur J Biochem. Blackwell Publishing Ltd 139(2):411–416
Kajikawa M, Yamaoka S et al (2003a) Functional analysis of a β-ketoacyl-CoA synthase gene, MpFAE2, by gene silencing in the liverwort Marchantia polymorpha L. Biosci Biotechnol Biochem 67(3):605–612
Kajikawa M, Yamato KT et al (2003b) MpFAE3, a β-ketoacyl-CoA synthase gene in the liverwort Marchantia polymorpha L., is preferentially involved in elongation of palmitic acid to stearic acid. Biosci Biotechnol Biochem 67(8):1667–1674
Kenrick P, Crane PR (1997) The origin and early evolution of plants on land. Nature 389:33–39
Kerr JB, McElroy CT (1993) Evidence for large upward trends of ultraviolet-B radiation linked to ozone depletion. Science (Washington) 262(5136):1032–1034
Kerstiens G (1996) Cuticular water permeability and its physiological significance. J Exp Bot. Oxford University Press 47(12):1813–1832
Kim H et al (2012) Characterization of glycosylphosphatidylinositol-anchored lipid transfer protein 2 (LTPG2) and overlapping function between LTPG/LTPG1 and LTPG2 in cuticular wax export or accumulation in arabidopsis thaliana. Plant Cell Physiol. Oxford University Press 53(8):1391–1403
Koch K et al (2006) Influences of air humidity during the cultivation of plants on wax chemical composition, morphology and leaf surface wettability. Environ Exp Bot 56(1):1–9
Kolattukudy PE (1970) Reduction of fatty acids to alcohols by cell-free preparations of Euglena gracilis. Biochemistry 9(5):1095–1102
Kolattukudy PE (1971) Enzymatic synthesis of fatty alcohols in Brassica oleracea. Arch Biochem Biophys. Academic Press 142(2):701–709
Kolattukudy PE (1980) Biopolyester membranes of plants: cutin and suberin. Science (New York, NY). American Association for the Advancement of Science 208(4447):990–1000
Kolattukudy PE, Liu TYJ (1970) Direct evidence for biosynthetic relationships among hydrocarbons, secondary alcohols and ketones in Brassica oleracea. Biochem Biophys Res Commun. Academic Press 41(6):1369–1374
Kolattukudy PE, Buckner JS, Liu TYJ (1973) Biosynthesis of secondary alcohols and ketones from alkanes. Arch Biochem Biophys. Academic Press 156(2):613–620
Kosma DK, Bourdenx B, Bernard A, Parsons EP, Lü S, Joubès J, Jenks MA (2009) The impact of water deficiency on leaf cuticle lipids of Arabidopsis. Plant Physiol 151(4):1918–1929
Kunst L, Samuels AL (2003) Biosynthesis and secretion of plant cuticular wax. Prog Lipid Res 42(1):51–80
Kurata T et al (2003) The YORE-YORE gene regulates multiple aspects of epidermal cell differentiation in Arabidopsis. Plant J. Blackwell Science Ltd 36(1):55–66

Le Provost G et al (2013) Soil water stress affects both cuticular wax content and cuticle-related gene expression in young saplings of maritime pine (*Pinus pinaster* Ait). BMC Plant Biol. BioMed Central 13(1):95

Leung KC et al (2006) Arabidopsis ACBP3 is an extracellularly targeted acyl-CoA-binding protein. Planta. Springer-Verlag 223(5):871–881

Li L et al (2013) The maize glossy13 gene, cloned via BSR-Seq and Seq-walking encodes a putative ABC transporter required for the normal accumulation of epicuticular waxes. PLoS One. Edited by M. Xu. Public Library of Science 8(12):e82333

Li-Beisson Y et al (2013) Acyl-lipid metabolism. Arabidopsis Book Am Soc Plant Biologists 11:e0161

Lü S et al (2009) Arabidopsis CER8 encodes LONG-CHAIN ACYL-COA SYNTHETASE 1 (LACS1) that has overlapping functions with LACS2 in plant wax and cutin synthesis. Plant J 59(4):553–564

Luo X et al (2013) Expression of wild soybean WRKY20 in Arabidopsis enhances drought tolerance. J Exp Bot. Oxford University Press 64(8):2155–2169

Lydon J, Teramura AH, Summers EG (1986) Effects of ultraviolet-B radiation on the growth and productivity of field grown soybean. In: Stratospheric ozone reduction, solar ultraviolet radiation and plant life. Springer, Berlin/Heidelberg, pp 313–325

Macková J et al (2013) Plant response to drought stress simulated by ABA application: changes in chemical composition of cuticular waxes. Environ Exp Bot 86:70–75

Magallon S, Hilu KW, Quandt D (2013) Land plant evolutionary timeline: Gene effects are secondary to fossil constraints in relaxed clock estimation of age and substitution rates. Am J Bot 100(3):556–573

Mao B et al (2012) Wax crystal-sparse leaf2, a rice homologue of WAX2/GL1, is involved in synthesis of leaf cuticular wax. Planta. Springer-Verlag 235(1):39–52

McFarlane HE et al (2014) Golgi- and trans-Golgi network-mediated vesicle trafficking is required for wax secretion from epidermal cells. Plant Physiol. American Society of Plant Biologists 164(3):1250–1260

McNevin JP et al (1993) Isolation and characterization of eceriferum (cer) mutants induced by T-DNA insertions in Arabidopsis thaliana. Genome Natl Res Counc Can = Génome Cons Natl Rech Can. NRC Research Press Ottawa, Canada 36(3):610–618

Merchant SS et al (2007) The Chlamydomonas genome reveals the evolution of key animal and plant functions. Science (New York, NY). American Association for the Advancement of Science 318(5848):245–250

Metz JG et al (2000) Purification of a jojoba embryo fatty acyl-coenzyme A reductase and expression of its cDNA in high erucic acid rapeseed. Plant Physiol 122(3):635–644

Millar AA, Kunst L (1997) Very-long-chain fatty acid biosynthesis is controlled through the expression and specificity of the condensing enzyme. Plant J Cell Mol Biol. Blackwell Science Ltd 12(1):121–131

Millar AA et al (1999) CUT1, an Arabidopsis gene required for cuticular wax biosynthesis and pollen fertility, encodes a very-long-chain fatty acid condensing enzyme. Plant Cell Online 11(5):825–838

Mizoi J, Shinozaki K, Yamaguchi-Shinozaki K (2012) AP2/ERF family transcription factors in plant abiotic stress responses. Biochim Biophys Acta (BBA) Gene Regul Mech 1819(2):86–96

Mouillon J-M, Eriksson SK, Harryson P (2008) Mimicking the plant cell interior under water stress by macromolecular crowding: disordered dehydrin proteins are highly resistant to structural collapse. Plant Physiol. American Society of Plant Biologists 148(4):1925–1937

Nawrath C (2006) Unraveling the complex network of cuticular structure and function. Curr Opin Plant Biol 9:281–287

Nodskov Giese B (1975) Effects of light and temperature on the composition of epicuticular wax of barley leaves. Phytochemistry. Pergamon 14(4):921–929

Ohlrogge J, Browse J (1995) Lipid biosynthesis. Plant Cell 7(7):957–970

Ohme-Takagi M, Shinshi H (1995) Ethylene-inducible DNA binding proteins that interact with an ethylene-responsive element. Plant Cell. American Society of Plant Biologists 7(2):173–182

Panikashvili D et al (2011) The Arabidopsis ABCG13 transporter is required for flower cuticle secretion and patterning of the petal epidermis. New Phytol. Blackwell Publishing Ltd 190(1):113–124

Pighin JA et al (2004) Plant cuticular lipid export requires an ABC transporter. Science (New York, NY). American Association for the Advancement of Science 306(5696):702–704

Pii Y, Pandolfini T, Crimi M (2010) Signaling LTPs – a new plant LTPs sub-family? Plant Signal Behav. Taylor & Francis 5(May):1–4

Pii Y, Molesini B, Pandolfini T (2013) The involvement of Medicago truncatula non-specific lipid transfer protein N5 in the control of rhizobial infection. Plant Signal Behav. Taylor & Francis 8(7):e24836

Prochnik SE et al (2010) Genomic analysis of organismal complexity in the multicellular green alga Volvox carteri. Science. American Association for the Advancement of Science 329(5988):223–226

Qin BX et al (2011) Rice OsGL1-1 is involved in leaf cuticular wax and cuticle membrane. Mol Plant 4(6):985–995

Qiu Y-L et al (1998) The gain of three mitochondrial introns identifies liverworts as the earliest land plants. Nature. Nature Publishing Group 394(6694):671–674

Raven JA (1997) The role of marine biota in the evolution of terrestrial biota: gases and genes. Biogeochemistry. Kluwer Academic Publishers 39(2):139–164

Raven JA (2002) Selection pressures on stomatal evolution. New Phytol. Blackwell Science Ltd 153(3):371–386

Reed DW, Tukey HB (1982) Light intensity and temperature effects on epicuticular wax morphology and internal cuticle ultrastructure of carnation and Brussels sprouts leaf cuticles. J Am Soc Hortic Sci 107:417–420

Reicosky DA, Hanover JW (1978) Physiological effects of surface waxes I. Light reflectance for glaucous and nonglaucous *Picea pungens*. Plant Physiol 62(1):101–104

Riedel M, Eichner A, Jetter R (2003) Slippery surfaces of carnivorous plants: composition of epicuticular wax crystals in Nepenthes alata Blanco pitchers. Planta. Springer-Verlag 218(1):87–97

Riederer M, Schneider G (1990) The effect of the environment on the permeability and composition of citrus leaf cuticles – II. Composition of soluble cuticular lipids and correlation with transport properties. Planta. Springer-Verlag 180(2):154–165

Riederer M, Schreiber L (1995) Waxes—the transport barriers of plant cuticles. In: Hamilton RJ (ed) Waxes: chemistry, molecular biology and functions, 6th edn. The Oily Press, West Ferry, Dundee, pp 130–156

Rowland O et al (2006) CER4 encodes an alcohol-forming fatty acyl-coenzyme a reductase involved in cuticular wax production in Arabidopsis. Plant Physiol 142(3):866–877

Sakuma Y et al (2002) DNA-binding specificity of the ERF/AP2 domain of Arabidopsis DREBs, transcription factors involved in dehydration- and cold-inducible gene expression. Biochem Biophys Res Commun. Academic Press 290(3):998–1009

Samuels L, Kunst L, Jetter R (2008) Sealing plant surfaces: cuticular wax formation by epidermal cells. Annu Rev Plant Biol. Annual Reviews 59(1):683–707

Schirmer A et al (2010) Microbial biosynthesis of alkanes. Science 329(5991):559–562

Schroeder JI, Kwak JM, Allen GJ (2001) Guard cell abscisic acid signalling and engineering drought hardiness in plants. Nature. Nature Publishing Group 410(6826):327–330

Seki M et al (2007) Regulatory metabolic networks in drought stress responses. Curr Opin Plant Biol 10:296–302

Seo PJ, Park C-M (2011) Cuticular wax biosynthesis as a way of inducing drought resistance. Plant Signal Behav 6(7):1043–1045

Seo PJ et al (2009) The MYB96 transcription factor mediates abscisic acid signaling during drought stress response in Arabidopsis. Plant Physiol 151(1):275–289

Seo PJ et al (2011) The MYB96 transcription factor regulates cuticular wax biosynthesis under drought conditions in Arabidopsis. Plant Cell 23(3):1138–1152

Sharma MK et al (2010) Identification, phylogeny, and transcript profiling of ERF family genes during development and abiotic stress treatments in tomato. Mol Gen Genomics. Springer-Verlag 284(6):455–475

Shepherd T, Robertson GW, Griffiths DW, Birch ANE, Duncan G (1995) Effects of environment on the composition of epicuticular wax from kale and swede. Phytochemistry 40(2):407–417

Shepherd T, Griffiths DW (2006) The effects of stress on plant cuticular waxes. New Phytol 171(3):469–499

Shinozaki K, Yamaguchi-Shinozaki K, Seki M (2003) Regulatory network of gene expression in the drought and cold stress responses. Curr Opin Plant Biol 6:410–417

Somerville CR, Browse J, Jaworski J, Ohlrogge J (2000) Lipids. In: Buchanan BB, Gruissem W, Jones RL (eds) Biochemistry and molecular biology of plants. American Society of Plant Physiologists, Rockville, pp 456–527

Song H et al (2016) Evolution of the KCS gene family in plants: the history of gene duplication, sub/neofunctionalization and redundancy. Mol Genet Genomics: MGG 291(2):739–752

Sterk P et al (1991) Cell-specific expression of the carrot EP2 lipid transfer protein gene. Plant Cell. American Society of Plant Biologists 3(9):907–921

Sturaro M et al (2005) Cloning and characterization of GLOSSY1, a maize gene involved in cuticle membrane and wax production. Plant Physiol. American Society of Plant Biologists 138(1):478–489

Suh MC (2005) Cuticular lipid composition, surface structure, and gene expression in Arabidopsis stem epidermis. Plant Physiol 139(4):1649–1665

Sun W et al (2015) The TsnsLTP4, a nonspecific lipid transfer protein involved in wax deposition and stress tolerance. Plant Mol Biol Report. Springer US 33(4):962–974

Tanaka H, Machida Y (2008) 10 the cuticle and cellular interactions. Ann Plant Rev Biol Plant Cuticle. John Wiley & Sons 23:312

Tanaka H et al (2002) ACR4, a putative receptor kinase gene of Arabidopsis thaliana, that is expressed in the outer cell layers of embryos and plants, is involved in proper embryogenesis. Plant Cell Physiol 43(4):419–428

Terzaghi WB, Cashmore AR (1995) Light-regulated transcription. Annu Rev Plant Physiol Plant Mol Biol. Annual Reviews USA 46(1):445–474

Tevini M, Steinmüller D (1987) Influence of light, UV-B radiation, and herbicides on wax biosynthesis of cucumber seedling. J Plant Physiol. Urban & Fischer 131(1):111–121

Tevini M, Braun J, Fieser G (1991) The protective function of the epidermal layer of rye seedlings against ultraviolet-B radiation. Photochem Photobiol. Blackwell Publishing Ltd 53(3):329–333

Thomas DA, Barber HN (1974) Studies on leaf characteristics of a cline of Eucalyptus urnigera from Mount Wellington, Tasmania: water repellency and the freezing of leaves. Aust J Bot 22:501–512

Thoma S, Kaneko Y (1991) A non-specific lipid transfer protein from. Plant J. Blackwell Publishing Ltd 3(1 993):427–436

Tran LP et al (2007) Plant gene networks in osmotic stress response: from genes to regulatory networks. Methods Enzymol 428:109–128

Uppalapati SR et al (2012) Loss of abaxial leaf epicuticular wax in Medicago truncatula irg1/palm1 mutants results in reduced spore differentiation of anthracnose and nonhost rust pathogens. Plant Cell. American Society of Plant Biologists 24(1):353–370

Vioque J, Kolattukudy PE (1997) Resolution and purification of an aldehyde-generating and an alcohol-generating fatty acyl-CoA reductase from pea leaves (Pisum sativum L.) Arch Biochem Biophys 340(1):64–72

Vogg G et al (2004) Tomato fruit cuticular waxes and their effects on transpiration barrier properties: functional characterization of a mutant deficient in a very-long-chain fatty acid ß-ketoacyl-CoA synthase. J Exp Bot 55(401):1401–1410

von Wettstein-Knowles P, Avato P, Mikkelsen JD (1980) Light promotes synthesis of the very long fatty acyl chains in maize wax. In: Mazliak P, Benveniste P, Costes C, Douce R (eds) Biogenesis and function of plant lipids. Elsevier/North-Holland Biomedical Press, New York, pp 271–274

von Wettstein-Knowles P (1995) Biosynthesis and genetics of waxes. In: Hamilton R (ed) Waxes. The Oily Press Ltd, West Ferry, Dundee, pp 91–129
Walton TJ (1990) Waxes, cutin and suberin. Methods in plant biochemistry 4.11. Edited by membranes and aspects of photobiology Lipids. pp 5–158
Wang Y et al (2012) An ethylene response factor OsWR1 responsive to drought stress transcriptionally activates wax synthesis related genes and increases wax production in rice. Plant Mol Biol. Springer Netherlands 78(3):275–288
Wang W et al (2014) Cucumber ECERIFERUM1 (CsCER1), which influences the cuticle properties and drought tolerance of cucumber, plays a key role in VLC alkanes biosynthesis. Plant Mol Biol. Springer Netherlands 87(3):219–233
Wang W et al (2015a) Cucumis sativus L. WAX2 plays a pivotal role in wax biosynthesis, influencing pollen fertility and plant biotic and abiotic stress responses. Plant Cell Physiol. Oxford University Press 56(7):1339–1354
Wang Y, Wang M, Sun Y, Wang Y et al (2015b) FAR5, a fatty acyl-coenzyme A reductase, is involved in primary alcohol biosynthesis of the leaf blade cuticular wax in wheat (Triticum aestivum L.) J Exp Bot. Oxford University Press 66(5):1165–1178
Wang Y, Wang M, Sun Y, Hegebarth D et al (2015c) Molecular characterization of TaFAR1 involved in primary alcohol biosynthesis of cuticular wax in hexaploid wheat. Plant Cell Physiol. Oxford University Press 56(10):1944–1961
Wang M et al (2016) Three TaFAR genes function in the biosynthesis of primary alcohols and the response to abiotic stresses in *Triticum aestivum*. Sci Rep. Nature Publishing Group 6:25008
Warui DM et al (2011) Detection of formate, rather than carbon monoxide, as the stoichiometric coproduct in conversion of fatty aldehydes to alkanes by a cyanobacterial aldehyde decarbonylase. J Am Chem Soc. American Chemical Society 133(10):3316–3319
Webb AR (1997) Monitoring changes in UV-B radiation. In: Seminar series-society for experimental biology. Cambridge University Press, pp 13–30
Wessler SR et al (2005) Homing into the origin of the AP2 DNA binding domain. Trends Plant Sci. Elsevier 10(2):54–56
Wollenweber E (1984) The systematique implication of flavonoids secreted by plants. In: Biology and chemistry of plant trichomes. Plenum Press, pp 53–69
Wollenweber E, Schneider H (2000) Lipophilic exudates of Pteridaceae – chemistry and chemotaxonomy. Biochem Syst Ecol 28(8):751–777
Wu XY, Moreau RA, Stumpf PK (1981) Studies of biosynthesis of waxes by developing jojoba seed: III. Biosynthesis of wax esters from Acyl-CoA and long chain alcohols. Lipids. Springer-Verlag 16(12):897–902
Xiao S, Chye ML (2009) An Arabidopsis family of six acyl-CoA-binding proteins has three cytosolic members. Plant Physiol Biochem 47:479–484
Yamaguchi-Shinozaki K, Shinozaki K (1993) Characterization of the expression of a desiccation-responsive rd29 gene of Arabidopsis thaliana and analysis of its promoter in transgenic plants. Mol Gen Genet MGG. Springer-Verlag 236(2–3):331–340
Yang J et al (2011) Induced accumulation of cuticular waxes enhances drought tolerance in Arabidopsis by changes in development of stomata. Plant Physiol Biochem. Elsevier Masson SAS 49(12):1448–1455
Yu N, Song Chao WX (2014) Investigation on response mechanism of epicuticular wax on Arabidopsis thaliana under cold stress. Sci Agric Sin 47(2):252–261
Zhang Q (2007) Strategies for developing green super rice. Proc Natl Acad Sci U S A. National Academy of Sciences 104(42):16402–16409
Zhang J-Y et al (2005) Overexpression of WXP1, a putative Medicago truncatula AP2 domain-containing transcription factor gene, increases cuticular wax accumulation and enhances drought tolerance in transgenic alfalfa (Medicago sativa). Plant J. Blackwell Science Ltd 42(5):689–707
Zhang JY et al (2007) Heterologous expression of two Medicago truncatula putative ERF transcription factor genes, WXP1 and WXP2, in Arabidopsis led to increased leaf wax accumulation

and improved drought tolerance, but differential response in freezing tolerance. Plant Mol Biol. Kluwer Academic Publishers 64(3):265–278

Zhang D et al (2010) OsC6, encoding a lipid transfer protein, is required for postmeiotic anther development in rice. Plant Physiol. Am Soc Plant Biol 154(1):149–162

Zheng H, Rowland O, Kunst L (2005) Disruptions of the Arabidopsis Enoyl-CoA reductase gene reveal an essential role for very-long-chain fatty acid synthesis in cell expansion during plant morphogenesis. Plant Cell 17(5):1467–1481

Zhou L et al (2013) Rice OsGL1-6 is involved in leaf cuticular wax accumulation and drought resistance. PLoS One. Edited by G. Bonaventure. Public Library of Science 8(5):e65139

Zhou X et al (2015) OsGL1-3 is involved in cuticular wax biosynthesis and tolerance to water deficit in rice. PLoS One 2015(10):e116676

Zhu X et al (2012) Overexpression of wheat lipid transfer protein gene TaLTP5 increases resistances to Cochliobolus sativus and Fusarium graminearum in transgenic wheat. Funct Integr Genomics. Springer-Verlag 12(3):481–488

Zhu X, Xiong L, Qifa Zhang B (2013) Putative megaenzyme DWA1 plays essential roles in drought resistance by regulating stress-induced wax deposition in rice. PNAS 110(44):17790–17795

Zuk-Golaszewska K, Upadhyaya MK, Golaszewski J (2003) The effect of UV-B radiation on plant growth and development. Plant Soil Environ. Institute of Agricultural and Food Information 49(3):135–140

Abiotic Stress Response in Plants: A *Cis*-Regulatory Perspective

6

Aditi Jain, Gauri Joshi, Chetan Chauhan, and Sandip Das

Abstract

The growth and development of plant are adversely affected due to various abiotic stresses such as cold, heat, drought, salinity, heavy metal, hypoxia, etc. Plant adaptation to the stress conditions is brought out by changes in its physiological parameters which have underlying molecular basis. Multiple signalling pathways are activated in response to particular stress which are often complex and work through a network of *trans*- and *cis*-regulatory factors. In the following chapter, we summarize the current understanding of *cis*-regulatory factors, viz. promoters and transcription factor binding motifs, which are responsible for relaying signals during anabiotic signalling cascade ultimately preventing and mitigating the ill effect of climatic conditions on plant growth and survival.

Keywords

Abiotic stress · Cis-regulation · Stress inducible promoters · Epigenetic regulation

6.1 Introduction

Abiotic stress such as cold, drought, salt, heat, heavy metals and hypoxia leads to enormous loss of crop yield annually. Plants being sessile must adapt to the changing environmental conditions to ameliorate the stresses and maintain their optimal growth, metabolism and physiology (Niklas 2009; Sultan 1995). Dedicated

A. Jain · G. Joshi · C. Chauhan · S. Das (✉)
Department of Botany, University of Delhi, New Delhi, India
e-mail: sdas@botany.du.ac.in; sandipdsas04@gmail.com

S. M. Zargar, M. Y. Zargar (eds.), *Abiotic Stress-Mediated Sensing and Signaling in Plants: An Omics Perspective*, https://doi.org/10.1007/978-981-10-7479-0_6

regulatory networks/networks of genetic pathways have evolved for the regulation of stress-responsive genes (Chapin et al. 1993).

A principle mode of regulation of protein and some RNA coding gene (such as miRNA) expression occurs at the *cis*-regulatory element where transcription factors or *trans*-factors interact with their relevant *cis*-regulatory motifs. Such protein and RNA coding genes that are differentially regulated (in contrast to tRNA and rRNA which are constitutively expressed) are driven by type II promoters as the major *cis*-element and require the active participation of RNA polymerase II (Butler and Kadonaga 2002). *Cis*-regulatory elements are generally located near the transcription start site (TSS) in the plants as compared to the animals, where they are present far from TSS (Katagiri and Chua 1992; Chandler 2001). Each type II promoter as *cis*-element consists of motifs for binding of multiple *trans*-factors to regulate and modulate (enhance or repress) gene expression (Kwok et al. 1994). Conservation of *cis*-regulatory element for the external stimulus within the promoter in different species shows universality of mechanisms/pathways of abiotic stress responses (Meier et al. 2008).

6.2 *Cis*-Regulatory Elements and Motif Involved in Abiotic Stresses

6.2.1 Cold-Responsive *Cis*-Elements and Motifs

Cold stress includes chilling (<20 °C) and/or freezing (<0 °C) which often affects plant growth and development and causes significant losses to crop. Cold stress inhibits the metabolic reaction through cold-induced osmotic, oxidative and other stresses. Plants respond and adapt to these stresses at molecular and cellular levels through a process termed as 'cold acclimation' which induces profound biochemical and physiological changes, by altering gene expression. In *Arabidopsis thaliana*, cold-regulated genes have been estimated to constitute 4–20% of the genome (Chinnusamy et al. 2010). Several *cis*- and *trans*-acting factors are known to be involved in cold stress-responsive transcription. DRE/C-repeat was the first low-temperature-responsive *cis*-element identified by Yamaguchi and Shinozaki in 1994 and is located on promoter of *RD29* gene in *Arabidopsis thaliana* (Yamaguchi and Shinozaki 1994). Another motif, termed as DRE (drought-responsive element), also known as C-repeat (CRT) is a 9-bp *cis*-element (5′-TACCGACAT-3′) with a core sequence of 5 bp (5′-CCGAC-3′) and is responsive to drought, cold and salt stress (Yamaguchi and Shinozaki 1994). *CBF/DREB* (CRT/DRE binding factor) is a transcription factor which binds to DRE/C-repeat and positively regulates the expression of cold stress-responsive gene (Stockinger et al. 1997).

ICE (inducer of CBF expression) recognizes a *cis*-regulatory motif known as ICE-box with the sequence 5′-CACATG-3′ (CANNNTG; Meshi and Iwabuchi 1995) which is present in the promoters of CBF genes. ICE protein is present in the cytoplasm in inactive form, which becomes active at low temperature, binds to ICE-box and activates CBF expression (Gilmour et al. 1998). The expression of COR

(Cold Responsive) genes are mediated by cold in *Arabidopsis thaliana* (Wilhelm and Thomashow 1993; Baker et al. 1994), and an analysis of its promoter region revealed that it contains CRT/DRE *cis*-motifs. In *C. bursa-pastoris*, two homologs of *COR* are present *COR15a* and *COR15b*, and their associated promoters contain two and one *cis*-motif of CRT/DRE, respectively. *CbCOR15* gene is expressed at normal level under normal growing conditions, but under cold stress, its expression is greatly increased in all tissues (Lin et al. 2016). Two other genes, *RAB18* (Lang and Palva 1992) and *KIN1/2* (Kurkela and Franck 1990), from *A. thaliana* also contain CRT/DRE (also known as LTRE—low-temperature-responsive element) in their promoter region. *BN115* from *Brassica napus*, a homolog of *A. thaliana COR15a*, also is transcriptionally induced in response to cold which can be ascribed to the presence of CRT/DRE motif in its promoter (Jiang et al. 1996). Similarly, in wheat, the cold inducibility of *WCS120* is because of two CRT/LTRE motifs in the promoter, which was confirmed by deleting these motifs that led to loss of induction (Vazquez-Tello et al. 1998).

Another motif that has been associated with cold-induced genes are myelocytomatosis viral oncogene homolog (MYC, CANNTG) and myeloblastosis viral oncogene homolog (MYB) binding motif (WAACCA, YAACKG or GGATA, where W (A/T), Y(C/T), K (G/T)) that had previously been identified in *A. thaliana* in response to dehydration. Subsequently, these were also reported to respond under chilling and freezing conditions (Abe et al. 2003; Agarwal et al. 2006). Characterization of the *Paeonia suffruticosa* MPT (mitochondrial phosphate transfer) gene promoter, which transfers inorganic phosphate to mitochondria where ATP is synthesized, revealed that it responds to many abiotic stresses and hormones. Various *cis*-elements present in the promoter include putative pyrimidine box, two GARE motifs, TAG element, SURE element and MYB and MYC recognition site. The chilling/cold response could be ascribed to the presence of one MYC element (CAATTG—412/408) that is sufficient to respond effectively to chilling, although four sites each of MYB and MYC binding sites are present (Zhang et al. 2016).

Another motif that is found in cold-inducible promoter and responsible for induction is a highly conserved EE motif which is bound by CCA1 (circadian clock-associated 1)/LHY (late elongated hypocotyl) transcription factor proteins (Harmer et al. 2000; Mikkelsen and Thomashow 2009). Interestingly, CCA/LHY proteins are classified as MYB transcription factors, related to circadian clock, and also bind to timing of CAB expression 1 (TOC1) promoters (Gendron et al. 2012). Dehydration-inducible promoter also contains EE motif but less frequently. Some novel motifs CGTACG and GTAGTA were also reported to be present in highly conserved manner in cold-inducible rice promoters (Maruyama et al. 2012), and their cognate *trans*-factors are yet to be fully characterized.

6.2.2 Drought-Responsive *Cis*-Elements and Motifs

Mesophilic plants exposed to drought conditions respond via abscisic acid (ABA)-dependent and ABA-independent signalling pathways. This induction is regulated at the level of *cis*-elements which contain specific stress-responsive motifs that interact with a plethora of *trans*-factors to govern the regulation of drought-responsive genes (Yamaguchi-Shinozaki and Shinozaki 2005; Kim et al. 2010).

Drought-responsive genes that are known to act independently of abscisic acid contain *drought-responsive elements* (DRE) and *C-repeat* (CRT) *cis*-elements in their promoter with a similar 9-bp conserved sequence TACCGACAT (consensus A/GCCGAC) both involved in drought stress; CRT/DRE *cis*-motifs have therefore also been termed as drought-responsive *cis*-element (Yamaguchi-Shinozaki and Shinozaki 2006, Baker et al. 1994; Jiang et al. 1996; Stockinger et al. 1997; Thomashow 1999; Yamaguchi-Shinozaki 1994). In contrast, genes that act via the ABA-dependent pathway harbour *abscisic acid-responsive element* (ABRE), an 8-bp conserved sequence with a consensus of PyACGTGG/TC and with a core ACGT sequence in their promoters (Fujita et al. 2011; Nakashima et al. 2009). During the analysis of *Osem* gene in rice, it was found that the ABRE (with an A/GCGT motif) occurs together with a coupling element, CE3 (Hobo et al. 1999). Similarly, the promoter of *rd29A* of *A. Thaliana* contains one ABRE motif and DRE/CRT motif, and the activation via *rd29A* promoter requires interdependence of both DRE/CRT and ABRE motifs. The DRE/CRT motif may be acting as a coupling element (CE) to the ABRE with binding of DREB1/CBF and DREB2 protein to DRE/CRT and AREB/ABF protein to ABRE (Narusaka et al. 2003).The *rd29B* gene on the other hand contains two ABRE sequence, and both are responsible for ABA-dependent dehydration responsiveness, in a slow induction manner (Yamaguchi-Shinozaki and Shinozaki 1994; Uno et al. 2000). Single copy of ABRE is not sufficient to drive the expression; it requires the presence of another copy of ABRE element or CRT/DRE element or coupling elements, CE1, with a motif of TGCCACCGG or CE3 sequence with a motif of ACGCGTGTCCTG. The coupling elements CE1 and CE3 function together with a ACGT-box, namely, ABRE3 element (GCCACGTACA), that was discovered during the analysis of promoter of *HVA1* and *HVA22* genes from barley (Shen and Ho 1995; Shen et al. 1996; Hobo et al. 1999), an 8-bp (CACGTGGC) ABRE studied in *Em* gene from wheat (Guiltinan et al. 1990). MYC (CANNTG) and MYB (C/TAACNA/G) *cis*-motifs are also found in the promoters of drought-responsive genes (Abe et al. 1997).

Other motifs that are known to induce the drought-responsive genes are CCAAT-box (Li et al. 2008; Laloum et al. 2013), Sph-box and GTGTC-box of maize *C1* gene (McCarty 1995), MYC-like sequence (CATGTG) and a 14-bp rps1 site 1-like sequence in the promoter of *early responsive to dehydration1* (*ERD1*) gene under dehydration stress (Tran 2004; Simpson et al. 2003). The promoter of *Rd22* gene in *Arabidopsis thaliana* does not contain any ABRE motif but respond in drought condition because of the presence of two MYC and MYB recognition sites (Abe et al. 1997; Iwasaki et al. 1995).The promoter of *ProDH* (proline dehydrogenase) contains the ACTCAT *cis*-motif that is involved in rehydration-responsive gene expression (Satoh 2002).

6.2.3 Heavy Metal-Responsive *Cis*-Elements and Motifs

Promoters of heavy metal-responsive genes contain metal-interacting *cis*-motifs that play a major role in their regulation. Several *cis*-acting motifs involved in responses to heavy metals, such as Cu^{2+}, Zn^{2+}, Al^{3+}, Cu^{2+} and Cd^{2+}, have been found. Metal-responsive element (MRE) like sequence 5′-TGCGCNC-3′ has been shown to be present in the promoter of Cd^{2+}-responsive genes (Karin et al. 1987). Subsequently, several researchers demonstrated the presence of MRE sequence responsible for heavy metal-responsive regulation in various other plants such as Cd^{2+} response in *Nicotiana tabacum* (Kusaba et al. 1996), *Oryza sativa* class I-4b metallothionein gene (OsMT-I-4b; Ren and Zhao 2009; Dong et al. 2010) and *Phaseolus vulgaris* stress-related gene 2 (PvSR2; Qi et al. 2007; Sun et al. 2015). Apart from protein-coding genes, Cd^{2+}-responsive elements have also been shown to be present in the promoter of miRNA genes of rice (Ding et al. 2011).

Fe^{2+} deficiency *cis*-motifs, IDE1 (5′-ATCAAGCATGCTTCTTGC-3′) and IDE2 (5′-TTGAACGGCAAGTTTCACGCTGTCACT-3′), have been shown to be responsible for the regulation of iron-responsive genes (Kobayashi 2005; Kobayashi et al. 2003; Kong and Yang 2010; Li et al. 2015; Ogo et al. 2008). In the green algae *Chlamydomonas reinhardtii*, a Cu^{2+} response element (CuRE) with the conserved core sequence 5′-GTAC-3′ that lies 120 bp upstream to the transcription start site of *Cyc6* gene was detected (Quinn and Merchant 1995; Quinn 2000; Kropat et al. 2005; Zhou et al. 2013). In plants, a Cu^{2+}-responsive *cis*-element conferred a strong GUS expression in seedling stage of transgenic *Arabidopsis thaliana* when used with rice MT gene promoter transcriptionally fused to *uidA* (Lü et al. 2007). Similar CuRE motifs have also been reported to be present in the promoters of *Pisum sativum* metallothionein genes (PsMTA; Fordham-Skelton et al. 1997), *Lycopersicon esculentum* metallothionein-like genes (LeMTA and LeMTB; Whitelaw et al. 1997), *Pseudotsuga menziesii* metallothionein-like gene (PmMT; Chatthai et al. 2004) and *Pisum sativum* stress-related gene 2 (PvSR2; Qi et al. 2007).

An Al^{3+}-responsive *cis*-acting motif 5′-GGN(T/g/a/C)V(C/ A/g)S(C/G)-3′ has been found in the promoter regions of 29 ARTI-regulated genes in *Oryza sativa* (Tsutsui et al. 2011) and in the promoter of *TaALMT1*(Ryan et al. 2010). Interestingly, a miniature inverted-repeat transposable element (MITE) responsible for Al^{3+} response was found in the promoter of *SbMATE* gene in *Sorghum bicolor* (Magalhaes et al. 2007).

Zn^{2+}-responsive *cis*-motif 5′-ACCYYNAAGGT-3′ present on the promoter of *ZRT1*, *ZRT2* and *ZAP1* genes is responsible for the expression under zinc stress (Zhao et al. 1998); another *cis*-motif called Zn^{2+} deficiency response element 5′(ZDRE; 5′-RTGTCGACAY-3′) present on the promoter of *ZIP4* gene can regulate this gene under zinc-deficient condition (Assunção et al. 2010). The regulation of Ni^{2+}-responsive genes is mediated by both copper response elements (CuRE) and copper response regulator (CRR1) *cis*-motifs (Quinn et al. 2003).

6.2.4 High Temperature-Responsive *Cis*-Elements and Motifs

Temperature, one of the primary factors affecting the rate of plant development, has been severely affected by global climate change. High temperature leads to a decrease in photosynthetic efficiency, grain number and grain weight leading to crop productivity losses. The importance of ambient temperature for plant development was realized as early as in the 1980s, and studies on heat shock response in several higher plants such as soybean, tobacco, maize, carrot, tomato and *Tradescantia paludosa* were conducted which showed that a thermal stress is followed by an increase in production/expression of heat shock proteins (Spena and Schell 1987). Heat shock proteins and ROS scavenging enzymes are key functional proteins that are produced in response to heat stress and are thus widely studied proteins to understand heat stress signalling in plants.

In 1982, Pelham and co-workers successfully isolated and characterized the first heat-inducible promoter, i.e. hsp70 (heat shock protein), from *Drosophila* (Pelham 1982) which in subsequent studies was shown to be also active in plant and animal tissues, demonstrating that the induction mechanism involved in heat shock response is conserved across the evolutionary boundaries (Spena et al. 1985; Spena and Schell 1987; Schmülling et al. 1989). Later, Gurley and co-workers characterized a heat-inducible promoter from heat shock protein gene (*HSP17.5*) from soybean and identified important regions required for thermal induction (Gurley et al. 1986). Further analysis of this promoter led to identification of four *cis*-motifs, viz. TATA domain, AT-rich region and two sites containing consensus heat shock elements (HSE), which are present between −179 and −40 positions (Czarnecka et al. 1989). Several other heat-induced promoters from plants were characterized by fusing them to GUS or GFP reporters and measuring the reporter activity under developmental stages and in response to various. One such example is the promoter of *HSP18.2* which was isolated and characterized from *Arabidopsis thaliana* and showed enhanced activity in all the tissues and at all developmental stages in response to heat stress as expected (Takahashi et al. 1992). The expression of this promoter was also seen in heterologous systems such as *Nicotiana plumbaginifolia* (Moriwaki et al. 1999) and *Nicotiana tabacum* (Lee et al. 2007). Other promoters which have been characterized for heat inducibility are *HSP81* from *A. thaliana* (Yabe et al. 1994), APX1 from *A. thaliana* (Storozhenko et al. 1998), *Hsp90-1* from *A. thaliana* (Haralampidis et al. 2002), *Hsp101* from *O. sativa* (Wu et al. 2009), *Hvhsp17* from *Hordeum vulgare* (Freeman et al. 2011) and *sHsp26* from *Triticum aestivum* (Khurana et al. 2013). Interestingly, a heat-inducible bidirectional promoter (intergenic region between heat shock protein 100 and choline kinase 2) has also been isolated (Mishra and Grover 2014).

Apart from the routine characterization for the heat-inducible promoters, these studies also focused on deciphering the characteristics of the *cis*-regulatory regions which are required for heat induction of such promoters. Early studies established that a relatively simple element termed as heat response element (HSE), if present on promoter, can confer heat induction. Following heat stress, HSPs form trimer and bind to HSE to relay signals for activation/expression of heat-induced genes. This element consists of a module of trinucleotide repeat 5′-GAA-3′ or 5′-TTC-3′

(inverse orientation) arranged in a palindromic fashion separated by 2 bp from one another (Pelham 1982; Amin et al. 1988). Later, a detailed analysis of *Hsp17.3* from soybean indicated the importance of these two bp flanking the trinucleotide repeat and established that 5′aGAAg-3′ is the transcriptional core sequence in plants (Barros et al. 1992). Further, AT-rich sequence has been shown to have moderate effect on thermo-induction of promoters, whereas CCAAT-box sequence was found to act cooperatively with HSEs to increase the promoter activity in response to heat (Rieping and Schöffl 1992). Study on regulatory network of heat stress showed that DREB2A (dehydration-responsive element-binding protein 2A), a major protein involved in dehydration and salt stress, is also a key player in relaying signals of heat stress. Therefore, the DRE element (DREB2A binding cis-element), with a core sequence A/GCCGAC, is also frequently located on heat-inducible promoters (Sakuma et al. 2006; Ohama et al. 2017).

Apart from deletion analysis of promoter regions, comparative analysis of the promoter regions of heat shock protein genes was carried out to elucidate the regions which are important for heat response and are thus conserved in otherwise variable promoters. One such study on promoter regions of small heat shock proteins in rice identified 18 highly conserved *cis*-regulatory motifs in these promoters, such as ARRIAT (5′NGATT 3′), BIHD1OS (5′ TGTCA 3′), CAATBOX (5′ CAAT 3′), DOFCOREZM (5'AAAG 3′), CACTFTPPCA1(5′ YACT3'), EBOXBNNAPA (CANNTG), GATABOX (GATA), GT1CONSENSUS (GRWAAW), GTGANTG10 (GTGA), MYBCORE (CNGTTR), MYBCOREATCYB1(AACGG), NODCON2GM (CTCTT), POLLEN1LELAT52 (AGAAA), RAV1AAT (CAACA), RHERPATEXPA7 (TTGAC) and WRK71OS (TGAC), which may have a role in thermal induction of promoters (Safdar et al. 2016). A similar study on promoter regions of heat shock-inducible genes from *Arabidopsis*, soybean, rice and maize led to identification of hexamers which are overrepresented in these genes. The results showed that sequence TCTAGA is abundant in *Arabidopsis* and soybean, TCCAGA in rice and maize and CTAGAA in the promoters of all the studied species. Further, these predictions were employed to design an optimal heat shock-inducible promoter which can contribute to future molecular breeding programmes in relation to climate change (Maruyama et al. 2017).

6.2.5 Anaerobic/Hypoxic Stress-Responsive *Cis*-Elements and Motifs

The global climate change is expected to perturb the water availability drastically creating drought-like conditions in some areas while flooding in others (Durack et al. 2012). The adverse conditions in terms of water availability are expected to affect farming in almost every region of the world. Soil waterlogging (a result of flooding) causes oxygen deficiency in plants affecting its normal metabolism ultimately leading to death of the plants. Under conditions of hypoxia (low oxygen concentration) and anoxia (total absence of oxygen), plants respire via fermentation and glycolytic pathways (anaerobic respiration) which are less energy efficient than aerobic respiration. Therefore, genes involved in fermentation pathway such as alcohol dehydrogenase, pyruvate decarboxylase and sucrose synthase are upregulated in response to anaerobic stress.

Maize alcohol dehydrogenase (*Adh1*) is the first and one of the most extensively studied genes in context of anaerobic stress signalling as its expression level increases 50-fold after 7 h of flooding (Olive et al. 1991). The analysis of promoter region of *Adh1* showed the presence of two anaerobic responsive elements (ARE, with consensus sequence 5′-GC(G/C)CC-3′) which are required in tandem to be bound by a protein for hypoxia-induced transcription of *Adh1* (Walker et al. 1987; Olive et al. 1991). The *Arabidopsis* counterpart of this gene, *Adh1*, was shown to express in immature seedlings and induced by hypoxic condition in roots of adult plant (Delisle and Ferl 1990). The promoter region of *Adh* genes also shows the presence of G-box sequence (5′-CCACGTGG -3′) which is responsible for induction of promoter in response to abscisic acid and other abiotic stresses such as low temperature and drought (McKendree and Ferl 1992; de Vetten and Ferl 1995; de Bruxelles et al. 1996). The G-boxes are bound by G-box binding factor 1 (GBF1) protein which is required for hypoxia responsiveness of *Adh* promoter (de Vetten and Ferl 1995). Recently, a hybrid *Adh* promoter has been constructed by fusing two fragments of *Adh* promoter from maize, and the characterization of this promoter through promoter/reporter lines confirmed its high expression in response to hypoxia (Ammara et al. 2016).

Interestingly, characterization of promoter of glyceraldehyde-3-phosphate dehydrogenase (gene involved in glycolytic pathway) of maize in heterologous systems, *A. thaliana* and tobacco, showed that its induction in response to anaerobic condition is light-dependent. The hypoxic induction of this gene is attributed to a TATA box and Myb binding site present in its promoter region (Kohler et al. 1996; Hänsch et al. 2003; Zeng et al. 2016). Alanine aminotransferase is an enzyme which catalyses the reversible conversion of pyruvate and glutamate into alanine and oxoglutarate and has been shown to increase its expression in response to anaerobic stress, light and nitrogen (Ricoult et al. 2006; Miyashita et al. 2007; Rocha et al. 2010). The promoter region of this gene from millet also showed the presence of anaerobic-responsive element, and its activity increases twofold in roots in response to hypoxia (Muench et al. 1998).

Hypoxia leads to build up of ethylene in submerged plant, which triggers ethylene signalling pathway that helps in adaptation of plants to low oxygen condition. A hypoxic plant can either employ a quiescent or escape strategy to adapt via ethylene signalling pathway. In quiescent strategy, *SUB1A*, a transcription factor belonging to VII group of the ethylene-responsive factor (ERF) family, decreases gibberellin responsiveness, thus leading to shoot elongation by a process involving ethylene. In escape strategy, SNORKEL1 and SNORKEL2 proteins (belonging to VII group of the ethylene-responsive factor) promote GA-mediated internode elongation as seen in the case of deepwater rice varieties. Promoter analysis of *SUB1A* gene from rice showed its expression (GUS expression) at the base of the leaf sheath and in the leaf collar region (Singh et al. 2010). SUB1A protein from rice has further been shown to bind to GCC box of *ubiquinol-cytochrome C chaperone* gene and regulate its expression (Prajapati et al. 2013).

A microarray study performed on wild-type and transgenic P_{sag12}-IPT of *A. thaliana* under normal and hypoxic conditions showed that expression of genes involved in glycolysis and fermentation pathways, ethylene synthesis and perception, calcium signalling, nitrogen utilization, trehalose metabolism and alkaloid synthesis were altered in response to hypoxia. Furthermore, the study showed that promoters of cluster of upregulated genes show the presence of motifs such as AtMYB2 binding motif (GT motif), a sugar response element-like motif and a G-box-related sequence (Liu et al. 2005). Mohanty and co-worker performed a bioinformatics study to find the common motifs which are overrepresented in a promoter region of the majority of anaerobically induced genes. The study identified five novel motifs: 5′-AAACAAA-3′, 5′-AGCAGC-3′, 5′-TCATCAC-3′, 5′-GTTT(A/C/T)GCAA-3′ and 5′-TTCCCTGTT-3′ which are shared by the majority of these genes (Mohanty and Swarup 2005).

6.2.6 Salinity Stress-Responsive *Cis*-Elements and Motifs

Salinity is one of the major factors that limit the growth and productivity of agricultural crops in many parts of the world by inhibiting seed germination and growth physiology of plant, ultimately leading to a decrease in plant vigour and crop yield. One of the major effects of salt stress is on plant-water relations which can be seen as a large overlap between drought, salt stress and ABA (abscisic acid)-induced genes, i.e. >50%. Salt stress alters both osmotic and ionic stress in plants, and tolerance/adaptation to salt stress involves multiple pathways. Osmotic stress signalling occurs via ABA-dependent or ABA-independent pathway, whereas the ionic stress response takes place either via ROS, SOS or Ca^{2+}-mediated signalling pathway (Kumar et al. 2013). *Rd29A* and *rd29B* are one of the first and extensively studied genes in response to abiotic stress. Expression patterns of these genes showed that *rd29A* is an early stress-responsive gene, whereas *rd29B* is a late stress-responsive gene. Analysis of promoter regions of these genes showed that *rd29A* contains DRE element (TACCGACAT), ABRE (YACGTGGC) and as-1-like motif, whereas *rd29B* showed the presence of two ABRE and one Myb recognition sequence (PyAACT/GG). It was further inferred that DRE element is responsible for early stress response of *rd29A* (ABA-independent pathway), and the slow response to stress is mediated via ABA pathway which requires either two ABRE element or single ABRE element together with a coupling element such as CE3 (ACGCGTGTCCTG) or A/GCGT (Yamaguchi-Shinozaki and Shinozaki 1994; Narusaka et al. 2003).

SOS (Salt Overly Sensitive) pathway plays an important role in maintaining ionic homeostasis by Na^+/K^+ influx-efflux. The pathway starts with stress generated Ca^{2+} signal that is perceived by SOS3 and results in the formation of the SOS3-SOS2 complex. This complex in turn phosphorylates SOS1 and activates its transport activity (plasma membrane Na^+/H^+ antiporter). The analysis of promoter region of *SOS1* gene from *Arabidopsis thaliana* showed its expression in epidermal cells at the root tip and in parenchyma cells at the xylem/symplast boundary of roots, stems and leaves (Shi et al. 2002). Further, in silico analysis of promoter of *SOS1* gene from an extreme halophile *Salicornia brachiate* showed the presence of various stress-specific motifs such as DOF motif (AAAG), GT elements (GAAAAA), ABRE-like sequence (ACGTG/ ACGT) and root-specific motifs (KCACGW, CAACA, CACCTG). Functional analysis of this promoter showed that it is induced by salt stress but not by ABA or cold (Goyal et al. 2013). Promoter analysis of *H⁺-translocating pyrophosphatases* (*H+-PPase*, a gene involved in proton transport across membranes by catalytic hydrolysis of inorganic pyrophosphate to provide energy) of maize identified a 71-bp segment (−219 to −148 bp) that responds to NaCl or PEG stress (Hou et al. 2016). The salt and PEG inducibility of this 71-bp fragment was attributed to GT1 element (GAAAAA) present in this promoter region.

Another important family of transcription factors studied in context of salt and other abiotic stresses is NAC transcription factor family. *ANAC019*, *ANAC055* and *ANAC072* (members of NAC transcription factor family) were shown to be responsive to drought, salinity and ABA. Promoter analysis of *ANAC092* as carried out by transcriptional fusion with *GUS* reporter showed that it is expressed in leaf, roots and young and older flowers. Levels of GUS reporter expression increased in response to salt indicating that the gene is involved in salinity-induced plant senescence (Balazadeh et al. 2010). Upon stress-mediated induction, ANAC proteins bind to CATGTG motif which is present on promoter region of *erd1* (*early responsive to dehydration stress 1*) gene (Tran et al. 2004), triggering multitude of downstream stress tolerance pathways.

A summary of the *cis*-motifs involved in abiotic stress responses found in promoter regions is listed in Table 6.1.

6.3 Epigenetic Changes and Abiotic Stress Response

In nature, plants are exposed to a variety of biotic and abiotic stresses that adversely affect crop productivity, growth and development. Plants have developed innate mechanisms to combat these environmental stresses. The most prevalent abiotic stresses in environment include high temperature, low temperature, drought, high salinity stress, heavy metal stress, etc. The growth and development of plants are regulated by several proteins and genes under abiotic stresses (Li et al. 2015). In the recent past, epigenetic mechanisms such as DNA methylation and histone modifications have been shown to play an important role in regulation of stress-related genes (Chinnusamy and Zhu 2009; Vanyushin and Ashapkin 2011).

DNA methylation is one of such epigenetic modifications that has been shown to be involved in abiotic stress response. DNA methylation includes both hyper- and hypomethylation (Boyko et al. 2010; Bilichak et al. 2012; Karan et al. 2012).

Demethylation of promoter region that could enhance the expression of stress-related gene was postulated as early as in 1998 (Finnegan et al. 1998). GAPC (glyceraldehyde-3-phosphate dehydrogenase cytosolic) plays an important role in stress conditions (Pillai et al. 2002; McLoughlin et al. 2013). In *Oryza sativa*, *OsGAPC2* responds to drought stress, salt stress, heat stress and ABA; *Zea mays*

Table 6.1 *Cis*-motifs involved in abiotic stress signalling in plants

Cis-element	Consensus sequence	Stress	Plant species	References
DRE (drought-responsive element)	TACCGACAT	Drought/cold/heat/salt	*Arabidopsis thaliana*	Yamaguchi and Shinozaki (1994)
CRT (C-repeat)	CCGAC	Drought/cold	*Arabidopsis thaliana*	Yamaguchi and Shinozaki(1994)
ICE-box	CACATG	Cold	*Arabidopsis thaliana*	Meshi and Iwabuchi (1995)
MYB binding site	CANNTG,CATGTG	Drought/cold	*Arabidopsis thaliana*	Abe et al. (1997)
MYC binding site	WAACCA, YAACKG,GGATA,CATGTG	Drought/cold	*Arabidopsis Thaliana*	Abe et al. (1997)
EE motif (evening element)	AAAATATCT	Cold	*Arabidopsis thaliana*, *Oryza sativa*, *Glycine max*	Harmer et al. (2000) and Mikkelsen and Thomashow (2009)
ABRE (ABA response element)	PyACGTGG/TC	Drought/salt	*Arabidopsis thaliana*	Fujita et al. (2011) and Nakashima et al. (2009)
CE1 (coupling element)	TGCCACCGG	Drought	Barley	Shen and Ho (1995), Shen et al. (1996) and, Hobo et al. (1999)
Sph-box	CATGCA	Drought	Maize	McCarty (1995)
RRE (rehydration responsive element)	ACTCAT	Rehydration	*Arabidopsis Thaliana*	Satoh (2002)
MRE (metal-responsive element)	TGCGCNC	Heavy metal	*Nicotiana tabacum*	Kusaba et al. (1996)
IDE1	ATCAAGCATGCTTCTTGC	Heavy metal	Barley	Kobayashi (2005), Kobayashi et al. (2003), Kong and Yang (2010), Li et al. (2015), and Ogo et al. (2008)
IDE2	TTGAACGGCAAGTTTCACGCTGTCACT	Heavy metal	Barley	Kobayashi (2005); Kobayashi et al. (2003); Kong and Yang (2010); Li et al. (2015); Ogo et al. (2008)
CuRE (copper response element)	GTAC	Heavy metal	*Chlamydomonas reinhardtii*	Quinn and Merchant (1995), Quinn (2000), Kropat et al. (2005), and Zhou et al. (2013)

(continued)

Table 6.1 (continued)

Cis-element	Consensus sequence	Stress	Plant species	References
Al-responsive element	GGN(T/g/a/C)V(C/A/g)S(C/G)	Heavy metal	*Oryza sativa*	Tsutsui et al. (2011)
Zn-responsive elements	ACCYYNAAGGT	Heavy metal	*Arabidopsis thaliana*	Assunção et al. (2010)
Zn deficiency-responsive elements	RTGTCGACAY	Heavy metal	*Arabidopsis thaliana*	Assunção et al. (2010)
Heat shock element (HSE)	aGAAg	Heat	Soybean	Barros et al. (1992)
CCAAT-box	CCAAT	Heat	Soybean	Rieping and Schöffl (1992)
ARE	GC(G/C)CC	Anaerobic/hypoxia	Maize	Walker et al. (1987) and Olive et al. (1991)
G-box	CCACGTGG	Anaerobic/hypoxia	Maize	De Vetten and Ferl (1995)
Myb binding site	CAACGG	Anaerobic/hypoxia	Maize	Zeng et al. (2016)
CE3	ACGCGTGTCCTG	Salt	Barley	Shen and Ho (1995)
GT1 element	GAAAAA	Salt	Maize	Hou et al. (2016)
CATGTG motif	CATGTG	Salt	*Arabidopsis thaliana*	Tran et al. (2004)

ZmGAPC1 respond to anoxia, while *Arabidopsis thaliana* GAPC1/GAPC2 is involved in heat, anaerobic and salt stresses and drought. In a study on GAPC1 from *Triticum aestivum* (TaGAPC1), it was revealed that it is induced and highly expressed by stress in drought-tolerant wheat genotype, whereas it is expressed at a very low level in drought-sensitive wheat genotype. Bisulphite sequencing of the promoter region revealed that osmotic and salt stress in drought-tolerant wheat genotype induces demethylation of promoter region at C_PG and C_PH_PG and enhances the expression of the gene (Fei et al. 2016).

In *Oryza sativa*, OsGAPC2 responds to drought stress, salt stress, heat stress and ABA. Analysis of this salt-responsive miRNA in salt-tolerant (FL478) and salt-responsive (IR29) rice genotype reveals that it behaves differently in both the varieties which can be accounted for by differences in their promoter methylation and differential presence of salt-responsive *cis*-elements in the promoter region. Detailed analysis revealed that OsamiR393a promoter is methylated to some extent in FL478 under salt stress resulting in lower level of transcript and higher level of TIR1 transcript, which is the target of miR393a for PTGS (Ganie et al. 2016).

Regulation of expression of the genes involved in flavonoid and antioxidative pathway is under the control of DNA methylation. Demethylation of both the promoter and coding region has been shown to increase the expression level of the enzymes involved in flavonoid biosynthetic and oxidative pathways under stress condition (Bharti et al. 2015). The enzymes of these pathways play a key role and protect the plant cells from oxidative damage by scavenging of free radicals (Bors et al. 1990; Fini et al. 2011; Ghasemzadeh and Ghasemzadeh 2011; Sharma et al. 2012). Bisulphite sequencing of the promoter region of genes involved as enzymes in this pathway during normal and stress conditions revealed that both promoter and gene undergo demethylation under salt stress condition.

Apart from DNA modification, histone proteins undergo epigenetic modifications including acetylation, phosphorylation, methylation, ubiquitylation and sumoylation (Xhemalce et al. 2011; Oki et al. 2007; Hershko and Ciechanover 1998). Histone modifications such as acetylation, phosphorylation and ubiquitination are associated with transcriptional activation (Sridhar et al. 2007; Zhang et al. 2012), while biotinylation and sumoylation with transcriptional repression (Nathan et al. 2006; Camporeale et al. 2007). Among histone methylation, H3K4 trimethylation activates transcription, whereas H3K9 and H3K27 dimethylation represses transcription (Zhang et al. 2012).

In rice, alcohol dehydrogenase (AD1) and pyruvate decarboxylase (PDC1) are submergence-responsive genes. Transcription of these genes increases due to increased binding of RNAPol II to promoter as well as 5′ and 3′ coding region of *AD1* and *PDC1* which correlates with conversion of lysine residue of histone from dimethylated to trimethylated state group upon induction. This is also accompanied by increased acetylation throughout histone H3 proteins present at *Adh1* and *PDC1* loci (Tsuji et al. 2006). Another study analysed *HOS15* gene which was shown to be responsible for repression of gene associated with abiotic stress tolerance through histone deacetylation in *A. thaliana*. It was found that the *rd29A* promoter is hyper-induced due to the higher level of acetylated H4 histone in *hos15* mutant under cold stress (Zhu et al. 2008).

Table 6.2 Epigenetic modification in promoter region contributing to stress tolerance in plants

Gene	Stress	Plant species	Site of modification	Epigenetic modification/ promoter status	References
Glyceraldehyde-3-phosphate dehydrogenase(GAPC2)	Salt/ osmotic	*T. aestivum*	CpG, CpHpG (H= A,C,T)	Demethylated	Fei et al. (2016)
MIR393a	Salt	*O. sativa*	CpG, CpHpG	Methylated	Ganie et al. (2016)
ROS1	Oxidative stress	*A. thaliana*	CpG, CpHpG	Demethylated	Bharti et al. (2015)
Alcohol dehydrogenase(AD1)/ pyruvate decarboxylase(PDC1)	Flooding stress	*O. sativa*	Lys residue of histone	Dimethylated to trimethylated	Tsuji et al. (2006)
rd29A	Cold stress	*A. thaliana*	H4 histone	Acetylated	Zhu et al. (2008)

Table 6.2 compiles the various epigenetic modifications that have been reported in promoter regions of genes associated with abiotic stress tolerance.

6.4 Application of Stress-Inducible Promoters

Currently, most of the studies employ constitutive promoters for generation of transgenics which leads to constitutive overproduction of proteins often causing growth defects under non-stress conditions (Sakuma et al. 2006). Use of stress-specific promoters for expression of stress-mitigating genes can help in combating the above problem.

Although the main aim of characterizing the heat-inducible promoter was to make transgenics which show tolerance to heat, there are limited reports in this respect. One of the early reports shows the use of heat-inducible promoter to drive *ipt* gene which is a cytokinin biosynthesis gene, where heat application enabled tobacco calli to grow in cytokinin-free media (Schmülling et al. 1989). Masclaux and group used promoter of *Hsp18.2*, isolated from *A. thaliana*, for gene silencing of phytoene desaturase using an RNAi construct (Masclaux et al. 2004). In rice, heat-induced expression of *OsWRKY11* was achieved by using *Hsp101*promoter to drive expression of *WRKY11*. Such transgenic plants showed significant heat and drought tolerance after heat treatment (Wu et al. 2009). One of the other applications of heat-inducible promoters has been in creation of marker-free plants. *Cre-Lox* system is a well-known system for marker excision where the transgene construct present between the *loxP* sites is excised when recombinase is produced. Heat-inducible promoters, such as those derived from GmHsp17.5, Hsp81, Hsp70m and Hsp 18.2, were used to drive the expression of recombinase allowing the generation of a limited percentage of marker-free plants (Lyznik et al. 1995; Hoff et al. 2001; Chen et al. 2003; Luo et al. 2008; Zheng et al. 2016).

Overexpression of the *TaDREB3* gene from *Triticum aestivum* in barley under constitutive promoter improved frost tolerance of transgenic plants but also led to stunted phenotypes and 3–6-week delays in flowering compared to control plants. These negative developmental phenotypes could be eliminated with the use of cold-inducible promoter of *WRKY71* and *COR39* gene to drive the expression of *TaDREB3*. Analysis of the transgenic plants revealed that in the absence of stress, plant characteristics and grain yield remain unchanged when compared with wild-type plants. A comparison of the promoter-gene efficiency revealed that *WRKY71:TaDREB3* promoter-transgene combination is better than *COR9::TaDREB3* and appears to be a promising tool for the enhancement of cold and frost tolerance in crop plants (Kovalchuk et al. 2013). In another application of cold-inducible promoter, *Solanum tuberosum* was transformed with three *Arabidopsis AtCBF* (1–3) genes driven by either CaMV35S constitutive promoter or *rd29A* promoter. *S. tuberosum* is a frost-sensitive species incapable of cold acclimation, and even a brief exposure to frost can significantly reduce its yield, while longer duration of frost causes greater losses. Constitutive expression of AtCBF genes showed detrimental effect on plant growth and phenotype including smaller leaves, stunted plants, delayed flowering and lack of tuber production. This detrimental, negative phenotype could be avoided by the use of *rd29A* promoter which showed freezing tolerance and restored tuber production to levels like wild-type plant under non-stressed conditions.

In order to confer salt tolerance to *Medicago truncatula*, transgenic plants constitutively overexpressing *DREB1*, whose protein product binds to DRE/CRT elements, were generated with such plant having drastically reduced growth (Liu et al. 1998; Dubouzet et al. 2003; Ito et al. 2006). This negative impact was overcome by the use of *rd29A* promoter from *Arabidopsis* to drive the expression of *DREB1* (isolated from soybean) in alfalfa. The transgenic plants thus derived showed high tolerance to salt stress (Jin et al. 2010).

6.5 Summary

Climate change is a major factor in the decline in productivity of plants and is a matter of serious concern worldwide. Plants, being sessile in nature, have developed robust and elaborate mechanisms, including altered gene expression patterns, to fight adverse climatic conditions that pose threat to their survival. Understanding of mechanisms of stress signalling would help us prepare additional strategies to overcome issues of low plant productivity. The study of promoter regions of stress-responsive genes serves twin objectives, i.e. understanding the stress signalling pathway and identifying promoters that can be utilized to drive transgenes in crop improvement programmes. Previous studies have unravelled overlapping molecular response machinery and adaptive mechanism across several abiotic stresses. The commonality existing across major stress pathways such as salt, drought and cold may allow us to counteract the threat using novel strategies such as the use of a single promoter-gene combination to generate transgenic plants. Further, the information derived from the study of *cis*-elements can help in devising synthetic promoters which can respond in a desired manner paving the path for making stress-tolerant crops for a better future.

Acknowledgements Research in our laboratory on plant promoters has received financial support from DBT, Government of India (grant number BT/PR14532/AGR/36/673/2010), and UGC, Government of India (grant number 40-170/2011(SR). Research grant received from the University of Delhi through Research and Development (R&D) grant is also acknowledged. AJ and GJ are thankful to UGC for financial support in the form of JRF and SRF; CC acknowledges DBT for financial support as JRF/SRF.

References

Abe H, Yamaguchi-Shinozaki K, Urao T, Iwasaki T, Hosokawa D, Shinozaki K (1997) Role of *Arabidopsis* MYC and MYB homologs in drought- and abscisic acid-regulated gene expression. Plant Cell 9:1859–1868

Abe H, Urao T, Ito T, Seki M, Shinozaki K, Yamaguchi-Shinozaki K (2003) *Arabidopsis* AtMYC2 (bHLH) and AtMYB2 (MYB) function as transcriptional activators in abscisic acid signalling. Plant Cell 15:63–78

Agarwal M, Hao Y, Kapoor A, Dong CH, Fujii H, Zang X, Zhu JK (2006) A R2R3 type MYB transcription factor is involved in the cold regulation of CBF genes and in acquired freezing tolerance. J Biol Chem 281(49):37636–37645

Amin J, Ananthan J, Voellmy R (1988) Key features of heat shock regulatory elements. Mol Cell Biol 8:3761–3769

Ammara M, Nadia I, Hira M, Rubab ZN, Asia K, Aftab B (2016) Cloning and expression analysis of alcohol dehydrogenase (Adh) hybrid promoter isolated from *Zea mays*. Afr J Biotechnol 15:2384–2393

Assunção AGL, Herrero E, Lin Y-F, Huettel B, Talukdar S, Smaczniak C, Immink RGH, van Eldik M, Fiers M, Schat H, Aarts MGM (2010) *Arabidopsis thaliana* transcription factors bZIP19 and bZIP23 regulate the adaptation to zinc deficiency. Proc Natl Acad Sci U S A 107:10296–10301

Baker SS, Wilhelm KS, Thomashow MF (1994) The 5′-region of *Arabidopsis thaliana* COR15a has cis-acting elements that confer cold, drought and ABA-regulated gene expression. Plant Mol Biol 24(5):701–713

Balazadeh S, Siddiqui H, Allu AD, Matallana-Ramirez LP, Caldana C, Mehrnia M, Zanor M-I, Köhler B, Mueller-Roeber B (2010) A gene regulatory network controlled by the NAC transcription factor ANAC092/AtNAC2/ORE1 during salt-promoted senescence. Plant J 62:250–264

Barros MD, Czarnecka E, Gurley WB (1992) Mutational analysis of a plant heat-shock element. Plant Mol Biol 19(4):665–675

Bharti P, Mahajan M, Vishwakarma AK, Bhardwaj J, Yadav SK (2015) AtROS1 overexpression provides evidence for epigenetic regulation of genes encoding enzymes of flavonoid biosynthesis and antioxidant pathways during salt stress in transgenic tobacco. J Exp Bot 66(19):5959–5969

Bilichak A, Ilnystkyy Y, Hollunder J, Kovalchuk I (2012) The progeny of *Arabidopsis thaliana* plants exposed to salt exhibit changes in DNA methylation, histone modifications and gene expression. PLoS One 7(1):e30515

Bors W, Heller W, Michel C, Saran M (1990) Flavonoids as antioxidants: determination of radical-scavenging efficiencies. Methods Enzymol 186:343–355

Boyko A, Blevins T, Yao Y, Golubov A, Bilichak A, Ilnytskyy Y, Hollander J, Meins F Jr, Kovalchuk I (2010) Transgenerational adaptation of *Arabidopsis* to stress requires DNA methylation and the function of dicer-like proteins. PLoS One 5(3):e9514

Butler JEF, Kadonaga JT (2002) The RNA polymerase II core promoter: a key component in the regulation of gene expression. Genes Dev 16:2583–2592

Camporeale G, Oommen AM, Griffin JB, Sarath G, Zempleni J (2007) K12-biotinylated histone H4 marks heterochromatin in human lymphoblastoma cells. J Nutr Biochem 18:760–768
Chandler VL (2001) Gene activation and gene silencing. Plant Physiol 125:145–148
Chapin FS, Stuart Chapin F, Autumn K, Pugnaire F (1993) Evolution of suites of traits in response to environmental stress. Am Nat 142:S78–S92
Chatthai M, Osusky M, Osuska L, Yevtushenko D, Misra S (2004) Functional analysis of a Douglas-fir metallothionein-like gene promoter: transient assays in zygotic and somatic embryos and stable transformation in transgenic tobacco. Planta 220:118–128
Chinnusamy V, Zhu JK (2009) Epigenetic regulation of stress responses in plants. Curr Opin Plant Biol 12(2):133–139
Chinnusamy V, Zhu JK, Ranrar S, Lee BH, Hong X, Aggarwal M, Zhu JK (2010) Gene regulation during cold stress acclimation in plants. Methods Mol Biol 639:39–55
Czarnecka EVA, Key J, Gurley WB (1989) Regulatory domains of the Gmhsp l7. 5-E heat shock promoter of soybean. Mol Cell Biol 9:3457–3463
De Bruxelles GL, Peacock WJ, Dennis ES, Dolferus R (1996) Abscisic acid induces the alcohol dehydrogenase gene in *Arabidopsis*. Plant Physiol 111:381–391
De Vetten NC, Ferl RJ (1995) Characterization of a maize G-box binding factor that is induced by hypoxia. Plant J 7:589–601
Delisle AJ, Ferl RJ (1990) Characterization of the *Arabídopsís* Adh G-box binding factor. Plant Cell Am Soc Plant Physiol 2:547–557
Ding Y, Chen Z, Zhu C (2011) Microarray-based analysis of cadmium-responsive microRNAs in rice (*Oryzasativa*). J Exp Bot 62:3563–3573
Dong C-J, Wang Y, Yu S-S, Liu J-Y (2010) Characterization of a novel rice metallothionein gene promoter: its tissue specificity and heavy metal responsiveness. J Integr Plant Biol 52:914–924
Dubouzet JG, Sakuma Y, Ito Y, Kasuga M, Dubouzet EG, Miura S, Seki M, Shinozaki K, Yamaguchi Shinozaki K (2003) OsDREB genes in rice, *Oryza sativa* L., encode transcription activators that function in drought, high salt and cold responsive gene expression. Plant J 33:751–763
Durack PJ, Wijffels SE, Matear RJ (2012) Ocean salinities reveal strong global water cycle intensification during 1950 to 2000. Science 336:455–458
Fei Y, Xue Y, Du P, Yang S, Deng X (2016) Expression analysis and promoter methylation under osmotic and salinity stress of TaGAPC1 in wheat (Triticum aestivum L). Protoplasma 254(2):987–996
Fini A, Brunetti C, Di Ferdinando M, Ferrini F, Tattini M (2011) Stress- induced flavonoid biosynthesis and the antioxidant machinery of plants. Plant Signal Behav 6(5):709–711
Finnegan EJ, Genger RK, Kovac K, Peacock WJ, Dennis ES (1998) DNA methylation and the promotion of flowering by vernalization. Proc Natl Acad Sci U S A 95(10):5824–5829
Fordham-Skelton AP, Lilley C, Urwin PE, Robinson NJ (1997) GUS expression in *Arabidopsis* directed by 5′ regions of the pea metallothionein-like gene PsMTA. Plant Mol Biol 34:659–668
Freeman J, Sparks CA, West J, Shewry PR, Jones HD (2011) Temporal and spatial control of transgene expression using a heat-inducible promoter in transgenic wheat. Plant Biotechnol J 9:788–796
Fujita Y, Fujita M, Shinozaki K, Yamaguchi-Shinozaki K (2011) ABA-mediated transcriptional regulation in response to osmotic stress in plants. J Plant Res 124:509–525
Ganie SA, Dey N, Mondal TK (2016) Promoter methylation regulates the abundance of osa-miR393a in contrasting rice genotypes under salinity stress. Func Integr Genomics 16:1–11
Gendron JM, Pruneda-Paz JL, Doherty CJ, Gross AM, Kang SA, Kay SA (2012) *Arabidopsis* circadian clock protein, TOC1, is a DNA-binding transcription factor. Proc Natl Acad Sci U S A 109(8):3167–3172
Ghasemzadeh A, Ghasemzadeh N (2011) Flavonoids and phenolic acids: role and biochemical activity in plants and human. J Med Plant Res 5(31):6697–6703
Gilmour SJ, Zarka DG, Stockinger EJ, Salazar MP, Houghton JM, Thomashow MF (1998) Low temperature regulation of the *Arabidopsis* CBF family of AP2 transcriptional activators as an early step in cold-induced COR gene expression. Plant J 16(4):433–442

Goyal E, Singh RS, Kanika K (2013) Isolation and functional characterization of salt overly sensitive 1 (SOS1) gene promoter from *Salicornia brachiata*. Biol Plant 57:465–473

Guiltinan MJ, Marcotte WR Jr, Quatrano RS (1990) A plant leucine zipper protein that recognizes an abscisic acid response element. Science 250:267–271

Gurley WB, Czarnecka E, Nagao RT, Key JL (1986) Upstream sequences required for efficient expression of a soybean heat shock genet. Mol Cell Biol 6:559–565

Hänsch R, Mendel RR, Cerff R, Hehl R (2003) Light-dependent anaerobic induction of the maize glyceraldehyde-3-phosphate dehydrogenase 4 (GapC4) promoter in *Arabidopsis thaliana* and *Nicotiana tabacum*. Ann Bot 91:149–154

Haralampidis K, Milioni D, Rigas S, Hatzopoulos P (2002) Combinatorial interaction of cis elements specifies the expression of the *Arabidopsis* AtHsp90-1 gene. Plant Physiol 129:1138–1149

Harmer SL, John B, Hogenesch JB, Straume M, Chang HS, Han B, Zhu T, Wang X, Krep JA (2000) Orchestrated transcription of key pathways in *Arabidopsis* by the circadian clock. Science 290(5499):2110–2113

Hershko A, Ciechanover A (1998) The ubiquitin system. Annu Rev Biochem 67:425–479

Hobo T, Asada M, Kowyama Y, Hattori T (1999) ACGT-containing abscisic acid response element (ABRE) and coupling element 3 (CE3) are functionally equivalent. Plant J 19:679–689

Hoff T, Schnorr KM, Mundy J (2001) A recombinase-mediated transcriptional induction system in transgenic plants. Plant Mol Biol 45:41–49

Hou J, Jiang P, Qi S, Zhang K, He Q, Xu C, Ding Z, Zhang K, Li K (2016) Isolation and functional validation of salinity and osmotic stress inducible promoter from the maize type-II H+-pyrophosphatase gene by deletion analysis in transgenic tobacco plants. PLoS One 11:1–23

Ito Y, Katsura K, Maruyama K, Taji T, Kobayashi M, Seki M, Shinozaki K, Yamaguchi-Shinozaki K (2006) Functional analysis of rice DREB1/CBF-type transcription factors involved in cold-responsive gene expression in transgenic rice. Plant Cell Physiol 47:141–153

Iwasaki T, Yamaguchi-Shinozaki K, Shinozaki K (1995) Identification of a *cis*-regulatory region of a gene in *Arabidopsis thaliana* whose induction by dehydration is mediated by abscisic acid and requires protein synthesis. Mol Gen Genet 247:391–398

Jiang C, Iu B, Singh J (1996) Requirement of a CCGAC cis-acting element for cold induction of the BN115 gene from winter *Brassica napus*. Plant Mol Biol 30(3):679–684

Jin T, Chang Q, Li W, Yin D, Li Z, Wang D, Liu B, Liu L (2010) Stress-inducible expression of GmDREB1 conferred salt tolerance in transgenic alfalfa. Plant Cell Tissue Organ Cult 100:219–227

Karan R, De Leon T, Biradar H, Subudhi PK (2012) Salt stress induced variation in DNA methylation pattern and its influence on gene expression in contrasting rice genotypes. PLoS One 7(6):e40203

Karin M, Haslinger A, Heguy A, Dietlin T, Cooke T (1987) Metal-responsive elements act as positive modulators of human metallothionein-IIA enhancer activity. Mol Cell Biol 7:606–613

Katagiri F, Chua NH (1992) Plant transcription factors: present knowledge and future challenges. Trends Genet 8:22–27

Khurana N, Chauhan H, Khurana P (2013) Wheat chloroplast targeted sHSP26 promoter confers heat and abiotic stress inducible expression in transgenic *Arabidopsis* plants. PLoS One 8:e54418

Kim T-H, Böhmer M, Hu H, Nishimura N, Schroeder JI (2010) Guard cell signal transduction network: advances in understanding abscisic acid, CO_2, and Ca^{2+}signaling. Annu Rev Plant Biol 61:561–591

Kobayashi T (2005) Expression of iron-acquisition-related genes in iron-deficient rice is co-ordinately induced by partially conserved iron-deficiency-responsive elements. J Exp Bot 56:1305–1316

Kobayashi T, Nakayama Y, Itai RN, Nakanishi H, Yoshihara T, Mori S, Nishizawa NK (2003) Identification of novel cis-acting elements, IDE1 and IDE2, of the barley IDS2 gene promoter conferring iron-deficiency-inducible, root-specific expression in heterogeneous tobacco plants. Plant J 36:780–793

Kohler U, Mendel RR, Cerff R, Hehl R (1996) A promoter for strong and ubiquitous anaerobic gene expression in tobacco. Plant J 10:175–183

Kong WW, Yang ZM (2010) Identification of iron-deficiency responsive microRNA genes and *cis*-elements in *Arabidopsis*. Plant Physiol Biochem 48:153–159

Kovalchuk M,Jia W,Eini O,Morran S,Pyvovarenko T,Fletcher S,Bazanova N,Harris J,Beck-Oldach K,Shavrukov Y,Langridge P,Lopato S (2013) Optimization of TaDREB3 gene expression in transgenic barley using coldinducible promoters.Plant Biotechnol J. 11(6):659–70.)

Kropat J, Tottey S, Birkenbihl RP, Depège N, Huijser P, Merchant S (2005) A regulator of nutritional copper signaling in *Chlamydomonas* is an SBP domain protein that recognizes the GTAC core of copper response element. Proc Natl Acad Sci U S A 102:18730–18735

Kumar K, Kumar M, Kim S-R, Ryu H, Cho Y-G (2013) Insights into genomics of salt stress response in rice. Rice (N Y) 6:27

Kurkela S, Franck M (1990) Cloning and characterization of a cold- and ABA-inducible *Arabidopsis* gene. Plant Mol Biol 15(1):137–144

Kusaba M, Takahashi Y, Nagata T (1996) A multiple-stimuli-responsive as-1-related element of parA gene confers responsiveness to cadmium but not to copper. Plant Physiol 111:1161–1167

Kwok RPS, Lundblad JR, Chrivia JC, Richards JP, Bächinger HP, Brennan RG, Roberts SGE, Green MR, Goodman RH (1994) Nuclear protein CBP is a coactivator for the transcription factor CREB. Nature 370:223–226

Laloum T, De Mita S, Gamas P, Baudin M, Niebel A (2013) CCAAT-box binding transcription factors in plants: y so many? Trends Plant Sci 18:157–166

Lang V, Palva ET (1992) The expression of a rab-related gene, rab 18, is induced by abscisic acid during the cold acclimation process of *Arabidopsis thaliana* (L.) Heyn. Plant Mol Biol 21(3):581–582

Lee KT, Chen SC, Chiang BL, Yamakawa T (2007) Heat-inducible production of -glucuronidase in tobacco hairy root cultures. Appl Microbiol Biotechnol 73:1047–1053

Li W-X, Oono Y, Zhu J, He X-J, Wu J-M, Iida K, Lu X-Y, Cui X, Jin H, Zhu J-K (2008) The *Arabidopsis* NFYA5 transcription factor is regulated transcriptionally and post-transcriptionally to promote drought resistance. Plant Cell 20:2238–2251

Li H, Wang L, Yang ZM (2015) Co-expression analysis reveals a group of genes potentially involved in regulation of plant response to iron-deficiency. Gene 554(1):16–24

Lin P, Wu L, Wei D, Chen H, Zhou M, Yao X, Linet J (2016) Promoter analysis of cold-responsive (COR) gene from *Capsella bursa-pastoris* and expression character in response to low temperature. Int J Agric Biol 18(2):346–352

Liu Q, Kasuga M, Sakuma Y, Abe H, Miura S, Yamaguchi-Shinozaki K, Shinozaki K (1998) Two transcription factors, DREB1 and DREB2, with an EREBP/AP2 DNA binding domain separate two cellular signal transduction pathways in drought- and low-temperature-responsive gene expression, respectively, in *Arabidopsis*. Plant Cell 10:1391–1406

Liu F, VanToai T, Moy PL, Bock G, Linford LD, Quackenbush J (2005) Global transcription profiling reveals comprehensive insights into hypoxic response in *Arabidopsis*. Plant Physiol 137:1115–1129

Lü S, Gu H, Yuan X, Wang X, Wu A-M, Qu L, Liu J-Y (2007) The GUS reporter-aided analysis of the promoter activities of a rice metallothionein gene reveals different regulatory regions responsible for tissue-specific and inducible expression in transgenic *Arabidopsis*. Transgenic Res 16:177–191

Luo K, Sun M, Deng W, Xu S (2008) Excision of selectable marker gene from transgenic tobacco using the GM-gene-deletor system regulated by a heat-inducible promoter. Biotechnol Lett 30:1295–1302

Lyznik LA, Hirayama L, Rao KV, Abad A, Hodges TK (1995) Heat-inducible expression of FLP gene in maize cells. Plant J 8:177–186

Magalhaes JV, Liu J, Guimarães CT, Lana UGP, Alves VMC, Wang Y-H, Schaffert RE, Hoekenga OA, Piñeros MA, Shaff JE, Klein PE, Carneiro NP, Coelho CM, Trick HN, Kochian LV (2007) A gene in the multidrug and toxic compound extrusion (MATE) family confers aluminum tolerance in sorghum. Nat Genet 39:1156–1161

Maruyama K, Todaka D, Mizoi J, Yoshida T, Kidokoro S, Matsukura S, Takasaki H, Sakurai T, Yamamoto YY, Yoshiwara K, Kojima M, Sakakibara H, Shinozaki K, Yamaguchi-Shinozaki K (2012) Identification of cis-acting promoter elements in cold and dehydration induced transcriptional pathways in *Arabidopsis*, rice, and soybean. DNA Res 19(1):37–49

Maruyama K, Ogata T, Kanamori N, Yoshiwara K, Goto S, Yamamoto YY, Tokoro Y, Noda C, Takaki Y, Urawa H, Iuchi S, Urano K, Yoshida T, Sakurai T, Kojima M, Sakakibara H, Shinozaki K, Yamaguchi-Shinozaki K (2017) Design of an optimal promoter involved in the heat-induced transcriptional pathway in *Arabidopsis*, soybean, rice, and maize. Plant J 89:671–680

Masclaux F, Charpenteau M, Takahashi T, Pont-Lezica R, Galaud JP (2004) Gene silencing using a heat-inducible RNAi system in *Arabidopsis*. Biochem Biophys Res Commun 321:364–369

McCarty DR (1995) Genetic control and integration of maturation and germination pathways in seed development. Annu Rev Plant Physiol Plant Mol Biol 46:71–93

McKendree WLM Jr, Ferl RJ (1992) Functional elements of the Arabidopsis Adh promoter include the G-box. Plant Mol Biol 19(5):859–862

McLoughlin F, Arisz SA, Dekker HL, Kramer G, De Koster CG, Haring MA, Munnik T, Testerink C (2013) Identification of novel candidate phosphatidic acid-binding proteins involved in the salt-stress response of *Arabidopsis thaliana* roots. Biochem J 450(3):573–581

Meier S, Bastian R, Donaldson L, Murray S, Bajic V, Gehring C (2008) Co-expression and promoter content analyses assign a role in biotic and abiotic stress responses to plant natriuretic peptides. BMC Plant Biol 8:24

Meshi T, Iwabuchi M (1995) Plant transcription factors. Plant Cell Physiol 36:1405–1420

Mikkelsen MD, Thomashow MF (2009) A role for circadian evening elements in cold-regulated gene expression in *Arabidopsis*. Plant J 60(2):328–339

Mishra RC, Grover A (2014) Intergenic sequence between *Arabidopsis* caseinolytic protease B-cytoplasmic/heat shock Protein100 and choline kinase genes functions as a heat-inducible bidirectional promoter. Plant Physiol 166:1646–1658

Miyashita Y, Dolferus R, Ismond KP, Good AG (2007) Alanine aminotransferase catalyses the breakdown of alanine after hypoxia in *Arabidopsis thaliana*. Plant J 49:1108–1121

Mohanty MI, Swarup S (2005) Detection and preliminary analysis of motifs in promoters of anaerobically induced genes of different plant species. Anna 96:669–68132

Moriwaki M, Yamakawa T, Washino T, Kodama T, Igarashi Y, Yoshikawa T (1999) Organ-specific expression of β-glucuronidase activity driven by the *Arabidopsis* heat-shock promoter in heat-stressed transgenic *Nicotiana plumbaginifolia*. Plant Cell Rep 19:92–95

Muench DG, Christopher ME, Good AG (1998) Cloning and expression of a hypoxic and nitrogen inducible maize alanine aminotransferase gene. Physiol Plant 103:503–512

Nakashima K, Ito Y, Yamaguchi-Shinozaki K (2009) Transcriptional regulatory networks in response to abiotic stresses in *Arabidopsis* and grasses. Plant Physiol 149:88–95

Narusaka Y, Nakashima K, Shinwari ZK, Sakuma Y, Furihata T, Abe H, Narusaka M, Shinozaki K, Yamaguchi-Shinozaki K (2003) Interaction between two cis-acting elements, ABRE and DRE, in ABA-dependent expression of *Arabidopsis* rd29A gene in response to dehydration and high-salinity stresses. Plant J 34:137–148

Nathan D, Ingvarsdottir K, Sterner DE, Bylebyl GR, Dokmanovic M, Jean A, Dorsey JA, Kelly A, Whelan KA, Mihajlo Krsmanovic M, William S, Lane WS, Meluh PB, Johnson ES, Berge SL (2006) Histone sumoylation is a negative regulator in *Saccharomyces cerevisiae* and shows dynamic interplay with positive-acting histone modifications. Genes Dev 20(8):966–976

Niklas KJ (2009) Functional adaptation and phenotypic plasticity at the cellular and whole plant level. J Biosci 34:613–620

Ogo Y, Kobayashi T, Nakanishi Itai R, Nakanishi H, Kakei Y, Takahashi M, Toki S, Mori S, Nishizawa NK (2008) A novel NAC transcription factor, IDEF2, that recognizes the iron deficiency-responsive element 2 regulates the genes involved in iron homeostasis in plants. J Biol Chem 283:13407–13417

Ohama N, Sato H, Shinozaki K, Yamaguchi-Shinozaki K (2017) Transcriptional regulatory network of plant heat stress response. Trends Plant Sci 22(1):53–65

Oki M, Aihara H, Ito T (2007) Role of histone phosphorylation in chromatin dynamics and its implications in diseases. Subcell Biochem 41:319–336

Olive MR, Peacock WJ, Dennis ES (1991) The anaerobic responsive element contain two GC rich sequences essential for binding to a nuclear protein and hypoxic activation of the maize Adh1 promoter. Nuclec Acids Res 19:7053–7060

Pelham HRB (1982) A regulatory upstream promoter element in the *Drosophila* Hsp 70 heat-shock gene. Cell 30:517–528

Pillai M, Lihuang Z, Akiyama T (2002) Molecular cloning, characterization, expression and chromosomal location of OsGAPDH, a submergence responsive gene in rice. Theor Appl Genet 105(1):34–42

Prajapati GK, Kashyap N, Kumar A, Pandey DM (2013) Identification of GCC box in the promoter region of ubiquinol cytochrome C chaperone gene using molecular beacon probe and its in silico protein-DNA interaction study in rice (*Oryza sativa* L.) Int J Comput Bioinforma Silico Model 2:213–222

Qi X, Zhang Y, Chai T (2007) Characterization of a novel plant promoter specifically induced by heavy metal and identification of the promoter regions conferring heavy metal responsiveness. Plant Physiol 143:50–59

Quinn JM (2000) Coordinate copper- and oxygen-responsive Cyc6 and Cpx1 expression in Chlamydomonas is mediated by the same element. J Biol Chem 275:6080–6089

Quinn JM, Merchant S (1995) Two copper-responsive elements associated with the *Chlamydomonas* Cyc6 gene function as targets for transcriptional activators. Plant Cell 7:623–628

Quinn JM, Kropat J, Merchant S (2003) Copper response element and Crr1-dependent Ni(2+)-responsive promoter for induced, reversible gene expression in *Chlamydomonas reinhardtii.* Eukaryot Cell 2:995–1002

Ren Y, Zhao J (2009) Functional analysis of the rice metallothionein gene OsMT2b promoter in transgenic *Arabidopsis* plants and rice germinated embryos. Plant Sci 176:528–538

Ricoult C, Echeverria LO, Cliquet JB, Limami AM (2006) Characterization of alanine aminotransferase (AlaAT) multigene family and hypoxic response in young seedlings of the model legume *Medicago truncatula*. J Exp Bot 57:3079–3089

Rieping M, Schöffl F (1992) Synergistic effect of upstream sequences, CCAAT box elements, and HSE sequences for enhanced expression of chimaeric heat shock genes in transgenic tobacco. Mol Gen Genet 231:226–232

Rocha M, Sodek L, Licausi F, Hameed MW, Dornelas MC, Van Dongen JT (2010) Analysis of alanine aminotransferase in various organs of soybean (Glycine max) and in dependence of different nitrogen fertilisers during hypoxic stress. Amino Acids 39:1043–1053

Ryan PR, Raman H, Gupta S, Sasaki T, Yamamoto Y, Delhaize E (2010) The multiple origins of aluminium resistance in hexaploid wheat include *Aegilops tauschii* and more recent cis mutations to TaALMT1. Plant J 64:446–455

Safdar W, Ahmed H, Bostan N, Zahra NB, Sharif HR, Haider J, Abbas S (2016) Comparative analysis of nine different small heat shock protein gene promoters in *Oryza sativa* L. subsp. indica. Plant Syst Evol 302:1195–1206

Sakuma Y, Maruyama K, Qin F, Osakabe Y, Shinozaki K, Yamaguchi-Shinozaki K (2006) Dual function of an *Arabidopsis* transcription factor DREB2A in water-stress-responsive and heat-stress-responsive gene expression. Proc Natl Acad Sci U S A 103:18822–18827

Satoh R (2002) ACTCAT, a novel cis-acting element for proline- and hypoosmolarity-responsive expression of the ProDH gene encoding proline dehydrogenase in *Arabidopsis*. Plant Physiol 130:709–719

Schmülling T, Beinsberger S, De Greef J, Schell J, Van Onckelen H, Spena A (1989) Construction of a heat-inducible chimaeric gene to increase the cytokinin content in transgenic plant tissue. FEBS Lett 249:401–406

Sharma P, Jha AB, Dubey RS, Pessarakliet M. al(2012) Reactive oxygen species, oxidative damage, and antioxidative defense mechanism in plants under stressful conditions. J Bot:1–26

Shen Q, Ho T-HD (1995) Functional dissection of an abscisic acid (ABA)-inducible gene reveals two independent ABA-responsive complexes each containing a G-box and a novel cis-acting element. Plant Cell 7:295

Shen Q, Zhang P, Ho TH (1996) Modular nature of abscisic acid (ABA) response complexes: composite promoter units that are necessary and sufficient for ABA induction of gene expression in barley. Plant Cell 8:1107–1119

Shi H, Quintero FJ, Pardo JM, Zhu J-K (2002) The putative plasma membrane Na+/H+ antiporter SOS1 controls long-distance Na+ transport in plants. Plant Cell 14:465–477

Simpson SD, Nakashima K, Narusaka Y, Seki M, Shinozaki K, Yamaguchi-Shinozaki K (2003) Two different novel cis-acting elements of erd1, a clpA homologous *Arabidopsis* gene function in induction by dehydration stress and dark-induced senescence. Plant J 33:259–270

Singh N, Dang TTM, Vergara GV, Pandey DM, Sanchez D, Neeraja CN, Septiningsih EM, Mendioro M, Tecson-Mendoza EM, Ismail AM, Mackill DJ, Heuer S (2010) Molecular marker survey and expression analyses of the rice submergence-tolerance gene SUB1A. Theor Appl Genet 121:1441–1453

Spena A, Schell J (1987) The expression of a heat-inducible chimeric gene in transgenic tobacco plants. Mol Gen Genet 206:436–440

Spena A, Hain R, Ziervogel U, Saedler H, Schell J (1985) Construction of a heat-inducible gene for plants. Demonstration of heat-inducible activity of the *Drosophila* hsp70 promoter in plants. EMBO J 4:2739–2743

Sridhar VV, Kapoor A, Zhang K, Zhu J, Zhou T, Hasegawa PM, Bressan RA, Zhu JK (2007) Control of DNA methylation and heterochromatic silencing by histone H2B de-ubiquitination. Nature 447(7145):735–738

Stockinger EJ, Gilmour SJ, Thomashow MF (1997) *Arabidopsis thaliana* CBF1 encodes an AP2 domain-containing transcriptional activator that binds to the C-repeat/DRE, a cis-acting DNA regulatory element that stimulates transcription in response to low temperature and water deficit. Proc Natl Acad Sci U S A 94(3):1035–1040

Storozhenko S, De Pauw P, Van Montagu M, Inzé D, Kushnir S (1998) The heat-shock element is a functional component of the *Arabidopsis* APX1 gene promoter. Plant Physiol 118:1005–1014

Sultan SE (1995) Phenotypic plasticity and plant adaptation. Acta Bot Neerl 44:363–383

Sun N, Liu M, Zhang W, Yang W, Bei X, Ma H, Qiao F, Qi X (2015) Bean metal-responsive element-binding transcription factor confers cadmium resistance in tobacco. Plant Physiol 167:1136–1148

Takahashi T, Naito S, Komeda Y (1992) The Arabidopsis HSP18.2 promoter/GUS gene fusion in transgenic *Arabidopsis* plants: a powerful tool for the isolation of regulatory mutants of the heat-shock response. Plant J 2:751–761

Thomashow MF (1999) Plant cold acclimation: freezing tolerance genes and regulatory mechanisms. Annu Rev Plant Physiol Plant Mol Biol 50:571–599

Tran L-SP (2004) Isolation and functional analysis of *Arabidopsis* stress-inducible NAC transcription factors that bind to a drought-responsive cis-element in the early responsive to dehydration stress 1 promoter. Plant Cell Online 16:2481–2498

Tran LP, Nakashima K, Sakuma Y, Simpson SD, Fujita Y, Maruyama K, Fujita M, Seki M, Shinozaki K, Yamaguchi-shinozaki K (2004) Isolation and functional analysis of *Arabidopsis* stress-inducible NAC transcription factors that bind to a drought-responsive *cis*-element in the early responsive to dehydration stress 1 promoter. Plant Cell 16:2481–2498

Tsuji H, Saika H, Tsutsumi N, Hirai A, Nakazono M (2006) Dynamic and reversible changes in histone H3-Lys4 methylation and H3 acetylation occurring at submergence-inducible genes in rice. Plant Cell Physiol 47(7):995–1003

Tsutsui T, Yamaji N, Feng Ma J (2011) Identification of a cis-acting element of ART1, a C2H2-type zinc-finger transcription factor for aluminum tolerance in rice. Plant Physiol 156:925–931

Uno Y, Furihata T, Abe H, Yoshida R, Shinozaki K, Yamaguchi-Shinozaki K (2000) Arabidopsis basic leucine zipper transcription factors involved in an abscisic acid-dependent signal transduction pathway under drought and high-salinity conditions. Proc Natl Acad Sci U S A 97:11632–11637

Vanyushin BF, Ashapkin VV (2011) DNA methylation in higher plants: past, present and future. Biochim Biophys Acta 1809:360–368

Vazquez-Tello A, Ouellet F, Sarhan F (1998) Low temperature- stimulated phosphorylation regulates the binding of nuclear factors to the promoter of Wcs120, a cold-specific gene in wheat. Mol Gen Genet 257:157–166

Walker JC, Howard EA, Dennis ES, Peacock WJ (1987) DNA sequences required for anaerobic expression of the maize alcohol dehydrogenase 1 gene. Biochemistry 84:6624–6628

Chen Ming, Wang li-Xa, Peng Xiang-Lei, Xu Hui-Jun Z-P Lin (2003) Gene expression controlled by heat-inducible site specific recombination in tobacco. Acta Bot Sin 45:1481–1488

Whitelaw CA, Le Huquet JA, Thurman DA, Tomsett AB (1997) The isolation and characterisation of type II metallothionein-like genes from tomato (*Lycopersicon esculentum* L.) Plant Mol Biol 33(3):503–511

Wilhelm K, Thomashow MF (1993) *Arabidopsis thaliana* corl5b, an apparent homologue of corl5a, is strongly responsive to cold and ABA, but not drought. Plant Mol Biol 23(5):1073–1077

Wu X, Shiroto Y, Kishitani S, Ito Y, Toriyama K (2009) Enhanced heat and drought tolerance in transgenic rice seedlings overexpressing OsWRKY11 under the control of HSP101 promoter. Plant Cell Rep 28:21–30

Xhemalce B, Dawson MA, Bannister AJ (2011) Histone modifications. In: Meyers R (ed) Encyclopedia of molecular cell biology and molecular medicine. Wiley, Weinheim

Yabe N, Takahashi T, Komeda Y (1994) Analysis of tissue-specific expression of *Arabidopsis thaliana* Hsp90-family gene HSP81. Plant Cell Physiol 35(8):1207–1219

Yamaguchi-Shinozaki K (1994) A novel cis-acting element in an *Arabidopsis* gene is involved in responsiveness to drought, low-temperature, or high-salt stress. Plant Cell 6:251–264

Yamaguchi-Shinozaki K, Shinozaki K (1994) A novel cis-acting element in an *Arabidopsis* gene is involved in responsiveness to drought, low-temperature, or high-salt stress. Plant Cell 6(2):251–264

Yamaguchi-Shinozaki K, Shinozaki K (2005) Organization of *c*is-acting regulatory elements in osmotic- and cold-stress-responsive promoters. Trends Plant Sci 10:88–94

Yamaguchi-Shinozaki K, Shinozaki K (2006) Transcriptional regulatory networks in cellular responses and tolerance to dehydration and cold stresses. Annu Rev Plant Biol 57:781–803

Zeng L, Deng R, Guo Z, Yang S, Deng X (2016) Genome-wide identification and characterization of Glyceraldehyde-3-phosphate dehydrogenase genes family in wheat (Triticumaestivum). BMC Genomics 17:240

Zhang L, Wang Y, Zhang X (2012) Dynamics of phytohormone and DNA methylation patterns changes during dormancy induction in strawberry (*Fragaria×ananassa* Duch.) Plant Cell Rep 31(1):155–165

Zhang Y, Sun T, Liu S, Dong L, Liu C, Song W, Lui J, Gai S (2016) MYC *cis*-elements in *PsMPT* promoter is involved in chilling response of *Paeonia suffruticosa*. PLoS One 11(5)

Zhao H, Butler E, Rodgers J, Spizzo T, Duesterhoeft S, Eide D (1998) Regulation of zinc homeostasis in yeast by binding of the ZAP1 transcriptional activator to zinc-responsive promoter elements. J Biol Chem 273:28713–28720

Zheng Y, Pan Y, Li J, Zhou Y, Pan Y, Ding Y (2016) Visible marker excision via heat-inducible Cre/LoxP system and Ipt selection in tobacco. Vitr Cell Dev Biol 52(5):492–499

Zhou M-L, Qi L-P, Pang J-F, Zhang Q, Lei Z, Tang Y-X, Zhu X-M, Shao J-R, Wu Y-M (2013) Nicotianamine synthase gene family as central components in heavy metal and phytohormone response in maize. Funct Integr Genomics 13:229–239

Zhu J, Jeong JC, Zhu Y et al (2008) Involvement of *Arabidopsis* HOS15 in histone deacetylation and cold tolerance. Proc Natl Acad Sci U S A 105(12):4945–4950

Multifarious Role of ROS in Halophytes: Signaling and Defense

7

G. C. Nikalje, S. J. Mirajkar, T. D. Nikam, and P. Suprasanna

Abstract

Halophytes are the native flora of saline soil. They are well adapted to saline soils and get benefit from salt ions. This ability makes them a potential candidates for understanding physiological and biochemical mechanisms of salinity stress tolerance. Oxidative stress signaling and detoxification of reactive oxygen species (ROS) are both important components of salt tolerance mechanisms in these plants. The reactive oxygen species (ROS) play dual role in plant growth, development, and stress tolerance. The ROS are toxic by-products of cellular metabolism, but their key role in stress signaling and development is now emerging. Different biotic and abiotic stress conditions lead to overaccumulation of different kinds of ROS which cause oxidative damage and cell death.

G. C. Nikalje (✉)
Department of Botany, R. K. Talreja College of Arts, Science and Commerce, Ulhasnagar, Thane 421003, Maharashtra, India

Nuclear Agriculture and Biotechnology Division, Bhabha Atomic Research Centre, Mumbai 400085, Maharashtra, India
e-mail: ganeshnikalje7@gmail.com

S. J. Mirajkar
Department of Agricultural Botany, Dr. Panjabrao Deshmukh Krishi Vidyapith, Akola, Maharashtra, India

T. D. Nikam
Department of Botany, Savitribai Phule Pune University, Pune 411 007, India

P. Suprasanna
Nuclear Agriculture and Biotechnology Division, Bhabha Atomic Research Centre, Mumbai, 400085 Maharashtra, India
e-mail: penna888@yahoo.com

S. M. Zargar, M. Y. Zargar (eds.), *Abiotic Stress-Mediated Sensing and Signaling in Plants: An Omics Perspective*, https://doi.org/10.1007/978-981-10-7479-0_7

However, they also act as signaling molecules in various cellular and developmental processes. The balance, between production and scavenging of ROS by antioxidants, defines their positive or negative role in plant growth and development. In this article, ROS- production, scavenging systems (enzymatic and non-enzymatic), and their role in stress signaling and defense in halophytes have been discussed.

Keywords
Halophytes · ROS · Antioxidants · Salt stress · Signaling

7.1 Introduction

Reactive oxygen species (ROS) are the by-products of the aerobic metabolism. At ground state, molecular oxygen is less reactive, but under abnormal metabolic and/or stressful conditions, it gives rise to harmful reactive free radicals (Mittler 2017). Initially ROS were identified as toxic to cells which are detoxified by enzymatic and nonenzymatic antioxidants. Recently, their beneficial role as signaling molecule in plants during growth, development, response to abiotic and biotic stresses, and programmed cell death have been reported (Bose et al. 2014). With the due course of evolutionary process, plant cells have developed unique and efficient antioxidant machinery which detoxifies excess ROS and tend to use ROS molecules as signal transducers. The ROS-producing enzymes such as plant homolog of respiratory-burst NADPH oxidases demonstrated that plant cells can amplify ROS production for cell signaling. Cell organelles like plastids, mitochondria, and peroxisomes also produce ROS for initiating signaling cascade (Bailey-Serres and Mittler 2006). ROS-mediated signaling is under control of proper balance between production and scavenging of ROS. During intercellular and intracellular ROS signal transduction, the localized and temporal production of ROS is highly critical. Similar to calcium signaling, ROS is under the control of regional production and scavenging. Now it is well understood about the enzymes involved in ROS scavenging, their compartmentalization but the initiation of ROS signaling, sensing, response and the control between production and scavenging of ROS is yet to be ascertained in detail (Bailey-Serres and Mittler 2006).

7.2 The ROS Hub

Excitation of O_2 results in the formation of singlet oxygen ($^1O_2$) (Triantaphylidès and Havaux 2009), and by transferring electrons, it forms superoxide radical ($O^{\bullet-2}$, one electron), hydrogen peroxide (H_2O_2, two electrons), or hydroxyl radicals ($OH^\bullet$, three electrons) (Mittler 2002). The chemistry of different ROS is given in Table 7.1. Among the different ROS, hydroxyl ions are highly unstable OH- (half-life period, 1 ns) and can travel only 1 nm distance from the site of generation. The superoxides

Table 7.1 Chemistry of reactive oxygen species (ROS)

ROS	Characteristics	References
Singlet oxygen (O • − 2)	Highly unstable	Storz et al. (1990)
	Unable to diffuse through membranes due to negative charge	
	High attraction, $O•^{-2}$ oxidizes Fe–S clusters at a rate that is almost diffusion limited	
Hydrogen peroxide (H_2O_2)	Poor oxidant	Imlay (2003), D'Autréaux and Toledano (2007), Paulsen and Carroll (2010)
	Reacts mildly with [Fe–S] (rate constant of 102–103 M^{-1} s^{-1})	
	Relatively stable (cellular half-life ~1 ms, steady-state levels ~10–7 M	
	Can diffuse through biological membranes because it is not charged	
	Mediate intracellular signal transduction through chemoselective oxidation of Cys residues in signaling proteins such as glutathione, thioredoxins, and peroxiredoxins	
Hydroxyl ion (OH•)	Highly toxic	Halliwell and Gutteridge (1999), D'Autréaux and Toledano (2007)
	Highly reactive which limits its diffusion to sites of production	
	Half-life 10–9 s, but seems to operate in H_2O_2 sensing	
Singlet oxygen ($^1O_2$)	Excited state molecule	Niedre et al. (2002), Kochevar (2004), Triantaphylidès and Havaux (2009)
	Half-lifetime of $^1O_2$ is very short (~100 ns)	
	Travel distance in cells <100 nm	
	Reacts very rapidly with amino acids, unsaturated lipids, and other cell constituents	
	Can react directly only with molecules in close proximity to its production	
	Could deliver specific signaling events mainly through spatial aspects of ROS production	

(O_2^-) and singlet oxygen ($^1O_2$) have moderate half-life period (1–4 μs) and traveling distance (30 nm). Hydrogen peroxide (H_2O_2) is highly stable (half-life period, >1 ms) and can travel >1 μm distance from its site of production (Bose et al. 2014). Their mode of action and scavenging system is shown in Fig. 7.1.

ROS generation and detoxification constitute the defense strategy adopted by plants exposed to abiotic stress conditions. Plants possess a set of enzymatic and nonenzymatic antioxidants that either detoxify or prevent further formation of ROS such as singlet oxygen, superoxide anion radicals, hydroxyl radicals, and H_2O_2 in different plant subcellular compartments. The enzymatic route has superoxide dismutase, catalase, peroxidases, and glutathione peroxidases. ROS can be categorized on the basis of site of production such as chloroplastic, mitochondrial, peroxisomal and apoplastic ROS (Table 7.2).

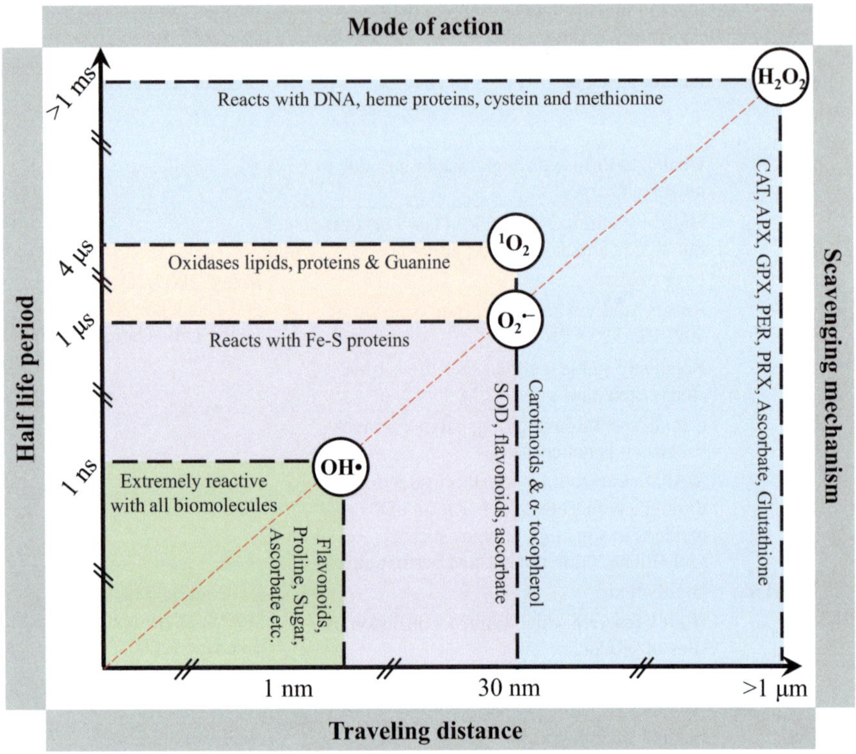

Fig. 7.1 Comparison between different properties of ROS (mode of action, half-life, scavenging mechanism, and traveling distance): The horizontal axis shows traveling distance, and vertical axis shows half-life period of ROS. The dotted line attached to horizontal axis shows scavenging system, while the dotted line attached to vertical axis shows mode of action of ROS. Singlet oxygen ($^1O_2$) and superoxide (O_2^-) have almost the same half-life period, scavenging mechanism, and traveling distance but different mode of action

7.2.1 Chloroplastic ROS

Both the photosystems (PSI and PSII) are the main sites of ROS production in chloroplast. Under illuminated condition, the antenna pigments of PSI forms $O_2^{\bullet -}$ by Mehler reaction (Asada and Takahashi 1987). These superoxides are quickly scavenged by SODs to form hydrogen peroxide that sequentially cleaved using catalase or ascorbate peroxidase (Asada 2006). At PSII, under high irradiance, the excited triplet state of chlorophyll transfers electrons to $^3O_2$ which is then converted into singlet oxygen ($^1O_2$). An enzyme alternative oxidase (AOX) is extensively studied in the management of ROS in halophytes. Some halophytes under salt treatment divert from C_3 to C_4 or CAM metabolism (Nikalje et al. 2017). The well-known halophyte, *Mesembryanthemum*, converts from C_3 to CAM which results in an increased activity of AOX and decreased ROS production (Sunagawa et al. 2010). In *Thellungiella,* in both control and salt treatment conditions, the increased

Table 7.2 Sites of production of reactive oxygen species (ROS) in plants (Modified and adapted from Sharma et al. 2012)

Sr. no.	Organelle	Site of production
1.	Chloroplast	PSI, electron transport chain (Fd, 2Fe-2S, and 4Fe-4S clusters); PSII, electron transport chain; QA and QB; chlorophyll pigments
2.	Mitochondria	Complex I, NADH dehydrogenase segment; complex II, reverse electron flow to complex I; complex III, ubiquinone-cytochrome region; enzymes, aconitase, 1-galactono-γ lactone dehydrogenase (GAL)
3.	Plasma membrane	Electron-transporting oxidoreductases; NADPH oxidase, quinone oxidase
4.	Peroxisome	Matrix, xanthine oxidase (XOD)
		Membrane, electron transport chain flavoprotein
		NADH and Cyt b; metabolic processes, glycolate oxidase, fatty acid oxidation, flavin oxidases, disproportionation of O2•− radicals
5.	Apoplast	Cell wall-associated oxalate oxidase amine oxidases
6.	Endoplasmic reticulum	NAD(P)H-dependent electron transport involving Cyt P450
7.	Cell wall	Cell wall-associated peroxidase diamine oxidases

electron flow gets (up to 30%) diverted to O_2 by plastid terminal oxidase via plastoquinone. This O_2 gets converted into water and not into toxic superoxides (Stepien and Johnson 2009). This type of diversion was not observed in a model glycophyte, *Arabidopsis*, which suggests that alternative electron sinks play an important role in decreasing salt-induced ROS production in halophytes. Apart from these basic strategies, halophytes increase abundance of 33 kDa Mn stabilizing proteins, a chlorophyll a/b protein (CP47), a PSI subunit protein, Rubisco activase, and sulfolipid content which stabilizes both the photosystems and prevents oxidative damage under salt stress (Ramani et al. 2004; Wang et al. 2008; Sengupta and Majumder 2009, 2010; Ozgur et al. 2013; Bose et al. 2014).

7.2.2 Mitochondrial ROS

Mitochondria are the powerhouse of the cell which generate energy in the form of ATPs. ATP formation takes place by oxidation of organic acids and electron transfer to O_2 by mitochondrial ETC (electron transport chain) in mitochondria and produces ROS (Huang et al. 2016). Mitochondria produce very low amount of ROS in leaves, almost 30–100 times lower than chloroplasts (Foyer and Noctor 2003). However, in non-photosynthetic tissues like roots, mitochondria must be the major source of ROS (Bose et al. 2014). In mitochondrial, electron transport chain (ETC) at complex I and III, $O_2^{\bullet -}$ are generated and immediately converted into H_2O_2 (Sweetlove and Foyer 2004). Similar to leaves, shifting of C_3–C_4 or CAM metabolism helps roots to control excessive load of ROS in halophytes. Basically, CAM plants have certain advantages like they close their stomata during the day and open in night which helps them to handle low water availability and reduces damage to photosynthetic apparatus (Amtmann 2009). Some halophytes switch from C_3 to C_4 mode of carbon metabolism under salt stress because due to photorespiration, C_3 plants are more vulnerable to oxidative damage. This switching strategy enables halophytes to control ROS load and to maintain photosynthesis under adverse environmental conditions. ROS production in mitochondria and chloroplasts affects endoplasmic reticulum (ER) stress, and organelle-originated ROS plays as a messenger molecule in the unfolded protein response (Ozgur et al. 2015) which could facilitate protein-folding and degradation of misfolded proteins, and in the vent of long-term stress, this can lead to excessive accumulation of misfolded proteins. The roles of mitochondrial ROS in cellular signaling, retrograde signaling, plant hormone signaling, and programmed cell death in plants have been reviewed by Huang et al. (2016).

7.2.3 Peroxisomal and Apoplastic ROS

Peroxisomes are the metabolically active subcellular organelles with highly variable enzyme content which differs with organism and environmental cue (Baker and Graham 2002). The organelle matrix and peroxisomal membrane are the main sites of ROS including nitric oxide (Corpas et al. 2001). In plant cells, apart from subcellular organelles, apoplastic space is also a major site for ROS generation. The NADPH oxidase and cell wall-associated peroxidase enzymes oxidize NADPH and transfer the electron to oxygen to generate $O_2^{\bullet-}$ radicals (Sagi and Fluhr 2006).

7.3 ROS Scavenging System

Antioxidants like ascorbate, glutathione, and tocopherol play a crucial role in defense and also influence plant growth and development (Foyer 2005). Antioxidants maximize cellular defense responses through cellular redox state and modulation of stress-related gene expression. It is suggested that for redox homeostasis, both reactive oxygen species (ROS) and antioxidant interaction act as a metabolic interface for signals derived from metabolism and from the environment (Foyer 2005). Basal level of ROS is essential for growth and development of all organisms. But after a certain limit, increased level of salt, 150 mM for most of the halophytes (Ozgur et al. 2013), enhances ROS production which leads to lipid peroxidation and array of chemical events leading to irreversible damage. To overcome oxidative stress, both enzymatic and nonenzymatic antioxidant defense mechanisms are activated which are discussed in the following sections.

7.3.1 Enzymatic Defence Machinery

The enzymatic antioxidant defense includes superoxide dismutase (SOD), catalase (CAT), peroxidases (POD), and glutathione peroxidases (GPX). All three types of SODs (chloroplastic Fe SOD, mitochondrial Mn SOD, and cytosolic Cu/Zn SOD) are positively correlated and are actively involved in salt tolerance mechanism in halophytes (Lokhande et al. 2013; Ozgur et al. 2013).

Catalase decomposes H_2O_2 into water and oxygen with high turnover rate, and the rate is almost millions of H_2O_2 molecules within a second. It is mainly localized in peroxisomes and mitochondria. APX is also involved in scavenging and degradation of H_2O_2 molecules into water and oxygen. It is localized in chloroplast membrane and stroma, mitochondria, cytosol, and peroxisomal membrane. Peroxidase family members are actively engaged in both scavenging of H_2O_2 in apoplast and signal transduction. They play an important role in stomata closure and cell elongation. Monodehydroascorbate reductase (MDHAR) plays a key role in regeneration of ascorbate from monodehydroascorbate for ROS scavenging (Ozgur et al. 2013). Glutathione reductase (GR) catalyzes recycling of GSSG to GSH at the expense of NADPH and maintains balance between GSH and ascorbate pools. It is predominantly found in chloroplast but also occurs in mitochondria and cytosol. Glutathione S-transferases are involved in cellular detoxification by conjugating GSH to electrophilic compounds found in apoplast, cytosol, chloroplast, mitochondria, and nucleus (Gechev and Hille 2005).

Respiratory electron chain provides electrons to cytochrome c oxidase for ATP synthesis and cyanide insensitive AOX for energy dissipation. Under salt stress, the electron flow is directed toward AOX pathway (Smith et al. 2009; Zhang et al. 2016). Thiol switch activates AOX1 which stimulates alternative respiration and decreases electron pressure. This prevents over-reduction of the ubiquinone (UQ) pool and controls excessive hydrogen peroxide and superoxide generation and minimizes ROS damage to cells under high salinity stress (Winger et al. 2007; Wang et al. 2010; Yoshida et al. 2011). Also, AOX defines threshold level for induction of programmed cell death by signaling (Van Aken et al. 2009) and modulates release of reactive nitrogen species (RNS). Reducing equivalent accumulation induces release of nitric oxide (NO) in mitochondria (Cvetkovska et al. 2014; Igamberdiev et al. 2014).

7.3.2 Nonenzymatic Defense Machinery

7.3.2.1 Redox Buffers

In halophytes, it has been observed that increased content of reduced ascorbate and glutathione improves salt tolerance in halophytes by ROS detoxification via ascorbate glutathione cycle (Gechev and Hille 2005). Ascorbate and glutathione acts as redox buffers during oxidative stress (Gechev and Hille 2005). They play multiple roles in several physiological processes such as regulation of sulfate transport, signal transduction during oxidative stress, and growth and development of plants

(Noctor et al. 1998; Ogawa et al. 2005). Although halophytes show different strategies and responses to salt stress, one common trait is observed in *Bruguiera parviflora* (Parida et al. 2004), *Cakile maritima* (Ben Amor et al. 2005), *Hordeum maritimum* (Hafsi et al. 2010), *Suaeda maritima* (Alhdad et al. 2013), and *Sesuvium portulacastrum* (Lokhande et al. 2011). They maintain GSH:GSSG ratio under oxidative stress and may regenerate oxidized GSH by using glutathione reductase or synthesis of new GSH to balance GSH redox state. Ascorbate is an electron donor of ascorbate peroxidase which reduces H_2O_2 to H_2O and generates monodehydroascorbate (MDHA). Ascorbate (Asc) along with DHA (ASC:DHA ratio) is an indicator of oxidative stress. They are actively involved in ROS scavenging. Ben Amor et al. (2006) have reported gradual decrease in ascorbate content by APX activity; Under low salinity ASC:DHA ratio increased but under high salinity, the activity decreased in a halophyte *Cakile maritima*. In contrast to this, *Suaeda salsa* showed coordinated upregulation of ROS scavenging systems such as ASC, GSH, APX, and GR under high salinity (Cai-Hong et al. 2005).

7.3.2.2 Osmolytes

Osmolytes are actively engaged in helping plants to combat with stress. The glycine betaine (GB), proline, polyols, phenolics, and sugars have been well studied for their role in salt stress tolerance in halophytes (Slama et al. 2015). It acts as a compatible solute and helps plant in osmotic adjustment. Under abiotic stress it stabilizes structure and function of PSII by preventing dissociation of regulatory intrinsic proteins (Nikalje et al. 2017). Many halophytes also accumulate high amount (>90 μmol dry weight) of glycine betaine. Betaine aldehyde dehydrogenase (BADH2) gene is responsible for GB synthesis. Halophytes show constitutive accumulation of GB; under control condition also, GB present in high amounts, for example, in *Sesuvium portulacastrum* and *Salvia fruticosa* (Khan et al. 1998; Lokhande et al. 2013).

Proline is one of the well-studied osmolytes. It plays duel role in plant metabolism; it is essential for normal metabolism as well as stress tolerance. It protects PSI and PSII from toxic effects of ROS, stabilizes mitochondrial respiration, maintains low NADPH to $NADP^+$ ratio, and prevents programmed cell death under stress (Suprasanna et al. 2015), in many halophytes such as *Lepidium crassifolium*, *Mesembryanthemum crystallinum*, *Sesuvium portulacastrum*, and *Thellungiella salsuginea* (Sanada et al. 1995; Taji et al. 2004; Lokhande et al. 2013; Slama et al. 2015). The polyols like mannitol, pinitol, myoinositol, ononitol, and sorbitol are involved in scavenging of highly toxic (ROS) such as hydroxyl ion, osmotic adjustment, and protection of cellular structures, proteins, and enzymes (Valluru and Van den Ende 2011). Their accumulation under different stresses has been reported in different halophytes. *Mesembryanthemum* accumulated pinitol showed a two fold high scavenging of 1,1-diphenyl-2-picrylhydrazyl as compared to *Lactuca sativa* (Agarie et al. 2009).

7.3.2.3 Polyamines

Polyamines possess antioxidative effect which helps plants in radical scavenging, inhibition of lipid peroxidation, and production of H_2O_2 under different abiotic stresses (Groppa and Benavides 2008). The conjugates of phenylpropanoid polyamine are known antioxidants of both ROS and reactive nitrogen species (RNS) under stressful conditions (Yamasaki and Cohen 2006). Polyamines activate antioxidant enzymes under stress conditions and are also involved in ROS degradation (Gupta et al. 2013). Free polyamine detoxifies superoxide anions, while conjugated polyamines actively participate in other ROS scavenging (Langebartels et al. 1991; Kubis 2005). In a halophyte *Prosopis strombulifera*, accumulation of putrescine was observed under salt stress (Raginato et al. 2012). The halophyte *T. halophila* accumulated all free polyamines, especially spermidine, during the first 2 days of stress. Simultaneously, the expression of the gene for S-adenosylmethionine synthetase, one of the enzymes of the pathway of aminopropyl synthesis, which is required for polyamine synthesis, was enhanced, and the amount of transcripts of the spermidine synthase gene were also increased (Radyukina et al. 2007).

7.3.2.4 Iron-Binding Proteins

Fenton reaction utilizes free iron to convert H_2O_2 to form toxic hydroxyl radicals (reviewed in Bose et al. 2014). To avoid this, iron-binding proteins like ferritin bind to free iron to make them unavailable to Fenton reaction, thereby preventing hydroxyl ion formation. This allows H_2O_2 to perform its signaling function. Ferritin has exceptional ability to store iron atoms. One molecule of ferritin can store up to 4500 iron atoms (Briat et al. 1999). In glycophytes such as maize, the role of ferritin in ROS homeostasis was shown (Lobreaux et al. 1995; Savino et al. 1997). After application of antioxidants like glutathione, it completely lowered the abundance of ferritin mRNA, and application of H_2O_2 enhanced the accumulation of ferritin transcript. Bose et al. (2014) outlined the role of this protein in both halophytes and non-halophytes and concluded that halophytes in early phase of stress prevent hydroxyl radical formation by using ferritin for proper functioning of H_2O_2-mediated signaling. In *Mesembryanthemum crystallinum* and *Avicennia marina*, abundance of ferritin and ferritin transcript (*fer 1*) were observed, respectively, under salt stress which are correlated with H_2O_2 formation rather than hydroxyl ions (Paramonova et al. 2004; Jitesh et al. 2006).

7.3.2.5 Carotenoids

The role of carotenoids in detoxifying $^1O_2$ and peroxyl radicals was for the first time confirmed by Knox and Dodge (1985) by applying norflurazon to inhibit carotenoid biosynthesis which adversely damages plants due to increased production of $^1O_2$. More than 600 members of carotenoid family have been identified for their role in ROS homeostasis specifically in quenching of singlet oxygen and peroxyl radicals formed during excess excitation of chlorophyll (Demmig-Adams and Adams 1996). In *Chloris virgata* and *Thellungiella halophila,* constitutive higher expression under control condition and very little decrease in carotenoid concentration upon exposure to salt stress confirmed its role in ROS homeostasis (Yang et al. 2009; Stepien and

Johnson 2009). It is apparent that halophyte species can rapidly induce H_2O_2 levels, having enough SOD "in stock," which gives them a certain adaptive advantage over glycophytes (Bose et al. 2014). It has also been suggested that similarities exist for Ca^{2+} and H_2O_2 stress-induced "signatures." In this context, the role of APX and CAT in the shaping of H_2O_2 signatures may be very similar to that of Ca^{2+} efflux systems (such as Ca^{2+}-ATPases and CAX Ca^{2+}/H^+ exchangers) which play in restoring the basal cytosolic Ca^{2+} levels (Bose et al. 2011).

7.4 Role of ROS in Signaling

ROS acts as secondary messenger/signaling molecule as well as oxidant (damaging molecules) (Fig. 7.2). Under optimum concentration they act as messengers in several intracellular signaling cascades which mediate response of plant cells to different developmental or stress conditions like programmed cell death, gravitropism, stomatal closure, and adaptation to biotic and abiotic stresses (Yan et al. 2007; Mittler 2002; Joo et al. 2001; Miller et al. 2008). Plants sense increased level of ROS and on the basis of their concentration translate signal into cellular response. Different ROS-responsive proteins, calcium mobility, phosphorylation of proteins, and expression of genes are involved in ROS-mediated signaling. The tyrosine phosphatase protein kinases and different transcription factors can directly sense

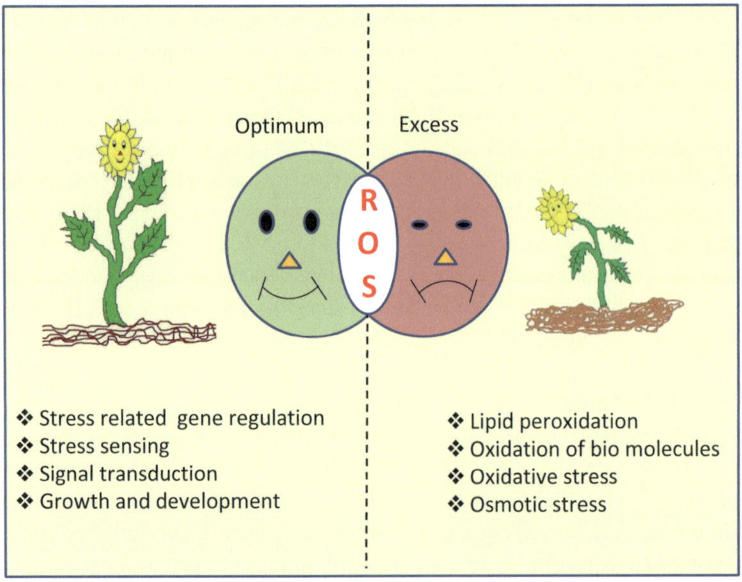

Fig. 7.2 Functional diversity of reactive oxygen species: Depending upon concentration in cell, ROS plays dual role in plants. Under optimum concentration they help in stress signaling and growth and development, while in excess concentration, they are toxic and lead to oxidation of biomolecules

ROS to activate downstream components of ROS signaling (for reviews see Xiong et al. 2002). The G proteins showed their role in ROS-mediated signaling. In *Arabidopsis*, G proteins are considered as first component of oxidative burst (Joo et al. 2005). This protein is also found in halophytes, like *Spartina alterniflora*, *S. maritima*, *Suaeda salsa*, and *Thellungiella halophila*, but its role has not been studied yet (reviewed in Ozgur et al. 2013). The balance between ROS production and removal by antioxidants determines strength, lifetime of ROS signaling pool, and consequently fate of cellular response. Miller et al. (2008) have analyzed mutants deficient in ROS scavenging enzymes and identified a signaling pathway which is activated in response to ROS accumulation. There are several pathways involved in ROS-mediated signaling. Among them calcium signaling, MAPK pathway, and SOS pathways are major genetic pathways (Fig. 7.3).

7.4.1 Genetic Pathways Involved in ROS-Mediated Signaling

7.4.1.1 Calcium Signaling

The ROS-mediated signaling mechanism is associated with Ca^{2+} and associated proteins like calmodulin, calcineurin, and GTP-binding proteins, abscisic acid, activation of MAPKs, and phospholipid signaling (Mittler 2004). Under normal physiological conditions, free calcium levels in cell organelles and cytoplasm are extremely low, but under osmotic stress, this level increased by factor of 10–20 within seconds (Medvedev 2005). The calcium sensor proteins (CAM, CMLs, CDPKs, CBL/CIPKs) decode this elevation and lead to specific physiological response (DeFalco et al. 2010). The CIPK (calcineurin B-like interacting protein kinase) protein interacts and forms complex with CBL (calcineurin B-like) protein and activates downstream proteins like AKT1 and SOS1 to regulate their physiological functions (Hashimoto et al. 2012). In a halophyte *H. brevisubulatum*, the HbCIPK2-interacting ferredoxin (HbCIPK2) protein was identified which acts as signal transducer and mediates the activities of its interacting partners (Zhang et al. 2015).

7.4.1.2 MAPK Pathway

The MAPKs (mitogen-activated protein kinases) are actively involved in many cellular responses under stressful conditions. They are evolutionarily conserved and play an important role in signaling in different stresses. Three MAPKs, MAPKKK, MAPKK, and MAPK, get activated by their phosphorylation under stress condition. The MAPKKK (serine/threonine kinase) phosphorylates S/T-X or -S/T motifs of MAPKK to activate it. The activated MAPKK further activates MAPK by phosphorylating threonine and tyrosine at conserved motif. These kinases have the ability to phosphorylate many substrates like transcription factors. In a halophyte *S. brachiata*, *SbMAPKK* gene showed high sequence similarity with MAPK genes from *N. benthamiana* and *L. esculentum*. The role of this gene under cold, dehydration, and salt tolerance has been already validated (Agarwal et al. 2010).

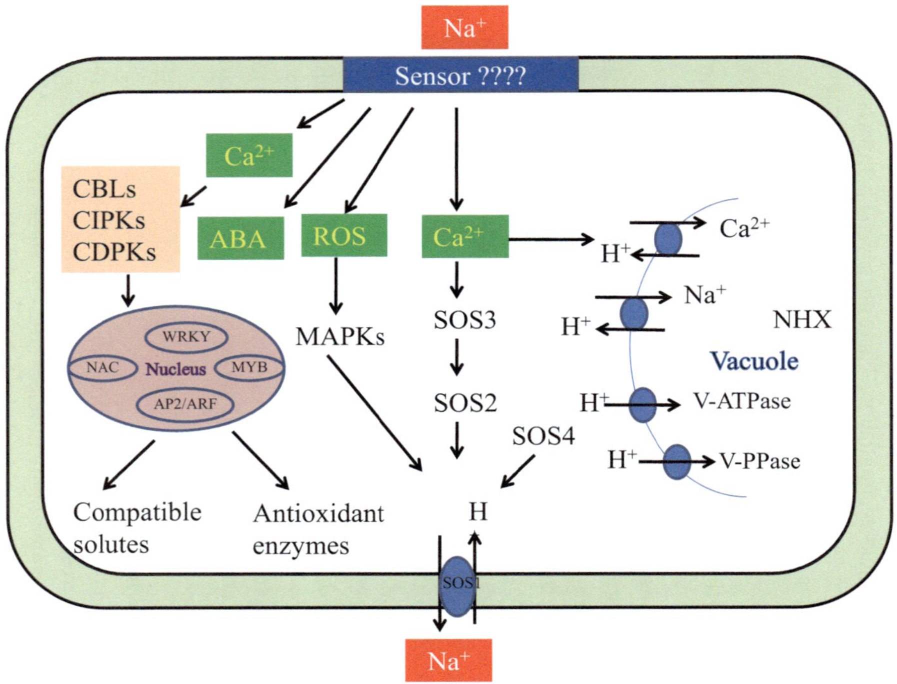

Fig. 7.3 Schematic representation of pathways involved in salt tolerance mechanism of plants (Source: Nikalje et al. 2017): Increased concentration of Na^+ activates membrane-bound sensors which activate the signaling molecules like Ca^{2+}, ABA, and ROS. These molecules interact and activate downstream genes like CBLs, CIPKs, CDPKs, different transcription factors, and membrane-bound ATPases and lead to synthesis of antioxidant enzymes and compatible solutes. Also, the MAPK and SOS pathways get activated, and they either sequester sodium into vacuole or exclude it outside the cell

7.4.1.3 SOS Pathway

Salt overly sensitive (SOS) is a major salt tolerance-related pathway which consists of three genes, *SOS3*, *SOS2*, and *SOS1*. The *SOS3* is a calcium sensor, *SOS2* is a serine/threonine kinase, and *SOS1* is a Na^+/H^+ antiporter (Zhu 2001). They play a key role in cellular signaling and maintaining ion homeostasis (Ji et al. 2013). Under stress condition when calcium level increases in cell, the *SOS3* gene gets activated and phosphorylates *SOS2* to form a complex. This complex activates *SOS1* which excludes toxic sodium ions outside the cell (reviewed in Nikalje et al. 2017). The *SOS1* gene is showed to be responsible for halotrophic nature of *T. halophila*. The 50% repression of *SOS1* gene in *Thellungiella* caused loss of halophytism (DH et al. 2009). Similarly, Yadav et al. (2012) showed major role of *SOS1* gene in salt adaptation mechanism of *Salicornia brachiata* and overexpression of *SOS1* in tobacco results into increased K^+ and Ca^{2+} content in roots and subsequently improves salt tolerance.

7.5 Conclusions

Depending on their concentration in the cell, ROS plays a multifarious role in plants. Under optimum concentrations, ROS play an important role in development and stress signaling while at high concentration they cause oxidation of many biomolecules. In halophytes, early sensing and signaling of salinity stress are the key component of their salt adaptation mechanism. The signaling molecules like Ca^{2+}, ROS and hormones play a vital role in signal transduction under stress. Although, ROS are toxic and highly reactive, halophytes may get benefit from them as signaling molecules. Being highly reactive and short lived, ROS rapidly activates their downstream interactors, and since halophytes are rich in enzymatic and nonenzymatic antioxidants, they maintain good balance between ROS production and scavenging. The intricacies of the mechanism of ROS signaling and ROS scavenging in halophytes are being understood, and more research is warranted to detail this mechanism. The advancements in different omics technologies like transcriptomics, proteomics, metabolomics, etc. are being applied to understand this early signaling mechanism, but it is largely limited to crop plants. Application of these omics techniques to halophytes will help to identify the molecular regulators involved in these stress sensing and signaling mechanisms which are responsible for high stress tolerance of halophytes. The genetic manipulation of these regulators will be useful to improve stress tolerance in crops.

References

Agarie S, Kawaguchi A, Kodera A et al (2009) Potential of the common ice plant, *Mesembryanthemum crystallinum* as a new high-functional food as evaluated by polyol accumulation. Plant Prod Sci 12:37–46

Agarwal PK, Gupta K, Jha B (2010) Molecular characterization of the *Salicornia brachiata* SbMAPKK gene and its expression by abiotic stress. Mol Biol Rep 37(2):981–986

Alhdad GM, Seal CE, Al-Azzawia MJ (2013) The effect of combined salinity and waterlogging on the halophyte Suaeda maritima: the role of antioxidants. Environ Exp Bot 87:120–125

Amtmann A (2009) Learning from evolution: *Thellungiella* generates new knowledge on essential and critical components of abiotic stress tolerance in plants. Mol Plant 22(1):3–12

Asada K (2006) Production and scavenging of reactive oxygen species in chloroplasts and their functions. Plant Physiol 141:391–396

Asada K, Takahashi M (1987) Production and scavenging of active oxygen in chloroplasts. In: Kyle DJ, Osmond CB, Arntzen CJ (eds) Photoinhibition. Elsevier, Amsterdam

Bailey-Serres J, Mittler R (2006) The roles of reactive oxygen species in plant cells. Plant Physiol 141(2):311

Baker A, Graham I (2002) Plant peroxisomes: biochemistry, cell biology and biotechnological applications. Kluwer Academic Publishers, Dordrecht

Ben Amor N, Ben Hamed K, Debez A et al (2005) Physiological and antioxidant responses of the perennial halophyte *Crithmum maritimum* to salinity. Plant Sci 168:889–899

Bose J, Pottosin II, Shabala SS et al (2011) Calcium efflux systems in stress signaling and adaptation in plants. Front Plant Sci 2:85

Bose J, Rodrigo-Moreno A, Shabala S (2014) ROS homeostasis in halophytes in the context of salinity stress tolerance. J Exp Bot 65(5):1241–1257

Briat JF, Lobréaux S, Grignon N et al (1999) Regulation of plant ferritin synthesis: how and why. Cell Mol Life Sci 56:155–166

Cai-Hong P, Su-Jun G, Zhi-Zhong W et al (2005) NaCl treatment markedly enhances H_2O_2-scavenging system in leaves of halophyte *Suaeda salsa*. Physiol Plant 125:490–499

Corpas FJ, Barroso JB, del Río LA (2001) Peroxisomes as a source of reactive oxygen species and nitric oxide signal molecules in plant cells. Trends Plant Sci 6:145–150

Cvetkovska M, Dahal K, Alber NA et al (2014) Knockdown of mitochondrial alternative oxidase induces the 'stress state' of signaling molecule pools in *Nicotiana tabacum*, with implications for stomatal function. New Phytol 203:449–461

D'Autréaux B, Toledano MB (2007) ROS as signalling molecules: mechanisms that generate specificity in ROS homeostasis. Nat Rev Mol Cell Biol 8(10):813–824

DeFalco TA, Bender KW, Snedden WA (2010) Breaking the code: Ca^{2+} sensors in plant signaling. Biochem J 425:27–40

Demmig-Adams B, Adams WW (1996) The role of xanthophylls cycle carotenoids in the protection of photosynthesis. Trends Plant Sci 1:21–26

DH O, Leidi E, Zhang Q et al (2009) Loss of halophytism by interference with SOS1 expression. Plant Physiol 151:210–222

Foyer CH (2005) Redox homeostasis and antioxidant signaling: a metabolic interface between stress perception and physiological responses. Plant Cell 17(7):1866–1875

Foyer CH, Noctor G (2003) Redox sensing and signaling associated with reactive oxygen in chloroplasts, peroxisomes and mitochondria. Physiol Plant 119:355–364

Gechev TS, Hille J (2005) Hydrogen peroxide as a signal controlling plant programmed cell death. J Cell Biol 168:17–20

Groppa MD, Benavides MP (2008) Polyamines and abiotic stress: recent advances. Amino Acids 34:35–45. https://doi.org/10.1007/s00726-007-0501-8

Gupta K, Dey A, Gupta B (2013) Plant polyamines in abiotic stress responses. Acta Physiol Plant 35:2015–2036

Hafsi C, Romero-Puertas LMC, del Rio A et al (2010) Differential antioxidative response in barley leaves subjected to the interactive effects of salinity and potassium deprivation. Plant Soil 334:449–460

Halliwell B, Gutteridge JM (1999) Free radicals in biology and medicine. Oxford University Press, Oxford

Hashimoto K, Eckert C, Anschutz U, Scholz M, Held K, Waadt R, Reyer A, Hippler M, Becker D, Kudla J (2012) Phosphorylation of calcineurin B-like (CBL) calcium sensor proteins by their CBL-interacting protein kinases (CIPKs) is required for full activity of CBL-CIPK complexes toward their target proteins. J Biol Chem 287:7956–7968

Huang S, Van Aken O, Schwarzlander M et al (2016) The roles of mitochondrial reactive oxygen species and stress responses in plants. Plant Physiol 171:1551–1155
Igamberdiev AU, Ratcliffe RG, Gupta KJ (2014) Plant mitochondria: source and target for nitric oxide. Mitochondrion 19:329–333
Imlay JA (2003) Pathways of oxidative damage. Annu Rev Microbiol 57:395–418
Ji H, Pardo JM, Batellic G, Van Oostend MJ, Bressane RA, Lia X (2013) The Salt Overly Sensitive (SOS) pathway: established and emerging roles. Mol Plant 6(2):275–286
Jitesh M, Prashanth S, Sivaprakash K et al (2006) Monitoring expression profiles of antioxidant genes to salinity, iron, oxidative, light and hyperosmotic stresses in the highly salt tolerant grey mangrove, *Avicennia marina* (Forsk.) Vierh. by mRNA analysis. Plant Cell Rep 25:865–876
Joo JH, Bae YS, Lee JS (2001) Role of auxin-induced reactive oxygen species in root gravitropism. Plant Physiol 126(3):1055–1060
Joo JH, Wang S, Chen JG et al (2005) Different signaling and cell death roles of heterotrimeric G protein alpha and beta subunits in the Arabidopsis oxidative stress response to ozone. Plant Cell 17:957–970
Khan MJ, Ungar IA, Showalter AM et al (1998) NaCl induced accumulation of glycine betaine in four subtropical halophytes from Pakistan. Physiol Plant 102:487–492
Knox JP, Dodge AD (1985) Singlet oxygen and plants. Phytochemistry 24:889–896
Kochevar IE (2004) Singlet oxygen signaling: from intimate to global. Sci Signal 2004(221):7. https://doi.org/10.1126/stke.2212004pe7
Kubis J (2005) The effect of exogenous spermidine on superoxide dismutase activity, H_2O_2 and superoxide radical level in barley leaves under water deficit conditions. Acta Physiol Plant 27:289–295
Langebartels C, Kerner K, Leonardi S et al (1991) Biochemical-plant responses to ozone and differential induction of polyamine and ethylene biosynthesis in tobacco. Plant Physiol 95:882–889
Lobreaux S, Thoiron S, Briat JF (1995) Induction of ferritin synthesis in maize leaves by an iron-mediated oxidative stress. Plant J 8:443–449
Lokhande VH, Srivastava AK, Srivastava S (2011) Regulated alterations in redox and energetic status are the key mediators of salinity tolerance in the halophyte Sesuvium portulacastrum (L.) L. Plant Growth Regul 65(2):287–298
Lokhande VH, Gor BK, Desai NS et al (2013) Sesuvium portulacastrum, a plant for drought, salt stress, and fixation, food and phytoremediation. A review. Agron Sustain Dev 33:329–348
Medvedev SS (2005) Calcium signaling system in plants. Russ J Plant Physiol 52:249–270
Miller G, Shulaev V, Mittler R (2008) Reactive oxygen signaling and abiotic stress. Physiol Plant 133(3):481–489
Mittler R (2002) Oxidative stress, antioxidants and stress tolerance. Trends Plant Sci 7(9):405–410
Mittler R (2004) Reactive oxygen gene network of plants. Trends Plant Sci 9(10):490–498
Mittler R (2017) ROS are good. Trends Plant Sci 22(1):11–19
Niedre MJ, Patterson MS, Boruvka N et al (2002) Measurement of singlet oxygen luminescence from AML5 cells sensitized with ALA-induced PpIX in suspension during photodynamic therapy and correlation with cell viability after treatment. Proc SPIE 4612:93–101
Nikalje GC, Nikam TD, Suprasanna P (2017) Looking at halophytic adaptations through genomic landscape. Curr Genomics 18:4
Noctor GN, Arisi ACM, Jouanin L et al (1998) Manipulation of glutathione and amino acid biosynthesis in the chloroplast. Plant Physiol 118(2):471–482
Ogawa T, Ueda Y, Yoshimura K et al (2005) Comprehensive analysis of cytosolic Nudix hydrolases in *Arabidopsis thaliana*. J Biol Chem 280:25277–25283
Ozgur R, Uzilday B, Sekmen AH et al (2013) Reactive oxygen species regulation and antioxidant defence in halophytes. Funct Plant Biol 40:832–847
Ozgur R, Uzilday B, Sekmen AH et al (2015) The effects of induced production of reactive oxygen species in organelles on endoplasmic reticulum stress and on the unfolded protein response in *Arabidopsis*. Ann Bot 116:541–553

Paramonova NV, Shevyakova NI, Kuznetsov V (2004) Ultrastructure of chloroplasts and their storage inclusions in the primary leaves of *Mesembryanthemum crystallinum* affected by putrescine and NaCl. Russ J Plant Physiol 51(1):86–96

Parida AK, Das AB, Mohanty P (2004) Defense potentials to NaCl in a mangrove, *Bruguiera parviflora*: differential changes of isoforms of some antioxidative enzymes. J Plant Physiol 161:531–542

Paulsen CE, Carroll KS (2010) Orchestrating redox signaling networks through regulatory cysteine switches. ACS Chem Biol 5(1):47–62. https://doi.org/10.1021/cb900258z

Radyukina NL, Kartashov AV, Ivanov YV et al (2007) Functioning of defense systems in halophytes and glycophytes under progressing salinity. Russ J Plant Physiol 54:806–815

Raginato MA, Abdala GI, Miersch O et al (2012) Changes in the levels of jasmonates and free polyamines induced by Na_2SO_4 and NaCl in roots and leaves of the halophyte *Prosopis strombulifera*. Biologia 67(4):689–697

Ramani B, Zorn H, Papenbrock J (2004) Quantification and fatty acid profiles of sulfolipids in two halophytes and a glycophyte grown under different salt concentrations. Z Naturforsch C 59:835–842

Sagi M, Fluhr R (2006) Production of reactive oxygen species by plant NADPH oxidases. Plant Physiol 141:336–340

Sanada Y, Ueda H, Kuribayashi K (1995) Novel light–dark change of proline levels in halophyte (*Mesembryanthemum crystallinum* L.) and glycophyte (*Hordeum vulgare* L. and *Triticum aestivum* L.) leaves and roots under salt stress. Plant Cell Physiol 36:965–970

Savino G, Briat JF, Lobréaux S (1997) Inhibition of the iron-induced ZmFer1 maize ferritin gene expression by antioxidants and serine/threonine phosphatase inhibitors. J Biol Chem 272:33319–33326

Sengupta S, Majumder AL (2009) Insight into the salt tolerance factors of wild halophytic rice, *Porteresia coarctata*: a physiological and proteomic approach. Planta 229:911–929

Sengupta S, Majumder AL (2010) Porteresia coarctata (Roxb.) Tateoka, a wild rice: a potential model for studying salt-stress biology in rice. Plant Cell Environ 33:526–542

Sharma P, Jha AB, Dubey RS et al (2012) Reactive oxygen species, oxidative damage, and antioxidative defense mechanism in plants under stressful conditions. Aust J Bot 2012:217037

Slama I, Abdelly C, Bouchereau A et al (2015) Diversity, distribution and roles of osmoprotective compounds accumulated in halophytes under abiotic stress. Ann Bot 115(3):433–447

Smith CA, Melino VJ, Sweetman C et al (2009) Manipulation of alternative oxidase can influence salt tolerance in Arabidopsis thaliana. Physiol Plant 137:459–472

Stepien P, Johnson GN (2009) Contrasting responses of photosynthesis to salt stress in the glycophyte *Arabidopsis* and the halophyte *Thellungiella*: role of the plastid terminal oxidase as an alternative electron sink. Plant Physiol 149:1154–1165

Storz G, Tartaglia LA, Ames BN (1990) Transcriptional regulator of oxidative stress-inducible genes: direct activation by oxidation. Science 248:189–119

Sunagawa H, Cushman JC, Agarie S (2010) Crassulacean acid metabolism may alleviate production of reactive oxygen species in a facultative CAM plant, the common ice plant *Mesembryanthemum crystallinum* L. Plant Prod Sci 13:256–260

Suprasanna P, Nikalje GC, Rai AN (2015) Osmolyte accumulation and implications in plant abiotic stress tolerance. In: Iqbal N et al (eds) Osmolytes and plants acclimation to changing environment: emerging Omics technologies. Springer, New Delhi. https://doi.org/10.1007/978-81-322-2616-1_1

Sweetlove LJ, Foyer CH (2004) Roles for reactive oxygen species and antioxidants in plant mitochondria. In: Day DA, Millar AH, Whelan J (eds) Plant mitochondria: from genome to function, 1. Advances in photosynthesis and respiration. Kluwer Academic Press, Dordrecht

Taji T, Seki M, Satou M et al (2004) Comparative genomics in salt tolerance between Arabidopsis and Arabidopsis-related halophyte salt cress using Arabidopsis microarray. Plant Physiol 135:1697–1709

Triantaphylidès C, Havaux M (2009) Singlet oxygen in plants: production, detoxification and signaling. Trends Plant Sci 14:219–228. https://doi.org/10.1016/j.tplants.2009.01.008

Valluru R, Van den Ende W (2011) Myo-inositol and beyond emerging networks under stress. Plant Sci 181:387–400

Van Aken OV, Giraud E, Clifton R et al (2009) Alternative oxidase: a target and regulator of stress responses. Physiol Plant 137:354–361

Wang M, Peng ZY, Li CL et al (2008) Proteomic analysis on a high salt tolerance introgression strain of *Triticum aestivum/Thinopyrum ponticum*. Proteomics 8:1470–1489

Wang H, Liang X, Huang J et al (2010) Involvement of ethylene and hydrogen peroxide in induction of alternative respiratory pathway in salt-treated *Arabidopsis* calluses. Plant Cell Physiol 51:1754–1765

Winger AM, Taylor NL, Heazlewood JL et al (2007) Identification of intra- and intermolecular disulphide bonding in the plant mitochondrial proteome by diagonal gel electrophoresis. Proteomics 7:4158–4170

Xiong L, Schumaker KS, Zhu JK (2002) Cell signalling during cold, drought, and salt stress. Plant Cell 14:165–183

Yadav NS, Shukla PS, Jha A et al (2012) The SbSOS1 gene from the extreme halophyte Salicornia brachiata enhances Na(+) loading in xylem and confers salt tolerance in transgenic tobacco. BMC Plant Biol 12:188. https://doi.org/10.1186/1471-2229-12-188

Yamasaki H, Cohen MF (2006) NO signal at the crossroads: polyamine-induced nitric oxide synthesis in plants? Trends Plant Sci 11:522–524

Yan J, Tsuichihara N, Etoh T et al (2007) Reactive oxygen species and nitric oxide are involved in ABA inhibition of stomatal opening. Plant Cell Environ 30(10):1320–1325

Yang CW, Zhang ML, Liu J et al (2009) Effects of buffer capacity on growth, photosynthesis, and solute accumulation of a glycophyte (wheat) and a halophyte (*Chloris virgata*). Photosynthetica 47:55–60

Yoshida K, Watanabe CK, Hachiya T et al (2011) Distinct responses of the mitochondrial respiratory chain to long- and short-term high-light environments in *Arabidopsis thaliana*. Plant Cell Environ 34:618–628

Zhang C, Ge R, Zhang J et al (2015) Identification and expression analysis of a novel HbCIPK2-interacting ferredoxin from halophyte H. brevisubulatum. PLoS One 10(12):e0144132

Zhang DW, Yuan S, Xu F et al (2016) Light intensity affects chlorophyll synthesis during greening process by metabolite signal from mitochondrial alternative oxidase in *Arabidopsis*. Plant Cell Environ 39:12–25

Zhu JK (2001) Plant salt tolerance. Trends Plant Sci 6:66–71

8 Enhancing Cold Tolerance in Horticultural Plants Using In Vitro Approaches

Samira Chugh, Shweta Sharma, Anjana Rustagi, Pratibha Kumari, Aayushi Agrawal, and Deepak Kumar

Abstract

The economic benefits of horticultural plants are steadily rising over the years owing to the fact that they provide high yield and returns per unit area as compared to other crops. However, the world has limited arable area which is further limited by various types of abiotic stresses, of which the extremes of temperature is very critical. Extreme temperatures lead to physiological, metabolic, and molecular damages to the plants causing a substantial loss to yield by lowering germination rate, killing seedlings, and inducing symptoms like surface lesions, chlorosis, necrosis, desiccation, wilting, etc. in mature plants. Conventional plant breeding has been employed for years to cross species and genera to and select varieties with abiotic stress tolerance. The use of this traditional method is however limited and has not achieved notable results in developing cold-tolerant varieties. In vitro tissue culture-based tools allow a deeper understanding of the physiology of plants growing under stress and also help in the development of abiotic stress-tolerant plants. In the last decade, ample research has also been done to develop cold-tolerant transgenic plants. This chapter reviews the role of biotechnology in the development of cold stress-tolerant horticultural plants.

S. Chugh · S. Sharma · A. Rustagi
Department of Botany, Gargi College, University of Delhi, New Delhi, India

P. Kumari
Leibniz Institute of Plant Biochemistry, Martin Luther University, Halle (Saale), Germany

A. Agrawal
Department of Biotechnology, Sharda University, Greater Noida, India

D. Kumar (✉)
Department of Plant Sciences, Central University of Jammu, Jammu, Jammu & Kashmir, India
e-mail: deepakkumar@cujammu.ac.in; deepakinjnu@gmail.com

S. M. Zargar, M. Y. Zargar (eds.), *Abiotic Stress-Mediated Sensing and Signaling in Plants: An Omics Perspective*, https://doi.org/10.1007/978-981-10-7479-0_8

Keywords
Abscisic acid · Cold stress · Horticultural crops · Somaclonal variation

8.1 Introduction

According to the International Society for Horticultural Science, horticultural crops include tree, bush, and perennial vine fruits; perennial bush and tree nuts; vegetables (roots, tubers, shoots, stems, leaves, fruits, and flowers of edible and mainly annual plants); aromatic and medicinal foliage, seeds, and roots (from annual or perennial plants); cut flowers, potted ornamental plants, and bedding plants (involving both annual or perennial plants); and trees, shrubs, turf, and ornamental grasses propagated and produced in nurseries for use in landscaping or for establishing fruit orchards or other crop production units (http://www.ishs.org/defining-horticulture). Horticultural crops constitute a significant portion of the total agricultural production and form an important aspect in the economy of any country. Cultivation of horticultural crops has numerous benefits like high export value, high per unit area yield, high returns per unit area, best utilization of wasteland, provision of raw materials for industries, production of more food energy per unit area than that of field crops, better use of undulating lands, and stabilization of women's empowerment by providing employment opportunities through processing, floriculture, seed production, mushroom cultivation, nursery preparation, etc. (Ravichandra 2014).

About three fourth of the total land area of our earth is not suitable for agriculture which is majorly because the abiotic stresses in one form or another limit the production. Each plant species has an optimum temperature range for its growth and metabolism. Extremes of both low and high temperature are major environmental constraints that affect plant growth and crop productivity, leading to substantial losses (Hasanuzzaman et al. 2013). Tropical and subtropical plants lack the mechanism of cold acclimation, thus rendering these plants more vulnerable to low-temperature stress (both chilling and freezing stress). In temperate regions, during winter, late autumn, and early spring, the vegetable plants, fruit trees, and ornamental plants are exposed to low temperatures leading to cold stress. The cold stress also affects the reproductive cycle and time of maturation of flowers and fruits, thereby decreasing their quality and quantity. Plants when challenged by cold stress show many physiological, biochemical, metabolic, and molecular changes in an effort to maximize growth and developmental processes and maintain cellular homeostasis (Fitter and Hay 1981).

Horticultural crops are a major chunk of the total agricultural produce and play a significant role worldwide. Lack of cold hardiness is a major limiting factor for production of fruit crops and other horticultural crops in many regions of the world. Low-temperature injuries to fruit trees are associated with either low autumn or winter temperatures, which kill trees or fruit buds, or with late spring frosts during bloom (Palonen and Buszard 1997). Improving cold tolerance of warm-season plants is important to expand its ecological range and extend the growing season throughout cooler climates (Liu et al. 2013). The main approaches to obtain cold-tolerant plants are in vitro cold acclimation of cells, tissues, and plantlets and treating them with growth regulators especially abscisic acid (ABA) and in vitro selection of cold-tolerant variants (cold-tolerant somaclones) and developing cold-tolerant transgenic plants.

In the present book chapter, we emphasize on the mitigation strategies using in vitro approaches used for enhancing cold tolerance in horticultural plants. This will provide important insights in understanding of strategies and mechanism of cold acclimatization in crop plants which could be useful in this field of research.

8.2 In Vitro Regeneration of Horticultural Crops

Micropropagation of horticulture plants is preferred over conventional propagation techniques as it offers rapid multiplication of uniform plants possessing desirable traits. Also large-scale production of plants is possible in the absence of seeds. Once established, the actively dividing cultures serve as a continuous source of microcuttings, thereby leading to plant production under greenhouse conditions without seasonal interruption (Akin-Idowu et al. 2009). Traditional methods of growing plants are too slow resulting in long gestation periods between the release of the variety and being commercially available (Varshney and Dhawan 1998). Biotechnology provides a powerful adjunct to existing plant breeding methods for the development of new plant varieties. In order to achieve commercial success, any new variety must be available in sufficient quantities to satisfy the demand from growers. Even though micropropagation is labor intensive, it can shorten the time needed to commercialize new varieties and permit the production of disease-free plants (Giles and Morgan 1987). Also, plants may acquire a new temporary characteristic through micropropagation which makes them more desirable from the conventionally raised stocks. The bushy habit in ornamental plants is one such characteristic. As a result, numerous plant tissue culture laboratories have come up worldwide, especially in the developing countries due to cheap labor costs (Rout et al. 2006). The establishment of an efficient in vitro selection system is an important aspect of protocols for increasing genetic variability of stress tolerance through somaclonal selection or genetic transformation (Liu et al. 2013).

8.3 Biological Implication of Abiotic Stresses Affecting Horticultural Plants

Abiotic stress is the adverse effect of nonliving factors from surrounding environment on the living organisms. Abiotic stresses are one of the primary factors responsible for significant reduction in agricultural yield worldwide (Bhatnagar-Mathur et al. 2008; Gao et al. 2007). Abiotic stresses can be of different types depending on the various environmental factors such as:

8.3.1 Salinity Stress

One of the very important abiotic stresses of the plants worldwide is the salinity stress. Salinity can be defined as the presence of elevated concentration of soluble salts, mostly chloride and sulfates of sodium in soil (Karan and Subudhi 2012). The main reasons for salinity are poor-quality irrigation water, inadequate drainage, saltwater flooding of coastal land, etc. (Kijne 2006). Salinity stress adversely affects plants as it inhibits seed germination and alters the physiology and anatomy of the plants. It reduces the rate of leaf expansion and causes stomata closure resulting in reduced rate of photosynthesis (Rahnama et al. 2010). Salinity stress severely reduces the yield of horticultural crops as it leads to water deficit due to relatively high solute concentration in soil (Yamaguchi and Blumwald 2005). In various tomato cultivars, NaCl concentrations of 80 mM have been observed to result in impaired seed germination, vegetative growth, leaf area, etc. (Van Ieperen 1996). Salt stress has also been reported to adversely affect fruit yield in tomato as a result of reduction in average fruit weight as well as fruit number. Exposure of horticultural crops to high salt concentrations has been shown to result in growth retardation, wilting, leaf curling, leaf abscission, tissue necrosis, etc. (Cuartero and Fernandez-Munoz 1999).

8.3.2 Heavy Metal Stress

The abiotic stress caused by heavy metals such as Cd, Cr, Cu, Ni, Pb, Se, etc. is proving to be extremely damaging for horticultural plants. These toxic heavy metals accumulate due to excessive use of chemical fertilizers and pesticides, irrigation with high levels of heavy metal wastewater, excessive mining, etc. (Das et al. 2004). Once the heavy metals enter into food chain, it potentially can accumulate in large amounts in plant and animal tissues. The toxic heavy metals adversely affect various functions of plant by damaging the protein structure, disrupting the essential metals present in molecules such as pigments or enzymes, damaging the integrity of membrane, leaf senescence, inhibition of overall growth, etc. This in turn harmfully impacts the various physiological processes in plants such as photosynthesis, respiration, transpiration, and enzymatic activities (Borowiak et al. 2012).

In *Helianthus annuus*, toxicity of heavy metals such as cadmium, copper, and nickel has resulted in genotoxic effects which include DNA base modifications, DNA strand breakage, rearrangements, depurination, etc. (Chakravarty and Srivastava 1992). Exposure to copper and cadmium results in responses such as thickening of root apices and fewer new roots in *Citrus* and *Pinus* (Zhu and Alva 1993; Arduini et al. 1995). High levels of lead inhibit root and shoot elongation in barley and *Raphanus sativus* (Juwarkar and Shende 1986). Iron toxicity has been reported to be responsible for "freckle leaf" of sugarcane and also produces brittle, brown to purple leaves with poor burning quality in tobacco (Foy 1978). Manganese toxicity has been reported to produce symptoms such as necrotic brown spotting on leaves, petioles, and stems (Wu 1994).

8.3.3 Water Stress

Water stress is reported both in terms of excess as well as the scarcity of water. Drought represents one of the major abiotic stresses responsible for huge yield losses across the world. Drought negatively affects plant growth by reducing the total available water content. In addition, it also leads to stomata closure for reducing water loss which in turn limits gaseous exchange and reduces transpiration as well as rate of photosynthesis (Chaves et al. 2003; Chaves and Oliveira 2004). Exposure to drought conditions leads to abscission of reproductive organs such as flower buds and flowers in vegetable crops, thus adversely affecting their yield (Wien et al. 1989). Water stress also includes waterlogging or flooding conditions. Flooding results in responses such as decreased root and shoot development, increased internal ethylene concentrations, low stomatal conductance, and reduced rate of photosynthesis. In horticultural crops such as avocado and pea, waterlogging has been shown to result in lower nitrogen absorption from the soil which in turn leads to reduced leaf concentration of N, P, K, Ca, and Mg (Rao and Li 2003). Deficiency of essential elements such as nitrogen results in yellowing of leaves due to reduced chlorophyll content. Tomato plants which are waterlogged undergo severe damages due to increased ethylene accumulation (Drew 1979). It has been reported that the symptom severity of waterlogged tomato plants increases with increase in temperature which results in wilting and death (Kuo et al. 1982).

8.3.4 Temperature Stress

An optimum range of temperature is very crucial for plant growth and development. Temperature extremes could prove to be damaging for plants. The presence of very high temperature may adversely affect plants due to denaturation of proteins, reduced permeability of membranes, etc. Exposure to high-temperature conditions results in postharvest deterioration of fruits and vegetables due to physiological changes (Kader 2002). Ledesma et al. (2008) reported that high-temperature stress negatively affects the number of inflorescence, flowers, fruits, etc. in strawberry. High-temperature stress also adversely affects pollen germination, pollen tube growth, and viability of ovule and pollen retention by stigma in tomato (Foolad 2005). Similarly, in papaya also reduced pollen viability was reported due to temperature stress (Galân-Saûco and Rodrîguez-Pastor 2007).

Similarly, low temperatures are also highly undesirable for plants and are responsible for major crop losses throughout the world. Chilling injury has most often been reported in horticultural crops of tropical and subtropical origin such as mango, papaya, sweet potato, eggplant, citrus, avocado, bananas, pineapple, etc. Major symptoms of chilling injury in economically important fruits and vegetables are pitting, water-soaked area, browning of pulp or core, browning of skin, rotting, red blotches, brown streaking on skin, etc. (Kader 2002).

8.4 Impact of Cold Stress: Chilling and Freezing Stress in Plants

Cold stress can be classified mainly into two categories:

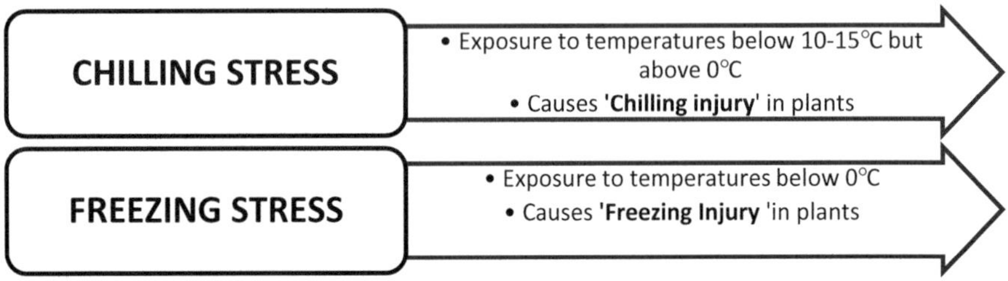

Plants develop various physiological and cytological damages upon exposure to low but nonfreezing temperatures, i.e., below 10–15 °C. The major symptoms observed in plants in response to cold stress include yellowing of leaves, i.e., chlorosis, reduced leaf expansion, wilting, poor germination, stunted seedlings, and necrosis. One of the primary effects of cold stress in plants is the loss of membrane integrity which is caused due to the formation of ice crystals. Formation of ice is the actual cause of damage as it leads to dehydration as well as cell rupture due to mechanical strain on cell wall (Olien and Smith 1997; Yadav 2010). Freezing injury refers to various changes in plants due to exposure to freezing temperatures, i.e., below 0 °C. Freezing injury in plants takes place mainly due to freezing of soil water which significantly lowers the amount of water available to plants. In addition, at sub-zero temperatures ice crystal formation takes place in the cell wall and intercellular spaces which causes severe damages (Olien and Smith 1997). The plants which are native of warm habitats are more prone to cold stress such as banana (*Musa* sp.), cotton (*Gossypium hirsutum*), maize (*Zea mays*), soybean (*Glycine max*), tomato (*Lycopersicon esculentum*), etc. The chilling injury symptom, which appears after 48–72 h after exposure to cold stress, varies from plant to plant (Hopkins 1999) (Fig. 8.1).

8.4.1 Cold Tolerance vs. Acclimation: Plants' Combat Strategies to Cold Stress

As plants are immobile in nature, the most effective way for them to survive in unfavorable environmental conditions is to adapt to the prevailing conditions (Altman 2003). Cold tolerance and cold acclimation are the ways by which plants respond to cold stress. The capability of a plant to tolerate low-temperature conditions without injury or damage is known as cold tolerance, while cold acclimation is the increase in the freezing tolerance of plants due exposure to chilling and nonfreezing temperatures (Guy 1990). Cold tolerance as well as cold acclimation involves a cascade of biochemical, molecular, and metabolic processes (Thomashow

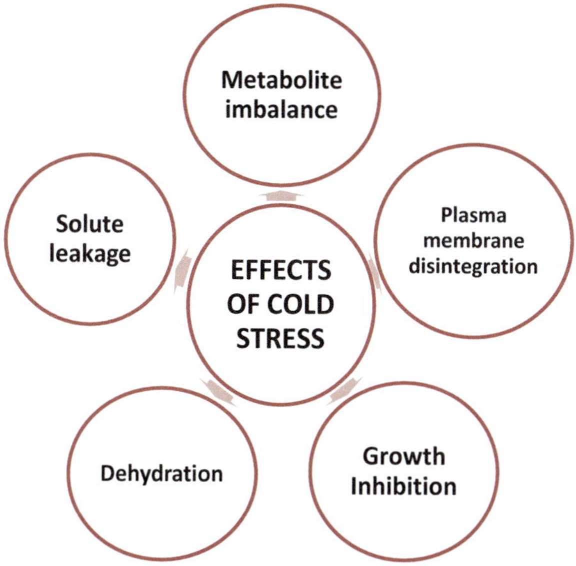

Fig. 8.1 Various physiological effects observed in plants due to cold stress

1999; Zhu et al. 2007). Recent studies have shown that a complex mechanism is involved in the cold stress tolerance. The cold stress is sensed by unknown receptors which are present on the membrane and triggers a signal transduction pathway, major components of which are calcium, reactive oxygen species, protein kinase, protein phosphatase, and lipid signaling cascades. ABA has also been shown to play an important role in cold stress response (Mahajan and Tuteja 2005). The signaling cascade regulates various transcription factors which are involved in the regulation of cold-responsive genes. The transcription factors which play an important role in cold stress response are mitogen-activated protein kinase (MAPK), C-repeat binding factors (CBF), and myeloblasts. The expression of cold-responsive genes leads to increase in level of several proteins such as chaperones, late embryogenesis abundant (LEA) proteins, osmotin, antifreeze proteins, etc. (Yadav 2010). These proteins and metabolites have been shown to have protective role in cold stress response.

8.4.2 Approaches to Enhance Cold Tolerance in Horticultural Plants

Conventional breeding methods involving interspecific or intergeneric hybridization have met with limited success in improving the cold tolerance of important crop plants (Sanghera et al. 2011). The conventional breeding approaches are limited by the complexity of stress tolerance traits, low genetic variance of yield components under stress conditions, and lack of efficient selection criteria. In vitro tissue culture-based tools have not only allowed a deeper understanding of the physiology and biochemistry in plants cultured under adverse environmental conditions but have also helped in the development of abiotic stress-tolerant plants (Pérez-Clemente and Gómez-Cadenas 2012). Tolerance to freezing can also be enhanced when plants are prior exposed to a period of low but nonfreezing temperatures, which is called cold acclimation (CA) (Weiser 1970). The freezing tolerance or cold acclimation of plants is enhanced over a period of time by temperatures below 10 °C and by a short photoperiod in certain species of trees and grasses (Gusta et al. 2005). Several studies have convincingly demonstrated that exogenously applied ABA can also increase cold tolerance. Application of ABA at room temperature increased cold resistance in callus explants of tobacco, cucumber, winter wheat, and alfalfa (Swamy and Smith 1999). There are also a few reports of in vitro induced variations to improve the abiotic stress tolerance.

8.4.2.1 In Vitro Low-Temperature and/or Abscisic Acid (ABA) Treatment to Enhance Cold Stress Tolerance in Cultures

Application of growth regulators is one possible way of enhancing cold tolerance in horticultural plants. Palonen and Buszard (1997) published an extensive review on the effect of growth regulators on the enhancement of cold hardiness in fruit trees and have concluded that effects of growth regulators vary between plant species and even between cultivars. They reported the effects of paclobutrazol, thidiazuron, and ethephon on enhancement of cold hardiness of apple, peach, and grapes. After reviewing the recent literature, we feel abscisic acid (ABA) has a more significant and consistent role in enhancing cold tolerance in horticultural plants. Adaptation and survival of plant cells and tissues to various environmental stresses may be enhanced by exogenous ABA (Rai et al. 2011). Irving and Lanphear (1968) reported plants of *Acer negundo*, sprayed with gibberellic acid (GA) under short days, did not acclimate, whereas plants sprayed with GA inhibitors showed a significant increase in cold hardiness. At that time a recently identified hormone, dormin, was shown to be a GA antagonist (Chrispeels and Varner 1967). The application of dormin to plants exposed to long days resulted in an increase in freezing tolerance equivalent to those exposed to short days. In lettuce seed germination assays, high levels of an inhibitor similar to dormin were produced in leaves exposed to short days. Dormin was later renamed abscisic acid (ABA). Gusta et al. (2005) have suggested Irving and Lanphear's work might be the first to demonstrate that ABA was involved in cold acclimation. Following this work many workers have reported an enhancement in cold tolerance of plants by exogenous ABA treatment by foliar sprays or under in vitro conditions. They pointed out in their extensive review on the role of ABA cold acclimatization that in contrast to limited results by foliar applications, ABA or ABA analogs supplied aseptically to tissue under in vitro conditions resulted in a significant increase in freezing tolerance. The leaf's cuticle is the major barrier to foliar ABA uptake (Blumenfeld and Bukovac 1972); also, ABA is readily inactivated by light.

In the next few decades, quite a few reports have been published dealing with cold tolerance in horticultural plants. Bornman and Jansson (1980) reported that pith explants of *Nicotiana tabacum* pretreated with 10^{-4} M ABA underwent approximately 50% less cellular leakage when chilled at 2 °C under short-day conditions for 10 days as compared to the non-treated tissue, and in ABA-treated tissue, some cells even survived sub-zero temperatures and regenerated callus again. Growth in terms of fresh and dry weights was also more than twice that of the non-ABA-treated material. The authors suggested that at least part of the promotory effects of ABA may have resulted from an increase in the level of proline. Chen and Gusta (1983) investigated the effect of abscisic acid (ABA) on the cold hardiness of cell suspension of bromegrass *Bromus inermis* Leyss. Their investigations revealed that cultures treated with 7.5×10^{-5} molar ABA for 4 days at 20 °C could tolerate −30 °C, whereas the control cultures tolerated only −7 to −8 °C. Lee et al. (1992) studied the induction of freezing tolerance by abscisic acid (ABA) or cold treatment in suspension cultured cells of *Solanum commersonii* and reported that both ABA (50–100 μM) at 23 °C and low temperature (4 °C) increased freezing tolerance in cultured *Solanum commersonii* cells from a LT_{50} (freezing temperature at which 50% cells were killed) of −5 °C (control) to −11.5 °C in 2 days.

Gusta et al. (2005) in their critical review of the role of ABA in cold acclimation have also concluded that ABA might regulate many though not all of the genes associated with cold tolerance. Cold hardiness is an extremely complex phenomenon; species and cultivars vary in their response to low temperatures, and genotype is the most critical factor in determining hardiness. Therefore, developing cold hardy cultivars for different climatic conditions is essential. Cultural practices can be used to further enhance hardiness only once the appropriate cultivar is available (Palonen and Buszard 1997).

8.4.2.2 Cold-Tolerant Somaclonal Variants of Horticultural Plants

Clonal propagation through tissue culture is an effective technique to realize propagation of significant plants rapidly within small space. The main advantage of tissue culture is the genetic uniformity it offers. However, genetic variation does occur in small amounts in the in vitro regenerated cells and plantlets. Tissue culture protocols are designed to produce large numbers of true-to-type plants. However, genetic variants occur and are problem in many micropropagation programs. From the point of commercial micropropagation, variation of any kind, in particular genetic variations, may be considered obstructive, since such variations may lead to loss of genetic fidelity (Krishna et al. 2016). The first report of variation observed in tissue culture plants was published in 1971 by Heinz and Mee in sugarcane. Following this report many such observations were made in other crops. Larkin and Scowcroft (1981) first suggested that this variation in in vitro regenerated plants could be exploited to raise new cultivars. They proposed the term somaclonal variation to describe the genetic variation in plants regenerated from any form of cell culture.

Somaclonal variations observed in in vitro clonally propagated plants can advantageously be utilized as a source of new variation in horticultural crops (Karp 1995). However, suitable tools for detection, evaluation, identification, and improvement of resistant clones have to be designed in order to realize the benefits of such variations (Sahijram et al. 2003). Crop improvement through somaclonal variation enables breeders to obtain plants tolerant to biotic or abiotic stress, such as drought, high salinity, high or low soil pH, and disease tolerance (Yusnita et al. 2005). A survey of available literature reveal previous reports on improved cold tolerance through somaclonal variants in agronomic crops like winter wheat (Kendall et al. 1990) and rice (Bertin and Bouharmont 1997). However, the potential of somaclonal variation has yet to be fully exploited by breeders of horticultural crops.

There are many limiting factors to the production of grapes for the multimillion dollar wine industry, but the most challenging one is that of winter cold damage. Cold-tolerant cell lines selected from suspension cultures of three cultivars of grape hybrids (*Vitis* spp.) survived temperatures as low as −9 °C (Zhang and Rajashekar 1994). Li et al. (2010) reported the development of freezing-tolerant variant of St. Augustine grass which is known to be the least cold tolerant of all turfgrasses. According to the authors, out of 7800 regenerated plants from tissue culture, somaclonal variant SVC3 showed significantly more freezing-tolerant than its parent "Raleigh." Liu et al. (2013) reported the establishment of an effective in vitro culture protocol for generating plants from calli using mature seeds of another warm turfgrass seashore paspalum (*Paspalum vaginatum*). Plant variants derived from cold-selected calli exhibited significant improvement in their tolerance to low temperature with higher turf quality, leaf chlorophyll content, and membrane stability as well as lower levels of lipid peroxidation compared with the control plants.

8.4.3 Transgenomics Approach Toward Generation of Cold-Resistant Horticultural Crops

Genetic engineering has been widely used for generation of stress-resistant crops. In the process of generating transgenic plant, a crucial initial step is to select genes involved in cold tolerance. Several genes have been used to transfer from source plant to host plant to develop transgenic plant resistant for cold stress. Nevertheless, very few examples are known in regard to horticultural crops. List of such genes and their source and host plants is listed in Table 8.1. Low temperature induces a number of genes which regulate the level of several proteins and metabolites, thereby providing protection to plants against stresses. Genes expressed during cold stress can be categorized into two groups. First group includes proteins that have potential role in stress tolerance. Examples of such genes are protein-encoding genes for late embryogenesis abundant proteins, osmolyte biosynthesis, chaperons, antifreeze proteins, water channel proteins, and detoxification enzymes. Some of the listed enzymes have been overexpressed in transgenic plants which results in overcoming cold stress. This indicates the role of these genes in stress tolerance. Second group include proteins which regulate gene expression and signal transduction such as transcription factors. Functional analysis of these transcription factors provides information about complex regulatory networks which become active due to different types of stresses. In addition to these transcription factors, there are other proteins involved in stress regulatory network of plants, such as kinases, protein phosphatases, calmodulin-binding proteins, enzymes involved in phosphate metabolism, and 14-3-3 proteins.

Table 8.1 Selective reports on cold stress-tolerant transgenic horticultural crops

Gene(s)/gene product	Origin	Transgenic host plant	References
COD/codA (choline oxidase)	*Arthrobacter globiformis*	Tomato	Park et al. (2003, 2004)
CBF1 (CRT/CRE binding factor)	*Arabidopsis*	Tomato	Hsieh et al. (2002)
		Canola	
		Strawberry	Lee et al. (2003)
	Canola	Canola	Jaglo-Ottosen et al. (2001)
			Owens et al. (2002)
			Gusta et al. (2002)
Fe-SOD (Fe-superoxide dismutase)	Tobacco	Alfalfa	McKersie et al. (2000)
cAPX (cytosolic ascorbate peroxidase)	*Pisum sativum L.*	Tomato	Wisniewski et al. (2007)
		Apple	
GRXS17	*Arabidopsis*	Tomato	Hu et al. (2015)
Lea-Gal (antisense β-galactosidase)	Tomato	Petunia	Pennycooke et al. (2003)
wft1/wft2 (Fructosyltransferase)	Wheat	Ryegrass	Hisano et al. (2004)
Mn-SOD (Mn-superoxide dismutase)	Tobacco	Alfalfa	McKersie et al. (1999)
	Wheat	Canola	Gusta et al. (2002)
MusabZIP53	Banana	Banana	Shekhawat and Ganapathi (2014)

Many cold-regulated (*COR*) genes are regulated by family of transcription factors known as C-repeat binding/dehydration-responsive element binding factors (*CBF/DREB*). The three *CBF/DREB1* genes induced by cold stress are *CBF1/DREB1b*, *CBF2/DREB1c*, and *CBF3/DREB1a*. They belong to AP2/DREBP family of DNA-binding proteins. These transcription factors bind to the promoter of *COR* genes and induce their expression, thus in turn enhancing abiotic stress tolerance (Stockinger et al. 1997. Overexpression of *CBF1/DREB1b* and *CBF3/DREB1a* resulted in enhanced cold tolerance by inducing expression of *COR* genes as well as other biochemical changes such as proline and sugar accumulation (Gilmour et al. 2000; Kasuga et al. 1999). The transcription factors bZIP have shown to be involved in diverse cellular processes in plants. Shekhawat and Ganapathi (2014) identified a *bZIP* gene, MusabZIP53, from banana and later overexpressed it in banana cultivar Rasthali. The transgenic plants were tolerant in response to cold and drought stress. Overexpression of *AtCBF1* shows improved chilling tolerance in transgenic tomato plants by producing higher level of catalase 1 (*CAT1*) (Hsieh et al. 2002). *CAT1* reduces H_2O_2 level in cell and thus protects plants from oxidative stress, which causes cellular damage in variety of abiotic stresses. Similarly *AtCBF1* overexpression increases cold tolerance in canola plants. Strawberry is an early blooming plant. Freezing temperature causes floral damage in such plants. *AtCBF1* expressed under CaMV 35S promoter in strawberry leads to cold tolerance in leaf tissue

(Owens et al. 2002). Different families of DRE-binding proteins are expressed during different types of stresses. *DREB1A* and its homologs (*DREB1B/CBF1* and *DREB1C/CBF2*) are induced by cold stress, whereas induction of *DREB2A* and its homologous genes is caused by drought stress (Sanghera et al. 2011).

It has been found that homeodomain leucine zipper interacts with DREB/CBFs. In hot pepper two *DREB1b/CBF1* such as cDNAs *CaCBF1A* and *CaCBF1B* are induced in response to low-temperature stress. During cold stress *CaCBFIB* interacts with homeodomain leucine zipper (Yadav 2010) suggesting the role of homeodomain leucine zipper in regulation of *DREB/CBFs*. *CBFs* expression is negatively regulated by upstream transcription factor, MYB domain 15 (*MYB15*) in *Arabidopsis*. *MYB15* is an *R2R3-MYB* family protein which is expressed even in the absence of cold stress. *MYB15* binds to the MYB recognition elements (MYBRS) in the promoter of *CBFs* (Sanghera et al. 2011).

Accumulation of reactive oxygen species (ROS) in response to variety of abiotic stresses activates respective signaling pathways to protect the plants from cold stress conditions by overproducing ROS scavenging enzymes such as superoxide dismutase (*SOD*), ascorbate peroxidase (*APX*), and glutathione reductase (*GR*). Overexpression of tobacco *Mn-SOD* in alfalfa confers freezing tolerance (McKersie et al. 1993). Transgenic alfalfa plants that expresses *Mn-SOD* show higher winter survival rates as well as increased yield in comparison to untransformed control plants (McKersie et al. 1999, 2000). ROS is converted into H_2O_2, which is further processed by APX into water and dehydroascorbate. Transgenic tomato and apple (Royal Gala) plants overexpressing cytosolic APX show enhanced tolerance to cold stress. In freezing temperature, LT50 of the transgenic apple leaves ranges from 1 to 3.2 °C above that of wild-type plants (Wisniewski et al. 2007). Glutaredoxins (GRXs) are small ubiquitous redox enzymes and belong to the thioredoxin (Trx) family. The GRXs maintain the intracellular redox state by catalyzing reversible reduction of substrate protein disulfide bonds in the presence of glutathione (GSH). Ectopic expression of *AtGRXS17* in tomato leads to chilling tolerance and suppresses chilling-induced oxidative damage (Hu et al. 2015).

Compatible solutes such as sugars, proteins, and related compounds are highly soluble compounds which act as cryoprotectant and osmolyte to protect the plants against freezing-induced dehydration and also stabilize membrane. Plants under low-temperature stress conditions accumulate high level of these solutes and upregulate genes for sugar biosynthesis. Both sucrose and raffinose have been reported to restore freezing stress. Exogenous application of sucrose restores freezing tolerance of sucrose-deficient 4 (*sfr4*) mutants (Uemura et al. 2003). However it has been reported that freezing tolerance is more closely related to raffinose accumulation than the sucrose level (Taji et al. 2002). Synthesis of raffinose is controlled by galactinol synthase (*GolS*) and raffinose synthase (*RafS*), whereas degradation is controlled by α-galactosidase (*α-Gal)*. Transgenic *Petunia* plants expressing antisense transcripts of tomato *Lea-Gal* show enhanced freezing tolerance (Pennycooke et al. 2003). Bacterial fructan polymerase as well as wheat fructosyltransferase (*wft1 and wft2)* expression in tobacco and ryegrass, respectively, enhances freezing tolerance (Hisano et al. 2004).

Heat shock proteins (HSPs) have been reported to play an important role in protein unfolding or assembly/disassembly reactions in an ATP-dependent manner. The HSPs prevent denaturation of protein during different stresses. Molecular chaperones, including HSP70 and HSP90 from maize and spinach and *Brassica napus*, respectively, confer freezing tolerance by stabilizing protein against stresses (Anderson et al. 1994; Krishna et al. 1995).

Glycine betaine (GB) is an example of osmolyte which helps in osmotic adjustment in plants. GB maintains membrane integrity and enzyme activity in plants grown under stress conditions. Exogenous application of GB to its non-accumulator plant species improves their growth capacity under stress conditions. GB biosynthetic gene could be the better candidate for generating transgenic plants with enhanced stress tolerance capacity. The gene *codA* encodes for choline oxidase (*COD*) that coverts choline to GB. Tomato plants transformed with *codA* using transit peptide sequence to target the *COD* into chloroplast show increased cold tolerance in vegetative and reproductive stage (Park et al. 2003, 2004).

Many stress-inducible genes have been identified using transcriptomic approaches. Techniques for transcriptome analysis include deoxyribonucleic acid (DNA) hybridization measurement, ribonucleic acid (RNA) hybridization measurement, subtractive hybridization, and microarray. Serial analysis of gene expression is used to identify novel genes together with their metabolic circuits and biochemical pathway and is a highly promising, efficient, and global approach for analyzing gene expression profiles under different physiological states. In addition to Affymetrix 22K GeneChip, ATH1 has been used recently to identify more stress-inducible genes (Yadav 2010).

8.5 Conclusions

The extremes of temperatures exceeding the limits of tolerance influence the different aspects of metabolism, viability, physiology, and yield of plants. The totality of damage induced by temperature stress is dependent on the duration of the adverse temperature, the genotypes of the exposed plants, and their stage of growth. The mitigation strategies to cold stress damage include exogenous application of cryoprotectants, somaclonal variations, and transgenomics technologies. The phytohormone ABA has shown its potential role for cryoprotecting horticultural crops, probably by playing a role in the regulation of genes associated with cold tolerance. Although some somaclonal variations might be epigenetic, there could also be genetic variants that can contribute to heritable improvement of cold stress tolerance. Recombinant DNA technology offers new strategies that can be used for efficient gene transfer for the generation of cold-tolerant transgenic lines. Although a number of genes have been characterized that are responsive to freezing stress, the plant responses to cold stress are complex owing to its multigenic characteristic. Thus improving cold tolerance by transgenic technology could be the thrust area in the future research.

References

Akin-Idowu PE, Ibitoye DO, Ademoyegun OT (2009) Tissue culture as a plant production technique for horticultural crops. African J Biotech 8:3782–3788

Altman A (2003) From plant tissue culture to biotechnology: scientific revolutions, abiotic stress tolerance, and forestry. In Vitro Cell Dev Biol Plant 39:75–84

Anderson JV, Li QB, Haskell DW, Guy CL (1994) Structural organization of the spinach endoplasmic reticulum-luminal 70-kilodalton heat shock cognate gene and expression of 70-kilodalton heat shock genes during cold acclimation. Plant Physiol 104:1359–1370

Arduini I, Godbold DL, Onnis A (1995) Influence of copper on root growth and morphology of Pinus pinea L. and Pinus pinaster Ait. seedlings. Tree Physiol 15:411–415

Bertin P, Bouharmont J (1997) Use of somaclonal variation and in vitro selection for chilling tolerance improvement in rice. Euphytica 96:135–142

Bhatnagar-Mathur P, Vadez V, Sharma KK (2008) Transgenic approaches for abiotic stress tolerance in plants: retrospect and prospects. Plant Cell Rep 27:411–424

Blumenfeld A, Bukovac MJ (1972) Cuticular penetration of ABA. Planta 107:261–268

Bornman CH, Jansson E (1980) *Nicotiana tabacum* callus studies X, ABA increases resistance to cold damage. Physiol Plant 48:491–493

Borowiak K, Drzewiecka K, Magdziak Z, Gasecka M, Mleczek M (2012) Effect of Ca/Mg ratio on copper uptake, photosynthesis activity and growth of Cu (II) –treated *Salix viminalis* L. "Cannabina". Photosynthetica 50:353–361

Chakravarty B, Srivastava S (1992) Toxicity of some heavy metals in vivo and in vitro in *Helianthus annuus*. Mutat Res 283:287–294

Chaves MM, Oliveira MM (2004) Mechanisms underlying plant resilience to water deficits: prospects for water-saving agriculture. J Exp Bot 55:2365–2384

Chaves MM, Maroco JP, Pereira JS (2003) Understanding plant responses to drought – from genes to the whole plant. Funct Plant Biol 30:239–264

Chen THH, Gusta LV (1983) Abscisic acid-induced freezing resistance in cultured plant cells. Plant Physiol 73:71–75

Chrispeels MJ, Varner JE (1967) Hormonal control of enzyme synthesis: on the mode of action of gibberellic acid and abscisin in aleurone layers of barley. Plant Physiol 41:1008–1016

Cuartero J, Fernandez-Munoz R (1999) Tomato and salinity. Sci Hortic 78:83–125

Das HK, Mitra AK, Sengupta PK, Hossain A, Islam F, Rabbani GH (2004) Arsenic concentrations in rice, vegetables, and fish in Bangladesh: a preliminary study. Environ Int 30:383–387

Drew MC (1979) Plant responses to anaerobic conditions in soil and solution culture. Curr Adv Plant Sci 36:1–14

Fitter AH, Hay RKM (1981) Environmental physiology of plants. Academic Press, New York

Foolad MR (2005) Breeding for abiotic stress tolerance in tomato. In: Ashraf M, Harris PJ (eds) Abiotic stresses: plant resistance through breeding and molecular approaches. The Hawarth Press, New York, pp 613–684

Foy CD (1978) The physiology of metal toxicity in plants. Annu Rev Plant Physiol 29:511–566

Galân-Saûco VG, Rodrîguez-Pastor MGR (2007) Greenhouse cultivation of papaya. Acta Hortic 740:191–195

Gao JP, Chao DY, Lin HX (2007) Understanding abiotic stress tolerance mechanisms: recent studies on stress response in rice. J Integr Plant Biol 49:742–750

Giles KL, Morgan WM (1987) Industrial-scale plant micropropagation. TIBTECH 5:35–39

Gilmour SJ, Sebolt AM, Salazar MP, Everard JD, Thomashow MF (2000) Overexpression of the *Arabidopsis CBF3* transcriptional activator mimics multiple biochemical changes associated with cold acclimation. Plant Physiol 1(24):1854–1865

Gusta LV, Nesbitt NT, Wu G, Luo X, Robertson AJ, Waterer D, Gusta ML (2002) Genetic engineering of cultivated plants for enhanced abiotic stress tolerance. In: Li PH, Palva T (eds) Plant cold hardiness: gene regulation and genetic engineering. Kluwer Academic/Plenum Publishers, Dordrecht, pp 237–248

Gusta LV, Trischuk R, Weiser C (2005) Plant cold acclimation: the role of abscisic acid. J Plant Growth Regul 24:308

Guy CL (1990) Cold acclimation and freezing stress tolerance: role of protein metabolism. Annu Rev Plant Physiol 41:187–223

Hasanuzzaman M, Nahar K, Fujita M (2013) Extreme temperature responses, oxidative stress and antioxidant defense in plants. In: Vahdati K (ed) Abiotic stress – plant responses and applications in agriculture. InTech. https://doi.org/10.5772/54833. Available from: https://www.intechopen.com/books/abiotic-stress-plant-responsesand-applications-in-agriculture/extreme-temperature-responses-oxidative-stress-and-antioxidant-defense-inplants

Hisano H, Kanazawa A, Kawakami A, Yoshida M, Shimamoto Y, Yamada T (2004) Transgenic perennial ryegrass plants expressing wheat fructosyltransferase genes accumulate increased amounts of fructan and acquire increased tolerance on a cellular level to freezing. Plant Sci 167:861–868

Hopkins WG (1999) The physiology of plants under stress. In: Introduction to plant physiology, 2nd edn. Wiley, New York, pp 451–475

Hsieh TH, Lee JT, Yang PT, Chiu LH, Charng YY, Wang YC, Chan MT (2002) Heterology expression of the *Arabidopsis* C-repeat/dehydration response element binding factor 1 gene confers elevated tolerance to chilling and oxidative stresses in transgenic tomato. Plant Physiol 129:1086–1094

Hu Y, Wu Q, Sprague SA, Park J, Oh M, Rajashekar CB, Koiwa H, Nakata PA, Cheng N, Hirschi KD, White FF, Park S (2015) Tomato expressing Arabidopsis glutaredoxin gene AtGRXS17 confers tolerance to chilling stress via modulating cold responsive components. Horticulture Research 2:15051

Irving RM, Lanphear FO (1968) Regulation of cold hardiness in *Acer negundo*. Plant Physiol 43:9–13

Juwarkar AS, Shende GB (1986) Interaction of Cd-Pb effect on growth yield and content of Cd, Pb in barley. Indian J Environ Health 28:235–243

Kader AA (2002) Postharvest biology and technology: an overview. In: Kader AA (ed) Postharvest technology of horticultural crop, publication number 3311. Regents of the University of California, Division of Agricultural and Natural Resources, Oakland, pp 39–48

Karan R, Subudhi PK (2012) Approaches to increasing salt tolerance in crop plants. In: Ahmad P, Prasad MNV (eds) Abiotic stress responses in plants metabolism, productivity and sustainability. Springer, New York, 978–1–4614-0633-4, pp 63–88

Karp A (1995) Somaclonal variation as a tool for crop improvement. Euphytica 85:295–302

Kasuga M, Liu Q, Miura S, Yamaguchi-Shinozaki K, Shinozaki K (1999) Improving plant drought, salt, and freezing tolerance by gene transfer of a single stress-inducible transcription factor. Nature Biotechnology 17:287–291

Kendall EJ, Qureshi JA, Kartha KK, Leung KN, Caswell CK, THH C (1990) Regeneration of freezing tolerant spring wheat (*Triticum aestivum* L.) plants from cryoselected callus. Plant Physiol 94:1756–1762

Kijne JW (2006) Abiotic stress and water scarcity: identifying and resolving conflicts from plant level to global level. Field Crop Res 97:3–18

Krishna P, Sacco M, Cherutti JF, Hill S (1995) Cold-induced accumulation of hsp90 transcripts in *Brassica napus*. Plant Physiol 107:915–923

Krishna H, Alizadeh M et al (2016) Somaclonal variations and their applications in horticultural crops improvement. 3 Biotech 6:54

Kuo CG, Tsay JS, Chen BW, Lin PY (1982) Screening for flooding tolerance in the genus *Lycopersicon*. Hortic Sci 17:76–78

Larkin P, Scowcroft W (1981) Somaclonal variation-a novel source of variability from cell cultures for plant improvement. Theor Appl Genet 60:197–214

Ledesma NA, Nakata M, Sugiyama N (2008) Effects of high temperature stress on the reproductive growth of strawberry CNS 'Nyoho' and 'Toyonoka'. Sci Hortic 116:186–193

Lee SP, Zhu B, Chen THH, Li PH (1992) Induction of freezing tolerance in potato (*Solanum commersonii*) suspension cultured cells. Physiol Plant 84:41–48

Lee JT, Prasad V, Yang P-T, Wu J-F, David Ho T-H, Charng Y-Y, Chan MT (2003) Expression of *Arabidopsis* CBF1 regulated by an ABA/stress inducible promoter in transgenic tomato confers stress tolerance without affecting yield. Plant Cell Environ 26:1181–1190

Li R, Qu R, Bruneau AH, Livingston DP (2010) Selection for freezing tolerance in St. Augustine grass through somaclonal variation and germplasm evaluation. Plant Breed 129:417–421

Liu J, Yang Z, Li W, Yu J, Huang B (2013) Improving cold tolerance through in vitro selection for somaclonal variations in seashore *Paspalum*. J Amer Soc Hort Sci 138(6):452–460

Mahajan S, Tuteja N (2005) Cold, salinity and drought stresses: an overview. Arch Biochem Biophys 444:139–158

McKersie BD, Chen Y, de Beus M, Bowley SR, Bowler C, Inze D, Halluin KD, Botterman J (1993) Superoxide dismutase enhances tolerance of freezing stress in transgenic alfalfa (*Medicago sativa* L.) Plant Physiol 10(3):1155–1116

McKersie BD, Bowley SR, Jones KS (1999) Winter survival of transgenic alfalfa overexpressing superoxide dismutase. Plant Physiol 119:839–849

McKersie BD, Murnaghan J, Jones KS, Bowley SR (2000) Iron-superoxide dismutase expression in transgenic alfalfa increases winter survival without a detectable increase in photosynthetic oxidative stress tolerance. Plant Physiol 122:1427–1437

Olien CR, Smith MN (1997) Ice adhesions in relation to freeze stress. Plant Physiol 60:499–503

Owens CL, Thomashow MF, Hancock JF, Iezzoni AF (2002) CBF1 orthologs in sour cherry and strawberry and the heterologous expression of CBF1 in strawberry. J Am Soc Hortic Sci 127:489–494

Palonen P, Buszard D (1997) Current state of cold hardiness research on fruit crops. Can J Plant Sci 77:399–420

Park EJ, Jeknic Z, Sakamoto A, DeNoma J, Murata N, Chen THH (2003) Genetic engineering of cold-tolerant tomato via glycine betaine biosynthesis. Cryobiol Cryotechnol 49:77–85

Park EJ, Jeknic Z, Sakamoto A, DeNoma J, Yuwansiri R, Murata N, Chen THH (2004) Genetic engineering of glycine betaine synthesis in tomato protects seeds, plants, and flowers from chilling damage. Plant J 40:474–487

Pennycooke JC, Jones ML, Stushnoff C (2003) Down-regulating α-galactosidase enhances freezing tolerance in transgenic petunia. Plant Physiol 133:901–909

Pérez-Clemente RM, Gómez-Cadenas A (2012) In vitro tissue culture, a tool for the study and breeding of plants subjected to abiotic stress conditions. In: Leva A (ed), Recent advances in plant in vitro culture, InTech, doi: https://doi.org/10.5772/50671

Rahnama A, Poustini K, Munns R, James RA (2010) Stomatal conductance as a screen for osmotic stress tolerance in durum wheat growing in saline soil. Funct Plant Biol 37:255–263

Rai MK, Shekhawat HNS, Gupta AK, Phulwaria M, Ram K, Jaiswal U (2011) The role of abscisic acid in plant tissue culture: a review of recent progress. Plant Cell Tissue Organ Cult 106:179–190

Rao R, Li YC (2003) Management of flooding effects on growth of vegetable and selected field crops. Hort Technol 13:610–616

Ravichandra NG (2014) Horticultural nematology. Springer, New Delhi

Rout GR, Mohapatra A, Jain SM (2006) Tissue culture of ornamental pot plant: a critical review on present scenario and future prospects. Biotechnol Adv 24:531–560

Sahijram L, Soneji J, Bollamma K (2003) Analyzing somaclonal variation in micropropagated bananas (*Musa* spp.) In Vitro Cell Dev Biol Plant 39:551–556

Sanghera GS, Wani SH, Hussain W, Singh NB (2011) Engineering cold stress tolerance in crop plants. Curr Genomics 12:30–43

Shekhawat UKS, Ganapathi TR (2014) Transgenic banana plants overexpressing MusabZIP53display severe growth retardation with enhanced sucrose and polyphenol oxidase activity. PlantCell Tiss Org Cult 116:387–402

Stockinger EJ, Gilmour SJ, Thomashow MF (1997) *Arabidopsis thaliana CBF1* encodes an AP2 domain-containing transcriptional activator that binds to the C-repeat/DRE, a cis-acting DNA regulatory element that stimulates transcription in response to low temperature and water deficit. Proc Natl Acad Sci U S A 94:1035–1040

Swamy PM, Smith B (1999) Role of abscisic acid in plant stress tolerance. Curr Sci 76:1220–1227
Taji T, Ohsumi C, Iuchi M, Seki M, Kasuga M, Kobayshi M, Shinozaki KY, Shinozaki K (2002) Important roles of drought- and cold-inducible genes for galactinol synthase in stress tolerance in *Arabidopsis thaliana*. Plant J 29:417–426
Thomashow MF (1999) Plant cold acclimation: freezing tolerance genes and regulatory mechanisms. Annu Rev Plant Physiol 50:571–599
Uemura M, Warren G, Steponkus PL (2003) Freezing sensitivity in the sfr4 mutant of *Arabidopsis* is due to low sugar content and is manifested by loss of osmotic responsiveness. Plant Physiol 131:1800–1807
Van Ieperen W (1996) Effects of different day and night salinity levels on vegetative growth, yield and quality of tomato. J Hortic Sci 71:99–111
Varshney A, Dhawan V (1998) Micropropagation of ornamental plants. In: Srivastava PS (ed) Plant tissue culture and molecular biology: applications and prospects. Narosa Publishing House, New Delhi, pp 402–528
Weiser C (1970) Cold resistance and injury in woody plants knowledge of hardy plant adaptations to freezing stress may help us to reduce winter damage. Science 169:1269–1278
Wien HC, Turner AD, Yang SF (1989) Hormonal basis for low light intensity induced flower bud abscission of pepper. J Am Soc Hortic Sci 114:981–985
Wisniewski M, Bassett C, Norelli JL, Artlip T (2007) Using biotechnology to improve resistance to environmental stress in fruit crops: the importance of understanding physiology. Biotechnol Temp Fruit Crops Trop Species 738
Wu S (1994) Effect of manganese excess on the soybean plant cultivated under various growth conditions. J Plant Nutr 17:993–1003
Yadav SK (2010) Cold stress tolerance mechanisms in plants. A review. Agron Sustain Dev 30(3):515–527
Yamaguchi T, Blumwald E (2005) Developing salt tolerant crop plants: challenges and opportunities. Trends Plant Sci 10:615–620
Yusnita Y, Widodo W, Sudarsono S (2005) In vitro selection of peanut somatic embryos on medium containing culture filtrate of *Sclerotium rolfsii* and plantlet regeneration. Hayati J Biosci 12(2):50–56
Zhang M, Rajashekar CB (1994) Selection of cold tolerant cells of grapes in suspension culture. Plant Sci 97:69–74
Zhu B, Alva AK (1993) Effect of pH on growth and uptake of copper by Swingle citrumelo seedlings. J Plant Nutr 16:1837–1845
Zhu J, Dong CH, Zhu JK (2007) Interplay between cold-responsive gene regulation, metabolism and RNA processing during plant cold acclimation. Curr Opin Plant Biol 10:290–295

9 Omics-Based Strategies for Improving Salt Tolerance in Maize (*Zea mays* L.)

Mohammed Shalim Uddin, Masum Billah, Neelima Hossain, Shamim Ara Bagum, and M. Tofazzal Islam

Abstract

Soil salinity is one of the most crucial abiotic stresses that limit global food production. Due to the impacts of climate change, salinity poses a serious threat to future food and nutritional security of many countries. Maize has become a staple food in many parts of the world, with total production surpassing that of wheat or rice. It is sensitive to soil salinity. Osmotic and ion toxicity are the two major physiological problems that maize plants face under salt stress conditions. The detrimental effects of salt stress on survival and growth of maize plants are very complex. It is a prerequisite to introduce novel breeding approaches for mitigating this specific peril. Plant breeders have been manipulating a wide range of strategies that definitely improve maize production under salt-affected soils. This chapter reviews current updates of omics approaches on maize under salt stress conditions to shed light on complex controlling networks involved in salt tolerance. Better understanding of the mechanisms of salt stress tolerance through omics approaches was found useful for engineering salt-tolerant maize varieties for sustainable production of this major food crop under saline soils.

Keywords

Maize · Abiotic stress · Salinity · Approaches · Omics

M. S. Uddin · M. Billah · N. Hossain · S. A. Bagum
Stress Breeding Lab, Bangladesh Agricultural Research Institute, Gazipur, Bangladesh

M. T. Islam (✉)
Department of Biotechnology, Bangabandhu Sheikh Mujibur Rahman Agricultural University, Gazipur, Bangladesh
e-mail: tofazzalislam@yahoo.com

S. M. Zargar, M. Y. Zargar (eds.), *Abiotic Stress-Mediated Sensing and Signaling in Plants: An Omics Perspective*, https://doi.org/10.1007/978-981-10-7479-0_9

9.1 Introduction

At present, the most distressing threat to the environment is the change of global climate, and thus, it is gaining substantial attention from researchers, policymakers and farmers for its foremost impacts on agriculture (Hasanuzzaman et al. 2014; Lobell et al. 2008; Ahuja et al. 2010). Due to environmental misfortunes, agriculture of the world leads to serious contests to fulfil demand, with augmented consumption and allotment of land use for other purposes (Curtis and Halford 2014). However, soil salinity is one of the most serious environmental abiotic stresses that affect agricultural efficiency and food security throughout the world (Hoque et al. 2015). It is accounted that more than 45 million hectare (ha) of irrigated land are spoiled by salt stress which account to 20% of total land of the world, and due to high salinity level, about 1.5 million ha are removed from production per year (Munns and Tester 2008). Mostly, establishment and reproduction stages of crops are affected by salt stress (Hassanzadehdelouei et al. 2013; Samineni et al. 2011). Biomass production of plants reduces due to salinity through influencing various important physiological and biochemical processes (Ahmad et al. 2012). The plant responds to salt stress through several processes that function to stabilize cellular hyper-osmolarity and ionic imbalance (Chandna et al. 2013). However, to maintain osmotic potential and ion homeostasis, the plant has several biochemical pathways that reduce the reactive oxygen species (ROS) which is produced under salt stress. Moreover, many intracellular substances, like amino acids, nucleic acids, proteins and carbohydrates, are influenced due to salt stress (Ahmad et al. 2010). It is by the means of the molecular strategies of biology into plant stress physiology that an opportunity to initiate the identification of stress-endurance genes was enhanced (Umezawa et al. 2006). Maize has become a staple food in many parts of the world, with the total production surpassing that of wheat or rice. Production of maize is significantly limited by soil salinity. Edward East in 1908 initiated the genetic studies on maize which reported on inbreeding depression and hybrid vigour, and the 1940s saw a cytogenetic innovation for transposable elements by Barbara McClintock (Walbot 2008). Thus, maize has been used as a typical species archaeologically for developmental biology, genetics, physiology and, more recently, genomic studies (Agrawal et al. 2013). In recent years, application of omics approaches significantly advanced our understanding on complex responses of maize plants to soil salinity. The collective morphological, physiological and molecular responses of maize to salt stress have been found useful for engineering and breeding maize for salinity tolerance. The objective of this chapter is to update our knowledge on complex interactions of maize to salinity that has been generated through various omics approaches. Effects of the application of these molecular mechanisms in breeding and engineering maize for salinity tolerance are also discussed in relation to sustainable maize production in saline soils.

9.2 Salt Stress in Maize

Salt tolerance is the capability of organisms to sense and adjust to high salt concentration in their cultural environment. It was well established that the efficiency of salt tolerance varies both at genotype and species level. Mechanism of salt tolerance of plants indicates the constraint of salt entrance into the plant cell, particularly reducing the gathering of salt in cytoplasm and photosynthetic tissues (Munns 2002).

9.2.1 Root Architecture System (RAS)

Roots have a crucial role in the establishment and enactment of plants. Through optimization of root system, researchers have initiated "the second green revolution" to search the opportunity of yield progresses (Lynch and Walsh 2007). Meanwhile, water and nutrients are not equally distributed in the soil, and the spatial organization of the root system is vital for optimum use of the available resources (Koevoets et al. 2016). Such kinds of special arrangement of the root and its components are denoted to as RAS (Koevoets et al. 2016). Therefore, on a macro-scale, RAS defines the association of the primary and lateral roots and also auxiliary roots as well as other root types observed in cereals where they play as a key factor of nutrient, space and water-use efficiency in plants (Smith and De Smet 2012). Thus, maize has seminal root, which is a distinct feature of RAS. However, salt has a unique impact on root growth (Galvan-Ampudia and Testerink 2011). Inhibition of maize root length has been detected under high salinity level (Ahmad et al. 2010; Hoque et al. 2015). Moreover, lateral root number specifically declines in the root zone established after being exposed to salt stress (Julkowska et al. 2014). No effect was observed in most of the studies of salt stress on lateral root density, signifying that the decline in number of lateral roots is associated with the inhibition of primary root growth (Julkowska et al. 2014). It was demonstrated that the relative growth rate of the primary root was more strongly affected than the growth rate of the lateral roots in *Arabidopsis* (Julkowska et al. 2014) which designated that the RAS is refashioned during salt stress. RAS can also adjust depending on their phenotypic plasticity and the trend of root progression to evade locally high salt levels (Galvan-Ampudia and Testerink 2011). In contrast, salt stress stimulates suberization of the hypodermis and endodermis, and the Casparian strip improves closer to the root tip than under non-salinity roots (Shannon et al. 1994).

9.2.2 Root-Shoot Ratio

The effect of salt (NaCl) on maize greatly reduces the shoot and root length (Usman et al. 2012). As a salt-sensitive crop, shoot growth in maize is sturdily inhibited in the first phase of salt stress. Water stress, ion toxicity, oxidative stress, nutritional disorders, alteration in the metabolic processes, membrane disorganization, reduction of cell division and expansion and genotoxicity are the salt stress factors that affect the plants (Munns 2002; Zhu 2001). These effects collectively decreased plant growth development and endurance (Carillo et al. 2011; Filippou et al. 2014). Thus, salt stress affected the root-shoot ratio of maize.

9.2.3 Root Hairs

Root hairs enhance the ability of roots to explore the rhizosphere by increasing surface area. Through increasing surface area, root hairs enhance the capability of roots to explore the rhizosphere (Zhu et al. 2010). Nearly 77% of the total root surface is covered by root hairs and thus is the major path of connection between plants and the rhizosphere. On a micro-scale, RAS embraces root hairs that upsurge the surface area, assisting with uptake of water and nutrients (Leitner et al. 2010; Tominaga-Wada et al. 2011; Zygalakis et al. 2011). The length of root hairs, their compactness and their life span are some very applicable factors for nutrient uptake, which may also be inhibited by plant water supply under saline soil environments.

9.2.4 Morpho-Physiological Responses Associated with Salt Stress in Maize

With the increase of salinity, the reaction of plants is too complex and included alterations in plant morphology, physiology and metabolism (Parida and Das 2005). Soil salinity decreases leaf size or area, stem extension and root explosion, interrupts plant-water relations, reduces water-use efficiency, disrupts photosynthetic pigments and reduces the gas exchange, which lead to a reduction in plant growth and yield (Table 9.1).

9.2.5 The Role of Wild Species of Maize or Related Species in Introgressiomics for Salt Tolerance in Maize

Introgressiomics is defined as a large-scale systematic development of plant materials and populations carrying introgressions of genome fragments of wild crop relatives into the genetic background of crops for developing a new generation of cultivars with desired traits (Prohens et al. 2017). Potential variability of the crop can be achieved through introgression of useful exotic genetic material into crop varieties that were formerly absent from the genome. A wide range of genotypes in natural hybrid zones consequence from many natural selection and generations of recombination, and all constituents of the obstacles to interspecific gene flow or can be beneficial simulations to study the effort of transgenes between species (Stewart et al. 2003).

However, the successful story of introgression in several crops for many traits, including both biotic and abiotic, have been introgressed from wild relatives into the cultivated without depression of oil yield and quality (Brar and Khush 2002; Rieseberg et al. 1999; Wang et al. 2001). While, maize has exposed little

Table 9.1 Genetic variability and effects of salt stress on morphological and physiological adaptations in maize

Si no.	Treatment	Genotypes (no.)	Growth conditions	Growth stage	Major effect or findings	References
Genetic variability study in maize for salt stress						
1.	Different levels of stress (control, 0.15 (EC = 13.6 dSm-1), 0.2 MNaCl (EC = 18 dSm-1)	14 maize pipe line hybrids	Hydroponics	Seedling	Variation was found in seedling emergence and number of adventitious roots	Maiti et al. (2012)
2.	Three levels of NaCl including control (0, 50 and 100 m Mol)	Haricon-11	Nutrition solution	Seedling	Maximum reduction in growth attributes was observed at 100 m M of NaCl	Hussain et al. (2010)
3.	100 mmol L^{-1} NaCl	Eight different maize genotypes	Vermiculture and nutrition solution	Seedling	Salt stress reduced plant growth of all genotypes	Azevedo Neto (2004)
4.	Soil salinity	Fifty-six genotypes	Field condition	Flowering	Differences observed in means of most traits studied were high	Umar et al. (2015)
5.	Four salinity levels (control, 40, 80 and 120 mM NaCl)	10 different maize hybrids	Solution culture	Seedling	Genetic variation for salt tolerance was assessed	Akram et al. (2010)
6.	Tow salinity levels (20–40 mM)	Single genotype	Pot culture	Seedling	Great variation exists for the morphological attributes	Usman et al. (2012)
7.	Received recommended cultural practices	Two hundred S1 generation selfed seeds	Field condition	Up to maturity	Highly significant differences for 18 characters	Praveenkumar et al. (2014)
Effect of salt stress on maize root growth						
8.	Three salinity levels (control, 1.35 dS M^{-1} and 2.70 dS M^{-1})	Single variety	Pot culture	Seedling	2.7 salinity water treatment decreased root length	Bilgin et al. (2008)

(continued)

Table 9.1 (continued)

Si no.	Treatment	Genotypes (no.)	Growth conditions	Growth stage	Major effect or findings	References
9.	Various concentrations of NaCl (50, 75, 100, 125, 150 mM)	Two genotypes	Pot culture	At germination	Root length reduced with increasing salinity	Saddiqe et al. (2016)
10.	Sodium chloride was applied at the rates of 0, 25, 50, 75 and 100 mM NaCl	Single genotype	Pot experiment	At seedling	NaCl significantly decreased root dry mass of maize plants	Turan et al. (2010)
11.	Five salinity levels	Single	Under field conditions	At seedling	Reduction was significant in all saline media	Abraha and Yohannes (2013)
Effect of salt stress on maize leaf area						
13.	Salinized culture solutions at different osmotic potentials	Two genotypes	Solution culture	Up to 30 days	Decreased leaf area	Cicek and Cakirlar (2002)
14.	Tow salinity level	Eight different maize genotypes	Hoagland's nutrient solution	At seedling	Reduced leaf area	Azevedo Neto (2004)
15.	Two treatments (0 mM NaCl and 100 mM NaCl)	Thirteen genotypes	Solution culture	At seedling	Leaf area discriminating	Collado et al. (2016)
16.	Four salinity treatments	Two maize cultivars	Pot experiment	At seedling	Partial root excision reduced leaf area remarkably under salinity	Zhang et al. (2012)
17.	Five salinity levels	Single genotype	Hoagland's nutrient solution	At seedling	Photosynthetic activity decreased with salinity treatments	Sayed (2011)
18.	Five salinity levels	Two cultivar	MS sugar-free liquid media with vermiculite	At seedling	Photosynthetic rate reduced with increased salinity	Cha-Um and Kirdmanee (2009)

advancement in splayed the cultivated gene pool with exotic sources for the disturbance of current logistical habits, temporary yield depression and potential for loss of quality (Warburton et al. 2017). Therefore, *T. dactyloides* was recommended as a source of salt stress resistance based on their evaluation of hybrids between *Z. mays* ssp. *mays* and *T. dactyloides* (Pesqueira et al. 2003).

9.3 Molecular Breeding Approaches for Salt Tolerance in Maize

9.3.1 Quantitative Trait Loci (QTL)

QTL mapping has been a crucial tool to study genetic design of complex characters (Kearsey and Farquhar 1998). Plant salt endurance is a quantitative trait governed by multiple genes and generally obstructed by the environmental dynamics. QTL analysis delivers an operative strategy to separate quantitative traits into single constituents to study their comparative impacts on any exact trait (Doerge 2002).

Defining the genetic comparisons between the phenotypic traits and salt endurance, and result the loci fretful with salt tolerance are quite difficult (Cui et al. 2015), while some traits have been utilized to identify QTL for salt tolerance; the constancy of such QTL is relatively low under diverse conditions. However, QTL analysis of conventional salt tolerance is usually designed at analysing appearance of phenotypic traits under salinity.

Using exclusively the effect of traits expressed under normal treatment, QTL mapping can detect the distinct loci under salt stress (Zhang et al. 2013). Therefore, QTL have been mapped for numerous imperative traits such as plant height (Zhang et al. 2007; Teng et al. 2013), popping ability (Babu et al. 2006; Li et al. 2008) and tolerance against other abiotic stresses such as salt stress (Cui et al. 2015), drought stress (Xiao et al. 2005; Prasanna et al. 2010) and waterlogging (Qiu et al. 2007). Numerous QTLs related to salt tolerance have been detected in rice (Lin et al. 2004), wheat (Xu et al. 2013; Thomson et al. 2010), barley (Mano and Takeda 1997) and *Arabidopsis* (DeRose-Wilson and Gaut 2011) at the seedling stage. Nevertheless, there have been few reports on QTL mapping for salt tolerance in maize (Table 9.2).

9.3.2 Marker-Assisted Breeding and Transportation of Gene for Salt Tolerance in Maize

Through marker-assisted selection (MAS), DNA markers have massive potential to develop the competence and precision of conventional plant breeding (Collard et al. 2008). Effective MAS applications have been described for introgression breeding in maize, as well as introgressions of transgenes (Ragot et al. 1995) and alterations including simple (Morris et al. 2003) or complex traits (Crosbie et al. 2008; Ragot et al. 2007). Maize head smut gene (*RsrR*) has been introduced into two elite

Table 9.2 QTL mapping reports on salt tolerance in maize

Population type/size	Parent performance	Traits	Map length (cm)	Marker no./ density (cm)	QTL no.	References
163/F3	Inbred line B73 (a salt-susceptible inbred line) and CZ-7 (a salt-tolerant inbred line)	SL, RL, RRLSL, SFW, RFW, PFW, SDW, RDW, PDW, RRDWSDW	1195.2	9.88	4	Hoque (2013)
171/F7	Huangzaosi × Mo17		1428.3	17.63	6	Shi and Xia (2012)
161/F2:5	F63 and F35 being salt tolerant and sensitive, respectively	FGFSTR field STR, SFW, SDW, TWC, SNC				Cui et al. (2015)

SL shoot length, *RL* root length, *RRLSL* ratio of root length and shoot length, *SFW* shoot fresh weight, *RFW* root fresh weight, *PFW* plant fresh weight, *SDW* shoot dry weight, *RDW* root dry weight, *PDW* plant dry weight, *RRDWSDW* ratio of root dry weight and shoot dry, *FGR* field germination rate, *FSTR* field salt tolerance ranking, *STR* salt tolerance ranking, *SFW* shoot fresh weight, *SDW* shoot dry weight, *TWC* tissue water content, *SNC* shoot Na + concentration, *SKC* shoot K $^+$ concentration, *SKN* shoot K^+/Na^+ ratio

maize inbred lines, "Huangzao4" and "Qi319", through conventional hybridization and MAS (Agrawal et al. 2013).

However, using genetic engineering tools, for transgenic plant development, has been anticipated as an enhanced solution for improving adverse environmental resistance variety. Though documents on the achievement of stress tolerance transgenic crops in environment are very insufficient, scientists must assemble to develop in this area (Ashraf and Akram 2009). Salt resistance is an intricate trait that is governed by a number of minor genes (Shahbaz and Ashraf 2013). Some genes for ion transport have been transferred to mend salt tolerance in wheat. In wheat, overexpression of the vacuolar Na^+/H^+ antiporter gene AtNHX1 has enhanced growth characteristics such as germination rate, plant biomass and yield (Ohta et al. 2002). This line displayed decreased leaf Na^+, improved leaf K^+ gathering and had 68% better shoot dry weight, as well as 26% higher root dry weight under salt stress in association with those of the wild kind (Ohta et al. 2002) (Table 9.3).

9.3.3 Association Mapping

Linkage disequilibrium (LD) analysis-based association mapping is a potential tool for the separation of complex agronomic traits and for the detection of alleles with major effect for the enhancement of a target trait (Yan et al. 2011). Through rapid advances in the fields of molecular biology and biotechnology, plant genome research has developed a focus on association mapping, particularly with the

Table 9.3 Development of salt tolerance in cereal crops through genetic engineering strategy

Name of cereal crop	Transferred gene	Organism source	Trait developed	Growth enriched	References
Rice	Vacuolar Na^+/H^+ antiporter *AgNHX1*	*Atriplex gmelini*	Activity of these antiporters was eightfold high	Seedling survival increased from 51% or 81 to 100%	Ohta et al. (2002)
	Vacuolar Na^+/H^+ antiporter gene *PgNHX1*	*Pennisetum glaucum* (L.) R. Br.	Well-developed root system	About 81% higher shoot and root lengths	Verma et al. (2007)
	Vacuolar $Na^{+/}H^+$ antiporter gene *OsNHX1*	Wild rice (*Oryza sativa* L.)	High accumulation of Na+ and low K^+	Endure salinity level up to 0.2 M where wild plants died	Su and Wu (2004)
	_1-pyrroline-5 carboxylate synthetase (*P5CS*)	Moth bean (*Vigna aconitifolia*)	Transgenic plants accumulated more proline under both saline and nonsaline conditions	Shoot fresh weight was improved from 30 to 93% and root fresh weight 37–74% under 200 mM NaCl as compared to those in wild type	Su and Wu (2004)
	Na^+/H^+ antiporter *SOD2*	Yeast	Transgenic plants accumulated higher K^+, Ca2+ and Mg^{2+} and lower Na+ in their shoots as compared to respective non-transformed controls	Transgenic plants showed good performance under saline conditions	Zhao et al. (2006)
	OPBP1 gene	Tobacco	Transgenic plants showed high resistance against salt and disease		Chen and Guo (2008)
Wheat	Vacuolar Na^+/H^+ antiporter gene *AtNHX1*	*Arabidopsis*	Germination rate, plant biomass and yield, low leaf Na^+ and high leaf K^+	Rise in shoot dry weight was 68% and root dry weight 26%	Ohta et al. (2002)
	1-pyrroline-5-carboxylate synthetase (*P5CS*)	Moth bean (*Vigna aconitifolia*)	Accumulated 2.5-fold more proline as compared to that in wild type	Transgenic plants remained unaffected up to 200 mM and showed slight reduction at 250 mM, while respective wild plants died at 100 m*M*	Sawahel and Hassan (2002)
Maize	Vacuolar Na^+/H^+ antiporter *AtNHX1*	*Arabidopsis thaliana*	Germination percentage	Germination capacity of transgenic plants was 80% while those of wild type 13–57% at 0.5% NaCl	Yin et al. (2004)
	Na^+/H^+ antiporter gene *OsNHX1*	*Oryza sativa*	Yield was improved	Exhibited higher biomass and yield even at 200 mM NaCl	Chen et al. (2007)

exploitation of molecular markers, exploring quantitative traits of plants (Zhao et al. 2016). Thus, association mapping has been extensively used in plants (Hao et al. 2012; Hu et al. 2013; Zhang et al. 2014). However, association mapping is generally used to identify molecular markers related with yield, quality and resistance. Therefore, association mapping analyses for salt tolerance have been described in wheat, rice, barley and *Arabidopsis* (DeRose-Wilson and Gaut 2011; Long et al. 2013).

9.3.4 Genome-Wide Association Studies

Genome-wide association studies (GWAS) are microarray technology that is used to detect relationship between definite results and genetic variants through the entire genome, rather than in an exact gene. A positive relationship ascends when a greater incidence occurs in the existence of a genetic variant in individuals with a disease or trait than in unaffected individuals. However, in diverse maize, GWAS is challenging, by way of linkage disequilibrium decays within 2000 bp (Gore et al. 2009). Using the recent innovation of 1.6 million SNPs by the maize HapMap project (Gore et al. 2009) and the development of a large combined linkage-relationship panel recognized as the nested association mapping (NAM) population (McMullen et al. 2009), currently, operative GWAS in maize have become possible (Tian et al. 2011). Producing about 5000 recombinants inbred with maize NAM panels, which were produced by crossing with 25 diverse lines of maize to one allusion line. Nevertheless, applied genome-wide association studies (GWAS) detected loci controlling salinity endurance in rice and identified 60 SNPs (loci) which were found to be significantly connected with the Na^+/K^+ ratio and other traits observed under stress conditions (Kumar et al. 2015).

9.3.5 Next-Generation Sequencing

A high-throughput technique is used to govern a portion of the nucleotide sequence of an individual's genome; this method uses DNA sequencing technologies that are proficient in dispensing multiple DNA sequences in parallel. The identification of novel unknown stress-tolerant genes can be problematic, as the conventional approaches such as real-time PCR or ELISA may be too definite to a particular species or even a trait. Next-generation sequencing (NGS) offers an alternative solution where sequence is generated in a non-specific fashion and identification is based on similarity searching against gene banks (Adams et al. 2013). Thus, sequencing recital is swiftly enlightening in read depth; technologies producing long-sequence fragments, for example, 454 Roche pyro-sequencing, have been established which are mostly valuable in de novo sequencing and progress of new possessions for non-model species, deprived of an accessible reference genome (Wheat 2010). However, the application of basic research is accelerated by NGS technologies to find out the breeding of re-sequencing abundance of tropical and temperate maize genotypes, thus initiating the germplasm pool available for maize development.

9.3.5.1 Epigenetic Regulators

Epigenetic refers to the study of changes in the regulation of gene activity and expression that are not dependent on gene DNA sequence (Chinnusamy and Zhu 2009). Epigenetic is the study of potentially heritable changes in gene expression that does not involve changes to the underlying DNA sequence, a change in phenotype without a change in genotype, which in turn affects how cells read the genes (Thomashow 1999).

Primary and secondary stress signals induce changes in the expression and/or activity of epigenetic regulators, namely, small RNAs, RdDM components, histone variants, histone modification enzymes and chromatin-remodelling factors (Iba 2002). These epigenetic regulators modify histone variants, histone modifications and DNA methylation (Tsuji et al. 2006). Some of these are heritable epigenetic modifications, while others are transient changes (Zhu et al. 2008). There are some genes, transposons and promoter fragments of genome differentially regulated through DNA methylation as an epigenetic reprogramming (Table 9.4).

9.3.6 Genomic Selection

Genomic selection (GS) is a novel technique for improving quantitative characters in the plant breeding germplasm that is based on whole-genome molecular markers. It is supposed that alternating progeny field-testing with selection based only on markers should increase the genetic gains per unit. Besides accelerating the selection cycles, GS offers the opportunity to increase the selection gains per unit of time. However, whole-genome prediction models estimate all marker impacts in all loci and capture small QTL effects. Relationship between markers and quantitative trait loci (QTL) is dignified by association analysis, and their estimation ignores genes with small effects that trigger underpinning quantitative traits (Desta and Ortiz 2014). Genome-wide selection estimates marker effects through the whole genome on the marked population grounded on a prediction model established in the drill population. Therefore, GS technique is used for root traits under conditions of nitrogen and phosphorus stress in tropical maize breeding programme (Fritsche-Neto et al. 2012). Nonetheless, phenotypic assortment or marker-assisted breeding procedures can be substituted by selection, dependent on whole-genome predictions in which phenotyping improve the model to form the prediction precision (Desta and Ortiz 2014).

9.3.7 Transcriptomics Analysis for Salt Tolerance in Maize

Efficient genomic strategies, such as transcriptomics, genomics, metabolomics, proteomics and ionomics, have been widely used to assess abiotic stress tolerance mechanisms in plants. These strategies can be exploited to advance our capability to determine the genes and paths that regulate definite characters in response to abiotic stress (Shelden and Roessner 2013). Microarrays have been used widely to produce

Table 9.4 Some of genes, transposons and promoter fragments in plant genome differentially regulated by biotic and abiotic stresses

Genome region	Plant	Methylation status	Stress	Mode of action	References
Transposons:					
MuDR	Maize	Hypomethylation	N^+ implantation	Increases the expression of mudrA and mudrB	Hashida et al. (2006)
Ac/Ds transposon	Maize	Demethylation	Cold stress	Cold-induced root-specific demethylation	Steward et al. (2000)
Gene/coding segment:					
ZmMI1	Maize	Demethylation	Cold stress	Cold-induced root-specific demethylation	Steward et al. (2000)
Sodium transporter gene (AtHKT1)	*Arabidopsis*	Hypomethylation	Salt tolerance	Loss in cytosine methylation in a putative small RNA target region	Baek et al. (2011)
Non-transposon Asr1	Tomato	Asymmetric CNN methylation	Water stress	Drought conditions brought about higher CG methylation levels in the first exon	González et al. (2011)
Promoter:					
Glyma11g 02400	Soybean	Hypomethylation	Salinity stress	Most of the cytosines were demethylated following exposure to salinity stress for 1–24 h	Song et al. (2012)
Glyma16g 27950	Soybean	Hypomethylation	Salinity stress	Hypomethylation at transcription	Song et al. (2012)
Glyma20g 30840	Soybean	Hypomethylation	Salinity stress	Hypomethylated cytosines at promoter region 1	Song et al. (2012)
RMG1 promoter	*Arabidopsis*	Demethylation	*Pseudomonas syringae*	RMG1 is targeted by RdDM and ROS1-dependent DNA demethylation	Yu et al. (2013)

transcriptional profiles in response to abiotic stresses in an array of plant species. Transcript studies on the basis of a specific cell type have been performed in many plant species such as the rice, maize, barley, soybean and *Arabidopsis* (Rogers et al. 2012). Very few have resulted to the detection of stress tolerance pathways, despite having a large number of microarray experiments (Deyholos 2010).

9.3.8 MicroRNAs of Salt Tolerance in Maize

MicroRNAs (miRNAs) are a subgroup of endogenous nearly 22-nucleotide small non-coding regulatory RNA molecules that operate gene expression post-transcriptionally, by interceding mRNA degradation or translational suppression in a sequence explicit manner. Plant miRNAs' role in various processes is connected with growth and development such as floral organ identity, reproductive development, developmental transitions, organ polarity, auxin signalling, boundary formation or organ separation and leaf and stem growth (Sun 2012).

Mature single-stranded miRNAs are produced from single-stranded primary miRNAs that are transcribed from miRNA loci and are administered by Dicer-like 1 (DCL1) which are laden into the RNA-induced silencing complex (RISC). MiRNA-loaded RISC targets cognate transcripts and induces their cleavage (Bartel 2004). About 1000 miRNAs have been detected in various plant species, with 20 miRNA families that are well preserved between dicots and monocots (Griffiths-Jones et al. 2008). Important progress has been completed in characterizing and analysing miRNAs in the maize genome as one of the world's most important crop species (Mica et al. 2006). Current studies have revealed that miRNAs are also complicated in abiotic and biotic stress responses. Salt stress is a significant yield-limiting factor in many regions of the world. Thus, understanding plant responses to salt stress through miRNA is important for developing crop efficiency.

9.3.9 Expression of Salt Stress-Responsive miRNAs in Maize Plant Parts

The data, collected from different used technologies, had shown changed expression profiles of many miRNAs in maize roots during salt stress. Hybridization of miRNA microarray showed evidence that from 27 plant miRNA families, 98 miRNAs had significantly changed their expression after being in salt treatment. They displayed different activities in salt stress (Ding et al. 2009). Reflection of expression of the family counterparts in maize was visible in miRNA profiles which were similarly expressed. A total of 77 out of the 98 miRNAs of 10 maize miRNA families aligned with 55 members, while from another 6 plant species, the other miRNAs corresponded to several members of them, and also, differentially expressed miRNAs were transcriptionally regulated during salt stress at different times (Ding et al. 2009) (Table 9.5).

Various modifications can be visible in salt-stressed maize plant. Some genes enhances auxin-responsive gene functions; some can ensure more root and shoot development and can decrease wastage of energy and lateral root elongation, accumulate more biomass to counteract consumption, protect photosynthesis and maintain energy supply; and some can help in morphological, physiological and biochemical adaptation as well as seedling survival under stress condition (Ding et al. 2009).

Table 9.5 Salt-responsive miRNA family members and their functions

Families of miRNA	Members of miRNA family	Functions
zma-miR156	ath-miR156a/g/h, bna-miR156a, osa-miR156l, ptc-mir156 k, smo-miR156c, sbi-miR156e, vvi-miR156e	From the flower developmental stage I and floral induction to specification of flower organ cell type and stage IV (Chuck et al. 2007; Gallavotti et al. 2010)
zma-miR159	ath-miR159a/b/c, osa-miR159a/c/d/e/f, pta-miR159a/b	Regulates both MYB and TCP transcription factors, which may control petal morphogenesis (Nag et al. 2009)
	ptc-miR159d/f, smo-miR159	
	sof-miR159e	
zma-miR160	ppt-miR160b	Targets six ARF genes to be involved in signalling pathways that mediate response to auxin (Liu et al. 2014)
zma-miR162	ath-miR162a	Targets Dicer-like 1 (DCL1) which is a homolog of DCL1 in *Arabidopsis* that is required for miRNA accumulation (Park et al. 2002)
zma-miR164	ath-miR164a/c, osa-miR164c/d/e	Involved in regulating ear development and encodes NAM proteins (Liu et al. 2014)
	ptc-miR164f, sbi-miR164c	
zma-miR166	ath-miR166a, osa-miR166e/k/m	Targets basic leucine zipper (bZIP) genes in maize (Zhang et al. 2009)
	ppt-miR166j, ptc-miR166n/p	
	sbi-miR166a, vvi-miR166b/d	
zma-miR167	ath-miR167a/c/d, bna-miR167a	Targets four auxin response factor (ARF) genes as for zma-miR160 (Liu et al. 2014)
	ppt-miR167, ptc-miR167e/f/h	
	vvi-miR167c	
zma-miR168	ath-miR168a	Responds to abscisic acid stress (Li et al. 2012)
	osa-miR168a/b	
	sof-miR168b	
zma-miR171	ath-miR171a, osa-miR171 g	Targets GRAS transcription factor that controlled gibberellic acid (GA) signalling and phytochrome (Kang et al. 2012)
	ptc-miR171c/e, smo-miR171a	
	zma-miR171c	
zma-miR319	ath-miR319a/c, gma-miR319a/c	Both MYB and TCP transcription factors are being regulated that leads to control of petal morphogenesis (Nag et al. 2009)
	osa-miR319a, ppt-miR319a	
	pta-miR319, ptc-miR319e	
	vvi-miR319b	
zma-miR395	ath-miR395b	Controls the activity of the ATP sulfurylase gene (APS) and the high-affinity sulphate transporter gene (SULTR2;1) (Allen et al. 2005; Buhtz et al. 2010; Kawashima et al. 2009)
	ppt-miR395	
zma-miR396	osa-miR396d	Regulates growth-regulating factors (GRFs), and in the leaf primordia, it expresses at low levels throughout the meristem, overlapping with the expression of its target, GRF2 (Rodriguez et al. 2010)

(continued)

Table 9.5 (continued)

Families of miRNA	Members of miRNA family	Functions
zma-miR399	ptc-miR399f/i	Regulates Pi homeostasis in *Arabidopsis* by binding to the 5′-UTR of PHO2 (a gene encoding ubiquitin-conjugating E2) (Aung et al. 2006 and Fujii et al. 2005)

Source: (Aung et al. 2006; Buhtz et al. 2010; Ding et al. 2009; Fujii et al. 2005; Kang et al. 2012; Li et al. 2012; Liu et al. 2014; Nag et al. 2009; Park et al. 2002; Rodriguez et al. 2010; Zhang et al. 2009)

9.3.10 Transcription Factors Can Induce Stress-Responsive Genes in Plants Under Drought or Water Stress

The multiplex signalling pathways in plants during water stress consist of several proteins, for example, TFs, enzymes, functional proteins and metabolites (Song et al. 2013). Genetic engineering works to overcome the challenge of drought stress which involve the overexpression of these TFs, enzymes and other metabolites (Thudi et al. 2014). In recent times, a huge number of TF families relevant to drought stress have been found out (Anbazhagan et al. 2015). At the time of signal transduction, TFs directly activate the expression of the genes, which are associated by means of molecular switches. These TFs interact specifically with *cis*-elements, which are located in the promoter region of the genes they regulate. A large proportion of genes in the plant (up to 10%) possibly encode TFs, which are categorized into different gene families such as *AREB*, *DREB*, *MYB*,*WRKY*, *NAC* and *b ZIP* because of their distinct structure (Franco-Zorrilla et al. 2014; Golldack et al. 2011). In the process of *Arabidopsis*, nearly 6% of the proteome is devoted to TFs, and among them, several TFs have showed response under drought stress through pathways dependent/independent of ABA (Joshi et al. 2016; Rayko et al. 2010).

9.3.11 Molecular Interaction Network

Molecular interaction network (MIN) represents the interaction between and within different biochemical families like protein, carbohydrate, fat DNA, RNA, lipid, microRNA-transcription factor, etc. Several evidences have been found about the molecular protein-protein interaction. Molecular biological networks are categorized in many ways (Liseron-Monfils and Ware 2015). A meta-gene network makes possible representing a factor of transcription and its promoter in a single node by simplifying the network view (Liseron-Monfils and Ware 2015). To understand and enucleate network behaviour for network view, visualization software is very much needed. For non-computational biologist, Cytoscape analyser, VisANT, etc. are very strongly useable (Bastian et al. 2009; Hu et al. 2009; Smoot et al. 2011). New major visualization tools, such as network analysis tools (NeAT), which only decipher protein-protein interaction network found from STRING database and igraph (which is a library of the R programming language), are being used vastly in the next-generation visualization of network view for molecular interactions. And, this type of methods is far more complex though faster than any other tools of this related aspect (Brohee et al. 2008; Csardi and Nepusz 2006). The new inventions give a wide range of network analyser applications to help to calculate node clusters or network centralities, and these can help in uniting two networks (Assenov et al. 2008).

9.3.12 Other miRNA Responses to Salt Stress

Various studies on different crops have revealed an important role on both up-regulation and downregulation for miRNAs in salt responses. Recently, the comparative observation between salt-tolerant maize genotype NC286 and salt-sensitive maize genotypes Huangzao4 showed that miR156, miR164, miR167 and miR396 relatives were downregulated, while miR162, miR168, miR395 and miR474 relatives were up-regulated in salt-stressed maize roots (Ahmad et al. 2013). Under salt stress, miR156, miR158, miR159, miR165, miR167, miR168, miR169, miR171, miR319, miR393, miR394, miR396 and miR397 were up-regulated, whereas the miR398 was downregulated in *Arabidopsis*, thus launching a role for miRNAs in the adaptive response to salt stress (Liu et al. 2008). Up-regulation of miRS1 and miR159.2 under salt stress condition was detected in *Phaseolus vulgaris* (Arenas-Huertero et al. 2009). Many studies revealed that several rice miRNAs are convoluted in salinity stresses (Zhao et al. 2009).

9.3.13 Proteomics of Salt Tolerance in Maize

Proteomics has been well-defined as "the systematic analysis of the protein population in a tissue, cell, or sub-cellular compartment" and is often related with two-dimensional electrophoresis (Agrawal et al. 2013). Currently, the study of proteomics research has been objected that the detecting new proteins in relation to their role and eventually at separating how their expression is managed within controlling networks (Agrawal et al. 2013). For neutralizing the harmful effects of salt stress, maize plants undergo a multiplicity of adaptive mechanisms at the molecular level. The technology of protein array is a highly parallel method to standard proteomics, and the idea enables the linking of recombinant proteins to clones identified (Agrawal et al. 2013). However, the gathering or reserve of several proteins and the up-regulation and downregulation of several gene transcripts are imperative (Menezes-Benavente et al. 2004). Countenance of antioxidant protection genes is activated in maize to defend the cells from salt stress-induced oxidative damage. Furthermore, fronting high concentration of salt stress, Sod1, Sod2, Sod4, Sod4A and all catalase transcripts were inhibited in maize plants. Fourteen proteins controlled by salt stress were deoxyribonucleic acid-directed: deoxyribonucleic acid polymerase (A1), ribosomal protein S4 (A2), cytochrome P450-like protein (A3), serine/threonine kinase (A4), adenosine kinase (A5), Rubisco large chain (B6),

Rubisco small chain (B7), fructose 1,6-bisphosphate aldolase (B8), glyceraldehyde-3-phosphate dehydrogenase (B9), β-glucosidase (B10), V-ATPase subunit α (B11), methionine synthase (C12), *S*-adenosyl methionine synthase (C13) and glutamate-ammonia ligase (C14). In maize under low salinity level, these 14 proteins fit to three dissimilar groups, i.e. proteins intricate in protein biosynthesis and alterations by kinases (A), enzymes of carbon metabolism (B) and nitrogen metabolism (C) (Zörb et al. 2004). Downregulation of *ZmExpB2*, *ZmExpB6* and *ZmExpB8* transcripts possibly contributed to the reduced protein abundance (Hoque 2013). Thus, the up-regulation of *ZmExpB2*, *ZmExpB6* and *ZmExpB8* can tolerate the steady expression of β-expansin protein under saline condition (Geilfus et al. 2010). Thus, gathering or inhibition of specific proteins and the up-regulation and downregulation of many gene transcripts help in salt resistance of maize.

9.4 Conclusion

Salt stress is one of the greatest limitations for the production of maize crop in salt-affected areas. On the other hand, conventional breeding programme is governed by the phenotype of crop plants, which is time-consuming and less efficient. Several approaches have recently been established to breed salt-tolerant cultivars to advance the productivity of crop under saline soils. Through high-throughput knowledge, the genomics-abetted manipulation was efficiently used to improve hybrids to overwhelm constraints of traditional breeding programmes. Various omics approaches, viz. phenomics, genomics, proteomics, breeding informatics, micromics, metabolomics, introgressiomics and transcriptomics, have shed light on underlying molecular mechanisms of plant responses and tolerance to soil salinity. This review updates state-of-the-art knowledge on complex responses of plants to salt stress which might be useful for novel targets for development of salinity-tolerant maize. Nevertheless, these information and associated mechanisms in maize offer new opportunities to develop salt-tolerant maize varieties through genetic engineering for sustainable maize production under saline environment.

Acknowledgments MTI is thankful to the World Bank for funding to this work by a sub-project CP#2071 of Higher Education Quality Enhancement Project of University Grants Commission of Bangladesh.

References

Abraha B, Yohannes G (2013) The role of seed priming in improving seedling growth of maize (*Zea mays* L.) under salt stress at field conditions. Agric Sci 04:666–672

Adams I, Miano D, Kinyua Z, Wangai A, Kimani E, Phiri N et al (2013) Use of next-generation sequencing for the identification and characterization of maize chlorotic mottle virus and sugarcane mosaic virus causing maize lethal necrosis in Kenya. Plant Pathol 62:741–749

Agrawal PK, Saini N, Babu BK, Bhatt JC (2013) Omics approaches in maize improvement. OMICS Applications in Crop Science: 73

Ahmad R, Waraich EA, Ashraf MY, Akram M, Mohsan M, Iqbal J (2010) Screening for salt tolerance in maize (*Zea mays* L.) hybrids at an early seedling stage. Pak J Bot 42:141–154

Ahmad P, Hakeem KR, Kumar A, Ashraf M, Akram NA (2012) Salt-induced changes in photosynthetic activity and oxidative defense system of three cultivars of mustard (*Brassica juncea* L.) Afr J Biotechnol 11:2694–2703

Ahmad P, Azooz MM, Prasad MNV (2013) Salt stress in plants: signalling, omics and adaptations. Springer, New York

Ahuja I, de Vos RCH, Bones AM, Hall RD (2010) Plant molecular stress responses face climate change. Trends Plant Sci 15:664–647

Akram M, Ashraf MY, Ahmad R, Waraich EA, Ishfaq M (2010) Screening for salt tolerance in maize (*Zea mays* L.) hybrids at an early seedling stage. Pak J Bot 42:141–154

Allen E, Xie Z, Gustafson AM, Carrington JC (2005) microRNA-directed phasing during trans-acting siRNA biogenesis in plants. Cell 121:207–221. https://doi.org/10.1016/j.cell.2005.04.004

Anbazhagan K, Bhatnagar-Mathur P, Vadez V, Dumbala SR, Kishor PB, Sharma KK (2015) DREB1A overexpression in transgenic chickpea alters key traits influencing plant water budget across water regimes. Plant Cell Rep 34:199–210. https://doi.org/10.1007/s00299-014-1699-z

Arenas-Huertero C, PÃrez B, Rabanal F, Blanco-Melo D, De la Rosa C, Estrada-Navarrete G et al (2009) Conserved and novel miRNAs in the legume Phaseolus vulgaris in response to stress. Plant Mol Biol 70:385–401

Ashraf M, Akram NA (2009) Improving salinity tolerance of plants through conventional breeding and genetic engineering: an analytical comparison. Biotechnol Adv 27:744–752

Assenov Y, Ramirez F, Schelhorn SE, Lengauer T, Albrecht M (2008) Computing topological parameters of biological networks. Bioinformatics 24:282–284. https://doi.org/10.1093/bioinformatics/btm554

Aung K, Lin S-I, Wu C-C, Huang Y-T, Su C-l, Chiou T-J (2006) pho2, a phosphate overaccumulator, is caused by a nonsense mutation in a MicroRNA399 target gene. Plant Physiol 141:1000–1011. https://doi.org/10.1104/pp.106.078063

Azevedo Neto AD, Prisco JT, Enéas-Filho J, Lacerda CF, Silva JV, Costa PHA, Gomes-Filho E (2004) Effects of salt stress on plant growth, stomatal response and solute accumulation of different maize genotypes. Braz J Plant Physiol 16(1):31–38

Babu R, Nair SK, Kumar A, Rao HS, Verma P, Gahalain A et al (2006) Mapping QTLs for popping ability in a popcorn × flint corn cross. Theor Appl Genet 112:1392–1399

Baek D, Jiang J, Chung J-S, Wang B, Chen J, Xin Z et al (2011) Regulated AtHKT1 gene expression by a distal enhancer element and DNA methylation in the promoter plays an important role in salt tolerance. Plant Cell Physiol 52:149–161

Bartel DP (2004) MicroRNAs: genomics, biogenesis, mechanism, and function. Cell 116:281–297

Bastian M, Heymann S, Jacomy M (2009) Gephi: an open source software for exploring and manipulating networks

Bilgin O, Baser I, Korkut KZ, Balkan A, Saglam N (2008) The impacts on seedling root growth of water and salinity stress in maize (*Zea mays* indentata sturt.) Bulgarian J Agr Sci 14:313–320

Brar D, Khush G (2002) Transferring genes from wild species into Rice. In: Kang MS (ed) Quantitative genetics, genomics and plant breeding. CABI, Oxford, pp 197–217

Brohee S, Faust K, Lima-Mendez G, Vanderstocken G, van Helden J (2008) Network analysis tools: from biological networks to clusters and pathways. Nat Protoc 3:1616–1629. https://doi.org/10.1038/nprot.2008.100

Buhtz A, Pieritz J, Springer F, Kehr J (2010) Phloem small RNAs, nutrient stress responses, and systemic mobility. BMC Plant Biol 10:64. https://doi.org/10.1186/1471-2229-10-64

Carillo P, Annunziata MG, Pontecorvo G, Fuggi A, Woodrow P (2011) Salinity stress and salt tolerance. Abiotic stress plants-mech. adapt. In: Prof. Arun Shanker (ed) InTech, https://doi.org/10.5772/22331

Chandna R, Azooz MM, Ahmad P (2013) Recent advances of metabolomics to reveal plant response during salt stress. Springer, New York, pp 1–14

Cha-Um S, Kirdmanee C (2009) Effect of salt stress on proline accumulation, photosynthetic ability and growth characters in two maize cultivars. Pak J Bot 41:87–98

Chen X, Guo Z (2008) Tobacco OPBP1 enhances salt tolerance and disease resistance of transgenic rice. Int J Mol Sci 9:2601–2613

Chen Z, Pottosin II, Cuin TA, Fuglsang AT, Tester M, Jha D et al (2007) Root plasma membrane transporters controlling K^+/Na^+ homeostasis in salt-stressed barley. Plant Physiol 145:1714–1725

Chinnusamy V, Zhu J-K (2009) Epigenetic regulation of stress responses in plants. Curr Opin Plant Biol 12:133–139

Chuck G, Cigan AM, Saeteurn K, Hake S (2007) The heterochronic maize mutant Corngrass1 results from overexpression of a tandem microRNA. Nat Genet 39:544–549. https://doi.org/10.1038/ng2001

Cicek N, Cakirlar H (2002) The effect of salinity on some physiological parameters in two maize cultivars. Bulg J Plant Physiol 28:66–74

Collado MnB, Aulicino MnB, Arturi MJ, Molina Md C (2016) Selection of maize genotypes with tolerance to osmotic stress associated with salinity. Agric Sci 07:82–92

Collard BC, Vera Cruz CM, McNally KL, Virk PS, Mackill DJ (2008) Rice molecular breeding laboratories in the genomics era: current status and future considerations. Int J Plant Genomics 2008:25

Crosbie TM, Eathington SR, Johnson GR, Edwards M, Reiter R, Stark S, et al (2008) Plant breeding: past, present, and future. Oxford Publishing, Oxford, UK:3–50

Csardi G, Nepusz T (2006) The igraph software package for complex network research. Inter J Complex Syst 1695:1695

Cui D, Wu D, Somarathna Y, Xu C, Li S, Li P et al (2015) QTL mapping for salt tolerance based on snp markers at the seedling stage in maize (*Zea mays* L.) Euphytica 203:273–283

Curtis T, Halford NG (2014) Food security: the challenge of increasing wheat yield and the importance of not compromising food safety. Ann Appl Biol 164:354–372

DeRose-Wilson L, Gaut BS (2011) Mapping salinity tolerance during arabidopsis thaliana germination and seedling growth. PLoS One. 6(8):e22832

Desta ZA, Ortiz R (2014) Genomic selection: genome-wide prediction in plant improvement. Trends Plant Sci 19:592–601

Deyholos MK (2010) Making the most of drought and salinity transcriptomics. PCE Plant Cell Environ 33:648–654

Ding D, Zhang L, Wang H, Liu Z, Zhang Z, Zheng Y (2009) Differential expression of miRNAs in response to salt stress in maize roots. Ann Bot 103:29–38. https://doi.org/10.1093/aob/mcn205

Doerge RW (2002) Mapping and analysis of quantitative trait loci in experimental populations. Nat Rev Genet 3:43–52

Filippou P, Fotopoulos V, Bouchagier P, Skotti E (2014) Proline and reactive oxygen/nitrogen species metabolism is involved in the tolerant response of the invasive plant species Ailanthus altissima to drought and salinity. Environ Exp Bot 97:1–10

Franco-Zorrilla JM, López-Vidriero I, Carrasco JL, Godoy M, Vera P, Solano R (2014) DNA-binding specificities of plant transcription factors and their potential to define target genes. Proc Natl Acad Sci 111:2367–2372

Fritsche-Neto R, DoVale JC, Éder Cristian Malta de Lanes, Marcos Deon Vilela de Resende, Miranda GV (2012) Genome-wide selection for tropical maize root traits under conditions of nitrogen and phosphorus stress – https://doi.org/10.4025/actasciagron.v34i4. 15884. Universidade Estadual de MaringÃ¡

Fujii H, Chiou TJ, Lin SI, Aung K, Zhu JK (2005) A miRNA involved in phosphate-starvation response in Arabidopsis. Curr Biol 15:2038–2043. https://doi.org/10.1016/j.cub.2005.10.016

Gallavotti A, Long JA, Stanfield S, Yang X, Jackson D, Vollbrecht E et al (2010) The control of axillary meristem fate in the maize ramosa pathway. Development 137:2849–2856. https://doi.org/10.1242/dev.051748

Galvan-Ampudia CS, Testerink C (2011) Salt stress signals shape the plant root. Curr Opin Plant Biol 14:296–302

Geilfus CM, Zorb C, Muhling KH (2010) Salt stress differentially affects growth-mediating β-expansins in resistant and sensitive maize (*Zea mays* L.) Plant Physiol Biochem 48:993–998

Golldack D, Lüking I, Yang O (2011) Plant tolerance to drought and salinity: stress regulating transcription factors and their functional significance in the cellular transcriptional network. Plant Cell Rep 30:1383–1391

González RM, Ricardi MM, Iusem ND (2011) Atypical epigenetic mark in an atypical location: cytosine methylation at asymmetric (CNN) sites within the body of a non-repetitive tomato gene. BMC Plant Biol 11:94

Gore MA, Chia JM, Elshire RJ, Sun Q, Ersoz ES, Hurwitz BL et al (2009) A first-generation haplotype map of maize. Science 326:1115–1117

Griffiths-Jones S, Saini HK, van Dongen S, Enright AJ (2008) miRBase: tools for microRNA genomics. Nucleic Acids Res 36:D154–D158

Hao D, Chao M, Yin Z, Yu D (2012) Genome-wide association analysis detecting significant single nucleotide polymorphisms for chlorophyll and chlorophyll fluorescence parameters in soybean (*Glycine max*) landraces. Euphytica 186:919

Hasanuzzaman M, Alam MM, Rahman A, Hasanuzzaman M, Nahar K, Fujita M (2014) Exogenous proline and glycine betaine mediated upregulation of antioxidant defense and glyoxalase systems provides better protection against salt-induced oxidative stress in two rice (Oryza sativa L.) varieties. Biomed Res Int 2014:17

Hashida S-N, Uchiyama T, Martin C, Kishima Y, Sano Y, Mikami T (2006) The temperature-dependent change in methylation of the antirrhinum transposon Tam3 is controlled by the activity of its transposase. Plant Cell 18:104–118

Hassanzadehdelouei M, Vazin F, Nadaf J (2013) Effect of salt stress in different stages of growth on qualitative and quantitative characteristics of cumin (Cuminum Cyminum L.) Cercet Agronomice Moldova 46:89–97

Hoque MMI (2013) Evaluation and mapping QTLs of maize salinity tolerance. Ph.D. thesis. Chinese Academy of Agricultural Sciences Dissertation, Chaina

Hoque MMI, Zheng J, Wang G (2015) Evaluation of salinity tolerance in maize (*Zea mays* L.) genotypes at seedling stage. J BioSci Biotechnol 4:39–49

Hu Z, Hung JH, Wang Y, Chang YC, Huang CL, Huyck M et al (2009) VisANT 3.5: multi-scale network visualization, analysis and inference based on the gene ontology. Nucleic Acids Res 37:W115–W121. https://doi.org/10.1093/nar/gkp406

Hu Z, Zhang H, Kan G, Ma D, Zhang D, Shi G et al (2013) Determination of the genetic architecture of seed size and shape via linkage and association analysis in soybean (*Glycine max* L. Merr.) Genet: Int J Genet Evol 141:247–254

Hussain K, Majeed A, Nawaz K, Nisar MF (2010) Changes in morphological attributes of maize (*Zea mays* L.) under NaCl salinity. Am-Eurasian J Agric Environ Sci 8:230–232

Iba K (2002) Acclimative response to temperature stress in higher plants: approaches of gene engineering for temperature tolerance. Annu Rev Plant Biol 53:225–245

Joshi R, Wani SH, Singh B, Bohra A, Dar ZA, Lone AA et al (2016) Transcription factors and plants response to drought stress: current understanding and future directions. Front Plant Sci 7:1029

Julkowska MM, Hoefsloot HC, Mol S, Feron R, de Boer GJ, Haring MA et al (2014) Capturing Arabidopsis root architecture dynamics with ROOT-FIT reveals diversity in responses to salinity. Plant Physiol 166:1387–1402

Kang M, Zhao Q, Zhu D, Yu J (2012) Characterization of microRNAs expression during maize seed development. BMC Genomics 13:360. https://doi.org/10.1186/1471-2164-13-360

Kawashima CG, Yoshimoto N, Maruyama-Nakashita A, Tsuchiya YN, Saito K, Takahashi H et al (2009) Sulphur starvation induces the expression of microRNA-395 and one of its target genes but in different cell types. Plant J 57:313–321. https://doi.org/10.1111/j.1365-313X.2008.03690.x

Kearsey M, Farquhar A (1998) QTL analysis in plants; where are we now? Heredity 80:137–142

Koevoets IT, Venema JH, Elzenga JT, Testerink C (2016) Roots withstanding their environment: exploiting root system architecture responses to abiotic stress to improve crop tolerance. Front Plant Sci 7:1335

Kumar V, Singh A, Mithra SVA, Parida SK, Jain S, Tiwari KK et al (2015) Genome-wide association mapping of salinity tolerance in rice (*Oryza sativa*). DNA Res 22:133–145

Leitner D, Klepsch S, Ptashnyk M, Marchant A, Kirk GJD, Schnepf A et al (2010) A dynamic model of nutrient uptake by root hairs. New Phytol 185:792–802

Li YL, Dong YB, Cui DQ, Wang YZ, Liu YY, Wei MG et al (2008) The genetic relationship between popping expansion volume and two yield components in popcorn using unconditional and conditional QTL analysis. Int J Plant Breed 162:345–351

Li W, Cui X, Meng Z, Huang X, Xie Q, Wu H et al (2012) Transcriptional regulation of Arabidopsis MIR168a and argonaute1 homeostasis in abscisic acid and abiotic stress responses. Plant Physiol 158:1279–1292. https://doi.org/10.1104/pp.111.188789

Lin H, Zhu M, Yano M, Gao J, Liang Z, Su W et al (2004) QTLs for Na+ and K+ uptake of the shoots and roots controlling rice salt tolerance. Theor Appl Genet 108:253–260

Liseron-Monfils C, Ware D (2015) Revealing gene regulation and associations through biological networks. Curr Plant Biol 3–4:30–39. https://doi.org/10.1016/j.cpb.2015.11.001

Liu HH, Tian X, Li YJ, Wu CA, Zheng CC (2008) Microarray-based analysis of stress-regulated microRNAs in *Arabidopsis thaliana*. RNA 14:836–843

Liu H, Qin C, Chen Z, Zuo T, Yang X, Zhou H, et al (2014) Identification of miRNAs and their target genes in developing maize ears by combined small RNA and degradome sequencing. BMC Genomics 15: 25–25. https://doi.org/10.1186/1471-2164-15-25

Lobell DB, Burke MB, Tebaldi C, Mastrandea MD, Falcon WP, Naylor RL (2008) Prioritizing climate chnage adaptation needs for food security in 2030. Science 319(5863):607–610

Long NV, Dolstra O, Malosetti M, Kilian B, Graner A, Visser RG et al (2013) Association mapping of salt tolerance in barley (*Hordeum vulgare* L.) Theor Appl Genet 126:2335–2351

Lynch M, Walsh B (2007) The origins of genome architecture Sinauer Associates Sunderland Associates, Inc:340

Maiti R, Rodríguez HG, Rajkumar D, Koushik S, Vidyasagar P (2012) Genotypic variability in salinity tolerance of maize pipe line hybrids at seedling stage. Int J Bio-resour Stress Manag 3:427–432

Mano Y, Takeda K (1997) Mapping quantitative trait loci for salt tolerance at germination and the seedling stage in barley *(Hordeum vulgare* L.) Euphytica 94:263–272

McMullen MD, Kresovich S, Villeda HS, Bradbury P, Li H, Sun Q et al (2009) Genetic properties of the maize nested association mapping population. Science 325:737–740

Menezes-Benavente L, Kernodle SP, Margis-Pinheiro M, Scandalios JG (2004) Salt-induced antioxidant metabolism defenses in maize (*Zea mays* L.) seedlings. Redox Rep: Commun Free Radic Res 9:29–36

Mica E, Gianfranceschi L, PÃ ME (2006) Characterization of five microRNA families in maize. J Exp Bot 57:2601–2612

Morris M, Dreher K, Ribaut J-M, Khairallah M (2003) Money matters (II): costs of maize inbred line conversion schemes at CIMMYT using conventional and marker-assisted selection. Mol Breed Mol Breed: New Strateg Plant Improv 11:235–247

Munns R (2002) Comparative physiology of salt and water stress. Plant Cell Environ 25:239–250

Munns R, Tester M (2008) Mechanisms of salinity tolerance. Annu Rev Plant Biol 59:651–681

Nag A, King S, Jack T (2009) miR319a targeting of TCP4 is critical for petal growth and development in Arabidopsis. Proc Natl Acad Sci U S A 106:22534–22539. https://doi.org/10.1073/pnas.0908718106

Ohta M, Hayashi Y, Nakashima A, Hamada A, Tanaka A, Nakamura T, Hayakawa T (2002) Introduction of a Na^+/H^+ antiporter gene from Atriplex gmelini confers salt tolerance to rice. FEBS Lett 532:279–282

Parida AK, Das AB (2005) Salt tolerance and salinity effects on plants: a review. Ecotoxicol Environ Saf 60:324–349

Park W, Li J, Song R, Messing J, Chen X (2002) CARPEL FACTORY, a Dicer homolog, and HEN1, a novel protein, act in microRNA metabolism in *Arabidopsis thaliana*. Curr Biol 12:1484–1495

Pesqueira J, García MD, Molina MC (2003) NaCl tolerance in maize (Zea mays ssp. mays) × Tripsacum dactyloides L. hybrid calli in regenerated plants. 2003 1: 5. https://doi.org/10.5424/sjar/2003012-21

Prasanna BM, Pixley K, Warburton ML, Xie CX (2010) Molecular marker-assisted breeding options for maize improvement in Asia. Mol Breed: New Strateg Plant Improv 26:339–356
Praveenkumar B, Salimath PM, Sridevi O (2014) Study of genetic variability in maize (*Zea mays* L.) inbred lines under stress condition. Plant Arch 14:679–685
Prohens J, Gramazio P, Plazas M, Dempewolf H, Kilian B, Díez MJ et al (2017) Introgressiomics: a new approach for using crop wild relatives in breeding for adaptation to climate change. Euphytica 213:158. https://doi.org/10.1007/s10681-017-1938-9
Qiu F, Yonglian Z, Zhang Z, Xu S (2007) Mapping of QTL associated with waterlogging tolerance during the seedling stage in maize. Annal Bot 99(6):1067–1081. https://doi.org/10.1093/aob/mcm055
Ragot M, Sisco P, Hoisington D, Stuber C (1995) Molecular-marker-mediated characterization of favorable exotic alleles at quantitative trait loci in maize. Crop Sci 35:1306–1315
Ragot M, Lee M, Guimaraes E (2007) Marker-assisted selection in maize: current status, potential, limitations and perspectives from the private and public sectors. Marker-assisted selection, Current status and future perspectives in crops, livestock, forestry and fish: 117–150
Rayko E, Maumus F, Maheswari U, Jabbari K, Bowler C (2010) Transcription factor families inferred from genome sequences of photosynthetic stramenopiles. New Phytol 188:52–66
Rieseberg LH, Whitton J, Gardner K (1999) Hybrid zones and the genetic architecture of a barrier to gene flow between two sunflower species. Genetics 152:713–727
Rodriguez RE, Mecchia MA, Debernardi JM, Schommer C, Weigel D, Palatnik JF (2010) Control of cell proliferation in *Arabidopsis thaliana* by microRNA miR396. Development 137:103–112. https://doi.org/10.1242/dev.043067
Rogers ED, Jackson T, Moussaieff A, Aharoni A, Benfey PN (2012) Cell type-specific transcriptional profiling: implications for metabolite profiling. Plant J 70:5–17
Saddiqe Z, Javeria S, Khalid H, Farooq A (2016) Effect of salt stress on growth and antioxidant enzymes in two cultivars of maize (*Zea mays* L.) Pak J Bot 48:1361–1370
Samineni S, Siddique KHM, Gaur PM, Colmer TD (2011) Salt sensitivity of the vegetative and reproductive stages in chickpea (*Cicer arietinum* L.): podding is a particularly sensitive stage. Environ Exp Bot 71:260–268
Sawahel WA, Hassan AH (2002) Generation of transgenic wheat plants producing high levels of the osmoprotectant proline. Biotechnol Lett 24:721–725
Sayed El (2011) Influence of salinity stress on growth parameters, photosynthetic activity and cytological studies of Zeamays L. plantusing hydrogel polymer. Agric Biol J N Am 2:907–920. https://doi.org/10.5251/abjna.2011.2.6.907.920
Shahbaz M, Ashraf M (2013) Improving salinity tolerance in cereals. Crit Rev Plant Sci 32:237–249
Shannon MC, Grieve CM, Francois LE (1994) Whole-plant response to salinity. Plant-Environ Interact:199–244
Shelden MC, Roessner U (2013) Advances in functional genomics for investigating salinity stress tolerance mechanisms in cereals. Front Plant Sci 4:123
Shi L, Xia W (2012) Photoredox functionalization of C–H bonds adjacent to a nitrogen atom. Chem Soc Rev 41:7687–7697
Smith S, De Smet I (2012) Root system architecture: insights from Arabidopsis and cereal crops. Philos Trans R Soc B: Biol Sci 367(1595):1441–1452
Smoot ME, Ono K, Ruscheinski J, Wang PL, Ideker T (2011) Cytoscape 2.8: new features for data integration and network visualization. Bioinformatics 27:431–432. https://doi.org/10.1093/bioinformatics/btq675
Song Y, Ji D, Li S, Wang P, Li Q, Xiang F (2012) The dynamic changes of DNA methylation and histone modifications of salt responsive transcription factor genes in soybean. PLoS One 7:e41274
Song X, Li Y, Hou X (2013) Genome-wide analysis of the AP2/ERF transcription factor superfamily in Chinese cabbage (*Brassica rapa* ssp. pekinensis). BMC Genomics 14:573. https://doi.org/10.1186/1471-2164-14-573
Steward N, Kusano T, Sano H (2000) Expression of ZmMET1, a gene encoding a DNA methyltransferase from maize, is associated not only with DNA replication in actively proliferating

cells, but also with altered DNA methylation status in cold-stressed quiescent cells. Nucleic Acids Res 28:3250–3259

Stewart CN, Halfhill MD, Warwick SI (2003) Transgene introgression from genetically modified crops to their wild relatives. Nat Rev Genet 4:806–817

Su J, Wu R (2004) Stress-inducible synthesis of proline in transgenic rice confers faster growth under stress conditions than that with constitutive synthesis. Plant Sci 166:941–948

Sun G (2012) MicroRNAs and their diverse functions in plants. Mol Genet Biochem 80:17–36

Teng F, Zhai L, Liu R, Bai W, Wang L, Huo D et al (2013) ZmGA3ox2, a candidate gene for a major QTL, qPH3.1, for plant height in maize. Plant J: Cell Mol Biol 73:405–416

Thomashow MF (1999) Plant cold acclimation: freezing tolerance genes and regulatory mechanisms. Annu Rev Plant Biol 50:571–599

Thomson MJ, Rahman MA, Rahman MA, Rahman MA, Sajise AG, Adorada DL et al (2010) Characterizing the Saltol quantitative trait locus for salinity tolerance in Rice. Rice 3:2–3

Thudi M, Gaur PM, Krishnamurthy L, Mir RR, Kudapa H, Fikre A et al (2014) Genomics-assisted breeding for drought tolerance in chickpea. Funct Plant Biol 41:1178–1190. https://doi.org/10.1071/FP13318

Tian F, Bradbury PJ, Brown PJ, Hung H, Sun Q, Flint-Garcia S et al (2011) Genome-wide association study of leaf architecture in the maize nested association mapping population. Nat Genet 43:159–162

Tominaga-Wada R, Ishida T, Wada T (2011) New insights into the mechanism of development of Arabidopsis root hairs and trichomes. Int Rev Cell Mol Biol 286:67–106

Tsuji H, Saika H, Tsutsumi N, Hirai A, Nakazono M (2006) Dynamic and reversible changes in histone H3-Lys4 methylation and H3 acetylation occurring at submergence-inducible genes in rice. Plant Cell Physiol 47:995–1003

Turan MA, Elkarim AHA, Taban N, Taban S (2010) Effect of salt stress on growth and ion distribution and accumulation in shoot and root of maize plant. Afr J Agric Res 5:584–588

Umar U, Ado S, Aba D, Bugaje S (2015) Studies on genetic variability in maize (*Zea mays* L.) under stress and non-stress environmental conditions. Int J Agron Agric Res 7:70–77

Umezawa T, Fujita M, Fujita Y, Yamaguchi-Shinozaki K, Shinozaki K (2006) Engineering drought tolerance in plants: discovering and tailoring genes to unlock the future. Curr Opin Biotechnol 17:113–122

Usman M, Haq AU, Ahsan T, Amjad S, Riasat Z, Umar M (2012) Effect of NaCl on morphological attributes of maize (*Zea mays* L.) World J Agric Sci 8:381–384

Verma D, Singla-Pareek SL, Rajagopal D, Reddy M, Sopory S (2007) Functional validation of a novel isoform of Na^+/H^+ antiporter from Pennisetum glaucum for enhancing salinity tolerance in rice. J Biosci 32:621–628

Walbot V (2008) Maize genome in motion. BioMed Central Ltd.

Wang Z, Zemetra RS, Hansen J, Mallory-Smith CA (2001) The fertility of wheat× jointed goatgrass hybrid and its backcross progenies. Weed Sci 49:340–345

Warburton ML, Rauf S, Marek L, Hussain M, Ogunola O, de Jesus Sanchez Gonzalez J (2017) The use of crop wild relatives in maize and sunflower breeding. Crop Sci. https://doi.org/10.2135/cropsci2016.10.0855

Wheat CW (2010) Rapidly developing functional genomics in ecological model systems via 454 transcriptome sequencing. Genetica 138:433–451

Xiao YN, Li XH, George ML, Li MS, Zhang SH, Zheng YL (2005) Quantitative trait locus analysis of drought tolerance and yield in maize in China. Plant Mol Biol Report 23:155–165

Xu Y, Li S, Li L, Zhang X, Xu H, An D (2013) Mapping QTLs for salt tolerance with additive, epistatic and QTL× treatment interaction effects at seedling stage in wheat. Plant Breed 132:276–283

Yan J, Crouch J, Warburton M (2011) Association mapping for enhancing maize (*Zea mays* L.) genetic improvement. Crop Sci 51:433–449

Yin XY, Yang AF, Zhang KW, Zhang JR (2004) Production and analysis of transgenic maize with improved salt tolerance by the introduction of AtNHX1 gene. Acta Bot Sin (English edition) 46:854–861

Yu A, Lepère G, Jay F, Wang J, Bapaume L, Wang Y et al (2013) Dynamics and biological relevance of DNA demethylation in Arabidopsis antibacterial defense. Proc Natl Acad Sci 110:2389–2394

Zhang ZM, Zhao MJ, Rong TZ, Pan GT, Original L (2007) SSR linkage map onstruction and QTL identification for plant height and ear height in maize (*Zea mays* L.) Zuo Wu Xue Bao 33:341–344

Zhang L, Chia JM, Kumari S, Stein JC, Liu Z, Narechania A et al (2009) A genome-wide characterization of MicroRNA genes in maize. PLoS Genet 5:e1000716. https://doi.org/10.1371/journal.pgen.1000716

Zhang H, Cui LN, Meng JJ, Zhang HY, Shi DY, Dong ST et al (2012) Effects of partial root excision on the growth, photosynthesis, and antioxidant enzyme activities of maize under salt stress. J Appl Ecol 23:3377–3384

Zhang H, Cui FA, Wang LIN, Li JUN, Ding A, Zhao C et al (2013) Conditional and unconditional QTL mapping of drought-tolerance-related traits of wheat seedling using two related RIL populations. J Genet 92:213–231

Zhang WJ, Niu Y, Bu SH, Li M, Feng JY, Zhang J, et al (2014) Epistatic association mapping for alkaline and salinity tolerance traits in the soybean germination stage. PLoS One 9.c

Zhao F, Wang Z, Zhang Q, Zhao Y, Zhang H (2006) Analysis of the physiological mechanism of salt-tolerant transgenic rice carrying a vacuolar Na^+/H^+ antiporter gene from Suaeda salsa. J Plant Res 119:95–104

Zhao B, Ge L, Liang R, Li W, Ruan K, Lin H, et al (2009) Members of miR-169 family are induced by high salinity and transiently inhibit the NF-YA transcription factor. BMC Mol Biol 10

Zhao YL, Wang HM, Shao BX, Chen W, Guo ZJ, Gong HY et al (2016) SSR-based association mapping of salt tolerance in cotton (Gossypium hirsutum L.) Genet Mol Res 15(2)

Zhu J-K (2001) Plant salt tolerance. Trends Plant Sci 6:66–71

Zhu J, Jeong JC, Zhu Y, Sokolchik I, Miyazaki S, Zhu J-K et al (2008) Involvement of Arabidopsis HOS15 in histone deacetylation and cold tolerance. Proc Natl Acad Sci 105:4945–4950

Zhu J, Brown KM, Lynch JP (2010) Root cortical aerenchyma improves the drought tolerance of maize (*Zea mays* L.) Plant Cell Environ 33:740–749

Zörb C, Schmitt S, Neeb A, Karl S, Linder M, Schubert S (2004) The biochemical reaction of maize (*Zea mays* L.) to salt stress is characterized by a mitigation of symptoms and not by a specific adaptation. Plant Sci 167:91–100

Zygalakis KC, Kirk GJD, Jones DL, Wissuwa M, Roose T (2011) A dual porosity model of nutrient uptake by root hairs. New Phytol 192:676–688

Drought Stress Tolerance in Wheat: Omics Approaches in Understanding and Enhancing Antioxidant Defense

10

Mirza Hasanuzzaman, Jubayer Al Mahmud, Taufika Islam Anee, Kamrun Nahar, and M. Tofazzal Islam

Abstract

Plants face various kinds of stresses in the changing environment. Among the environmental stresses, drought is one of the most devastating stressors due to its diverse negative effects on crop plants. Drought stress in plants is very complex as it occurs due to varying environmental conditions such as soil water scarcity, soil salinity, and high temperature. The latter ones are termed as physiological drought. Bread wheat (*Triticum aestivum* L.) ranks first in the world's grain production and is consumed as staple food by more than 36% of the world population. Wheat plant is highly sensitive to drought, especially at flowering and grain filling stages. Growth, photosynthesis, metabolic processes, nutrient assimilation, and yield of wheat plants remarkably decrease under drought. The responses of wheat to drought are varied at morphological, physiological, molecular, and biochemical levels. One of the most common consequences of drought is the disturbance of the balance between production of reactive oxygen species (ROS) and antioxidant defense causing overaccumulation of ROS which induces

M. Hasanuzzaman (✉) · T. I. Anee
Department of Agronomy, Faculty of Agriculture, Sher-e-Bangla Agricultural University, Sher-e-Bangla Nagar, Dhaka, Bangladesh
e-mail: mhzsauag@yahoo.com

J. A. Mahmud
Department of Agroforestry and Environmental Science, Faculty of Agriculture, Sher-e-Bangla Agricultural University, Sher-e-Bangla Nagar, Dhaka, Bangladesh

K. Nahar
Department of Agricultural Botany, Faculty of Agriculture, Sher-e-Bangla Agricultural University, Sher-e-Bangla Nagar, Dhaka, Bangladesh

M. T. Islam
Department of Biotechnology, Bangabandhu Sheikh Mujibur Rahman Agricultural University, Gazipur, Bangladesh

S. M. Zargar, M. Y. Zargar (eds.), *Abiotic Stress-Mediated Sensing and Signaling in Plants: An Omics Perspective*, https://doi.org/10.1007/978-981-10-7479-0_10

oxidative stress. This happens due to closure of the stomata, CO_2 influx, and decrease of leaf internal CO_2 which direct more electrons to form ROS and enhance photorespiration. These ROS can incur direct damage to protein, lipid, and nucleic acid which can ultimately cause plant cell death. Enhancing the antioxidant defense system to mitigate the oxidative stress is one of the effective strategies to make the wheat plants tolerant to drought. It appears that plants synthesize or activate several molecules like osmoprotectants, phytohormones, signaling molecules, and antioxidants to protect themselves from drought-induced oxidative damages. Novel approaches for enhancing the antioxidant defense system to minimize the impacts of drought-induced damage in plants are prime targets of plant biologists. Several genes and their overexpression were found to confer drought tolerance in plants. Application of plant probiotic bacteria also enhances tolerance of wheat plants to drought. Recent advances in genomic, transcriptomic, proteomic, and metabolomic studies on wheat under varying levels of drought generate useful information for designing drought-tolerant wheat. This chapter comprehensively reviews and updates our understanding on molecular mechanisms of adaptation of wheat plants to drought stress with special emphasis to antioxidant defense systems.

Keywords
Abiotic stress · Antioxidant defense · Cereal crops · Reactive oxygen species · Water stress

10.1 Introduction

Wheat (*Triticum aestivum* L.) is one of the major cereals feeding one-fifth of the total world population (FAO 2011). Wheat production needs to be doubled by 2050 to fulfill the demand of increasing population (Foresight 2011). Because of the decreasing cultivable land, in spite of increasing the demand for wheat, it is more important to increase the production of wheat per unit area (Araus et al. 2003). But the production of all crops including wheat is severely affected by abiotic stresses. More than 50% of the land covered by wheat cultivation is affected by periodic drought (Rajaram 2001; Pfeiffer et al. 2005). Using statistical methods, Lesk et al. (2016) recently demonstrated that losses of production of 9–10% of cereal crops are caused by drought and extreme weather. Furthermore, drought is expected to increase in frequency and severity in the future as a result of climate change, mainly as a consequence of decrease in precipitation but also because of increasing evaporation driven by global warming (Battisti and Naylor 2009; Dai 2010). Therefore, understanding the drought-induced damages in wheat plants and approaches to improve drought tolerance is crucial to increase the productivity of wheat. Severity of drought-induced damage on plants varies depending on plant genotype and growth stage. Drought stress imposes damaging effects on plant developmental processes and in its different growth stages such as germination, growth, reproductive, and maturity. Uptake and transportation of water and nutrient, photosynthesis, respiration, transpiration, and translocation of assimilates are restricted under drought stress conditions. Drought-induced disruption of membrane structure and properties, disorganization of ultra-structural cellular components, enzyme activities, and anion and cationic imbalance are some of the major reasons for disturbing plant physiological processes (Szegletes et al. 2000; Yordanov et al. 2000; Zhu 2002; Lawlor and Cornic 2002; Chaves et al. 2003; Hasanuzzaman et al. 2014). Drought stress usually leads to the production of reactive oxygen species (ROS). Hydrogen peroxide (H_2O_2), superoxide ($O_2^{\bullet-}$), singlet oxygen ($^1O_2$), and hydroxyl radicals ($OH^{\bullet}$) are the most common species which are generated due to iron-catalyzed Fenton reaction due to the activities of lipoxygenases, peroxidases (POX), NADPH oxidase, and xanthine oxidase. The ROS in any form causes substantial damage to components of the cell and even can cause cell death (Hasanuzzaman et al. 2014; Mattos and Moretti 2015). Plants have a well-developed antioxidant defense system to scavenge and maintain a balanced state of ROS generation to prevent cells from oxidative damage. Antioxidant defense systems possess some nonenzymatic antioxidants (ascorbic acid, AsA; glutathione, GSH; phenolic compounds; alkaloids; nonprotein amino acids; and α-tocopherols) and some antioxidant enzymes (superoxide dismutase, SOD; catalase, CAT; ascorbate peroxidase, APX; glutathione reductase, GR; monodehydroascorbate reductase, MDHAR; dehydroascorbate reductase, DHAR; glutathione peroxidase, GPX; and glutathione-*S*-transferase, GST) which work coordinately to scavenge ROS in an efficient way (Gill and Tuteja 2010).

Drought stress tolerance in plants is a complex physiological trait where enhancement of the plant antioxidant defense system is vital to prevent oxidative damage to cell components. Prevention of oxidative damage is a prerequisite for running cellular activities and physiological processes. Several lines of evidence suggest that enhancing the antioxidant defense system improved plant performance under drought stress (Djibril et al. 2005; Manivannan et al. 2008; Hasanuzzaman and Fujita 2011; Nahar et al. 2015). Transgenic approaches also proved that enhancing the antioxidant system is vital to confer oxidative as well as abiotic stress tolerance (Gaber et al. 2006; Simon-Sarkadi et al. 2006; Molinari et al. 2007; Kim et al. 2008). Performances of drought-affected plants are remarkably improved by exogenous application of osmolytes (Wang et al. 2010a; Shahbaz et al. 2011), hormones (Kang et al. 2012; Yavas and Unay 2016), nutrients (Wei et al. 2013; Yasmeen et al. 2013), antioxidants (Farooq et al. 2013; Hafez and Gharib 2016), and signaling molecules (Hasanuzzaman et al. 2013, 2016; Shan et al. 2015). This chapter reviews and discusses on (i) common effects of drought stress on the wheat plant and (ii) the roles of osmolytes, phytohormones, nutrients, antioxidants, and signaling molecules to improve drought stress tolerance.

10.2 Responses of Wheat Plants to Drought Stress

Unlike other crops, it is not so simple to tell how drought affects the vulnerability of wheat production in combination with several co-varying factors (i.e., phenological phases, agroclimatic regions, soil texture) because of its complex nature (Hasanuzzaman et al. 2014; Daryanto et al. 2016). Drought stress often combines and coincides with others stress factors like high temperature, and the ultimate effects are even more devastating than we expect. A recent estimation showed that only 40% water reduction in soil may result in about 21% yield reduction in wheat (Daryanto et al. 2016). The causes for drought-induced yield reduction in wheat are due to the decreased cell division and growth, decreased photosynthesis, membrane damage, impairment of water and nutrient uptake and transport, abnormal reproductive growth, and oxidative stress (Fig. 10.1; Tables 10.1 and 10.2; Hasanuzzaman et al. 2014).

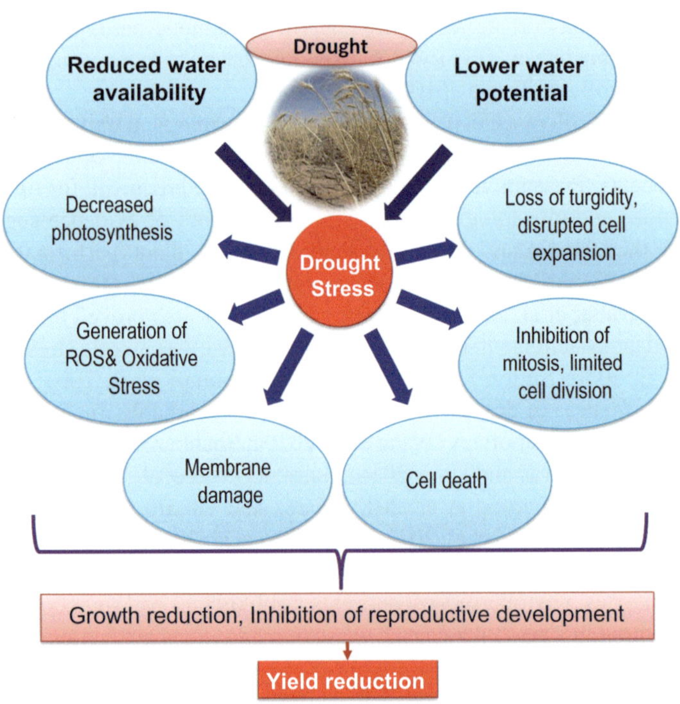

Fig. 10.1 Possible drought stress responses in plants (Adapted from Hasanuzzaman et al. 2014, with permission from Wiley)

Table 10.1 Effects of drought stress on seed germination, plant growth, water relation, nutrient uptake, and photosynthesis of *T. aestivum*

Stress treatments and duration	Effects	References
Withheld water at 10 DAS, 10 days	30% reduction of P_n	Timmusk et al. (2014)
	Reduced DW of both shoot and root	
	24% decrease in WUE	
	Decreased NAR and g_s	
35% water holding capacity	Delayed MET by 14% and $E_{50}$by 29%, also decreased CUE by 35%	Farooq et al. (2013)
	35% decrease in seedling DW	
180, 240, and 300 gL^{-1} PEG 6000, 8 days	FGP, MDG, GI, CVG, and MGT decreased with increasing concentration of PEG	Almaghrabi (2012)
	Length of shoot, root, and seedling and FW and DW of both shoot and root decreased with increasing concentration of PEG	
50% irrigation	34% decrease in plant height	Abdel-Motagally and El-Zohri (2016)
	26% and 18% decrease in chl*a* and chl*b*, respectively	
Withheld water at 17 DAS, 16 days	Reduced FW of both cultivars	Hassan et al. (2015)
	Dry weight of susceptible variety and leaf area of tolerant variety were decreased	
Withheld water at the grain filling stage, 10 days	Plant height decreased	Yavas and Unay (2016)
	20% and 26% decrease in RWC of tolerant and susceptible varieties, respectively	
	Reduced chl content	
60% FC	34% reduction of Ψ_w and 22% of RWC	Nawaz et al. (2015)
	43%, 30%, and 27% decrease of P_n, T_r, and g_s, respectively	
	Reduced Fe uptake, but increased Zn, Na, and Ca uptake	
26% PEG (−1.5 MPa), 2 days	Decreased leaf RWC and RDIR	Grzesiak et al. (2013)
20% PEG 6000, 7 days	40% decrease of plant height and 38% of root length	Wei et al. (2013)
	Decreased leaf water potential	
	More than 40% reduction of K^+ contents in both shoot and root	
	Decrease P_n and g_s	
	Reduced chl content	
50% FC	Reduced leaf area, chl*a*, and chl*b* contents	Yasmeen et al. (2013)
	Leaf K^+ content decreased	
15% PEG 6000, 3 days	Reduced FW, DW, and absolute water content (AWC)	Kang et al. (2012)

(continued)

Table 10.1 (continued)

Stress treatments and duration	Effects	References
60% FC, 49 days	Reduced FW and DW of both shoot and root and shoot length also	Shahbaz et al. (2011)
	Decreased P_n and g_s	
	Higher WUE	
	Lower contents of shoot and root P, N, and K	
25% and 35% FC	Decreased leaf area and plant height	Khakwani et al. (2011a)
	Reduced RWC	
Withheld irrigation at different growth stages	Lower CGR and LAI	Akram (2011)
	Reduced RWC, Ψ_w, Ψ_s, and Ψ_p	
Withheld water, 14 days	Reduced leaf RWC	Prasad et al. (2011)
	Decreased chl content and P_n	
50% and 30% FC	Decreased P_n in drought-sensitive cultivars	Akhkha et al. (2011)
	Reduced chl content	
	Increased dark respiration	
Withheld water onset of stem elongation stage and booting stage	Reduced plant height by 35% and 23% for water stress from stem elongation stage and booting stage, respectively	Shamsi and Kobraee (2011)
60% and 30% FC, 21 days	Decreased DW and water content of both shoot and root	Azooz and Youssef (2010)
	Reduced chl*a*, chl*b*, and carotenoid contents	
30% FC, 14 days	35% decrease of root DW	El Tayeb and Ahmed (2010)
	29% decrease of shoot DW in tolerant cultivar and 40% of sensitive one	
	Lower contents of K^+, Ca^{2+}, and Mg^{2+} in both shoot and root	
−0.4 MPa PEG, 3 days	37% decrease of shoot FW	Alavi et al. (2014)
	25% decrease of total chl content	
60% FC	Reduced the water relations, i.e., Ψ_w, Ψ_p, Ψ_s, and RWC	Bukhari et al. (2015)
	Reduced P, K, Mn, Zn, and Si uptake	
Withheld water after stem elongation stage	13% decrease of RWC	Hafez and Gharib (2016)
	Reduced leaf area and chl content	
	Lower N uptake in both grain and straw	
Water scarcity, 7 days	Decrease the rate of plant survival	Kasim et al. (2012)
	Reduced DW and water content	
Water scarcity, 6–7 days	Decreased chl and total carotenoid contents, P_n, T_r, g_s, and C_i of flag leaves	Wang et al. (2010b)
	Decreased leaf RWC, Ψ_w, Ψ_s and OA	

Table 10.2 Effects of drought on yield and yield contributing attributes of *T. aestivum*

Stress level and duration	Effects	References
50% FC	Number and length of spike decreased by 19.8% and 14.5%, respectively	Abdel-Motagally and El-Zohri (2016)
	1000-seed weight decreased by 18.2%	
	Grain yield decreased by 28.3%	
Withholding water at tillering and anthesis stage	Severe reduction in yield (21.5%)	Akram (2011)
40% water reduction	Yield reduced by 20.6%	Daryanto et al. (2016)
Withholding irrigation, 10 days	Spike number plant^{-1}, grain number spike^{-1}, and grain yield decreased by 5.88%, 4.44%, and 6.31%, respectively	Dhayal et al. (2012)
Withholding irrigation after the heading stage	Grain yield decreased by 25%	Taheri et al. (2011)
Withholding irrigation after anthesis	1000-kernel weight, seed, and straw yield decreased by 10.6%, 26.3%, and 28.5%, respectively	Johari-Pireivatlou (2010)
Water withdrawn at initiation of stem elongation stage	Grain yield decreased by 46.1%	Shamsi et al. (2010)
Withholding irrigation at anthesis, 5–7 days	Grain and biological yield decreased by 51.5% and 41.4%, respectively	Gupta et al. (2001)
35% FC	Grain yield reduced by 13–32% in different cultivars	Khakwani et al. (2011b)
Natural drought conditions	Grain yield decreased by 50–66% in 14 tested varieties	Kiliç and Yağbasanlar (2010)
Withholding irrigation at anthesis	Spike length decreased by 8–16% and grain yield decreased by 19–42% in three tested varieties	Maqbool et al. (2015)
No irrigation at pre-flowering or post-tillering stage	Reduced length of spike, spikelet spike^{-1}, grains spike^{-1}, and 1000-grain weight of all tested varieties	Mirbahar et al. (2009)
Irrigation skipped at tillering	Grain yield decreased by 28%	Naveed et al. (2014)
Withholding irrigation about 80% after anthesis	Kernel number and kernel weight ear^{-1} decreased by 3% and 28%, respectively	Plaut et al. (2004)
Created drought onset of stem elongation stage	Grain and biological yield reduced by 47% and 23%, respectively	Shamsi and Kobraee (2011)
Withholding water at anthesis, 7–10 days	Grain yield decreased	Khan et al. (2012)
60% FC	Grain weight, yield, and biological yield decreased by 39%, 40%, and 38%, respectively	Nawaz et al. (2015)
Withholding water, 22 days	Spikelet fertility rate and grain number decreased by 22% and 38%, respectively	Prasad et al. (2011)

10.2.1 Seed Germination

Seed germination is the first and most important stage of a plant life cycle. It determines the subsequent growth features of a seed to survive as a healthy seedling. If this vital stage is interrupted, it can result in mass damage to the plant growth, development, and yield. There are a number of factors that can influence the germination of a seed. Like other cereals, germination of wheat seeds also depends on the availability of moisture, air, light, temperature, and others. Under water-scarce condition, seed faces unavailability of moisture that hampers the seed imbibition which is the main prerequisite for successful germination. A number of studies revealed the negative impacts of drought stress on wheat seed germination (Table 10.1). Some important parameters that signify germination characteristics are mean germination time (MGT), time to 50% emergence (E_{50}), coefficient of uniformity of emergence (CUE), mean daily germination (MDG), germination index (GI), coefficient velocity germination (CVG), and final germination percentage (FGP) which has been studied by many researchers (Almaghrabi 2012; Farooq et al. 2013; Timmusk et al. 2014). In an experiment, two wheat varieties (Lasani-2008 and Mairaj-2008) grown at 35% water holding capacity of soil showed delayed mean emergence time (MET) and E_{50} (Farooq et al. 2013). Lasani-2008 recorded to delay MET by 14% and E_{50} by 29%, whereas Mairaj-2008 delayed MET by 10% and E_{50} by 19%. Both varieties reduced CUE by 35% and 33% in Lasani-2008 and Mairaj-2008, respectively. They indicated the possible causes as interrupted completion of pre-germination metabolic activities and lack of active germination metabolites due to drought stress. Timmusk et al. (2014) recorded that at drought condition germination rate of Stava cultivar was only 50%, whereas in control (irrigated) plants, it was 72%. These studies reveal that drought stress interrupts the wheat seed germination process significantly.

10.2.2 Plant Growth

Drought stress has always been considered to exert harmful effects on the growth of plants except xerophytes. Wheat, being sensitive to water availability, shows drastic changes in growth when exposed to drought. Moreover, the duration, type, and magnitude of drought and the stage of plant growth also regulate the possible changes. A large body of literature is available on the growth stage and tolerance level of wheat cultivars under drought stress (Table 10.1). Plant growth is also varied with duration and type of drought. Shamsi and Kobraee (2011) conducted a two-factor experiment with three wheat cultivars and three different stages of wheat growth. Drought stress was imposed at stem elongation, booting, and grain filling stages and continued up to harvest. Results showed that plants facing water stress from stem elongation stage suffered more compared to other two stages of plant growth. Plant height was reduced by 35% and 23% in plants facing drought from stem elongation stage and booting stage, respectively, but only by 7% in plants exposed to drought at grain filling stage. Almost similar findings were reported by Akram (2011), who initiated drought at two stages (stem elongation and anthesis) individually or at both. Two wheat cultivars (Inqlab-91 and Uqab-2000) were used, and both of them showed lower leaf area index (LAI) and crop growth rate (CGR) when faced drought starting from stem elongation stage alone or including anthesis stage compared to the plants exposed to drought at anthesis stage alone. On the other hand, Khakwani et al. (2011a) conducted an experiment with six wheat varieties (Damani, Hashim-8, Gomal-8, DN-73, Zam-04, and Dera-98), and these were subjected to two levels of field capacity (FC) conditions (25% and 35%) for the whole growing period. On an average, leaf area of six varieties was decreased by 30% and 54% at 35% and 25% FC conditions, respectively. Similar reductions were recorded in case of plant height also. Shahbaz et al. (2011) also showed a marked reduction of shoot length, shoot fresh weight (FW), shoot dry weight (DW), root FW, and root DW when wheat varieties were grown under drought condition for 7 weeks. Reduction in root and shoot growth by long-duration drought exposure to different wheat varieties has been reported by many other researchers (Azooz and Youssef 2010; El Tayeb and Ahmed 2010; Farooq et al. 2013; Yasmeen et al. 2013; Hassan et al. 2015; Abdel-Motagally and El-Zohri 2016). Kang et al. (2012) recorded seedlings FW and DW were decreased after exposure of wheat seedlings to 15% PEG for 3 days (d). Similar reductions in plant height, FW, DW, relative dry weight increase rate (RDIR), and root length were recorded in wheat plant exposed to PEG for 2–8 days (Almaghrabi 2012; Grzesiak et al. 2013; Wei et al. 2013; Alavi et al. 2014).

10.2.3 Plant Water Relations

Water is the vital component required for the survival of all living beings including plants. Plant water relation is important in many ways, such as the difference in water relation characteristics indicates the difference of tolerance and adaptability to drought stress between species and cultivars (Ashraf et al. 1994). According to FAO (2015), the wheat plant requires 450–650 mm water for better growth and yield depending on climate and length of growing period. Under drought condition, it becomes difficult for plants to meet this large requirement of water. As a result, damage occurs in plant at cellular levels, which leads to reduced yield or even death of plants. Drought-induced alterations in water relations of wheat crop have been documented in several studies (Table 10.1). Akram (2011) measured the following parameters, relative water contents (RWC), water potential (Ψ_w), osmotic potential (Ψ_s), and turgor potential (Ψ_p), in two wheat cultivars. Drought-stressed plants were recorded to have reduced RWC and other parameters, but Inqlab-91 maintained higher RWC and Ψ_w which signifies it was more tolerant to drought. Similarly, after short period (6–7 days) exposure of wheat seedlings to drought, leaf RWC, Ψ_w, and Ψ_s were recorded to decrease (Wang et al. 2010b). In addition, they measured osmotic adjustment (OA) which is the difference of water potential at full turgor between stressed seedling and unstressed seedlings. The transgenic wheat line (99T6) showed higher OA than its wild type (Shi4185; Wang et al. 2010b). Recently, Bukhari et al. (2015) studied the same parameters in two different wheat cultivars, and they also demonstrated that the tolerant cultivar (Chakwal-50) shows higher RWC, Ψ_w, Ψ_s, and Ψ_p compared to the sensitive one (Sehar-06). Similar results were recorded in several other experiments with different wheat cultivars and duration of drought (Azooz and Youssef 2010; Khakwani et al. 2011a; Prasad et al. 2011; Kang et al. 2012; Kasim et al. 2012; Grzesiak et al. 2013; Wei et al. 2013; Nawaz et al. 2015; Hafez and Gharib 2016; Yavas and Unay 2016). However, in case of water use efficiency (WUE), short period (10 days) exposure to drought caused a reduction (Timmusk et al. 2014), while longer period (49 days) resulted in higher WUE (Shahbaz et al. 2011). This is because WUE is the difference between total plant dry mass and total utilized water.

10.2.4 Nutrient Uptake

Plants require sufficient amount of water in their rhizosphere for the uptake of essential nutrient ions from soil. Therefore, water shortage remarkably affects nutrient uptake by plant roots (Table 10.1; El Tayeb and Ahmed 2010; Shahbaz et al. 2011; Bukhari et al. 2015; Nawaz et al. 2015). Shahbaz et al. (2011) measured the contents of major primary essential elements (N, P, and K) in shoot and root of wheat seedlings after 7 weeks of drought stress. Their results showed that both P and K contents reduced significantly in shoot and root under water deficit conditions, whereas N content decreased significantly in root only. But in a recent study Hafez and Gharib (2016) reported that N content decreased in both grain and straw due to drought stress. Reduction of P and K uptake under drought stress was also reported by Bukhari et al. (2015). El Tayeb and Ahmed (2010) studied secondary essential elements, Ca^{2+} and Mg^{2+}, along with K^+ and reported the reduction of these elements in both shoot and root after 14 days of 30% FC condition. Similar results were reported for Mn^+, Zn^{2+}, and Si (Bukhari et al. 2015) and also for Na^+ and Ca^{2+} (Nawaz et al. 2015). However, Nawaz et al. (2015) reported reduced uptake of Fe^{2+}/Fe^{3+} under drought condition.

10.2.5 Photosynthesis

Photosynthesis is a unique light harvesting process in plants, which is directly linked to the growth and survival of plants. Availability of water and CO_2 is the main prerequisite for photosynthesis. Drought stress causes inadequate water supply to plants which restricts production of carbohydrates by photosynthesis. Other factors that also influence the photosynthesis are also affected by drought stress. There are a number of studies which demonstrated the negative effects of water deficit on photosynthesis process of wheat plants (Table 10.1). Wang et al. (2010b) observed reduced levels of some photosynthetic gas exchange parameters including net photosynthesis rate (P_n), transpiration rate (T_r), stomatal conductance (g_s), and intercellular CO_2 concentration (C_i) in the flag leaves of wheat seedlings subjected to drought stress for 6–7 days. Similar reductions were reported in case of P_n (Akhkha et al. 2011; Prasad et al. 2011), g_s (Timmusk et al. 2014), or both P_n and g_s (Shahbaz et al. 2011; Wei et al. 2013) by other researchers. Recently Nawaz et al. (2015) recorded 43%, 30%, and 27% decrease in P_n, T_r, and g_s, respectively, in wheat seedlings grown under 60% FC for the whole growing period. When the wheat crop was grown at 50% and 30% FC conditions, increase of dark respiration was recorded (Akhkha et al. 2011). Net assimilation rate (NAR) reduced significantly in both sensitive and tolerant cultivars when they faced water scarcity for 10 days, from 10 days after sowing (DAS) to 20 DAS (Timmusk et al. 2014). Photosynthetic pigments like chlorophyll (chl)*a*, chl*b*, and carotenoids are also essential for the occurrence of photosynthesis, and any reduction in chl contents may upset the photosynthetic process. In a recent experiment, Abdel-Motagally and El-Zohri (2016) reported 26% and 18% reduction of chl*a* and chl*b*, respectively, when wheat plants experienced drought stress for the whole growing period. On the other hand, under artificial drought (by polyethylene glycol, PEG, exposure) for only 3 days, wheat seedlings also resulted in 25% reduction of chl contents (Alavi et al. 2014). There are many other studies which reported the reduction of photosynthetic pigments under water deficit condition (Azooz and Youssef 2010; Akhkha et al. 2011; Prasad et al. 2011; Wei et al. 2013; Yasmeen et al. 2013; Hafez and Gharib 2016; Yavas and Unay 2016).

10.2.6 Reproductive Growth

Drought stress in wheat shortens the life cycle and duration of grain filling. Reduced photosynthesis, accelerated leaf senescence, and reduced sink activity decrease the grain filling period of wheat under limited water conditions which decreased grain size (Madani et al. 2010; Wei et al. 2010). The young microspore stage of pollen development is one of the most sensitive stages often affected by drought stress and causes pollen sterility, thereby reducing grain number (Ji et al. 2010). Meiosis and anthesis failure due to drought affects grain number and reduces grain yield (Cattivelli et al. 2008). Drought stress at meiosis of pollen mother cells interrupts microsporogenesis that caused pollen sterility and a 40–50% reduction in grain set (Manjarrez-Sandoval et al. 1989). At booting to maturity stages of growth, drought caused 27–37% decrease in yield of wheat (Shamsi et al. 2010; Shamsi and Kobraee 2011). At heading stage, drought stress caused 57% yield reduction in wheat (Balla et al. 2011). Raza et al. (2012) reported that drought stress imposed at the tillering, flower initiation, and milking stage reduced N, K, and Na uptake and photosynthesis and hampered grain development. Drought stress decreases invertase activity (Dorion et al. 1996) and starch accumulation (Lalonde et al. 1997) which cause pollen sterility (Dorion et al. 1996). Abnormal degeneration of the tapetum in anthers and disorganization of reproductive cells under drought stress are responsible for early abortion of microspores (Lalonde et al. 1997). Drought-induced abscisic acid (ABA) biosynthesis in anthers causes pollen sterility (Ji et al. 2011). Drought after anthesis decreased grain filling duration which decreased grain dry weight (Plaut et al. 2004). Grain filling is a process of starch biosynthesis and accumulation where adenosine diphosphate glucose pyrophosphorylase, sucrose synthase, starch branching enzyme, and starch synthase play a vital role (Hurkman et al. 2003). Reduced soluble starch synthase activity caused drought-induced reductions in grain growth rate (Ahmadi and Baker 2001). Prolonged mild stress decreased yield by 58–92% during heading and grain filling period (Dhanda and Sethi 2002). Pre-anthesis drought stress decreased yield by 18–53% (Majid et al. 2007). During anthesis, Akram (2011) reported an 8% reduction of grain yield, Sangtarash (2010) reported 19% reduction, Jatoi et al. (2011) reported 11–39%, and Gupta et al. (2001) reported a reduction of 43–51% in grain yield. Post-anthesis mild stress decreased grain yield by 1–30% (Eskandari and Kazemi 2010). But Majid et al. (2007) reported that post-anthesis mild stress decreased yield by 13–38%. Grain filling drought spell may lose crop yield by 9–78% depending upon severity and duration of stress (Gúoth et al. 2009).

10.2.7 Yield

Scarcity of water or drought stress is one of the vital environmental constraints affecting crop yield worldwide. Yield loss through drought-induced stress possibly goes beyond losses from other environmental constraints, because both duration and intensity of drought stress are very critical for crop production. Water scarcity during vegetative and reproductive stages of wheat plant caused severe reduction in grain yield (Johari-Pireivatlou 2010; Taheri et al. 2011; Khan et al. 2012; Naveed et al. 2014; Maqbool et al. 2015). Decreasing photosynthesis rate as well as lower growth of plant induced by drought stress is one of the major reasons for yield loss of plant (Yordanov et al. 2000). Water insufficiency at vegetative phase is very dangerous because assimilates of vegetative portion of plants contribute in yield. On the other hand, drought stress at reproductive phase directly hampers the yield of plant by hindering the tiller surviving rate, number of spike, length of spike, number of fertile spikelet, grain ear^{-1}, grain size, and weight (Hasanuzzaman et al. 2013). Grain yield of wheat varies among diverse wheat genotypes under different levels of stress; even tolerant or resistant genotypes showed significant yield reduction. Yield reduction due to water shortage or withholding water at different growth stages of wheat was recorded by different plant biologists in diverse growing conditions. In some cases, more than 50% grain yield reduction was recorded in a number of wheat genotypes. Gupta et al. (2001) carried out an experiment by withholding water supply to wheat cultivar at anthesis stage and discovered that the grain and biological yield of wheat decreased by 51.5% and 41.4%, respectively. Creation of water deficit up to 80% after anthesis reduced kernel number and kernel weight ear^{-1} by 3% and 28%, respectively (Plaut et al. 2004), which is the ultimate reason of yield reduction. Mirbahar et al. (2009) recorded that no irrigation at pre-flowering or post-tillering stage reduced length of spike, number of spikelets, grains spike^{-1}, and 1000-grain weight of all tested wheat varieties. Kiliç and Yağbasanlar (2010) conducted an experiment in drought condition with 14 wheat genotypes and observed that the grain yield of all genotypes decreased by 50–66% than the well-watered plant. Withholding water after anthesis reduced 1000-seed weight, grain, and straw yield by 10.6%, 26.3%, and 28.5%, respectively (Johari-Pireivatlou 2010). Water withdrawn at initiation of stem elongation stage also reduced the yield of wheat cultivar. A 46% reduction of grain yield was recorded by Shamsi et al. (2010) at this condition. Spike length, spike number plant^{-1}, grain number spike^{-1}, number of fertile grain, etc. are the major attributes which determined the yield of wheat. Hampering yield contributing factors due to withholding water at tillering and anthesis stage finally reduced the grain yield by 21.5% in wheat plant (Akram 2011). Prasad et al. (2011) withdrew water for 22 days from wheat plant and found spikelet fertility rate decreased by 22% as well as grain number decreased by 38%. Creation of drought onset of stem elongation stage reduced grain and biological yield by 47% and 23%, respectively (Shamsi and Kobraee 2011). Irrigation skipped at tillering, heading, or anthesis stage considerably decreased the yield of grain in different wheat cultivars by hampering the yield contributing factors (Taheri et al. 2011; Khan et al. 2012; Naveed et al. 2014). Maqbool et al. (2015) also reported that spike length decreased by 8–16% and grain yield decreased by 19–42% in three tested varieties under water scarcity at anthesis stage. Dhayal et al. (2012) recorded that withholding irrigation only for 10 days reduced the spike number plant^{-1}, grain number spike^{-1}, and grain yield by 5.88%, 4.44%, and 6.31%, respectively. Nawaz et al. (2015) observed that grain weight, yield, and biological yield decreased by 39%, 40%, and 38%, respectively, due to scarcity of water (60% FC) in *T. aestivum* L. cv Pasban-90. Similar findings were found by Khakwani et al. (2011b) in other genotypes of wheat. They stated that grain yield reduced by 13–32% in different cultivars under 35% FC. Recently, Abdel-Motagally and El-Zohri (2016) and Daryanto et al. (2016) carried out two different experiments under different water deficit conditions and recorded yield reduction. Abdel-Motagally and El-Zohri (2016) raised wheat genotypes in soil which had 50% FC and recorded that grain yield decreased by 28.3% along with reduction of number of spike (19.8%), length of spike (14.5%), and 1000-seed weight (18.2%). Daryanto et al. (2016) also observed that grain yield reduced by about 21% due to 40% water reduction. So, it can be concluded that yield reduction of wheat depends on genotype, stress intensity, stress duration, and age of plants.

10.3 Drought-Induced Oxidative Stress in Wheat

Drought is one of the most vital expressions of environmental stress which obstruct almost all feature of plants' biochemistry and physiology. The root system of the plant is the first organ which is affected by water shortage under drought stress conditions. The plant roots not only supply water and mineral elements to the leaves through the xylem sap but also send different stress signals (Cruz de Carvalho 2008). Under drought stress condition, plant needs to avoid excess water loss and reduce the g_s. In that condition, ABA carries stress signal from root to leaves. Whenever the signal of drought stress arrives at leaves, it activates closure of stomata (Reddy et al. 2004; Cruz de Carvalho 2008). As a result, plant attains water-saving approach by regulating stomatal opening which reduces the rate of transpiration as well as the entrance of CO_2. Therefore, internal CO_2 concentration decreased, and reduction of CO_2 by the Calvin cycle becomes very slow which

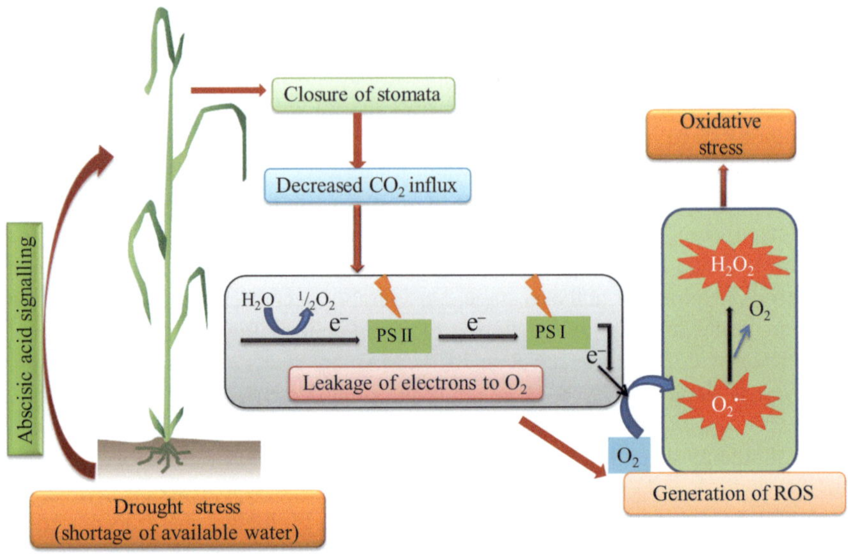

Fig. 10.2 Drought-induced oxidative stress in *T. aestivum*

decreases regeneration of $NADP^+$, thus provoking excess reduction of the photosynthetic electron transport chain (Fig. 10.2; Reddy et al. 2004; Cruz de Carvalho 2008; Hasanuzzaman et al. 2013). That means drought stress reduces the availability of CO_2 and obstruct fixation of carbon. As a result, the chloroplasts in leaf cells get exposed to excessive excited energy from photosystem (PS) I and enhance the production of different toxic ROS (Gill and Tuteja 2010; Hasanuzzaman et al. 2013). This excess generation of ROS throughout drought stress is a consequence of impaired electron transport procedures in the chloroplasts and mitochondria of the plant cell. On the other hand, cut of the activity in PSII results in a disproportion between the generation and utilization of electrons, resulting in an alteration in yield of quantum. These types of modification in the chloroplast photochemistry in the leaves of drought-stressed plants result in the rakishness of excess light energy in the PSII and generate different free radicals or ROS like $O_2^{\bullet-}$, $^1O_2$, H_2O_2, and $OH^\bullet$, which are potentially hazardous and create oxidative stress to plants (Fig. 10.2; Hasanuzzaman et al. 2013; Apel and Hirt 2004).

Reactive oxygen species are very reactive in nature and impede with plants' regular metabolism by causing the peroxidation of lipid, oxidation of protein, and nucleic acid and DNA damage (Fig. 10.3; Apel and Hirt 2004). The $OH^\bullet$ and their dismutation product, H_2O_2, can directly attack membrane lipids which results in higher malondialdehyde (MDA) content (Hossain et al. 2009). Overproduction of H_2O_2 can also hamper the actions of Calvin cycle (Hasanuzzaman et al. 2013). As a result, cell death increases considerably which leads to a reduction in growth and reproductive development (Fig. 10.3). When the presence of ROS causes only a little change in the cell redox potential, the cell's antioxidant system is stimulated

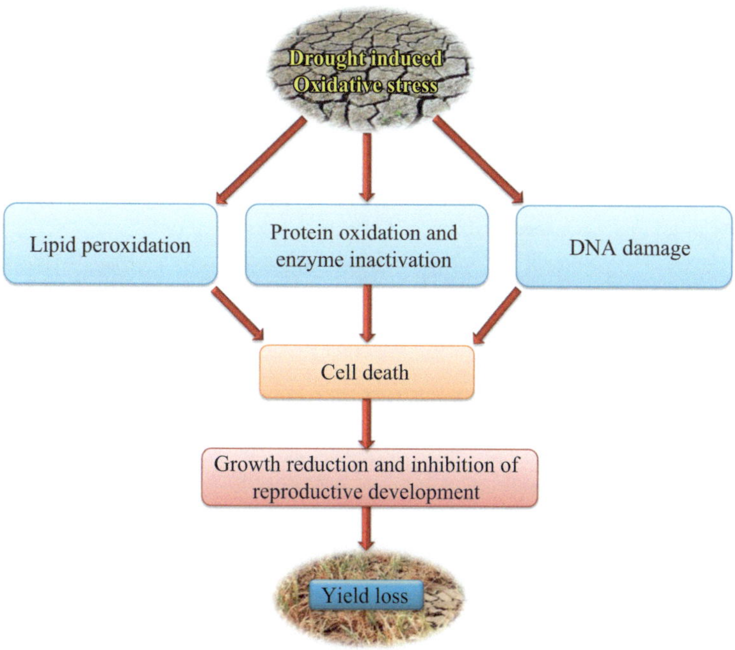

Fig. 10.3 Consequence of oxidative stress in *T. aestivum*

and defends the plant from the injury caused by ROS. Plants are well equipped with nonenzymatic antioxidants which give protection against oxidative stress (Gill and Tuteja 2010; Hasanuzzaman et al. 2012). Besides ROS, methylglyoxal (MG), a cytotoxic compound, can create oxidative stress by overproduction of $O_2^{\bullet-}$ (Yadav et al. 2005a, b). Chloroplast, mitochondrion, and cytosol are the potential sources of MG. Drought stress increases the production of MG in plants and finally increases the ROS (Kaur et al. 2015). Plants have a well-established MG detoxification system to detoxify MG, but severe water shortage condition creates a huge oxidative load for the plant (Yadav et al. 2005a, b). Therefore, the antioxidant defense and glyoxalase systems cannot work properly against toxic ROS and MG. As a result, the plant suffers a lot and can even die.

Several lines of evidence suggest the production of ROS and action of antioxidant under diverse drought stress conditions in wheat plant. Drought stress considerably increases oxidative damage and alters the antioxidant defense system based on drought intensity and plant growth stages (Table 10.3). Under mild stress condition, the oxidative stress marker (MDA, H_2O_2, etc.) increased along with upregulation of some antioxidant component of the plant. But severe water shortage downregulates the antioxidant activity. Short-term water shortage for the plant in controlled environment rapidly alters plant physiology (Tian and Lei 2007; Tan et al. 2008; Khan et al. 2012). Imposition of 15–20% PEG in wheat genotypes only for 24 h considerably increased oxidative stress by increasing the membrane

Table 10.3 Changes in metabolites of *T. aestivum* by drought-induced oxidative stress

Stress level and duration	Changes in metabolites, state of oxidative stress, and antioxidant components	References
50% FC	Root H_2O_2 content decreased by 7%	Abdel-Motagally and El-Zohri (2016)
	Leaf pro content increased by 153%	
Withholding water at the reproductive stage, 30 days	Increased lipid peroxidation and reduced stability of biomembrane	Aldesuquy and Ghanem (2015)
	Increased activities of phenylalanine ammonia lyase (PAL), ascorbic acid oxidase, and POX	
10% PEG	Increased content of H_2O_2	Alexieva et al. (2001)
	Decreased the activities of SOD, CAT, and POX	
Withholding water, 6 days	Increased MDA, H_2O_2, Pro, and phenol contents	Chakraborty and Pradhan (2012)
	Increased activities of POX and GR	
	Decreased activities of SOD and CAT	
Withholding irrigation, 20 days	Contents of MDA, H_2O_2, and Pro increased by 194%, 93%, and 193%, respectively	Ibrahim (2014)
	Membrane stability index and root viability decreased by 40% and 58%, respectively	
	The activities of CAT and SOD decreased by 46% and 42%, respectively, whereas POX activity increased by 178%	
	Decreased AsA content but increased GSH and α-tocopherol contents	
20% PEG, 42 days	Electrolyte leakage and Pro content increased by 291% and 262%, respectively	Karmollachaab and Gharineh (2015)
	Decreased soluble sugar by 22%	
Irrigation skipped at flowering	Lipid peroxidation and Pro content increased by 216% and 22%, respectively	Naveed et al. (2014)
	Total phenol content and total soluble sugar increased by 79% and 18%, respectively	
	Increased CAT, GR, and APX activities by 75%, 145%, and 235%, respectively	
PEG (−1.5 MPa), 7 days	Increased MDA and H_2O_2 content in both leaf and root	Nayyar and Gupta (2006)
	Stress injury increased by 330% in roots and 495% in leaves	
	Decreased activities of AsA, GSH, GR, APX, and DHAR in both leaf and root	
15% PEG, 6 days	The TBARS and H_2O_2 contents increased by 8.12- and 3.77-fold, respectively	Tian and Lei (2006)
	The activities of CAT and SOD decreased, but GPX activity increased by 1.52-fold	
Withholding water at initiation of grain filling stage	Increased Pro and soluble carbohydrates contents	Shamsi et al. (2010)

(continued)

Table 10.3 (continued)

Stress level and duration	Changes in metabolites, state of oxidative stress, and antioxidant components	References
15% PEG, 24 h	Increased MDA content, $O_2^{\bullet-}$ production, and Pro content	Tan et al. (2008)
15% PEG (about −0.5 MPa), 7 days	H_2O_2 content increased by 98%	Tian and Lei (2007)
	TBARS and Pro increased by 2.44 and 1.33 times, respectively	
	Increased activities of SOD, CAT, APX, and GPX	
15% PEG, 6 days	Increased Pro by 1.51 times	Lei et al. (2007)
	Increased activity of P5C synthetase	
20% PEG, 24 h	Membrane stability index, total soluble sugar, and soluble protein increased	Khan et al. (2012)
60% FC	Total soluble sugar and total free amino acid increased by 82% and 83%, respectively	Nawaz et al. (2015)
	Activities of CAT, APX, and POX increased by 16%, 15%, and 15%, respectively	
20% PEG, 7 days	Increased electrolyte leakage and MDA content	Wei et al. (2013)
	Increased activities of SOD, CAT, and POX	
30% FC, 21 days	Increased plant soluble carbohydrate and protein	Azooz and Youssef (2010)
	Increased Pro content	
Maintaining 35% of water holding capacity	Membrane stability index decreased by 23% and MDA content increased by 37%	Farooq et al. (2013)
	Soluble phenolics and leaf free Pro content increased by 30% and 57%, respectively	
	Enhanced content of AsA by 61%	
Maximum 30% FC	Increased Pro content	El Tayeb and Ahmed (2010)
	Soluble protein content of root and shoot increased by 97% and 29%, respectively	
	Soluble sugar content of root and shoot increased by 269% and 51%, respectively	

stability index, MDA content, and $O_2^{\bullet-}$ production (Tan et al. 2008; Khan et al. 2012). Stress induced by 10% PEG in *T. aestivum* L. (cv. Centauro) plant showed increased H_2O_2 content along with decreased activities of SOD, CAT, and POX (Alexieva et al. 2001). Nayyar and Gupta (2006) observed that under PEG (−1.5 MPa) for 7 days, wheat plant (cv. C306) experienced significantly more injury to its roots (49.4%) and leaves (79.1%) than maize (roots, 38.4; leaves, 59.7%). Besides these, drought stress decreased the activities of AsA, GSH, GR, APX, and DHAR in both leaf and root. On the other hand, 15% PEG for 6 days increased thiobarbituric acid-reacting substance (TBARS) and H_2O_2 contents by 8.12- and 3.77-fold in *T. aestivum* L. (cv. W7) (Tian and Lei 2006). They also found that CAT and SOD activities decreased, but GPX activity increased by 1.52-fold than the control. Besides short-term extreme drought stress, long-term low-range water shortage showed similar kind of oxidative damage for plants. Ibrahim (2014) withheld irrigation from late tillering to the early flowering stage of *T. aestivum* L. (Giza 168). He observed that contents of MDA, H_2O_2, and Pro increased by 194%, 93%, and 193%, respectively; membrane stability index and root viability decreased by 40% and 58%, respectively; activities of CAT and SOD decreased by 46% and 42% respectively, whereas (guaiacol peroxidase) POD activity increased by 178%, decreased AsA content, but increased GSH and α-tocopherol contents. Similarly Naveed et al. (2014) reported that lipid peroxidation and Pro content increased by 216% and 22%, respectively, under water scarcity stress (irrigation skipped at flowering). They also stated CAT, GR, and APX activities increased by 75%, 145%, and 235%, respectively. Changes in proline (Pro), phenol, protein, free amino acid, and soluble sugar contents are another kind of drought stress indicators for the wheat plant (Shamsi 2010; Chakraborty and Pradhan 2012; Naveed et al. 2014; Nawaz et al. 2016). Lei et al. (2007) investigated that Pro content increased by 1.51 times with the increase of P5C synthetase activity under 15% PEG for 6 days in wheat genotypes. Similarly, El Tayeb and Ahmed (2010) recorded increased Pro, protein, and soluble sugar contents in wheat cultivar at maximum 30% FC compared with well-watered plant. In addition, the soluble protein content of root and shoot increased by 97% and 29%, respectively, and soluble sugar content of root and shoot increased by 269% and 51%, respectively, by the effects of drought stress. But, Karmollachaab and Gharineh (2015) showed that long-time drought stress with PEG (20%, 42 days) decreased the leaf soluble sugar; hence electrolyte leakage and Pro content increased by 291% and 262%, respectively. In 2013, Farooq et al. reported that maintenance of 35% water holding capacity in soil decreased the membrane stability index by 23% and increased MDA content by 37%, compared to control plant. They also observed soluble phenolics and leaf free Pro content increased by 30% and 57%, respectively. So, oxidative damage in the wheat plant caused by drought stress mainly depends on the level of water scarcity, plant age, and duration of stress.

10.4 Approaches of Enhancing Antioxidant Defense in Wheat Under Drought Stress

Various approaches have been employed by the plant biologists to mitigate the deleterious effects of ROS through enhancing the antioxidant defenses. These include the genetic manipulation and/or use of exogenous protectants such as plant nutrients, phytohormones, antioxidants, osmolytes, signaling molecules, etc. (Tables 10.4 and 10.5).

10.4.1 Exogenous Osmolytes Enhance Drought Tolerance

Exogenous application of osmolytes protects wheat plants from drought-induced oxidative stress (Table 10.4). For example, exogenous glycine betaine (GB, 100 mM) treatment increased net photosynthetic rate, preserved the photochemical

Table 10.4 Protective effects of exogenous plant nutrients, antioxidants, and signaling molecules in mitigating oxidative stress in *T. aestivum*

Stress level	Protectants	Antioxidant defense	References
Withholding water	100 mM GB, co-treatment	Preserved the photochemical activity of PSII, recovered photoinhibition, maintained higher antioxidant enzyme (SOD and APX) activities, and reduced oxidative stress	Ma et al. (2006)
60% FC	0, 50, and 100 mM of GB, foliar spray	Improved photosynthesis g_s	Shahbaz et al. (2011)
Water deficit condition	50 mM Tre, co-treatment	Decreased H_2O_2, formation rate of $O_2^{\bullet-}$, and MDA content	Ma et al. (2013)
		Increased activities of SOD and GR and contents of AsA and GSH	
15% PEG	0.5 mM SA, co-treatment	Increased the GPX and APX activities	Horváth et al. (2007)
20% PEG	0.1 mM SA, seed soaking	Increased protein content and membrane stability index	Khan et al. (2012)
15% PEG	0.5 mM SA, co-treatment	Glutathione-*S*-transferases, APX, and 2-cysteine peroxiredoxin were enhanced under drought stress. Enhancement of antioxidant defense system worked against the oxidative damage	Kang et al. (2012)
15% PEG	0.5 mM SA, co-treatment	Enhanced the transcription of GSH synthetase (*GSHS*), *GST1, GST2, GR, MDHAR, DHAR,* and *GPX*	Kang et al. (2013)
		Decreased MDA content and increased AsA and GSH levels	
−0.8 MPa PEG	1.0 and 2.0 mM SA	Increased the activities of SOD, APX, GR, CAT	Agarwal et al. (2005a)
		Decreased H_2O_2 and TBARS	
		Increased membrane stability index	
−0.8 MPa PEG	0.5 and 1.0 mM ABA	Increased the activities of SOD, APX, GR, CAT	Agarwal et al. (2005b)
		Decreased H_2O_2 and TBARS	
		Increased membrane stability index	
		Prevented breakdown of chl and carotenoid	

activity of PSII, and increased the maximum photochemistry efficiency of PSII (Fv/Fm) and recovered drought-affected plants more rapidly from photoinhibition. Application of GB increases the activities of SOD and APX. The higher antioxidant enzyme activities decrease the oxidative stress. GB might have protective effect on PSII complex from damage through accelerating D1 protein turnover. In addition, GB prevented and maintained higher antioxidative enzyme activities under drought stress condition (Ma et al. 2006). Wheat plants transformed with *P5CS* (D1-pyrroline-5-carboxylate synthetase) encoding enzyme for Pro biosynthesis were exposed to drought stress (Vendruscolo et al. 2007). Transgenic wheat plants accumulated higher amount of Pro and showed tolerance to oxidative stress, compared to

Table 10.5 Protective effects of exogenous plant nutrients, antioxidants, and signaling molecules in mitigating oxidative stress in *T. aestivum*

Stress level	Protectants	Antioxidant defense	References
−0.7 MPa PEG solution	Spraying with 200 mM H_2O_2, 20 min	Increased the capacity of AOX pathway and *AOX1* expression	Feng et al. (2008)
		Kept the endogenous H_2O_2 level lower	
−0.7 MPa PEG solution	Soaking with 20–140 mM H_2O_2, 6 h	Higher expression of CAT and APX	He et al. (2009)
		Decreased MDA content	
15% PEG solution	Pretreatment with 1 mM NaHS, 12 h	Significant increase in AsA and GSH contents	Shan et al. (2011)
		32%, 27%, and 33% increase in APX, DHAR, and GR activities	
−8 bar PEG solution	Soaking with 0.05, 0.1 mM H_2O_2, 4 h	Higher activities of SOD, CAT, and APX	Agarwal et al. (2005a)
		Detoxification of H_2O_2	
PEG-induced osmotic stress	SNP co-treatment	Slight decrease of lipid peroxidation	Demiral et al. (2014)
		Suppression of SOD, CAT, and GR activities	
		Reduced H_2O_2 level	
25% PEG	0.5 mM SNP co-treatment	Increased activities of CAT, APX, POX	Zhang et al. (2003)
		Decreased LOX activity	
		Increased Pro content	
15% PEG	0.2 mM SNP	Increased activities of SOD, CAT, and GPX	Tian and Lei (2006)
		Decreased MDA and H_2O_2 content	
−0.4 MPa PEG	Pretreatment with 100 μM SNP	Reduced MDA content	Alavi et al. (2014)
		Decreased LOX activity	
		Increased activities of SOD, GR, and GPX	
Water scarcity	Foliar spray with 200 mg L^{-1}AsA	Increased activities of CAT and POX	Hafez and Gharib (2016)
35% FC	Osmopriming with 2 mM AsA	18–22% decrease of MDA content	Farooq et al. (2013)
		4–8% increase of AsA content	
		26–42% increase of Pro content	
Irrigation only at tillering and heading	2% KCl, 500 L ha^{-1}	Increased activities of SOD, CAT, and POX	Nawaz et al. (2016)
50% FC	2% sulfate of potash (SOP) at 25 ml $plant^{-1}$	Increased activities of SOD, POX, and CAT	Yasmeen et al. (2013)
		Increased AsA and phenolic content	

(continued)

Table 10.5 (continued)

Stress level	Protectants	Antioxidant defense	References
20% PEG	2.5–10 mM K_2CO_3	Increased SOD, POX, CAT, and APX activities and their gene expression	Wei et al. (2013)
		Decreased MDA content	
60% FC	Foliar spray of NPK (1.5%:2%:3%)	35% increase of TSS	Shabbir et al. (2016)
		66% increase of Pro content	
		19% and 8% increases of CAT and POX activities	
Deficit irrigation	0.3%	Increases of SOD, POX, and CAT activities	Yavas and Unay (2016)
	Zn as $ZnSO_4$		
55% FC	204 mg Na_2SiO_3 kg^{-1} soil	1.5–3-fold increases in expression levels of *TaSOD*, *TaCAT*, *TaGR*, *TaDHAR*, *TaMDHAR*, and *TaGS* genes	Ma et al. (2016)
		Decreased MDA and H_2O_2 content	
−1.0 MPa	2 mM Na_2SiO_3	11%, 10%, 13%, and 13% increase of CAT, POX, SOD, and APX activities	Ahmad and Haddad (2011)
60% FC	4 mMNa_2SiO_3	Increased activities of APX, POX, and CAT	Bukhari et al. (2015)
50% FC	2.11 mM Na_2SiO_3	Increased activities of SOD, CAT, and GR	Gong et al. (2005)
		Decreased H_2O_2 content	
		Increased total thiol content	
50% FC	Foliar spray of 50 ppm B	Decreased H_2O_2 and Pro contents	Abdel-Motagally and El-Zohri (2016)
		Increased carotenoid content	

non-transgenic one. Due to drought stress, the photosynthetic rate decreased by 59% in the P5CS-transformed line but decreased by 97% in control plants, compared to transgenic plants. Membrane stability index decreased approximately by 73% in control plants, while only 36% reduction was observed in transgenic plants under drought stress. In *P5CS*-transformed plants, 50% of cellular membranes remained intact after 14 days of withholding water, but in control plants this value was 13%. As a result, in control plants MDA level was 65% higher, compared to transgenic plant. Thus, the protective role of Pro has been proved in maintaining antioxidant system and decreasing oxidative damage (Vendruscolo et al. 2007). Exogenous trehalose (Tre) application increased the endogenous Tre content and significantly mitigated biomass reduction under water deficit condition. Dry weight increased by 18% and 22.4% after 5 and 10 days, respectively. Water deficit also created oxidative stress to the callus. Exogenous Tre (50 mM) application enhanced the activities of SOD and GR and increased the contents of AsA and GSH which decreased level of H_2O_2, formation rate of $O_2^{\bullet-}$, and MDA content. The overall effect of Tre was better on callus development process under water deficit condition (Ma et al. 2013). Raza et al. (2012) reported that application of GB effectively alleviated adverse effects of drought stress in wheat plant by increasing the rate of transpiration and photosynthesis and uptake of P and Ca, whereas GB reduced N, K, and Na uptake. Again, Raza et al. (2014) reported that different doses of GB (50, 100, and 150 mM) were beneficial at milking stage which improved plant water potential and increased plant height, spike length, number of spikelet spike^{-1}, number of grains spike^{-1}, and grain yield. Wheat plants overexpressing *P5CS* not only accumulated Pro but also reduced oxidative stress. After 8 days of withholding water, the lipid peroxidation levels increased in both transgenic and control plants with a higher MDA content in non-transgenic line. MDA contents in control plants were 65% higher, compared to transgenic plants. So, Pro helped to reduce oxidative damage in transgenic wheat plants (Vendruscolo et al. 2007). Contents of ethylene, osmoprotectants, levels and forms of polyamines (PAs), and activities of antioxidant enzymes were examined in different wheat cultivars under drought stress (−1.5 MPa). The roots and shoots of cv. Nawra, Parabola, and Manu (drought-tolerant cultivars) increased the osmoprotectors (Pro and soluble carbohydrates, mainly glucose, saccharose, and maltose), free PAs (putrescine, Put; spermidine, Spd; and spermine, Spm), and Spd-conjugated levels. In drought-sensitive Radunia and Raweta, the accumulation of those metabolites was lower. Activities of antioxidant enzymes, CAT and POXs, were significantly higher in tolerant wheat plants (Grzesiak et al. 2013). Seedlings of transgenic wheat line T6 overaccumulating GB and its wild-type Shi4185 were compared by imposing drought (30%, PEG-6000), heat (40 °C), and their combination. Ultrastructural damage to the chloroplast and thylakoid lamellae and withered phenotype were evident under those stressful conditions. Reduction of oxygen evolving complex function caused by stress had been noticed with higher damage in wild type. Overaccumulation of GB in T6 protected lipids of thylakoid membrane and stabilized the index of unsaturated fatty acids during stress condition. Glycine betaine overaccumulation also reduced the photoinhibition of PSII caused by stresses (Wang et al. 2010a). Five wheat cultivars (SARC-I, Inqlab-91, MH-97, Bhakkar, and S-24) were sprayed with three levels (0, 50, and 100 mM) of GB, and those were compared under well-watered and water stress (60% FC) growing conditions. Drought stress decreased shoot and root fresh and dry biomass, shoot length, leaf area, grain yield, photosynthetic and transpiration rates, g_s, and shoot and root P and K^+ contents. But foliar application of GB enhanced plant biomass, shoot length, transpiration rate, and root and shoot P and N contents and enhanced K^+ in both cultivars. The GB of 50 mM and cultivars SARC-I and Inqlab-91 performed better as compared to the others under drought stress condition (Shahbaz et al. 2011).

10.4.2 Enhancement of Drought Tolerance by Phytohormones and Growth Regulators

Apart from the direct role in growth regulation, phytohormones have an important role in stress signaling and mitigation of oxidative stress (Table 10.4). Kang et al. (2013) demonstrated that exogenous salicylic acid (SA) supplementation enhanced the transcription of *GST1*, *GST2*, *GR*, and *MDHAR* genes during almost the entire drought period. Increase of *DHAR* was noticed at 12 h, *GPX1* at 48 h, phospholipid hydroperoxide glutathione peroxidase (*GPX2*) at 12 and 24 h, and glutathione synthetase (*GSHS*) at 12, 24, and 48 h of drought stress. Upregulation of transcript level of AsA-GSH cycle enzymes together with increased AsA and GSH levels decreased MDA contents or oxidative stress and contributed to drought tolerance (Kang et al. 2013). Exogenous SA (0.5 mM) had been conferred tolerance against drought induced by 15% PEG. The drought tolerance of Chinese Spring increased the GPX and APX activities (Horváth et al. 2007). Spray application of SA on leaves of wheat genotypes C 306 and Hira at 25 showed alleviation of damaging effects under drought stress (−0.8 MPa). Increased activities of SOD, APX, GR, and CAT decreased H_2O_2 and TBARS, compared to unsprayed plants indicating the role of SA in enhancing oxidative stress tolerance under drought stress. Increase of chl and carotenoid contents, RWC, membrane stability index, leaf area, and total biomass over control plants also proved the overall protective effects of SA under drought stress (Agarwal et al. 2005a). Two wheat varieties, viz., Wafaq-2001 and Punjab-96, were subjected to drought stress. Drought decreased membrane stability index and reduced yield. Salicylic acid supplementation increased 37% soluble sugars in Wafaq-2001 cultivar which was higher, compared to Punjab-96 cultivar. Salicylic acid also increased protein content and membrane stability index in both cultivars with a higher increase in Wafaq-2001. Wafaq-2001 also showed better performance after SA application in terms of yield (Khan et al. 2012). Pretreatment with SA (0.5 mM) also enhanced drought tolerance. Pretreatment with SA upregulated 37 protein spots under drought stress which has been investigated through proteomics. Glutathione-*S*-transferases, APX, and 2-cysteine peroxiredoxin were enhanced under drought stress. Enhancement of antioxidant defense system worked against the oxidative damage. Signal transduction, photosynthesis, carbohydrate metabolism, protein metabolism, and energy production have been considered as SA-induced growth and drought tolerance in wheat seedlings (Kang et al. 2012). Foliar application of SA regulated growth and physiological processes in plants. Plant height, spike length, the number of grains spike^{-1}, 1000-grain weight, chl content, RWC, and activities of SOD, POX, and CAT under drought stress alleviated the adverse effects (Yavas and Unay 2016). Drought stress (−0.8 MPa, imposed by adding PEG-6000) resulted in a disrupted antioxidant defense system which increased the level of H_2O_2 and TBARS in different wheat genotypes. Drought stress also caused breakdown of photosynthetic pigments and negatively affected the growth of wheat genotypes. Abscisic acid (0.5 and 1.0 mM) spray on drought-affected plants increased the activities of SOD, APX, GR, and CAT compared to untreated control plants. ABA application with drought stress reduced H_2O_2 and TBARS indicating decreased lipid peroxidation, compared to untreated plants. Increased chl and carotenoid contents, RWC, membrane stability index, leaf area, and total biomass over control plants were also recorded from ABA-treated drought-affected plants, compared to non-sprayed showing the crucial role of ABA under drought stress (Agarwal et al. 2005b). Wheat plants (cv. Gabo) were grown in nutrient solution and supplied with gibberellic acid (GAs) which decreased Pro accumulation and alleviated growth inhibition at the apex and recovered inhibition of shoot elongation resulting from drought stress (Singh et al. 1973).

10.4.3 Application of Plant Nutrients Enhances Drought Tolerance

Under drought stress the uptake of plant nutrient becomes limited, and hence supplemental plant nutrients often help the plant to survive better. Various plant nutrients helped in improving physiological functions, antioxidant metabolism, and cell signaling as reported in *T. aestivum* studies mentioned in Table 10.5 (Fig. 10.4; Yasmeen et al. 2013; Sapre and Vakharia 2016; Yavas and Unay 2016).

Wei et al. (2013) studied the effects of K application in two winter wheat cultivars (drought-tolerant SN16 and susceptible JM22) grown under drought (20% PEG, 7 days) and found that antioxidant enzyme activities were upregulated by osmotic stress while external K_2CO_3 reduced those changes. However, the effects were largely depended on the dose of K. Interestingly, only 7.5 mM K supply drastically decreased the activities of SOD and CAT, but significant effects of 2.5–7.5 mM K supply were noted on POX activities in both cultivars. External K supply significantly upregulated the expression levels of SOD-related genes of both SN16 and JM22 in control condition but downregulated them in drought condition. As a consequence, K application under drought stress markedly reduced the MDA content where 7.5 mM was found to be the most effective (Wei et al. 2013). Two *T. aestivum* cultivars (AARI-11 and Millat-11) were grown under different irrigation regimes (canceling watering at different critical stages), and it was observed that activities of SOD, POX, and CAT increased under drought stress (Nawaz et al. 2016). However, application of K (2% KCl at 500 L ha^{-1}) further increased the activities of this enzyme which increased endogenous AsA content. The activities of antioxidant

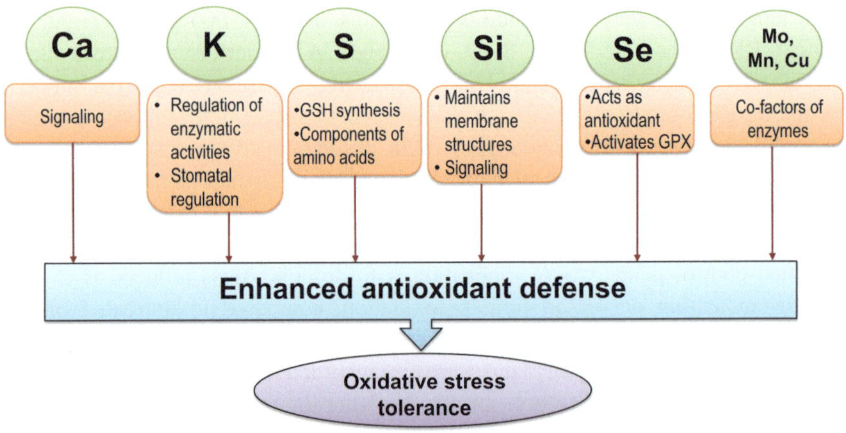

Fig. 10.4 Possible ways of oxidative stress tolerance by supplemental plant nutrients

enzymes (CAT, SOD, POX) and AsA content in *T. aestivum* slightly increased with the decrease of water regimes in soil (Yasmeen et al. 2013). However, application of 2% sulfate of potash (SOP) at 25 ml $plant^{-1}$ markedly increased the activities of these enzymes as well as soluble protein, AsA, and phenolic contents, compared to drought alone (Yasmeen et al. 2013). Combined application of N, P, and K also found to enhance antioxidant defense and drought stress tolerance in wheat (Shabbir et al. 2016). Under 60% FC, the plants showed 14%, 5%, and 55% increases in CAT, POX, and APX activities, respectively, with 24 and 29% increases in total soluble sugar (TSS) and Pro contents. Moreover, foliar application of N, P, and K resulted in further increases in CAT and POX activities by 19% and 8%, respectively, while no changes in APX activity were noticed. Supplemental N, P, and K under drought stress also increased TSS and Pro by 13% and 60%, respectively, compared to drought stress alone. Yavas and Unay (2016) investigated two wheat genotypes (drought susceptible, Basribey 95, and drought resistant, Ziyabey 98) and found that activities of antioxidant enzymes were increased in both cultivars under drought stress conditions (deficit irrigation). In salt susceptible, Basribey 95, the activities of SOD, POX, and CAT were increased by 74%, 2%, and 16%, respectively, while in drought resistant, Ziyabey 98, the changes were minimal. However, Zn supplementation resulted in further increases in the enzyme activities (Yavas and Unay 2016). The protective effect of B under drought stress in wheat was reported by Abdel-Motagally and El-Zohri (2016). Application of B decreased the content of H_2O_2 in *T. aestivum* leaves and roots under limited irrigation condition. However, B application at booting stage was more effective than anthesis stage. Interestingly, Pro content was also lower under drought stress when supplemented with B. However, B could increase the carotenoid content in plants.

Silicon (Si) is now considered as an essential nutrient in some plants such as rice. This element is rapidly absorbed by wheat plants from solution cultures initially containing Si at 0.5 mM, a concentration realistic in terms of the concentrations of the element in soil solutions (Sapre and Vakharia 2016). Thus, exogenous application of Si works well in exerting its beneficial effects. Ma et al. (2016) reported Si-mediated transcriptional regulation of multiple antioxidant defense pathways and subsequent drought stress mitigation in *T. aestivum*. Under moderate drought (55% FC) stress, exogenous Si reduced the H_2O_2 content by 6.9% and 18.5% at 10 and 20 DAS, respectively, compared to untreated control. Exogenous Si also decreased MDA content in the same manner. After 20 days of moderate stress, the relative expression levels of *TaSOD* and *TaCAT* were almost 2.0- and 1.5-fold greater than in the drought alone which were even higher (threefold) under severe drought condition. Silicon application also increased the relative expression levels of the *TaGR*, *TaDHAR*, *TaMDHAR*, and *TaGS* genes. Ahmad and Haddad (2011) found the exogenous Si could enhance the activities of the major antioxidant enzyme (SOD, CAT, APX, and POX) and prevented membrane damage in *T. aestivum* grown under drought. Drought stress significantly increased the activities of CAT, SOD, POX, and APX by 11%, 10%, 13%, and 13%, respectively, compared to drought alone (Ahmad and Haddad 2011). Bukhari et al. (2015) also found enhanced activities of APX, POX, and CAT upon Si supplementation which improved drought tolerance in wheat. However, the action was largely dependent on the application methods, and the foliar spray was more effective than other methods (fertigation and seed soaking) in alleviating the adverse effects of drought. In every case, enzymatic activities were higher under drought stress, but these activities were further increased when they were supplemented with Si. In pot culture, Si was found to enhance the antioxidant defense system in wheat plants exposed to drought stress (Gong et al. 2005). This stress tolerance was confirmed by reduced contents of MDA and H_2O_2 which were associated with higher activities of SOD, CAT, and GR. Exogenous Si also increased total thiol content under drought stress.

10.4.4 Enhanced Tolerance of Wheat Plants by Growth-Promoting Probiotic Bacteria

Plants harbor diverse classes of microorganisms including bacteria in their inside (endophytes) and on the surface (epiphytes) of their body tissues. The plant-associated bacteria which exert beneficial effects to plants are known as plant probiotic bacteria or plant growth-promoting bacteria (PGPB). Several lines of evidence suggest that application of these probiotic bacteria under the genera of *Bacillus, Pseudomonas, Paraburkholderia, Enterobacter, Micrococcus, Lysobacter*, etc. enhances plant tolerance to various kinds of stresses including drought. In a recent report, Gonitia-Mishra et al. (2016a) demonstrated that plant growth-promoting bacteria *Enterobacter ludwigii* IG10 and *Flavobacterium* sp. IG15 enhance drought tolerance in wheat. In a similar study, *Gammaproteobacteria Citrobacter* sp. (IG22) and *Empedobacter cloacae* (IG19) isolated from the wheat rhizosphere enhance tolerance of wheat to drought stress likely through enhanced activity of 1-aminocyclopropane-1-carboxylate (ACC) deaminase and other plant growth-promoting (PGP) traits (Gonitia-Mishra et al. 2016b). Mechanisms of enhancement of plant tolerance to drought induced by the application of plant probiotic bacteria include production of phytohormones, ACC deaminase, trehalose, volatile organic compounds, and exopolysaccharide (Forni et al. 2016). Recently, PGPB *Dietzia natronolimnaea* has been found to modulate abiotic stress-responsive genes in host plant, wheat (Bharti et al. 2016). Further gene expression studies confirmed the involvement of ABA signaling cascades such as TaABARE and TaOPR1 which were upregulated in PGPB-inoculated wheat plants leading to the induction of *TaMYB* and *TaWRKY* gene expression followed by stimulation of higher expressions of a plethora of abiotic stress (such as salinity and drought)-related genes. Enhanced expression of genes of various antioxidant enzymes such as *APX, MnSOD, CAT, POD, GPX*, and *GR* and higher Pro content were found in the PGPB-inoculated wheat plants. These findings suggest that PGP probiotic rhizobacteria could be used as enhancer of plant tolerance to drought stress. Drought-induced oxidative stress in *T. aestivum* was reduced by the application of bacterial biofertilizers (Khalilzadeh et al. 2016). When wheat plants were grown with only 50% water at heading or booting stage, it showed an adverse effect of the antioxidant defense system and resulted in oxidative stress. However, seed inoculation by *Azotobacter chrocoocum* strain 5, *Pseudomonas putida* strain 186, and their combination provided better protections of plants from drought stress. Water stress decreased the carotenoid but increased the activities of CAT, POX, polyphenol oxidase (PPO), and Pro content. However, these activities were increased further in the plants whose seeds were inoculated with biofertilizers. Inoculated plants also showed increased level of Pro under limited irrigation which provided protection against water stress. Further studies on molecular cross talks between these beneficial bacteria and their host wheat would reveal important information useful for an industrial use of these probiotic bacteria as a growth promoting as well as an enhancer of wheat plants tolerant to drought and other stresses.

10.4.5 Role of Antioxidants

Apart from the antioxidant enzymes, plant possesses various nonenzymatic antioxidants in their cellular components to protect themselves from oxidative stress by direct scavenging or detoxifying the ROS. The major antioxidants are AsA, GSH, tocopherol, and some phenolic compounds. Some of these antioxidants showed advanced protection against drought-induced oxidative stress when applied exogenously. However, these are mostly dose-dependent (Table 10.5).

Among the antioxidants AsA plays a vital role in maintaining equilibrium between the production and elimination of ROS, which in turn helps to avoid oxidative burst in cells caused by ROS and its products (Anjum et al. 2014). AsA was also found to be involved with the alteration of expression of hormones, biosynthetic genes, and/or signaling intermediates (Khan et al. 2011). Exogenous application of antioxidants has been reported to markedly improve the inhibitory effects of water stress on plant growth and metabolism (Malik and Ashraf 2012). In *T. aestivum*, exogenous AsA resulted in significant increases in the activities of CAT and POX under water stress (Hafez and Gharib 2016). Compared to water stress alone, AsA-supplemented plants grown under water stress showed 39–45% increase in CAT activity and 42–47% increase in POX activity in two consecutive seasons (Hafez and Gharib 2016). As a seed priming agent, AsA has been studied widely and found its beneficial effects in improving plant physiology including enhancement of antioxidant defense system. In their experiment, Farooq et al. (2013) showed that MDA content of *T. aestivum* seedlings was substantially increased under drought stress (35% FC) but osmopriming with AsA (2 mM) substantially decreased the MDA contents which established the role of AsA as ROS eliminator because this parameter was negatively correlated with other physiological parameters. Osmopriming with AsA also increased soluble phenolics, Pro, and endogenous AsA contents under drought. However, the changes were genotype-dependent. Glutathione is another strong antioxidant that acts as a substrate or cofactor for a number of biochemical reactions; it interacts with hormones and redox molecules and participates in stress-induced signal transduction (Foyer and Noctor 2005a, b; Szalai et al. 2009). Herbinger et al. (2002) showed that the concentration of GSH increased in flag leaf tissues of drought-sensitive wheat (*T. aestivum* Desf. cv. Nandu) and drought-resistant *T. durum* cultivars grown in open-top chambers using a water regime equivalent to 40% soil water capacity. Although the role of cellular GSH in

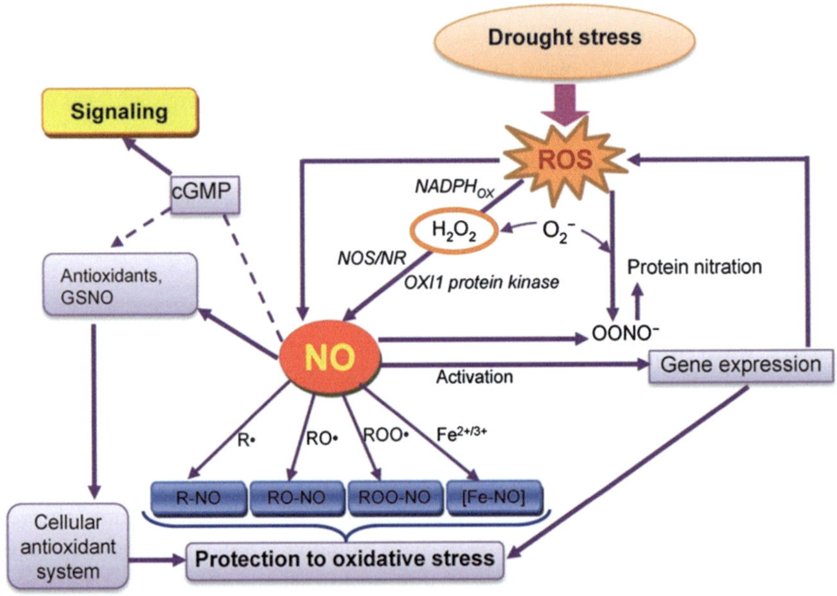

Fig. 10.5 Nitric oxide signaling and its interaction with other signaling molecules in conferring oxidative stress mitigation (Hasanuzzaman et al. 2016, with Permission from Wiley)

mitigating ROS is well known, however, the role of exogenous GSH in mitigating drought stress tolerance in wheat is yet to explore.

10.4.6 Role of Signaling Molecules

The role of signaling molecules such as ABA, SA, Ca, and H_2O_2 and nitric oxide (NO) on antioxidant enzyme induction in wheat both have been well reported in many plant studies (Agarwal et al. 2005a; Hasanuzzaman et al. 2013, 2016).

Nitric oxide is a bioactive signaling molecule that plays a key role as signaling molecule in plants, and several plant studies have shown the protective effect of NO against abiotic stress which is closely related to the reduction of ROS in plants (Corpas et al. 2011; Hasanuzzaman et al. 2013; Khan et al. 2014). This protective effect is mediated by "(i) reaction with lipid radicals, which stops the propagation of lipid oxidation; (ii) scavenging the $O_2^{\bullet -}$ and formation of peroxynitrite ($ONOO^-$) that can be neutralized by other cellular processes (iii) activation of antioxidant enzymes (SOD, CAT, APX, GPX, GR, POX etc.) and (iv) functioning as a signaling molecule in the cascade of events leading to changes in gene expression" (Fig. 10.5; Hasanuzzaman et al. 2016). Exogenous NO also helps the plant to activate efficient mechanisms against the deleterious effects of drought stress through hormonal regulation, water retention, and signal transduction and antioxidant defense (Table 10.5; García-Mata and Lamattina 2002, 2015; Hasanuzzaman et al. 2013, 2016). Besides these, NO also interacts with other signaling molecules and plant hormones. For instance, it activates H_2O_2, SA, ABA, and jasmonic acid (JA) and ethylene signaling pathways (Fig. 10.5). Moreover, all these signals can induce the generation of antioxidant activity that ameliorates oxidative stress (Neill 2007). In addition, NO may also protect plant cells against oxidative processes by stimulating nonenzymatic antioxidant such as GSH (Innocenti et al. 2007).

Exogenous NO was found effective in mitigating short-term (10 days) drought stress (induced by PEG) in *T. aestivum* genotype, Karahan 99 (Demiral et al. 2014). Nitric oxide could revert PEG-induced elevation of lipid peroxidation and endogenous H_2O_2 while suppressing the activities of SOD, CAT, and GR enzymes and H_2O_2 accumulation more under osmotic stressed condition. In another experiment, hydroponically grown *T. aestivum* cv. Kaplica and Goksu showed increased lipid peroxidation and H_2O_2 accumulation and downregulation of the activity of GR and upregulation of CAT activity. However, exogenous NO decreased the H_2O_2 level and lipid peroxidation, while enzymatic activities were not increased further. Importantly, NO resulted in higher accumulation of Pro in both genotypes, even under drought condition. Efficient ROS metabolism and subsequent osmotic stress tolerance in *T. aestivum* were also reported by Zhang et al. (2003). Sodium nitroprusside (SNP)-treated seedlings showed higher activities of CAT and APX than PEG treatment without any apparent influence on POX activity. However, SNP treatment decreased the LOX activity which might cause lower lipid peroxidation under osmotic stress (Zhang et al. 2003). Apart from its direct role, NO also participated in the JA-induced regulation of the AsA-GSH cycle in *T. aestivum* under drought stress as reported by Shan et al. (2015). In plants grown under 15% PEG treatment, visible signs of oxidative stress were observed, but exogenous JA alleviated the stress by upregulating APX, MDHAR, DHAR, and GR with a concomitant increase in the redox state of AsA and GSH which was correlated with the increase in endogenous NO content. However, these effects of JA were inhibited by NO scavenger (cPTIO, 2-4-carboxyphenyl-4,4,5,5-tetramethylimidazoline-1-oxyl-3-oxide) indicating that NO induced by exogenous JA was responsible for enhancing oxidative stress tolerance in wheat under drought condition. Tian and Lei (2006) found marked increase in MDA and H_2O_2 after drought stress (15% PEG) in a time-dependent manner with differential responses in the activities of antioxidant enzymes. In contrary, exogenous NO, at low concentration (0.2 mM SNP), helped the plants in activating antioxidant enzymes and reaction with active oxygen and lipid radicals directly. However, higher concentration of SNP (2 mM) inhibited the activities. Alavi et al. (2014) investigated the interactive effects of NO and SA in mitigating osmotic stress-induced oxidative damages in *T. aestivum*. Osmotic stress slightly upregulated the antioxidant defense system. However, the seedlings pretreated with SA or SNP further increased the activity of antioxidant systems. Relative to controls, SA or SNP increased the activity of SOD (33% and 34%, respectively), GPX (18% and 20%, respectively), and GR (40% and 45%, respectively) which in turn reduced the content of MDA (49% and 51%, respectively), activity of LOX (59% and 58%, respectively), and the level of H_2O_2 (32% and 30%, respectively) (Alavi et al. 2014).

Plant cross-tolerance and signaling paradigm though have been changed in recent decades. Previously the generation of $H_2O_2$was thought to be a harmful event. It is well established that H_2O_2 is one of the stress markers in plants exposed to any abiotic stress including drought (Hasanuzzaman and Fujita 2011). It was previously thought to be toxic substances and a major ROS but recently been considered as signaling molecules too. The dual role of H_2O_2 is not an interesting topic of research of many plant scientists. However, as exogenous application, most of the experiments were conducted using H_2O_2 as priming agents or pretreatments rather than using as co-treatment. Signaling cross-talk of H_2O_2 with NO is also well established since the last two decades (Neill et al. 2002).

Feng et al. (2008) found that H_2O_2 could partially modulate an alternative respiratory pathway under drought which protected wheat leaves from oxidative stress by upregulating antioxidant system. Exogenous H_2O_2 increased the capacity of alternative oxidase (AOX) pathway and *AOX1* expression without significant alteration of endogenous H_2O_2 contents. Exogenous H_2O_2also kept the electrolyte leakage lower. He et al. (2009) observed higher expression of antioxidant enzymes, viz., CAT and APX, under drought stress when pretreated with exogenous H_2O_2 which helped in minimizing the content of MDA. These results indicated that H_2O_2, a stress signal, could trigger the activation of antioxidants in seeds, which persisted in the seedlings to alleviate the oxidative damage (He et al. 2009). Exogenous hydrogen sulfide (H_2S) improved AsA and GSH contents in water-stressed wheat seedlings as reported by Shan et al. (2011). In *T. aestivum*, 1 mM sodium hydrogen sulfate (NaHS) pretreatment significantly increased the contents of AsA, total AsA, GSH, and total GSH under water stress which were mainly due to the enhanced activities of APX, DHAR, and GR. After NaHS pretreatment, the activities of APX, DHAR, and GR increased by 32%, 27%, and 33%, compared to drought alone, while no change in MDHAR activity was observed. In *T. aestivum*, osmotic stress was greatly reduced when the seedlings were pretreated with exogenous ABA (0.5, 1 mM), SA (1, 2 mM), Ca^{2+} (5, 10 mM), or H_2O_2 (0.05, 0.1 mM) for 4 h where higher concentrations were less effective (Agarwal et al. 2005a). The activities of SOD, CAT, and APX were enhanced by the exogenous application of these signaling agents which minimized ROS-induced damages. They also suggested a direct role of H_2O_2 in final signaling event, as Ca^{2+} treatment which affected antioxidant enzyme activity and also had pronounced effect on cellular H_2O_2 content (Agarwal et al. 2005a).

10.4.7 Omics Technology for Dissecting Drought-Induced Plant Responses

Application of various omics techniques such as genomics, transcriptomics, proteomics, and metabolomics has shed light on understanding underlying molecular mechanisms of plant responses to drought stress. It appears that the major responses of plants to drought stress include perception of stress signals and their transduction that activates various stress-related genes and synthesis of proteins with diversified functions (Zhu 2002; Seki et al. 2003; Hu et al. 2006). Drought stress-related well-characterized major proteins are molecular chaperones, osmotic adjustment proteins (Tamura et al. 2003), ion channels (Ward and Schroeder 1994), transporters (Klein et al. 2004), and antioxidant proteins (Bartels 2001). Specific transcription factors regulate the expression of their functional proteins (Singh et al. 2002; Zhu 2002; Hu et al. 2006). Discovered first in tomato (Iusem et al. 1993), ABA stress- and ripening-induced (ASR) genes found in many plants are thought to be linked with downstream components of a common signal transduction pathway involved in plant cell responses to environmental factors including drought (Saumonneau et al. 2012). Three major signaling pathways which are involved in drought response and defense are ABA, ROS, and Ca^{2+}signaling. In signaling pathway, processing of a signal generally involves protein phosphorylation and dephosphorylation by protein kinases and phosphatases when plants are subjected to drought stress. Recently, a dynamic phosphoproteome analysis of seedling leaves in *Brachypodium distachyon* L. reveals central phosphorylated proteins involved in the drought stress response (Yuan et al. 2016). Based on their study, a putative pathway of drought stress response in *B. distachyon* seedlings leaves through phosphorylation modification has been proposed (Yuan et al. 2016). Kim et al. (2012) demonstrated that rice ASR protein, OsASR1, functions as an effective ROS scavenger and its expression in yeast cell enhanced acquired tolerance of ROS-induced oxidative stress through induction of various cell rescue proteins. In another study, OsASR5-enhanced drought tolerance was found to be linked with a stomatal closure pathway associated with ABA and H_2O_2 signaling in rice (Li et al. 2016c). Gene orthologous to *OsASR1*, *TaASR1* conferred drought/osmotic stress tolerance through regulating the expression of stress- and defense-associated genes and enhancing the antioxidant system, thus preventing plants from oxidative damage (Hu et al. 2012; Gonzalez and Insem 2014). Advances in research on ASR genes and proteins are reviewed (Gonzalez and Insem 2014). Further studies are required for identifying the targets of *TaASR1* for better characterizing.

The ROS scavenging-mediated drought tolerance has been a subject of many omics studies. Hu et al. (2013) demonstrated that *TaASR1*, a transcription factor gene in wheat, confers drought stress tolerance in transgenic tobacco. Overexpression of *TaASR1* in transgenic tobacco resulted in lesser MDA, iron leakage (IL), and ROS but higher relative water content and SOD and CAT activities than wild type under drought stress. Proteomic studies shed light on the complexity of drought resistance mechanisms in wheat (Cheng et al. 2016). Proteomic signature of drought adaptability in varying drought-tolerant wheat varieties has been analyzed quantitatively. The iTRAQ proteomic analyses revealed that the Spd-induced physiological effects are associated with drought tolerance in plants (Li et al. 2016b). The expression of *BdWRKY36* gene is upregulated by drought treatment. Overexpression of *BdWRK36* in transgenic tobacco plants resulted in enhanced tolerance to drought stress through control of ROS homeostasis and regulating transcription of stress-related genes (Sun et al. 2015). On the other hand, ectopic overexpression of the aldehyde dehydrogenase, ALDH21, from *Syntrichia caninervis* in tobacco conferred salt and drought stress tolerance (Yang et al. 2015). Moreover, enhanced gene expression rather than natural polymorphism in coding sequence OsbZIP23 was found accountable for improved drought tolerance and yield performance in rice genotypes (Dey et al. 2016). The C_{1-2i} subclass of *TaZFP* (zinc finger protein) is responsive to the various abiotic stresses including oxidative stress generator H_2O_2 and drought tolerance in wheat (Chenk and Houde 2016). *Agrobacterium*-mediated transformation of *AtHDG11* (*Arabidopsis thaliana* homeodomain glabrous 11) genes conferred drought tolerance in wheat through decreasing stomatal density and rate of water loss and increasing proline accumulation and activities of CAT and SOD (Li et al. 2016a).

Plants cope up with drought stress by two major molecular and cellular responses: (i) accumulation of various osmolytes such as Pro, glutamate, GB, and sugars (mannitol, sorbitol, and trehalose), which play a key role in preventing membrane disintegration and activation of drought-induced enzymes (Mahajan and Tuteja 2005; Bhushan et al. 2007), and (ii) induction of a large number of drought-responsive genes and specific protective proteins under drought tolerance (Reddy et al. 2004; Zang and Komatsu 2007). These drought stress-related transcripts and proteins were induced for drought tolerance. A large body of literature revealed that drought stress-responsive genes and specific protective proteins are mostly involved in signal transduction and activation or regulation of transcription, antioxidants, and ROS scavengers (Cui et al. 2009). In wheat, expression of *TaDREE2* and *TaDREB3* genes demonstrated substantial resistance to drought stress (Morran et al. 2011). *TaMYBsdu1* from wheat is unregulated under drought stress and differentially expressed in tolerant and sensitive genotypes (Rahaie et al. 2010). Five families of transcription factors and regions belonging to five multigene families (*AP2/EREBP, bZIP, MYB/MYC, NAC,* and *WRKY*) have been described, and their role in drought tolerance in improvement of wheat is reviewed (Gahlaut et al. 2016).Identifying the physiological and genetic basis of stress tolerance in plants has proven to be critical to understanding adaptation in both agricultural and natural systems (Lovell et al. 2016). One of the major goals of modern plant biology is to better understand the stress tolerance including drought stress responses to crop plants under the changing climate (Ahuja et al. 2010). Although several genes and transcription factors associated with drought tolerance of plants, precise underlying molecular mechanisms of plant responses to drought stress still remain to be clarified through advanced molecular techniques including genome-wide expression of phenotypes.

10.5 Conclusion and Future Perspectives

Drought is one of the most limiting abiotic stress factors for crop production. The severity and frequency of drought may increase in the future due to the impacts of climate change. Wheat is a highly sensitive to drought stress especially at flowering and grain filling stages. Drought causes high yield loss in wheat by affecting the balance between productions of reactive oxygen species in plant cells and induces overproduction of ROS which results in oxidative stress. The drought-induced ROS directly damages various macromolecules including proteins and nucleic acids, which ultimately results in death of plant cells. Enhancement of antioxidative defense system is thought to be a plausible approach for tolerance of wheat to drought. In fact, drought in combination with other stresses disrupts photosynthesis and increases photorespiration by altering the normal homeostasis of plant cells and causes an increased production of ROS. In fact, ROS impacted on plants in two ways – (i) functioning as toxic by-products of stress metabolism and (ii) important signal transduction molecules (Miller et al. 2010). On the other hand, various metabolites are also biosynthesized in plant cells under drought stress such as osmoprotectants, phytohormones, antioxidant signaling substances for minimizing the detrimental effects of drought in plants. Several plant genes were found to be upregulated during drought stress, and overexpression of some plant genes has been found to confer improvement in drought tolerance in plants. In addition, application of plant probiotic bacteria also enhances remarkable drought stress tolerance in wheat and other plants through upregulation of drought tolerance-related plant genes. Better understanding of drought-induced damages as well as plants' inherent protective mechanisms through genomic, transcriptomic, proteomic, and metabolomic approaches would help us to design new strategies for development of drought-tolerant wheat varieties.

Acknowledgments We are highly thankful to Tasnim Farha Bhuiyan and Mazhar Ul Alam, Laboratory of Plant Stress Responses, Faculty of Agriculture, Kagawa University, Japan, for their critical reading and formatting of the manuscript draft. The first author acknowledges Japan Society for the Promotion of Science (JSPS) for funding in his research. M. Tofazzal Islam is thankful to World Bank for funding through a HEQEP CP # 2071 to the Department of Biotechnology of BSMRAU, Bangladesh. We are also highly thankful to Mr. Md. Mosfeq-Ul-Hasan, Zhejiang University, Hangzhou, China, for providing us several supporting articles. As page limitation precluded us from citing a large number of studies, we apologize to those whose original publications are therefore not directly referenced in this chapter. Special thanks to Tahsin Islam Sakif, Banani, Dhaka, Bangladesh, for linguistic editing of the manuscript.

References

Abdel-Motagally FMF, El-Zohri M (2016) Improvement of wheat yield grown under drought stress by boron foliar application at different growth stages. J Saudi Soc Agric Sci. https://doi.org/10.1016/j.jssas.2016.03.005

Agarwal S, Sairam RK, Srivastava GC, Tyagi A, Meena RC (2005a) Role of ABA, salicylic acid, calcium and hydrogen peroxide on antioxidant enzymes induction in wheat seedlings. Plant Sci 169:559–570

Agarwal S, Sairam RK, Srivastava GC, Meena RC (2005b) Changes in antioxidant enzymes activity and oxidative stress by abscisic acid and salicylic acid in wheat genotypes. Biol Plant 49:541–550

Ahmad ST, Haddad R (2011) Study of silicon effects on antioxidant enzyme activities and osmotic adjustment of wheat under drought stress. Czech J Genet PlantBreed 47:17–27

Ahmadi A, Baker DA (2001) The effect of water stress on the activities of key regulatory enzymes of the sucrose to starch pathway in wheat. Plant Growth Regul 35:81–91

Ahuja I, de Vos RCH, Bones AM, Hall RD (2010) Plant molecular stress responses face climate change. Trends Plant Sci 15:664–674

Akhkha A, Boutraa T, Alhejely A (2011) The rates of photosynthesis, chlorophyll content, dark respiration, proline and abscisic acid (ABA) in wheat (*Triticum durum*) under water deficit conditions. Int J Agric Biol 13:215–221

Akram M (2011) Growth and yield components of wheat under water stress of different growth stages. Bangladesh J Agril Res 36:455–468

Alavi SMN, Arvin MJ, Kalantari KM (2014) Salicylic acid and nitric oxide alleviate osmotic stress in wheat (*Triticum aestivum* L.) seedlings. J Plant Interact 9:683–688

Aldesuquy H, Ghanem H (2015) Exogenous salicylic acid and trehalose ameliorate short-term drought stress in wheat cultivars by up-regulating membrane characteristics and antioxidant defense system. J Hortic 2:139. https://doi.org/10.4172/2376-0354.1000139

Alexieva V, Sergiev I, Mapelli S, Karanov E (2001) The effect of drought and ultraviolet radiation on growth and stress markers in pea and wheat. Plant Cell Environ 24:1337–1344

Almaghrabi OA (2012) Impact of drought stress on germination and seedling growth parameters of some wheat cultivars. Life Sci J 9:590–598

Anjum NA, Gill SS, Gill R, Hasanuzzaman M, Duarte AC, Pereira E, Ahmad I, Tuteja R, Tuteja N (2014) Metal/metalloid stress tolerance in plants: role of ascorbate, its redox couple, and associated enzymes. Protoplasma 251:1265–1283

Apel K, Hirt H (2004) Reactive oxygen species: metabolism oxidatives and signal transduction. Ann Rev Plant Mol Biol 55:373–399

Araus JL, Bort J, Steduto P, Villegas D, Royo C (2003) Breeding cereals for Mediterranean conditions: ecophysiology clues for biotechnology application. Ann Appl Biol 142:129–141

Ashraf MY, Azmi AR, Khan AH, Naqvi SSM (1994) Water relation in different wheat (*Triticum aestivum* L.) genotypes under water deficit. Acta Physiol Plant 3:231–240

Azooz MM, Youssef MM (2010) Evaluation of heat shock and salicylic acid treatments as inducer of drought stress tolerance in Hassawi wheat. Am J Physiol 5:56–70

Balla K, Rakszegi M, Li Z, Béekés F, Bencze S, Veisz O (2011) Quality of winter wheat in relation to heat and drought shock after anthesis. Czech J Food Sci 29:117–128

Bartels D (2001) Targeting detoxification pathways: an efficient approach to obtain plants with multiple stress tolerance? Trends Plant Sci 6:284–286

Battisti DS, Naylor RL (2009) Historical warnings of future food insecurity with unprecedented seasonal heat. Science 323:240–244

Bharti N, Pandey SS, Barnawal D, Patel VK, Karla A (2016) Plant growth promoting rhizobacteria *Dietzia natronolimnaea* modulates the expression of stress responsive genes providing protection of wheat from salinity. Sci Rep 6:34768

Bhushan D, Pandey A, Choudhary MK, Datta A, Chakraborty S, Chakraborty N (2007) Comparative proteomics analysis of differentially expressed proteins in chickpea extracellular matrix during dehydration stress. Mol Cell Proteomics 6:1868–1884

Bukhari MA, Ashraf MY, Ahmad R, Waraich EA, Hameed M (2015) Improving drought tolerance potential in wheat (*Triticum aestivum* L.) through exogenous silicon supply. Pak J Bot 47:1641–1648

Cattivelli L, Rizza F, Badeck F, Mazzucotelli E, Mastrangelo AM, Francia E, Marè C, Tondelli A, Stanca AM (2008) Drought tolerance improvement in crop plants: an integrated view from breeding to genomics. Field Crops Res 105:1–14

Chakraborty U, Pradhan B (2012) Oxidative stress in five wheat varieties (*Triticum aestivum* L.) exposed to water stress and study of their antioxidant enzyme defense system, water stress responsive metabolites and H_2O_2 accumulation. Braz J Plant Physiol 24:117–130

Chaves MM, Maroco JP, Pereira JS (2003) Understanding plant responses to drought from genes to the whole plant. Funct Plant Biol 30:239–264

Cheng L, Wang Y, He Q, Li H, Zhang X, Zhang F (2016) Comparative proteomics illustrates the complexity of drought resistance mechanisms in two wheat (*Triticum aestivum* L) cultivars under dehydration and rehydration. BMC Plant Biol 16:188. https://doi.org/10.1186/s12870-016-0871-8

Chenk A, Houde M (2016) Genome wide identification of $C_{1\text{-}2i}$ zinc finger proteins and their response to abiotic stress in hexaploid wheat. Mol Genet Genomics 291:873–890

Corpas FJ, Leterrier M, Valderrama R, Airaki M, Chaki M, Palma JM, Barroso JB (2011) Nitric oxide imbalance provokes a nitrosative response in plants under abiotic stress. Plant Sci 181:604–611

Cruz de Carvalho MH (2008) Drought stress and reactive oxygen species: production, scavenging and signaling. Plant Signal Behav 3:156–165

Cui SX, Hu J, Yang B, Shi L, Huang F, Tsai SN, Nqai SM, He Y, Zhang J (2009) Proteomic characterization of Phragmites communis in ecotypes of swamp and desert dune. Proteomics 9:3950–3967

Dai A (2010) Drought under global warming: a review. Wiley Interdiscip Res Clim Change 2:45–65

Daryanto S, Wang L, Jacinthe P (2016) Global synthesis of drought effects on maize and wheat production. PLoS One 11:e0156362. https://doi.org/10.1371/journal.pone.0156362

Demiral T, Hamurcu M, Avsaroglu ZZ, Calik M, Almas S, Hakki EE, Topal A, Gezgin S, Bell RW (2014) Implication of nitric oxide on growth and development of wheat under drought conditions. J Biotechnol 185:S33. https://doi.org/10.1016/j.jbiotec.2014.07.111

Dey A, Samanta MK, Gayen S, Sen SK, Maiti MK (2016) Enhanced gene expression rather than natural polymorphism in coding sequence of the *OsbZIP23* determines drought tolerance and yield improvement in rice genotypes. PLoS One 11(3):eo150763

Dhanda SS, Sethi GS (2002) Tolerance to drought stress among selected Indian wheat cultivars. J Agric Sci 139:319–326

Dhayal SS, Bagdi D, Kakralya B, Saharawat YS, Jat ML (2012) Brassinolide induced modulation of physiology, growth and yield of wheat (*Triticum aestivum* L.) under water stress condition. Crop Res 44:14–19

Djibril S, Mohamed OK, Diaga D, Diégane D, Abaye BF, Maurice S, Alain B (2005) Growth and development of date palm (*Phoenix dactylifera* L.) seedlings under drought and salinity stresses. Afr J Biotechnol 4:968–972

Dorion S, Lalonde S, Saini HS (1996) Induction of male sterility in wheat (*Triticum aestivum* L.) by meiotic-stage water deficit is preceded by a decline in invertase activity and changes in carbohydrate metabolism in anthers. Plant Physiol 111:137–145

El Tayeb MA, Ahmed NA (2010) Response of wheat cultivars to drought and salicylic acid. Am-Euras J Agron 3:1–7

Eskandari H, Kazemi K (2010) Response of different bread wheat (*Triticum aestivum* L.) genotypes to post-anthesis water deficit. Not Sci Biol 2:49–52

FAO (2011) Crop prospects and food situation. Food and Agriculture Organization, Rome

FAO (2015) http://www.fao.org/nr/water/cropinfo_wheat. Accessed 15 Dec 2016

Farooq M, Irfan M, Aziz T, Ahmad I, Cheema SA (2013) Seed priming with ascorbic acid improves drought resistance of wheat. J Agron Crop Sci 199:12–22

Feng H, Duan J, Li H, Liang H, Li X, Han N (2008) Alternative respiratory pathway under drought is partially mediated by hydrogen peroxide and contributes to antioxidant protection in wheat leave. Plant Prod Sci 11:59–66

Foresight (2011) The future of food and farming: challenges and choices for global sustainability. Final project report. The Government Office for Science, London

Forni G, Duca D, Glick BR (2016) Mechanisms of plant response to salt and drought stress and their alteration by rhizobacteria. Plant Soil. https://doi.org/10.1007/s11104-016-3007-x

Foyer CH, Noctor G (2005a) Redox homeostasis and antioxidant signaling: a metabolic interface between stress perception and physiological responses. Plant Cell 17:1866–1875

Foyer CH, Noctor G (2005b) Oxidant and antioxidant signalling in plants: a re-evaluation of the concept of oxidative stress in a physiological context. Plant Cell Environ 28:1056–1071

Gaber A, Yoshimura K, Yamamoto T, Yabuta Y, Takeda T, Miyasaka H, Nakano Y, Shigeoka S (2006) Glutathione peroxidase-like protein of Synechocystis PCC 6803 confers tolerance to oxidative and environmental stresses in transgenic *Arabidopsis*. Physiol Plant 128:251–262

Gahlaut V, Jaiswal V, Kumar A, Gupta PK (2016) Transcription factors involved in drought tolerance and their possible role in developing drought tolerant cultivars with emphasis on wheat (*Triticum aestivum* L). Theor Appl Genetics 129:2019–2042

García-Mata C, Lamattina L (2015) Nitric oxide induces stomatal closure and enhances the adaptive plant responses against drought stress. Plant Physiol 126:1196–1204

Garcíia-Mata C, Lamattina L (2002) Nitric oxide and abscisic acid cross talk in guard cells. Plant Physiol 128:790–792

Gill SS, Tuteja N (2010) Reactive oxygen species and antioxidant machinery in abiotic stress tolerance in crop plants. Plant Physiol Biochem 48:909–930

Gong HJ, Zhu XY, Chen KM, Wang SM, Zhang CL (2005) Silicon alleviates oxidative damage of wheat plants in pots under drought. Plant Sci 169:313–321

Gonitia-Mishra I, Sapre S, Kachare S, Tiwari S (2016a) Molecular diversity of 1-aminocyclopr opane-1-carboxylate (ACC) deaminase producing PGPR from wheat (*Triticum aestivum* L.) rhizosphere. Plant Soil. https://doi.org/10.1007/s11104-016-3119-3

Gonitia-Mishra I, Sapre S, Sharma A, Tiwari S (2016b) Amelioration of drought tolerance in wheat by the interaction of plant growth-promoting rhizobacteria. Plant Biol 18:992–1000

Gonzalez RM, Insem ND (2014) Twenty years of research on Asr (ARA-stress-ripening) genes are proteins. Planta 239:941–949

Grzesiak M, Filek M, Barbasz A, Kreczmer B, Hartikainen H (2013) Relationships between polyamines, ethylene, osmoprotectants and antioxidant enzymes activities in wheat seedlings after short-term PEG- and NaCl-induced stresses. Plant Growth Regul 69:177–189

Gúoth A, Tari I, Gallé A, Csiszár J, Pécsvaradi A, Cseuz L, Erdei L (2009) Comparison of the drought stress responses of tolerant and sensitive wheat cultivars during grain filling: changes in flag leaf photosynthetic activity, ABA levels, and grain yield. J Plant Growth Regul 28:167–176

Gupta NK, Gupta S, Kumar A (2001) Effect of water stress on physiological attributes and their relationship with growth and yield of wheat cultivars at different stages. J Agron Crop Sci 186:55–62

Hafez EM, Gharib HS (2016) Effect of exogenous application of ascorbic acid on physiological and biochemical characteristics of wheat under water stress. Int J Plant Prod 10:579–596

Hasanuzzaman M, Fujita M (2011) Selenium pretreatment upregulates the antioxidant defense and methylglyoxal detoxification system and confers enhanced tolerance to drought stress in rapeseed seedlings. Biol Trace Elem Res 143:1758–1776

Hasanuzzaman M, Hossain MA, Teixeira da Silva JA, Fujita M (2012) Plant responses and tolerance to abiotic oxidative stress: antioxidant defense is a key factor. In: Bandi V, Shanker AK, Shanker C, Mandapaka M (eds) Crop stress and its management: perspectives and strategies. Springer, Berlin, pp 261–316

Hasanuzzaman M, Gill SS, Fujita M (2013) Physiological role of nitric oxide in plants grown under adverse environmental conditions. In: Tuteja N, Gill SS (eds) Plant acclimation to environmental stress. Springer, New York, pp 269–322

Hasanuzzaman M, Nahar K, Gill SS, Fujita M (2014) Drought stress responses in plants, oxidative stress and antioxidant defense. In: Tuteja N, Gill SS (eds) Climate change and plant abiotic stress tolerance. Wiley, Weinheim, pp 209–250

Hasanuzzaman M, Nahar K, Mahmud J, Ahmad P, Fujita M (2016) Nitric oxide: a jack of all trades of drought stress tolerance in plants. In: Ahmad P, Wani MR (eds) Water stress and crop plants: a sustainable approach. Wiley, London, pp 628–648

Hassan NM, El-Bastawisy ZM, El-Sayed AK, Ebeed HT, Alla MMN (2015) Roles of dehydrin genes in wheat tolerance to drought stress. J Adv Res 6:179–188

He L, Gao Z, Li R (2009) Pretreatment of seed with H_2O_2 enhances drought tolerance of wheat (*Triticum aestivum* L.) seedlings. Afr J Biotechnol 8:6151–6157

Herbinger K, Tausz M, Wonisch A, Soja G, Sorger A, Grill D (2002) Complex interactive effects of drought and ozone stress on the antioxidant defence systems of two wheat cultivars. Plant Physiol Biochem 40:691–696

Horváth E, Pál M, Szalai G, Páldi E, Janda T (2007) Exogenous 4-hydroxybenzoic acid and salicylic acid modulate the effect of short-term drought and freezing stress on wheat plants. Biol Plant 51:480–487

Hossain Z, López-Climent MF, Arbona V, Pérez-Clemente RM, Gómez-Cadenas A (2009) Modulation of the antioxidant system in citrus under waterlogging and subsequent drainage. J Plant Physiol 166:1391–1404

Hu HH, Dai MQ, Yao JL, Xiao BZ, Li XH, Zhang QF, Xiong LZ (2006) Overexpressing a NAM, ATAF, and CUC (NAC) transcription factor enhances drought resistance and salt tolerance in rice. Proc Natl Acad Sci U S A 103:12987–12992

Hu W, Huang C, Deng X, Zhon S, Chen L, Li Y, Wang C, Ma Z, Yuan Q, Wang Y, Cai R, Liang X, Yang G, He G (2013) TaASRI, a transcription factor gene in wheat confers drought stress tolerance in transgenic tobacco. Plant Cell Environ 37:1449–1464

Hu L, Li H, Pangb H, Fua J (2012) Responses of antioxidant gene, protein and enzymes to salinity stress in two genotypes of perennial ryegrass (*Lolium perenne*) differing in salt tolerance. J Plant Physiol 169:146–156

Hurkman WJ, McCue KF, Altenbach SB, Korn A, Tanaka CK, Kothari KM, Johnson EL, Bechtel DB, Wilson JD, Anderson OD, DuPont FM (2003) Effect of temperature on expression of genes encoding enzymes for starch biosynthesis in developing wheat endosperm. Plant Sci 164:873–881

Ibrahim HM (2014) Selenium pretreatment regulates the antioxidant defense system and reduces oxidative stress on drought-stresses wheat (*Triticum aestivum* L.) plants. Asian J Plant Sci 13:120–128

Innocenti G, Pucciariello C, LeGleuher M, Hopkins J, De Stefano M, Delledonne M, Puppo A, Baudouin E, Frendo P (2007) Glutathione synthesis is regulated by nitric oxide in *Medicago truncatula* roots. Planta 225:1597–1602

Iusem ND, Bartholomew DM, Hitz WD, Scolnik PA (1993) Tomato (*Lycopersicon esculentum*) transcript induced by water deficit and ripening. Plant Physiol 102:1353–1354

Jatoi WA, Baloch MJ, Kumbhar MB, Khan NU, Kerio MI (2011) Effect of water stress on physiological and yield parameters at anthesis stage in elite spring wheat cultivars. Sarhad J Agric 27:59–65

Ji X, Shiran B, Wan J, Lewis DC, Jenkins CLD, Condon AG, Richards RA, Dolferus R (2010) Importance of pre-anthesis anther sink strength for maintenance of grain number during reproductive stage water stress in wheat. Plant Cell Environ 33:926–942

Ji X, Dong B, Shiran B, Talbot MJ, Edlington JE, Hughes T, White RG, Gubler F, Dolferus R (2011) Control of abscisic acid catabolism and abscisic acid homeostasis is important for reproductive stage stress tolerance in cereals. Plant Physiol 156:647–662

Johari-Pireivatlou M (2010) Effect of soil water stress on yield and proline content of four wheat lines. Afr J Biotechnol 9:36–40

Kang G, Li G, Xu W, Peng X, Han Q, Zhu Y, Guo T (2012) Proteomics reveals the effects of salicylic acid on growth and tolerance to subsequent drought stress in wheat. J Proteome Res 11:6066–6079

Kang GZ, Li GZ, Liu GQ, Xu W, Peng XQ, Wang CY, Zhu YJ, Guo TC (2013) Exogenous salicylic acid enhances wheat drought tolerance by influence on the expression of genes related to ascorbate-glutathione cycle. Biol Plant 57:718–724

Karmollachaab A, Gharineh MH (2015) Effect of silicon application on wheat seedlings growth under water-deficit stress induced by polyethylene glycol. Iran Agric Res 34:31–38

Kasim WA, Osman ME, Omar MN, Abd El-Daim IA, Bejai S, Meijer J (2012) Control of drought stress in wheat using plant-growth promoting bacteria. J Plant Growth Regul 32:122–130

Kaur C, Sharma S, Singla-Pareek SL, Sopory SK (2015) Methylglyoxal, triose phosphate isomerase and glyoxalase pathway: implications in abiotic stress and signaling in plants. In: Pandey GK (ed) Elucidation of abiotic stress signaling in plants. Springer, New York, pp 347–366

Khakwani AA, Dennett MD, Munir M (2011a) Drought tolerance screening of wheat varieties by inducing water stress conditions. Songklanakarin J Sci Technol 33:135–142

Khakwani AA, Dennett MD, Munir M (2011b) Early growth response of six wheat varieties under artificial osmotic stress condition. Pak J Agri Sci 48:119–123

Khalilzadeh R, Sharifi RS, Jalilian J (2016) Antioxidant status and physiological responses of wheat (*Triticum aestivum* L.) to cycocel application and bio fertilizers under water limitation condition. J Plant Interact 11:130–137

Khan TA, Mazid M, Mohammad F (2011) A review of ascorbic acid potentialities against oxidative stress induced in plants. J Agrobiol 28:97–111

Khan SU, Bano A, Jalal-ud-din GA (2012) Abscisic acid and salicylic acid seed treatment as potent inducer of drought tolerance in wheat (*Triticum aestivum* L.) Pak J Bot 44:43–49

Khan MN, Mobin M, Mohammad F, Corpas FJ (2014) Nitric oxide in plants: metabolism and role in stress physiology. Springer, New York

Kiliç H, Yağbasanlar T (2010) The effect of drought stress on grain yield, yield components and some quality traits of durum wheat (*Triticum turgidum* ssp. durum) cultivars. Not Bot Hort Agrobot Cluj 38:164–170

Kim YH, Kim CY, Song WK, Park DS, Kwon SY, Lee HS, Bang JW, Kwak SS (2008) Overexpression of sweetpotato *swpa4* peroxidase results in increased hydrogen peroxide production and enhances stress tolerance in tobacco. Planta 227:867–881

Kim IS, Kim YS, Yoon HS (2012) Rice ASR1 protein with reactive oxygen species scavenging and chaperone-like activities enhances acquired tolerance to abiotic stresses in *Saccharomyces cerevisiae*. Mol Cells 33:285–293

Klein M, Geisler M, Suh SJ, Kolukisaoglu HU, Azevedo L, Plaza S, Curtis MD, Richter A, Weder B, Schulz B, Martinoia E (2004) Disruption of AtMRP4, a guard cell plasma membrane ABCC-type ABC transporter, leads to deregulation of stomatal opening and increased drought susceptibility. Plant J 39:219–236

Lalonde S, Beebe D, Saini HS (1997) Early signs of disruption of wheat anther development associated with the induction of male sterility by meiotic-stage water deficit. Sex Plant Reprod 10:40–48

Lawlor DW, Cornic G (2002) Photosynthetic carbon assimilation and associated metabolism in relation to water deficits in higher plants. Plant Cell Environ 25:275–294

Lei Y, Yin C, Ren J, Li C (2007) Effect of osmotic stress and sodium nitroprusside pretreatment on proline metabolism of wheat seedlings. Biol Plant 51:386–390

Lesk C, Rowhani P, Ramankutty N (2016) Influence of extreme weather disasters on global crop production. Nature 529:84–87. https://doi.org/10.1038/nature16467

Li L, Zheng M, Deng G, Liang J, Zhang H, Pan Z, Long H, Yu M (2016a) Overexpression of AtHDG11 enhanced drought tolerance in wheat (*Triticum aestivum* L.) Mol Breed 36:23. https://doi.org/10.1007/s11D32-016-0447-1

Li Z, Zhang Y, Xu Y, Zhang X, Peng Y, Ma X, Huang X, Yi Y (2016b) Physiological and: TRAQ proteomic analyses reveal the function of spermidine on improving drought tolerance in white clover. J Proteome Res 15:1563–1579

Li J, Li Y, Yin Z, Jiang J, Zhang M, Guo X, Ye Z, Zhao Y, Kiang C, Zhang H, An G, Paek NC, Ali J, Li Z (2016c) OsASR5 enhances drought tolerance through a stomatal closure pathway associated with ABA and H_2O_2 signaling in rice. Plant Biotech J. https://doi.org/10.1111/pbi.12601

Lovell JT, Shakirov EV, Schwarte S, Lowry DB, Aspinwall MJ, Taylor SH, Bonnette J, Palacio-Mejia JD, Hawkes CV, Fay PA, Juenger T (2016) Promises and challenges of ecophysiological genomics in the field: tests of drought responses in switchgrass. Plant Physiol 172:734–748

Ma Q, Wang W, Li Y, Li D, Zou Q (2006) Alleviation of photoinhibition in drought-stressed wheat (*Triticum aestivum*) by foliar-applied glycine betaine. J Plant Physiol 163:165–175

Ma C, Wang Z, Kong B, Lin T (2013) Exogenous trehalose differentially modulate antioxidant defense system in wheat callus during water deficit and subsequent recovery. Plant Growth Regul 70:275–285

Ma D, Sun D, Wang C, Qin H, Ding H, Li Y, Guo T (2016) Silicon application alleviates drought stress in wheat through transcriptional regulation of multiple antioxidant defense pathways. J Plant Growth Regul 35:1–10

Madani A, Rad AS, Pazoki A, Nourmohammadi G, Zarghami R (2010) Wheat (*Triticum aestivum* L.) grain filling and dry matter partitioning responses to source: sink modifications under post-anthesis water and nitrogen deficiency. Acta Sci Agron 32:145–151

Mahajan S, Tuteja N (2005) Cold, salinity and drought stresses: an overview. Arch Biochem Biophys 444:139–158

Majid SA, Asghar R, Murtaza G (2007) Yield stability analysis conferring adaptation of wheat to pre- and post-anthesis drought conditions. Pak J Bot 39:1623–1637

Malik S, Ashraf M (2012) Exogenous application of ascorbic acid stimulates growth and photosynthesis of wheat (*Triticum aestivum* L.) under drought. Soil Environ 31:72–77

Manivannan P, Jaleel CA, Somasundaram R, Panneerselvam R (2008) Osmoregulation and antioxidant metabolism in drought stressed *Helianthus annuus* under triadimefon drenching. C R Biol 331:418–425

Manjarrez-Sandoval P, Gonzales-Hernandez VA, Mendoza-Onofre LE, Engleman EM (1989) Drought stress effects on the grain yield and panicle development of sorghum. Can J Plant Sci 69:631–641

Maqbool MM, Ali A, Haq T, Majeed MN, Lee DJ (2015) Response of spring wheat (*Triticum aestivum* L.) to induced water stress at critical growth stages. Sarhad J Agric 31:53–58

Mattos LM, Moretti CL (2015) Oxidative stress in plants under drought conditions and the role of different enzymes. Enz Eng 5:136. https://doi.org/10.4172/2329-6674.1000136

Miller G, Suzuki N, Cifci-Yilmaz S, Miller R (2010) Reactive oxygen species homeostasis and signaling during drought and salinity stress. Plant Cell Environ 33:453–467

Mirbahar AA, Markhand GS, Mahar AR, Abro SA, Kanhar NA (2009) Effect of water stress on yield and yield components of wheat (*Triticum aestivum* L.) varieties. Pak J Bot 41:1303–1310

Molinari HBC, Marur CJ, Daros E, de Campos MKF, de Carvalho JFRP, Filho JCB, Pereira LFP, Vieira LGE (2007) Evaluation of the stress-inducible production of proline in transgenic sugarcane (*Saccharum* spp.): osmotic adjustment, chlorophyll fluorescence and oxidative stress. Physiol Plant 130:218–229

Morran S, Eini O, Pyvovarenko T, Parent B, Singh R, Ismagul A, Eliby S, Shirley N, Langridge P, Lopato S (2011) Improvement of stress tolerance of wheat and barley by modulation of expression of DREB/CBF factors. Plant Biotechnol J 9:230–249

Nahar K, Hasanuzzaman M, Alam MM, Fujita M (2015) Exogenous glutathione induced drought stress tolerance in mung bean (*Vigna radiata* L.) seedlings: coordinated roles of the antioxidant defense and methylglyoxal detoxification systems. AoB Plants. https://doi.org/10.1093/aobpla/plv069

Naveed M, Hussain MB, Zahir ZA, Mitter B, Sessitsch A (2014) Drought stress amelioration in wheat through inoculation with *Burkholderia phytofirmans* strain PsJN. Plant Growth Regul 73:121–131

Nawaz F, Ahmad R, Ashraf MY, Waraich EA, Khan SZ (2015) Effect of selenium foliar spray on physiological and biochemical processes and chemical constituents of wheat under drought stress. Ecotoxicol Environ Saf 113:191–200

Nawaz H, Yasmeen A, Anjum MA, Hussain N (2016) Exogenous application of growth enhancers mitigate water stress in wheat by antioxidant elevation. Front Plant Sci 7:597. https://doi.org/10.3389/fpls.2016.00597

Nayyar H, Gupta D (2006) Differential sensitivity of C_3 and C_4 plants to water deficit stress: association with oxidative stress and antioxidants. Environ Exp Bot 58:106–113

Neill S (2007) Interactions between abscisic acid, hydrogen peroxide and nitric oxide mediate survival responses during water stress. New Phytol 175:4–6

Neill SJ, Desikan R, Clarke A, Hurst RD, Hancock JT (2002) Hydrogen peroxide and nitric oxide as signalling molecules in plants. J Exp Bot 53:1237–1247

Pfeiffer WH, Trethowan RM, van Ginkel M, Ortiz MI, Rajaram S (2005) Breeding for abiotic stress tolerance in wheat. In: Ashraf M, Harris PJC (eds) Abiotic stresses: plant resistance through breeding and molecular approaches. Haworth Press, New York, pp 401–489

Plaut Z, Butow BJ, Blumenthal CS, Wrigley CW (2004) Transport of dry matter into developing wheat kernels and its contribution to grain yield under post-anthesis water deficit and elevated temperature. Field Crop Res 86:185–198

Prasad PVV, Pisipati SR, Momĉilović I, Ristic Z (2011) Independent and combined effects of high temperature and drought stress during grain filling on plant yield and chloroplast *ef-tu* expression in spring wheat. J Agron Crop Sci 197:430–441

Rahaie M, Xue GP, Naghavi MR, Alizadeh H, Schenk PM (2010) A MYB gene from wheat (*Triticum aestivum* L.) is up-regulated during salt and drought stresses and differentially regulated between salt-tolerant and sensitive genotypes. Plant Cell Rep 29:835–844

Rajaram S (2001) Prospects and promise of wheat breeding in the 21st century. Euphytica 119:3–15

Raza MAS, Saleem MF, Ashraf MY, Ali A, Asghar HN (2012) Glycine betaine applied under drought improved the physiological efficiency of wheat (*Triticum aestivum* L.) plant. Soil Environ 31:67–71

Raza MAS, Saleem MF, Shah GM, Khan IH, Raza A (2014) Exogenous application of glycine betaine and potassium for improving water relations and grain yield of wheat under drought. J Soil Sci Plant Nutr 14:348–364

Reddy AR, Chaitanya KV, Vivekanandan M (2004) Drought-induced responses of photosynthesis and antioxidant metabolism in higher plants. J Plant Physiol 161:1189–1202

Sangtarash MH (2010) Responses of different wheat genotypes to drought stress applied at different growth stages. Pak J Biol Sci 13:114–119

Sapre SS, Vakharia DN (2016) Role of silicon under water deficit stress in wheat: (biochemical perspective): a review. Agric Rev 37:109–116

Saumonneau A, Laloi M, Lallemand M, Rabot A, Atanassova R (2012) Dissection of the transcriptional regulation of grape ASR and response to glucose and abscisic acid. J Exp Bot 63:1495–1510

Seki M, Kamei A, Yamaguchi-Shinozaki K, Shinozaki K (2003) Molecular responses to drought, salinity and frost: common and different paths for plant protection. Curr Opin Biotech 14:194–199

Shabbir RN, Waraich EA, Ali H, Nawaz F, Ashraf MY, Ahmad R, Awan MI, Ahmad S, Irfan M, Hussain S, Ahmad Z (2016) Supplemental exogenous NPK application alters biochemical processes to improve yield and drought tolerance in wheat (*Triticum aestivum* L.) Environ Sci Pollut Res 23:2651–2662

Shahbaz M, Masood Y, Perveen S, Ashraf M (2011) Is foliar-applied glycine betaine effective in mitigating the adverse effects of drought stress on wheat (*Triticum aestivum* L.)? J Appl Bot Food Qual 84:192–199

Shamsi K (2010) The effects of drought stress on yield, relative water content, proline, soluble carbohydrates and chlorophyll of bread wheat cultivars. J Anim Plant Sci 8:1051–1060

Shamsi K, Kobraee S (2011) Bread wheat production under drought stress conditions. Ann Biol Res 2:352–358

Shamsi K, Petrosyan M, Noor-Mohammadi G, Haghparast R (2010) The role of water deficit stress and water use efficiency on bread wheat cultivars. J Appl Biosci 35:2325–2331

Shan CJ, Zhang S, Li D, Zhao Y, Tian X, Zhao X, Wu Y, Wei X, Liu R (2011) Effects of exogenous hydrogen sulfide on the ascorbate and glutathione metabolism in wheat seedlings leaves under water stress. Acta Physiol Plant 33:2533

Shan C, Zhou Y, Liu M (2015) Nitric oxide participates in the regulation of the ascorbate-glutathione cycle by exogenous jasmonic acid in the leaves of wheat seedlings under drought stress. Protoplasma 252:1397–1405

Simon-Sarkadi L, Kocsy G, Várhegyi A, Galiba G, De Ronde JA (2006) Stress induced changes in the free amino acid composition in transgenic soybean plants having increased proline content. Biol Plant 50:793–796

Singh TN, Aspinall D, Paleg LG (1973) The influence of (2-chloroethyl)trimethylammonium chloride and gibberellic acid on the growth and proline accumulation of wheat plants during water stress. Aust J Biol Sci 26:77–86

Singh K, Foley RC, Onate-Sanchez L (2002) Transcription factors in plant defense and stress responses. Curr Opin Plant Biol 5:430–436

Sun J, Hu W, Zhou R, Wang L, Wang X, Feng Z, Li Y, Qiu D, He G, Yang G (2015) The Brachypodium distachyon BdRKY36 gene confers tolerance to drought stress in transgenic tobacco plants. Plant Cell Rep 34:23–35

Szalai G, Kellős T, Galiba G, Kocsy G (2009) Glutathione as an antioxidant and regulatory molecule in plants under abiotic stress conditions. J Plant Growth Regul 28:66–80

Szegletes ZS, Erdei L, Tari I, Cseuz L (2000) Accumulation of osmoprotectants in wheat cultivars of different drought tolerance. Cereal Res Commun 28:403–410

Taheri S, Saba J, Shekari F, Abdullah TL (2011) Effects of drought stress condition on the yield of spring wheat (*Triticum aestivum*) lines. Afr J Biotechnol 10:18339–18348

Tamura T, Hara K, Yamaguchi Y, Koizumi N, Sano H (2003) Osmotic stress tolerance of transgenic tobacco expressing a gene encoding a membrane-located receptor-like protein from tobacco plants. Plant Physiol 131:454–462

Tan J, Zhao H, Hong J, Han Y, Li H, Zhao W (2008) Effects of exogenous nitric oxide on photosynthesis, antioxidant capacity and proline accumulation in wheat seedlings subjected to osmotic stress. World J Agric Sci 4:307–313

Tian X, Lei Y (2006) Nitric oxide treatment alleviates drought stress in wheat seedlings. Biol Plant 50:775–778

Tian XR, Lei YB (2007) Physiological responses of wheat seedlings to drought and UV-B radiation. Effect of exogenous sodium nitroprusside application. Russ J Plant Physl 54:676–682

Timmusk S, Abd El-Daim IA, Copolovici L, Tanilas T, Ka¨nnaste A, Behers L, Nevo E, Seisenbaeva G, Stenstro¨m G, Niinemets U (2014) Drought-tolerance of wheat improved by rhizosphere bacteria from harsh environments: enhanced biomass production and reduced emissions of stress volatiles. PLoS One 9:e96086. https://doi.org/10.1371/journal.pone.0096086

Vendruscolo ECG, Schuster I, Pileggi M, Scapim CA, Molinari HBC, Marur CJ, Vieira LGE (2007) Stress-induced synthesis of proline confers tolerance to water deficit in transgenic wheat. J Plant Physiol 164:1367–1376

Wang GP, Li F, Zhang J, Zhao MR, Hui Z, Wang W (2010a) Overaccumulation of glycine betaine enhances tolerance of the photosynthetic apparatus to drought and heat stress in wheat. Photosynthetica 48:30–41

Wang GP, Hui Z, Li F, Zhao MR, Zhang J, Wang W (2010b) Improvement of heat and drought photosynthetic tolerance in wheat by overaccumulation of glycine betaine. Plant Biotechnol Rep 4:213–222

Ward JM, Schroeder JI (1994) Calcium-activated k+ channels and calcium-induced calcium release by slow vacuolar ion channels in guard cell vacuoles implicated in the control of stomatal closure. Plant Cell 6(5):669–683

Wei TM, Chang XP, Min DH, Jing RL (2010) Analysis of genetic diversity and trapping elite alleles for plant height in drought-tolerant wheat cultivars. Acta Agron Sin 36:895–904

Wei J, Li C, Li Y, Jiang G, Cheng G (2013) Effects of external potassium (K) supply on drought tolerances of two contrasting winter wheat cultivars. PLoS One 8:e69737. https://doi.org/10.1371/journal.pone.0069737

Yadav SK, Singla-Pareek SL, Reddy MK, Sopory SK (2005a) Methylglyoxal detoxification by glyoxalase system: a survival strategy during environmental stresses. Physiol Mol Biol Plants 11:1–11

Yadav SK, Singla-Pareek SL, Reddy MK, Sopory SK (2005b) Transgenic tobacco plants overexpressing glyoxalase enzymes resist an increase in methylglyoxal and maintain higher reduced glutathione levels under salinity stress. FEBS Lett 579:6265–6271

Yang H, Zhang D, Li H, Dong L, Lan H (2015) Ectopic overexpression of the aldehyde dehydrogenase ALDH21 from *Syntrichia caninervis* in tobacco confers salt and drought tolerance. Plant Physiol Biochem 96:83–91

Yasmeen A, Basra SMA, Wahid A, Farooq M, Nouman W, Rehman HU, Hussain N (2013) Improving drought resistance in wheat (*Triticum aestivum)* by exogenous application of growth enhancers. Int J Agric Biol 15:1307–1312

Yavas I, Unay A (2016) Effects of zinc and salicylic acid on wheat under drought stress. J Anim Plant Sci 26:1012–1018

Yordanov I, Velikova V, Tsonev T (2000) Plant responses to drought, acclimation, and stress tolerance. Photosynthetica 38:171–186

Yuan LL, Zhang M, Yan X, Bian YW, Zhen SM, Yan YM (2016) Dynamic phosphoproteome analysis of seedling leaves in Brachypodium distachyon L reveals central phosphorylated proteins involved in the drought stress response. Sci Rep 6:35280. https://doi.org/10.1038/srep35280

Zang X, Komatsu S (2007) A proteomic approach for identifying osmotic-stress-related proteins in rice. Phytochemistry 68:426–437

Zhang H, Shen W, Xu L (2003) Effects of nitric oxide on the germination of wheat seeds and its reactive oxygen species metabolisms under osmotic stress. Acta Bot Sin 45:901–905

Zhu JK (2002) Salt and drought stress signal transduction in plants. Annu Rev Plant Biol 53:247–273

Signalling During Cold Stress and Its Interplay with Transcriptional Regulation

11

Pushpika Udawat and Priyanka Deveshwar

Abstract

World population (7.4 billion) is rapidly increasing (1.11% per year) and is projected to reach more than nine billion by 2050. Conversely, increase in crop yield and productivity is declining because of deleterious environmental effects including abiotic stresses (cold, salinity and drought). As a result, a major area of concern throughout the world is to minimize the losses caused by these stresses to cope up with increasing food necessity. Particularly, low temperature stress (freezing and cold) overall leads to a mechanical constraint on cellular membrane. Cold acclimation requires accurate sensing, signalling and regulation of the transcriptional cascade. Cold stress signals are sensed by change in membrane fluidity, Ca^{2+} channels and several kinases and phospholipases and also by photosynthetic apparatus. Afterwards, cytosolic Ca^{2+} concentration increases, and this Ca^{2+} influx is identified by Ca^{2+} sensor (calmodulin and calcineurin B-like proteins) and Ca^{2+} responder proteins (CDPKs and CIPKs). Signal is then conversed downstream to induce the activity of C-repeat binding factors (CBFs) and hence COR gene expression. A MYC (myelocytomatosis)-type bHLH TF (basic helix-loop-helix transcription factor) activator of CBF expression 1 (*ICE1*) controls expression of CBFs. Cold acclimation is perceived in CBF-dependent or CBF-independent way, which regulates different set of TFs. In this chapter, we will emphasize on cold stress, its signalling, downstream effectors (dehydrins, ROS scavengers, cryoprotectants and proteins involved in transfer of lipids) and candidate genes responsible for cold stress tolerance. Several factors responsible for cold stress tolerance have been addressed including cold stress-responsive regulatory/promoter elements, and different transcription factors and down-

P. Udawat · P. Deveshwar (✉)
Department of Plant Molecular Biology, University of Delhi, New Delhi, India
e-mail: pushpika.udawat@gmail.com; deveshwar.priyanka@gmail.com

S. M. Zargar, M. Y. Zargar (eds.), *Abiotic Stress-Mediated Sensing and Signaling in Plants: An Omics Perspective*, https://doi.org/10.1007/978-981-10-7479-0_11

stream signalling pathways have been covered. The process of cold stress sensing, signalling and TFs involved in cellular response requires further understanding.

Keywords
Cold signaling · Cold tolerance · Cold-regulated genes · Cold acclimation · ICE1 · CBF

11.1 Introduction

Stress in biological terms is defined as an undesirable condition that hinders the regular functioning and wellbeing of any biological system, for instance, plants. Plants, being sessile in nature, have to cope up with biotic (fungi, viruses, bacteria, insects, parasites and weeds) and abiotic stresses (cold, heat, salinity, drought, UV and heavy metal), a route without escape. Abiotic stresses are a main threat to crop yield and therefore responsible for losses worth millions of dollars every year. Feeding the increasing world population will require increase in crop yield and thus productivity. Therefore, to comprehend the consequence of various abiotic stresses on crop productivity, several studies have been undertaken to define the influence of changing climate on crop yield (Osbornea et al. 2013). Cold stress is amongst key environmental cues that affects yield and productivity of important crop plants. Around 70.8% of peripheral region of the earth is occupied by oceans and polar regions. Merely one-third of the entire terrestrial region (29.2%) is suitable for agriculture and 42% of which is frequently exposed to temperatures below −20 °C (Solanke and Sharma 2008). To grow under such unfavourable environments, plants undergo several adaptive modifications at different levels: morphological, physiological, biochemical and molecular level. Depending upon plant response to cold stress, they are alienated as susceptible and tolerant species. Plants of tropical climates are susceptible, while those growing in temperate regions have developed tolerance to cold stress through cold acclimation. Cold acclimation is a practice of developing tolerance to cold stress (both freezing and nonfreezing temperature) in temperate plants. Several agronomically important crops are unable to acquire appropriate level of cold acclimation. The extent of plant tolerance to cold stress is determined by the duration of plant exposure to cold stress, genotype, developmental stage and physiology of the plant (Janska et al. 2009). Hence, identification of key genes responsible for higher levels of cold stress tolerance has become a major area of research, with a view to manipulate gene expression in crop plants. Promoters of several cold-responsive (COR) genes contain *cis*-elements like dehydration-responsive or C-repeat elements (DRE/CRE), abscisic acid-responsive elements (ABRE) and myeloblastosis (MYB) or myelocytomatosis (MYC) recognition sequences (Yamaguchi-Shinozaki and Shinozaki 2005). Genetic screens of *AtPRD29A*::LUC and *AtPCBF3*::LUC luciferase reporter have directed the recognition of upstream regulatory elements that have given new perspective about cold stress sensing, signalling and expression of COR genes (Chinnusamy et al. 2002).

The regulatory components of plant cold stress sensing and signalling are largely unknown and understudied. In this chapter, we have first emphasized on different genes responsible for cold stress signalling, and their involvement in membrane stabilization has been discoursed. Besides, the present state of the art in reference to model plants as well as crop plants in the field of genomics and transcriptomics under cold stress is discussed.

In plant cells, cold stress is perceived by membrane and its receptors, initiating downstream signalling that leads to generation of secondary messengers, viz. Ca^{2+} and reactive oxygen species (ROS). This change in cold stress-induced cytosolic Ca^{2+} level is sensed by cellular Ca^{2+} sensors, viz. calmodulins (CaMs), calmodulin B-like proteins (CMLs) and calcineurin B-like proteins (CBLs), which subsequently undergo conformational change in Ca^{2+}-dependent manner (Defalco et al. 2010). Afterwards, these non-enzymatic Ca^{2+} sensory proteins bind to their interacting protein kinases (Ca^{2+}-dependent protein kinases and CBL-interacting protein kinases), initiating phosphorylation of ICE1 and inducing ICE-CBF-COR transcriptional cascade. The artefacts of these COR genes then lead to cellular homeostasis and adaptations which assist plants to endure and surpass the stressful conditions. Hence, plants respond to stress as a single cell and concomitantly as a complete organism. Therefore, it is crucial to recognize the process of cold acclimation step by step, from signal perception to initial signal transduction that ultimately results into cellular and metabolic changes.

Abiotic stress response in plants is an intricate, multigenic process and a quantitative trait. Therefore, several signal transduction pathways are activated, and certain pathways are common with other abiotic stress signalling, suggesting crosstalk amongst them. Cold, salinity and drought stress activate expression and activity of MAP (mitogen-activated protein) kinases which imply that MAPK cascade operate as connecting point in abiotic stress signalling (Liang et al. 2011). In *Arabidopsis*, ROS generated during cold stress have been shown to induce MAP kinase module. Several genes of abiotic stress response are generally classified into two categories: "early" (transient expression) and "late" (sustained expression) genes (Fig. 11.1). The genes in early category are usually transcription factor-encoding genes and expressed immediately after onset of stress, while late genes are majorly stress-inducible genes, i.e. responsive to dehydration and/or cold induction, which encodes for osmolytes, ROS scavengers (enzymatic and non-enzymatic), membrane stabilizers and late embryogenesis abundant (LEA)-like proteins. In plants, several effectors of cold stress tolerance have been characterized, viz. cryoprotectants (glycine betaine, proline, polyamines, mannitol, trehalose and galactinol), detoxifying enzymes (SOD, APX and CAT), transporters, TFs (DREB, NAC, MYB, MYC and AP2), protein kinases (diacylglycerol kinase) and phosphatases (CDPKs) (Shi et al. 2015). Sasaki et al. (2007) reported involvement of cold shock domain proteins (CSDPs) amid cold acclimation. Recently, *AtCSP3* was characterized in *Arabidopsis*, and it was demonstrated that it provides freezing tolerance (Kim et al. 2009). Thus, in-depth knowledge of transcriptional control of cold-responsive genes, CBF-dependent and CBF-independent regulon, will help interpretation of the ways by which plants develop cold stress tolerance.

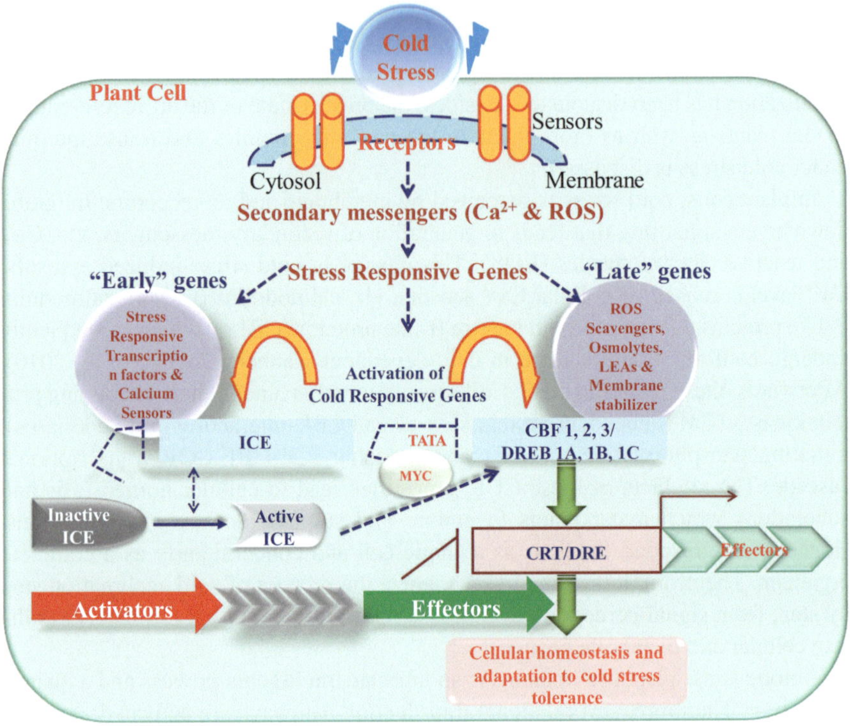

Fig. 11.1 An overview of cold stress sensing, signalling and cold acclimation

11.2 Cold Stress

Around two-thirds of the world's terrestrial region experience rounds of freezing or permafrost, and approximately 42% of cultivated area endures cold stress up to −20 °C (Ramankutty et al. 2002). Plant exposure to temperature in the range of 0–10 °C leads to chilling stress, while temperature below 0 °C leads to freezing stress. Sensitivity to cold stress differs from one plant to another; for example, certain set of temperature conditions can possibly be ideal for growth and development of one while stressful for the other. Plants like *Glycine max*, *Zea mays*, *Gossypium hirsutum*, *Lycopersicon esculentum* and *Musa* sps. display marks of cold damage even when treated with nonfreezing temperatures (Mahajan and Tuteja 2005). Plant response to cold stress comprises several phenotypic changes such as lowered leaf growth, ethylene production, wilting, surface lesion, tissue damage, accelerated senescence, chlorosis and necrosis (Sharma et al. 2005).

The key detrimental consequence associated with cold stress is that it causes membrane impairment, owing to severe dehydration (Steponkus 1984). Cellular membrane is usually made up of two kinds of fatty acids, viz. saturated and unsaturated. The fluidity of membrane is determined by the relative proportion of saturated to unsaturated fatty acids (Steponkus et al. 1993). Cold stress-resistant and cold stress-sensitive plant species vary in composition of their cellular membrane with higher or lower proportion of saturated to unsaturated fatty acids. Cold-tolerant plants contain more amounts of unsaturated fatty acids than cold-susceptible plants that have greater proportion of saturated fatty acids. The real cause of damage brought upon by freeze stress is formation of ice crystals in apoplastic space (because it has lesser amount of solute) at low temperature (Ramankutty et al. 2002). It is a well-known fact that ice has lower vapour pressure than water; thus ice crystal formation generates a vapour pressure gradient between surrounding cells and apoplast. Therefore, water moves down the pressure gradient towards apoplast and causes enlargement of preformed ice crystals that subsequently leads to mechanical damage to the cell wall and membrane causing cellular disintegration (Mahajan and Tuteja 2005). Furthermore, ROS generated during cold stress causes cellular and organelle membrane damage, and characteristically plant shows metabolic dysfunctioning, reduced rate of photosynthesis and protein synthesis. Considerable improvement has been done in classifying components of transcriptional regulators of cold-responsive activation of cold-induced (KIN) genes (Chinnusamy et al. 2006).

11.3 Cold Acclimation

The process of developing enhanced tolerance against cold stress in plants is known as "cold acclimation", where cellular membrane is stabilized against freeze injury by enhancing the proportion of unsaturated fatty acids (Yamaguchi-Shinozaki and Shinozaki 2006). The increased proportion of unsaturated fatty acids in membrane avoids hexagonal II phase lipid and cell lysis due to ice crystal formation (Steponkus et al. 1993). Cold acclimation also results in increased solute (sucrose and several other sugars) concentration in apoplast that change freezing point of water more towards negative, thereby avoiding ice crystal. Several alterations are associated with cellular components during cold stress, such as ratio of unsaturated to saturated fatty acid; proportion of glycerolipids, proteins and carbohydrates; and opening of several ion channels (Huang et al. 2012).

11.4 Cold Stress Sensors

In plants, temperature sensing depends upon cooling rate (dT/dt) that implicates gradual reduction in temperature instead of total temperature (T). The components of plant cold stress signalling pathway have been thoroughly studied; nonetheless, few reports are available on cold stress sensors. Numerous reports have highlighted the function of membrane-associated proteins in cold stress sensing, as they sense the change in physical structure of membrane from liquid to crystalline to semi-liquid state (Los and Murata 2004). The divalent cation, Ca^{2+}, plays a crucial role in sensing and signalling during low temperature stress, as cellular Ca^{2+} level increases during cold stress (Shi et al. 2015). Generally, it is hypothesized that the reduced membrane fluidity is considered as a primary event in cold stress sensing, which subsequently activates Ca^{2+} channels. Several families of Ca^{2+} channels have been identified in plants that regulate the Ca^{2+} influx from different cellular compartments. Ligand-gated receptors (1,4,5-trisphosphate and cyclic ADP-ribose), two pore channel 1 (TPC1), cyclic nucleotide-gated channels (CNGCs) as well as glutamate receptor-like channels (GLRs) possess Ca^{2+} influx activities (Finka et al. 2012), whereas Ca^{2+} pumps and exchangers that transfer cytosolic Ca^{2+} back to internal stores are P-type Ca^{2+} ATPase and Ca^{2+} exchangers (CAXs), which are located on the membrane of different organelles (McAinsh and Pittman 2009). The information encoded by Ca^{2+} signatures is decoded by Ca^{2+} sensors and Ca^{2+}-binding proteins. Plant Ca^{2+} sensors are divided into two categories: sensor relays (CaM, CBLs and CaM-related proteins) and sensor responders (CIPKs and CDPKs), respectively (Kanwar et al. 2014). The Ca^{2+} sensor relays and responders undergo Ca^{2+}-mediated conformational modification which leads to changes in their structure and activity (phosphorylation or dephosphorylation). Subsequently, the secondary messengers such as phytohormone ABA and ROS further increase the levels of calcium signatures that lead to cold stress signalling. Henriksson and Trewavas (2003) found positive correlation between increase in Ca^{2+} concentration to increase in cold-induced transcripts in *Arabidopsis*. In plants, changes in Ca^{2+} level are perceived as a cold stress response within 40s, as identified through aequorin-based luminescence imaging (Zhu et al. 2013). Knight and Knight (2000) used spatial imaging technique to demonstrate the role of aequorin (a Ca^{2+}-sensitive luminescent protein) in *Arabidopsis* plants expressing 35S-aequorin and found cold-induced Ca^{2+} influx in whole plant.

In plants, the two components-based histidine kinases (HKs) are considered to be low temperature sensors. When plant faces cold stress, it leads to autophosphorylation of membrane-based HKs, which in turn acts as low temperature sensor. Interestingly, plant HKs are ethylene and non-ethylene receptors, and about ten putative HKs are present in *Arabidopsis* (Wulfetange et al. 2011). Furthermore, these HKs are separated into diverse subfamilies on the basis of their kinase activity. The members of subfamily I (ETR1; ethylene response 1 and ERS1; ethylene response sensor 1) possess kinase domain; however, subfamily II members (ETR2, ERS2 and EIN4; ethylene insensitive 4) lack kinase domain. In *Arabidopsis*, *AtHK1* gene gets activated under cold stress, water deficit and dehydration stress (Urao et al. 2000). Upon activation, *AtHKs* (*AtHK2*, *AtHK3* and *AtHK4*) transduce the signal through *AHPs* (*Arabidopsis* histidine phosphor-transfer proteins) towards *ARRs* (*Arabidopsis* response regulators) (Kakimoto 2003) which subsequently activates cold stress signalling.

The receptor-like kinases (RLKs) have a membrane-spanning domain that transmits extracellular signal to intracellular targets. Ruell et al. (2002) showed accumulation of phospholipase C and D immediately after onset of cold stress; therefore phospholipases are considered to be involved in low temperature stress signalling. Phospholipases increase the assembly of phosphatidic acid through hydrolysis of phospholipids present in membrane that further act as membrane localized secondary signalling molecule. Phospholipase D attaches microtubule and actin filament to cellular membrane; thus their rearrangement leads to changes in physical state of cytoskeleton (Dhonukshe et al. 2003). Low temperature leads to imbalance between light energy absorbed and energy utilized by photosystem II; this subsequently alters the redox state of photosynthetic machinery. The energy imbalance (absorbed to utilize) is sensed by altered excitation coefficient of photosystem II. The redox state, which is generated through plastoquinone pool, controls expression of several chloroplast and nuclear genes. Therefore, oxidized-to-reduced status of photosynthetic machinery may turn into a temperature sensor, which further intermingles with downstream regulatory networks and is accountable in cold acclimation (Ensminger et al. 2006).

11.5 Regulators of Cold Acclimation

Numerous efforts have been made towards understanding the components of cold acclimation in plants. Cold stress tolerance is a multigenic trait, as it activates several COR genes, such as genes encoding for dehydrins, ROS scavengers, cryoprotectants (polyamines), proteins involved in transfer of lipids, elongation factors (EFs) of translation machinery, MAP kinases (mitogen-activated protein), chaperons, calmodulin-related proteins and LEA-like proteins (Shi et al. 2015). Gibson et al. (1994) reported involvement of *AtFAD8* gene, which encodes fatty acid desaturase, in providing cold stress tolerance by changing the lipid composition of cellular membrane. Likewise, Miquel et al. (1993) studied *Arabidopsis Fad2* mutant lethality at low temperature due to defect in oleate desaturase pathway, which leads to membrane rigidity and irregular composition. Furthermore, Chen and Thelen (2013) showed the significance of *AtADS2* in providing cold stress tolerance by altering the ratio of membrane lipids. The *Atfro1* (frostbite1) gene encodes Fe-S component of mitochondrial NADH dehydrogenase complex I of ETC (electron transport chain); in addition, mutation in it results into elevated reactive oxygen species accumulation. As studied in *Arabidopsis Fro1* mutants, which continuously generated elevated levels of ROS, showed altered regulation of COR genes, therefore, hypersensitivity to cold stress (Lee et al. 2002).

Agricultural crops with improved cold stress tolerance can be developed with effective biotechnological tools and transgenic approaches. Amongst the cold-induced genes identified for cold acclimation, transcription factor-encoding genes are one of the central players in providing cold acclimation, and most of them are known in *Arabidopsis* and rice. Overexpression of *AtCRT*-1/3 (dehydration-responsive factor or C-repeat binding factor 1 or 3) gene in crop plants, viz. *Brassica* (Jaglo et al. 2001), tomato (Hsieh et al. 2002), wheat (Pellegrineschi et al. 2004) and rice (Oh et al. 2005), provided low temperature, desiccation and salt stress tolerance. Overexpression of *AtHOS9* and *AtHOS10* (osmotically responsive genes) in *Arabidopsis* provided basal level of cold stress tolerance (Zhu et al. 2004, 2005). Recently, Ma et al. (2015) identified a rice QTL, COLD1 (chilling tolerance divergence 1); SNP in this locus makes *japonica* subspecies chilling tolerant while *indica* subspecies chilling sensitive. COLD1 functions as regulator of G-protein signalling,

Table 11.1 Positive regulators of cold stress tolerance (freezing and low nonfreezing)

Gene	Gene product	Host plant	Trait incorporated	References
ATcor15a	Cold-regulated gene	*A. thaliana*	Improved freezing tolerance of protoplast and chloroplast	Artus et al. (1996)
codA	Glycine betaine biosynthesis	*A. thaliana*	Improved cold and salt stress tolerance	Hayashi et al. (1998)
dreb1 and dreb2	Transcription factor	*A. thaliana*	Improved freezing and dehydration stress tolerance	Liu et al. (1998)
abi3	Transcription factor	*A. thaliana*	Improved freezing tolerance by upregulation of *RAB18, LTI129, LTI130 and LTI178*	Tamminen et al. (2001)
Osmyb4	Transcription factor	*A. thaliana*	Improved freezing and chilling tolerance	Vannini et al. (2004)
ZAT12	Transcription factor	*A. thaliana*	Improved cold acclimation	Vogel et al. (2005)
OsMYB3R-2	Transcription factor	*A. thaliana*	Improved freezing, drought and salinity tolerance	Dai et al. (2007)
AtESK1	Novel regulator	*A. thaliana*	Improved freezing tolerance	Xin et al. (2007)
AtCSP3	RNA chaperon	*A. thaliana*	Improved freezing tolerance	Kim et al. (2009)
mybc1	Transcription factor	*A. thaliana*	Improved freezing tolerance	Zhai et al. (2010)
ThpI	Transcription factor	*A. thaliana*	Improved chilling tolerance	Zhu et al. 2010)
AtHAP5A	Transcription factor	*A. thaliana*	Improved freezing tolerance	Shi et al. (2014)
OsCOLD1	Regulator of G-protein signalling	*Oryza sativa*	Improved chilling tolerance	Ma et al. (2015)
AtBZR1	Transcription factor	*A. thaliana*	Improved freezing tolerance	Li et al. (2017)

by interacting with G-protein α-subunit and stimulating GTPase activity. Additionally, transcriptome analysis of CBFs (C-repeat binding factors) overexpressing in transgenic *Arabidopsis* showed that 12% of the COR genes are expressed by CBFs (Fowler and Thomashow 2002), which implies the likelihood of uncharacterized/novel TFs and regulators, responsible for cold acclimation (Table 11.1).

Result of several genetic and transgenic studies showed calmodulins, *AtPP2CA* and *SOS3*-like or *CBL1* as negative regulators of cold-responsive gene expression (Tahtiharju and Palva 2001; Townley and Knight 2002; Cheong et al. 2003). Benzyl alcohol (BA), which increases the fluidity of membrane, inhibits activation of COR genes at 4 °C, while at 25 °C same events were reversed when dimethylsulfoxide (DMSO) was added to the cell suspension culture of alfalfa (Bjorn Larus Orvar et al. 2000; Sangwan et al. 2002). Interestingly, transgenic *Arabidopsis* overexpressing *AtZAT12* disclosed that *ZAT12* regulon is made up of 24 cold-responsive genes, out of which 9 are positive regulators and 15 are negative regulators of cold acclimation (Vogel et al. 2005). In another remarkable study, *Arabidopsis* microarray data gave insight into the role of *AtAP2* (related to *ABI3/VP1* or *RAV1*) in providing cold stress tolerance. The activation of *AtRAV1* gene is downregulated by epibrassinolide which subsequently affects lateral root and rosette growth in transgenic *Arabidopsis* (Fowler and Thomashow 2002). All these adaptations facilitate the plant to survive and beat the stern water deficit condition coupled with low temperature stress.

11.6 CBF-ICE1 Transcriptional Cascade in Response to Freezing Stress

A number of genes have been identified to have roles in cold acclimation. Whole genome microarray expression profiling has identified 939 genes that show altered expression in *Arabidopsis* on exposure to cold treatment. Of these, 655 and 284 genes showed up- and downregulation by cold treatment in comparison to untreated plants, respectively (Lee et al. 2005). These COR genes help in cold acclimation by affecting metabolism, protein stability and cell structure. A number of the highly induced COR genes have CCGAC *cis*-element in their promoters. This element is also known as C-repeat or DRE element and is the binding site of CBFs (C-repeat binding factors). CBFs are also called as DREBs, and their binding to DRE of COR gene promoters induces COR gene expression (Gilmour et al. 1998). *Arabidopsis* genome has three CBFs (*CBF1*, *CBF2* and *CBF3*) whose expression is induced by cold. Induction of CBFs is followed by other COR genes which together comprises the CBF regulon. Recently, Zhao et al. (2016) demonstrated the role of three contiguously aligned genes: *CBF1*, *CBF2* and *CBF3* by generating *cbf* single, double and triple mutants. Transcriptome profile identified 414 COR genes to be regulated by *CBF* regulon; out of that, 346 remain to be upregulated, while 68 are downregulated by CBFs (Zhao et al. 2016).

ICE1 (INDUCER OF CBF EXPRESSION 1) is a MYC-type basic helix-loop-helix transcription factor identified from *Arabidopsis*. It is a main regulator of CBF regulon as it regulates the expression of *CBF3* by binding to a *cis*-regulatory element in its promoter. *CBF3* promoter has MYC recognition elements that show binding with ICE1. The *ice1* mutants have impaired expression of *CBF3* and hence impaired COR gene expression and henceforth show hypersensitivity to chilling stress and inability to undergo cold acclimation (Chinnusamy et al. 2003). A total of 933 cold-regulated genes are altered by *ice1* mutation (Lee et al. 2005). Contrasting to CBFs and COR genes, *ICE1* shows constitutive expression, i.e. expresses in warm temperatures also. Overexpression of ICE1 in transgenic plants showed better chilling tolerance with increased expression of *CBF3*, *CBF2* and COR genes but only under cold conditions. ICE1 could not induce CBF regulon under warm conditions which suggests that a post-translational modification occurs to ICE1 on cold treatment that converts it to an active form that thereafter activates downstream genes (Chinnusamy et al. 2003).

ICE1 has a paralog ICE2 that shows ~60% sequence homology of the amino acid sequence with identical DNA-binding bHLH domain. *ICE2* is also a positive regulator of cold acclimation-dependent freezing tolerance pathway. Overexpression of ICE2 showed better freezing tolerance and greater induction of *CBF1*, one of the important regulators of CBF regulon (Fursova et al. 2009). Calmodulin-binding transcription activators (CAMTAs) are a family of transcription factors. Similar to *ICE1*, *CAMTAs* are also not cold induced but expressed at warm temperatures. CAMTA3 has been shown to positively regulate *CBF2* expression by binding to its promoter region during cold conditions (Doherty et al. 2009). A CAMTA-binding region, namely, CG-1 element (vCGCGb), has been identified in *CBF2* promoter, but it is absent from *CBF3* promoter; hence *CBF3* expression is not regulated by CAMTA3. Moreover, *camta1 camta3* double mutants show sensitivity to freezing stress (Doherty et al. 2009). CAMTAs appear to be a link between cold-induced calcium signal and expression of *CBF*s. Combining together the above results, it seems that ICE2 and CAMTA3 along with ICE1 control the expression of CBFs by binding to their promoter regions and hence regulate cold-induced freezing tolerance.

Recently, Park et al. (2015) redirected the query of composition of *Arabidopsis* CBF transcriptional cascade by utilizing the Affymetrix ATH1 DNA chip that analyses 24,000 genes induction during cold stress. They have defined a criterion to assign gene to be part of CBF regulon; (1) expression of gene is necessarily being affected in plants overexpressing *CBF1*, *CBF2* or *CBF3* at higher temperature (22 °C); also (2) expression of gene must remain, respectively, affected in wild-type plants at 4 °C. Thus, differentially expressed genes were those whose transcript showed twofold change with respect to the treatment. This criterion leads to the identification of 2637 COR genes, of which 1256 are upregulated (11% of them were upregulated by overexpression of CBF TFs), and 1381 are downregulated in wild-type *Arabidopsis* plants (Park et al. 2015). In a recent study, *cbf 1,2,3* triple mutants were formed using CRISPR/Cas9 technology that showed extreme sensitivity to freezing tolerance after cold acclimation but no tolerance difference without acclimation. Expression profiling of the triple mutants using RNA seq showed that 112 and 22 genes were positively and negatively regulated by CBFs, respectively (Jia et al. 2016). Amongst several ERFs (ethylene-responsive factors) identified, *AtERF105*, a TF gene, is vital for freezing tolerance and cold acclimation (Bolt et al. 2016).

11.7 Regulation of ICE1 Activity by Post-translational Modifications

Post-translational modifications are known to regulate function of many proteins involved in cold acclimation (Barrero-Gil and Salinas 2013). Three different post-translational modifications of ICE1 have been identified till date that regulates its ability to activate *CBF* expression on exposure to cold. These modifications include phosphorylation, ubiquitination and SUMOylation (Dong et al. 2006; Miura et al. 2011; Ding et al. 2015).

OST1 (open stomata 1)/*SnRK*2.6 is a Ser/Thr protein kinase, whose mutation leads to defective activation of CBF regulon and hence decreased freezing tolerance. On the other hand, overexpression of *OST1* increased the cold tolerance ability by stronger induction of CBF regulon. *OST1* showed cold induction of its kinase activity and was found to interact with ICE1. Both in vitro and in vivo experiments confirmed that ICE1 is phosphorylated by *OST1*. Ding et al. (2015) showed that serine positioned at 278th residue of ICE1 is phosphorylated by *OST1*. Further, they showed that *in planta* phosphorylation was only under cold stress and was important for ICE1 ability to bind to the promoter region of *CBF3*. Thus, cold-induced, *OST1*-mediated, phosphorylated form of ICE1 is the functional form that activates the transcription of *CBF3*.

Contrary to *OST1*, HOS1 (high expression of osmotically responsive gene 1), a RING-type E3 ubiquitin ligase, is a negative regulator of ICE1 activity. Mutation of *HOS1* increases the freezing tolerance, and expectedly its overexpression decreases cold tolerance. HOS1 has been shown to interact with polyubiquitinate ICE1 for its proteasome-dependent degradation in response to cold stress (Dong et al. 2006). Cold stress has been shown to induce degradation of ICE1 in plants. *HOS1* is expressed constitutively, but cold stress induces its movement to the nucleus (Lee et al. 2001). This nuclear co-localization of ICE1 and HOS1 during cold stress triggers ICE1 degradation via proteasome pathway. Thus, HOS1 attenuates ICE1-CBF-mediated cold stress response. Interestingly, OST1 is also shown to interact with HOS1. Both HOS1 and OST1 compete with each other to interact with ICE1 (Ding et al. 2015). OST1 activity counteracts HOS1, as OST1-mediated phosphorylation of ICE1 inhibits the interaction of HOS1 and ICE1. Thus, OST1 counters HOS1-mediated proteolysis of ICE and provides stability to ICE1 protein under cold stress (Ding et al. 2015; Zhan et al. 2015). In ICE1 protein, a serine positioned at 403 (S^{403}) residue is shown to be important for HOS1-mediated ubiquitination. Overexpression of a recombinant ICE1 wherein S^{403} is substituted with alanine showed retention of ICE1 protein after cold treatment in comparison to wild-type protein and greater induction of *CBF* and COR genes. Furthermore, polyubiquitination was halted in recombinant ICE1 (S to A)^{403} protein (Miura et al. 2011). These results infer that serine^{403} of ICE1 is one of the key residues controlling the HOS1-mediated degradation of ICE1 and hence the negative regulation of cold stress response.

Arabidopsis genome has three SUMO (small ubiquitin-related modifier) E3 ligases, and mutation of one of them, viz. *SIZ1* (SAP and Miz1), leads to sensitivity to freezing stress and downregulation of CBF regulon especially expression of *CBF3* and its downstream genes (Miura et al. 2007). Interestingly, *siz1* plants did not show any effect on *ICE1* expression. It was shown that SIZ1 adds the SUMO conjugates to ICE1 protein principally at 393 lysine residue, wherein mutation of K^{393} blocked the ICE1 SUMOylation (Miura et al. 2007). Further, the ICE1 SUMOylation was induced by cold. HOS1-mediated polyubiquitination of ICE1 was drastically reduced in SUMOylated ICE1 proteins, whereas plants carrying ICE1 with mutated K^{393} (i.e. SUMOylation decapacitated) showed decreased cold acclimation ability with reduce induction of CBF regulon. Thus, it can be said that

SIZ1-dependent SUMOylation of ICE1 prevents its proteasome-mediated degradation and enables CBF-dependent cold response.

The above discussion suggests that ICE1 protein level and stability are determined by a balance between its phosphorylation, ubiquitination and SUMOylation status mediated by OST1, HOS1 and SIZ1, respectively. Further, three key ICE1 residues, namely, serine at 278, lysine at 393 and serine at 403, determine the phosphorylation, SUMOylation and ubiquitination of ICE1 by OST1, SIZ1 and HOS1, respectively. All these three post-translational modifications are reversible; thus these adjustable changes in the ICE1 protein may regulate its functionality.

11.8 Interplay Between ICE1, ICE2, CBF1, CBF2, CBF3 and Other Regulators

Recent reports about interaction amongst various regulators of cold stress response have helped in developing an interacting regulatory pathway governing the freezing stress regulon. ICE1, ICE2 and CAMTAs induce *CBF3*, *CBF1* and *CBF2* by binding to their promoter regions, respectively (Chinnusamy et al. 2003; Doherty et al. 2009; Fursova et al. 2009; Kim et al. 2015). Similar to ICE1, ICE2 is also polyubiquitinated by HOS3 (Kim et al. 2015). *ICE2* function in conferring freezing tolerance appeared redundant with *ICE1*, as the *ICE2*-defective mutant *ice2-2* was not freezing sensitive; however, the *ice1 ice2* double mutant showed apparently absolute freezing sensitivity. The double mutant showed downregulation of all the three CBFs (Kim et al. 2015).

Besides ICE1- and ICE2-binding Myc recognition sequences, promoters of CBFs also have Myb recognition sequences. A cold-induced Myb transcription factor MYB15 is shown to bind to the *CBF* promoter regions and negatively regulates their expression (Agarwal et al. 2006). The loss-of-function mutation of MYB15 causes increased expression of *CBF* genes (*CBF1*, *CBF2* and *CBF3*), whereas its overexpression decreased their expression, thus establishing it as a negative regulator of cold stress response. Furthermore, MYB15 is shown to physically interact with ICE1. Accumulation of *MYB15* transcript in *ICE1* mutants suggests that *ICE1* negatively regulates *MYB15*; however relevance of their interaction is not clear (Chinnusamy et al. 2003). Additionally, SIZ1-mediated SUMOylation of ICE1 appeared to be important for negative regulation of *MYB15* as the ICE1-K^{393} mutation resulted in *MYB15* transcript accumulation (Miura et al. 2007). Another negative regulator of *CBF* genes is a cold-inducible C2H2 zinc finger transcription factor, ZAT12. Plants overexpressing *ZAT12* had increased freezing tolerance and downregulation of *CBF* gene expression (Vogel et al. 2005). These results indicated that *MYB15* and *ZAT12* are also involved in cold stress response via a negative regulatory circuit of CBF expression regulation.

CBF1, *CBF2* and *CBF3* are all induced by cold stress, but induction of *CBF1* and *CBF3* precedes that of *CBF2*. *CBF1* and *CBF3* transcripts appear rapidly (15 min) after cold stress and increase for another 1.5 h and then decrease rapidly. On the other hand, *CBF2* transcript is induced slowly after cold stress and reaches a transcript maximum after 2.5 h of cold treatment wherein the expression of *CBF1* and *CBF3* is almost disappeared (Novillo et al. 2004). This sequential appearance and disappearance of *CBF1/3* and *CBF2* suggest that *CBF1/3* induces the expression of *CBF2* in response to cold stress which in return negatively regulated by *CBF2*. This proposal is further confirmed by the observation that loss-of-function mutant *cbf2* is highly freezing tolerant and shows increased expression of *CBF1* and *CBF3* (Novillo et al. 2004).

In addition, JAZ1 and JAZ4 (negative regulators of jasmonate signalling) physically interact with ICE1 and ICE2 to downregulate their transcription activity, thus acting as negative regulator of CBF regulon (Hu et al. 2013). Dong and Pei (2014) demonstrated the role of *miR397* in providing freezing tolerance. Northern blot analysis of *Arabidopsis* overexpressing *miR397* revealed induction of cold-directed CBF genes, as well as downstream COR genes (Dong and Pei 2014). In another remarkable study, MYB96-HHP module was recognized as upregulator of CBF-COR regulon, thus ensuring plant acclimatization to low temperature stress (Lee and Seo 2015). MYB96 binds to promoter region of HHP (heptahelical protein) genes, which interacts with ICE1, ICE2 and CAMTA3, thereby activating CBF regulon (Lee and Seo 2015). *Arabidopsis* mutants of *rdm4* (RNA-dependent DNA methylation: codes for a protein which associate with RNA Pol II as well as RNA Pol V) were sensitive to freezing stress, whereas *AtRDM4* transgenic plants showed enhanced freezing tolerance through the CBF-mediated pathway (Chan et al. 2015). Catala et al. (2014) identified *Arabidopsis* RARE COLD INDUCIBLE 1A (*AtRCI1A*) gene that codes for 14-3-3 protein, which downregulates expression of COR genes, thus leaving transgenic plants freezing sensitive. Consistent with this study, recently, it was found that cold stress activates a plasma membrane localized *AtCRPK1* (cold-responsive protein kinase 1), which interacts with and phosphorylates 14-3-3 proteins. This phosphorylation event translocates 14-3-3 proteins from cytosol to nucleus, where they degrade CBFs, hence acting as negative regulator of freezing stress tolerance (Liu et al. 2017).

11.9 CBF-Independent Pathway of Cold Acclimation and Freezing Tolerance

Cold stress triggers the induction of several COR genes, and many of them are controlled through CBF regulon. However, numerous COR genes show CBF-independent expression as evident from transgenic analysis and are governed by a CBF-divergent regulatory pathways (Chinnusamy et al. 2010). A freezing-tolerant mutant *eskimo1* (*esk1*) was identified in *Arabidopsis* that showed constitutive cold stress tolerance, even when plants have not acquired appropriate level of cold acclimation. The *esk1* mutants exhibited further improved cold stress tolerance and later cold acclimation. High levels of proline accumulated in *esk1* mutant that showed differential activation of genes responsible for proline synthesis and degradation (Xin et al. 2007). Interestingly, the genes controlled by *ESK1* as identified by transcriptome analysis of *esk1* mutant were different from those regulated by CBF3/

ICE1 regulon (identified by transcript profile of *CBF2*-overexpressing plants). Thus, *ESK1* that encodes a DUF23 domain protein and is expressed constitutively is a downregulator of COR genes, controlling a signalling mechanism different from the CBF3-ICE1 transcriptional cascade (Xin et al. 2007). *GIGANTEA* (*GI*) is a well-characterized *Arabidopsis* gene known to regulate circadian rhythms and photoperiodic flowering. Interestingly, it is also activated by low temperature stress and has roles in providing cold stress resistance. The *gi* mutants were susceptible to cold stress; however the transcript induction of *CBF1*, *CBF2* and *CBF3* along with their respective *COR* genes (*RD29A*, *COR15A*, *KIN1*, *KIN2*) remained unaltered in the mutant even after cold stress. Accordingly, it can be said that *GI* is an inducer of CBF-independent pathway of cold acclimation, hence providing cold stress resistance in *Arabidopsis* (Cao et al. 2005).

Confirmation of roles of different TFs revealed that many of them are regulating the cold stress tolerance mechanism in a CBF-independent way. One such gene, *HOS9*, is a CBF-independent regulator of cold tolerance. HOS9 is a homeodomain TF and is constitutively expressed. The *hos9* mutants were susceptible to cold stress, both with and without cold acclimation. The mutant *hos9* transcriptome analysis showed that genes afflicted as a result of mutation are not common with that regulated by CBFs. A different set of genes existed, to be controlled through HOS9 that were responsible for conferring CBF-independent low temperature stress tolerance of *Arabidopsis* (Zhu et al. 2004). Likewise, another TF, *AtHAP5A* (heme-associated protein), known to bind specifically to CCAAT box element, is responsible for enhanced cold tolerance. *AtHAP5A* was found to associate with CCAAT element in the regulatory region of *AtXTH2*. Overexpression plants of both *AtHAP5A* and *AtXTH21* were further tolerant to cold stress, whereas the mutants of both genes showed sensitivity to freezing stress. Furthermore, freezing tolerance/sensitivity changes brought in the plants by mutation and overexpression of *AtHAP5A* were reversed by complementing with overexpression and mutant of *AtXTH21*, respectively, thereby establishing that *AtHAP5A* functions upstream to *AtXTH21* via CBF-independent regulatory pathway towards regulation of cold stress resistance (Shi and Chan 2014; Shi et al. 2014). Yet another transcription factor, namely, MYBC1, a R3-MYB factor, is shown to negatively regulate freezing tolerance independently of the CBF regulon. The *mybc1* mutant and MYBC1 overexpresser transgenics were further tolerant and susceptible to freezing stress, respectively. However, the induction of *CBF1*, *CBF2* and *CBF3* along with the CBF-responsive COR genes (*KIN1*, *ADC1*, *ADC2* and *ZAT12*) remained unaltered in both the mutant and the MYBC1 overexpressing plants (Zhai et al. 2010). Furthermore, BZR1 is a TF which appears to regulate the low temperature stress response in both CBF-dependent and CBF-independent manner. BZR1 is shown to control the induction of *CBF1* and *CBF2* and at the same time impasse to the regulatory region of the COR genes that are disconnected from the CBF regulon (*WKRY6*, *PYL6*, *SOC1*, *JMT* and *SAG21*). Therefore, *BZR1* is a positive regulator of plant freezing tolerance that acts via two distinct pathways of regulation (CBF-dependent and CBF-independent) and is a connecting link between them (Li et al. 2017).

However, after cold acclimation, the precise control and degree of de-acclimation are correspondingly essential for normal growth, as well as subsistence of plant. The fundamentals of plant cold acclimation have been studied in depth, while few facts are obtainable on de-acclimation after exposure to ambient temperature. Therefore, Zuther et al. (2015) studied time-dependent de-acclimation after cold acclimation in ten different accessions of *Arabidopsis*. It was found that the correlation (as studied by levels of several cryoprotectants) between cold acclimation and expression of COR genes was lost during de-acclimation. This study shows that after de-acclimation, plant growth needed to be entirely returned to normal, i.e. non-acclimated state (Zuther et al. 2015).

11.10 Cold Stress and Omics Perspective

Amongst diverse abiotic stresses, cold stress is a key menace to crop productivity and yield worldwide that further confines the geological dispersion and sowing time of several crops. Transcriptome data reveals the intricacy of plant adaptation to cold stress tolerance. Indeed, several newly identified genes have been interpreted as "unidentified function" and "novel pathways" signifying that present knowledge of transcriptional control of COR genes is restricted and the transcriptional regulation is far more intricate than formerly assumed. Information of cold stress-responsive transcriptome, proteome and metabolome is likely to grow in forthcoming time using different omics perspective (Sinha et al. 2015). Omics approaches will enable to deepen our knowledge of the mechanisms of cold stress sensing and thus cold acclimation. Die and Rowland (2014) identified cold acclimation pathway of blueberry using transcriptome profiling and found 454 genes representative of first and second stage of flower bud cold acclimation. Likewise, Beike et al. (2015) studied cold-induced transcriptome of *Physcomitrella patens* and found that transcripts of 3220 genes were considerably affected by cold stress. Wang et al. (2013) studied total transcriptome profile of cold acclimation in *Camellia sinensis*, and out of total 216,831 transcripts, 1170 were differentially expressed, 1168 were upregulated and 602 were downregulated. This information will enrich our knowledge of underlying cold-induced complex regulatory networks. A well-decisive strategy combining all the omics perspective of cold stress tolerance remains prerequisite in connecting the knowledge gaps flanked by the early as well as late induced genes and their products with respect to their molecular and cellular expression at whole plant phenotype level. Therefore, we anticipate improvement of existing cold-sensitive plant species towards ecological food productivity in evolving nations and to combat losses caused by abiotic stresses.

11.11 Future Prospects

This chapter has enclosed the current advancements in plant cold stress signal transduction, including transcription cascade of COR genes, such as positive and negative regulators of CBF pathway. Plants of temperate, tropical and subtropical regions show difference in tolerance and sensitivity towards freezing and low nonfreezing temperature. Temperate plants develop freezing tolerance on pre-exposure to chilling temperature, thus adapting cold acclimation, whereas tropical plus subtropical plants are susceptible to nonfreezing temperature and, hence, do not acquire cold acclimation. Cold stress signal stays to be first received by plasma membrane, and change in membrane fluidity is sensed by different membrane-bound sensor proteins. This subsequently leads to accumulation of Ca^{2+} signatures, activation of several kinases and phosphatases and induction of transcription cascade of COR genes. Therefore, reprogramming of COR genes results in the increase in concentration, not only of cryoprotectants but also of several ROS scavengers and transporters. Developing transgenic plants, having enhanced cold stress tolerance by introgression of COR genes, gives a viable option in developing tolerant plants. Instinctively, plant genetic engineering will be a quicker means of integrating gene of interest compared to conventional or molecular breeding. Application of omics and gene knockout approaches is progressing rapidly which helps in understanding complex multigenic traits associated with acquired cold stress tolerance. The information of whole genome sequence of cold-tolerant model as well as cold-sensitive crop plants has facilitated genome-wide expression profiling of COR genes liable for cold acclimation. By using genomics and transcriptomics approaches, numerous COR genes have been identified till date, and novel insight incessantly emerges to improve current *CBF/DREB1* cold-responsive pathway. Therefore, *CBF/DREB1* regulon is an activator of several integral elements of cold acclimation pathway by which plants develop cold stress tolerance. In summary, the *CBF/DREB1* regulon has been effectively characterized and was engineered in several crop and model plants to develop cold stress tolerance. Therefore, in-depth knowledge of the TFs controlling COR genes, the product of these genes and crosstalk amid diverse signalling pathways should be part of potential research in forthcoming time.

References

Agarwal M, Hao Y, Kapoor A et al (2006) A R2R3 type MYB transcription factor is involved in the cold regulation of CBF genes and in acquired freezing tolerance. J Biol Chem 281:37636–37645

Artus NN, Uemura M, Steponkus PL et al (1996) Constitutive expression of the cold-regulated *Arabidopsis thaliana* COR15a gene affects both choroplast and protoplast freezing tolerance. Proc Natl Acad Sci U S A 93:13404–13409

Barrero-Gil J, Salinas J (2013) Post-translational regulation of cold acclimation response. Plant Sci 205–206:48–54

Beike AK, Lang D, Zimmer AD, Wüst F, Trautmann D, Wiedemann G, Beyer P, Decker EL, Reski R (2015) Insights from the cold transcriptome of : global specialization pattern of conserved transcriptional regulators and identification of orphan genes involved in cold acclimation. New Phytol 205(2):869–881

Bolt S, Zuther E, Zintl S et al (2016) *ERF105* is a transcription factor gene of *Arabidopsis thaliana* required for freezing tolerance and cold acclimation. Plant Cell Environ 40:108–120

Cao S, Ye M, Jiang S (2005) Involvement of GIGANTEA gene in the regulation of the cold stress response in *Arabidopsis*. Plant Cell Rep 24:683–690

Catala R, Lopez-Cobollo R, Castellano M et al (2014) The *Arabidopsis* 14-3-3 protein RARE COLD INDUCIBLE 1A links low-temperature response and ethylene biosynthesis to regulate freezing tolerance and cold acclimation. Plant Cell 26:3326–3342

Chan Z, Wang Y, Cao M et al (2015) RDM4 modulates cold stress resistance in *Arabidopsis* partially through the CBF-mediated pathway. New Phytol 209:1527–1539

Chen M, Thelen JJ (2013) ACYL-LIPID DESATURASE2 is required for chilling and freezing tolerance in *Arabidopsis*. Plant Cell 25:1430–1444

Cheong YH, Kim KN, Pandey GK et al (2003) CBL1, a calcium sensor that differentially regulates salt, drought, and cold responses in *Arabidopsis*. Plant Cell 15:1833–1845

Chinnusamy V, Stevenson B, Lee B-H et al (2002) Screening for gene regulation mutants by bioluminescence imaging. Sci STKE 140:PL10

Chinnusamy V, Ohta M, Kanrar S et al (2003) ICE1: a regulator of cold-induced transcriptome and freezing tolerance in *Arabidopsis*. Genes Dev 17:1043–1054

Chinnusamy V, Zhu J, Zhu JK (2006) Gene regulation during cold acclimation in plants. Physiol Plant 126:52–61

Chinnusamy V, Zhu JK, Sunkar R (2010) Gene regulation during cold stress acclimation in plants. Methods Mol Biol 639:39–55

Dai X, Xu Y, Ma Q et al (2007) Overexpression of an R1R2R3 MYB gene, *OsMYB3R*-2, increases tolerance to freezing, drought and salt stress in transgenic *Arabidopsis*. Plant Physiol 143:1739–1751

DeFalco TA, Chiasson D, Munro K, Kaiser BN, Snedden WA (2010) Characterization of GmCaMK1, a member of a soybean calmodulin-binding receptor-like kinase family. FEBS Lett 584(23):4717–4724

Dhonukshe P, Laxalt AM, Goedhart J et al (2003) Phospholipase D activation correlates with microtubule reorganization in living plant cells. Plant Cell 15:2666–2679

Die JV, Rowland LJ (2014) Elucidating cold acclimation pathway in blueberry by transcriptome profiling. Environ Exp Bot 106:87–98

Ding Y, Li H, Zhang X et al (2015) *OST1* kinase modulates freezing tolerance by enhancing ICE1 stability in *Arabidopsis*. Dev Cell 32:278–289

Doherty CJ, Van Buskirk HA, Myers SJ et al (2009) Roles for *Arabidopsis* CAMTA transcription factors in cold-regulated gene expression and freezing tolerance. Plant Cell 21:972–984

Dong CH, Pei H (2014) Over-expression of *miR397* improves plant tolerance to cold stress in *Arabidopsis thaliana*. Plant Biol 57:209–217

Dong CH, Agarwal M, Zhang Y et al (2006) The negative regulator of plant cold responses, HOS1, is a RING E3 ligase that mediates the ubiquitination and degradation of ICE1. Proc Natl Acad Sci U S A 103:8281–8286

Ensminger I, Busch F, Huner NPA (2006) Photostasis and cold acclimation: sensing low temperature through photosynthesis. Physiol Plant 126:28–44

Finka A, Cuendet AF, Maathuis FJ et al (2012) Plasma membrane cyclic nucleotide gated calcium channels control and plant thermal sensing and acquired thermotolerance. Plant Cell 24:3333–3348

Fowler S, Thomashow MF (2002) *Arabidopsis* transcriptome profiling indicated that multiple regulatory pathways are activated during cold acclimation in addition to the CBF cold response pathway. Plant Cell 14:1675–1690

Fursova OV, Pogorelko GV, Tarasov VA (2009) Identification of ICE2, a gene involved in cold acclimation which determines freezing tolerance in *Arabidopsis thaliana*. Gene 429:98–103

Gibson S, Arondel V, Iba K et al (1994) Cloning of a temperature-regulated gene encoding a chloroplast Omega-3 desaturase from *Arabidopsis thaliana*. Plant Physiol 106:1615–1621

Gilmour SJ, Zarka DG, Stockinger EJ, Salazar MP, Houghton JM, Thomashow MF (1998) Low temperature regulation of the Arabidopsis CBF family of AP2 transcriptional activators as an early step in cold-induced COR gene expression. Plant J 16(4):433–442

Hayashi H, Sakamoto A, Nonaka H et al (1998) Enhanced germination under high-salt conditions of seeds of transgenic *Arabidopsis* with a bacterial gene (*codA*) for choline oxidase. J Plant Res 111:357–362

Henriksson NK, Trewavas AJ (2003) The effect of short-term low-temperature treatments on gene expression in Arabidopsis correlates with changes in intracellular Ca2+ levels. Plant Cell Environ 26:485–496

Henriksson NK, Trewavas AJ (2003) The effect of short-term low-temperature treatments on gene expression in Arabidopsis correlates with changes in intracellular Ca2+ levels. Plant, Cell & Environ 26: 485–496

Hsieh TH, Lee JT, Yang PT et al (2002) Heterology expression of the *Arabidopsis* C repeat/dehydration response element binding factor 1 gene confers elevated tolerance to chilling and oxidative stresses in transgenic tomato. Plant Physiol 129:1086–1094

Hu Y, Jiang L, Wang F et al (2013) Jasmonate regulates the inducer of CBF expression-C-repeat binding factor/DRE binding factor1 cascade and freezing tolerance in *Arabidopsis*. Plant Cell 25:2907–2924

Huang G-T, Ma S-L, Bai L-P, Zhang L, Ma H, Jia P, Liu J, Zhong M, Guo Z-F (2012) Signal transduction during cold, salt, and drought stresses in plants. Mol Biol Rep 39(2):969–987

Jaglo KR, Kleff S, Amundsen KL, Zhang X, Haake V, Zhang JZ, Deits T, Thomashow MF (2001) Components of the Arabidopsis CRepeat/dehydration-responsive element binding factor cold-response pathway are conserved in Brassica napus and other plant species. Plant Physiol 127(3):910–917

Janska A, Marsik P, Zelenkova S et al (2009) Cold stress and acclimation – what is important for metabolic adjustment? Plant Biol 12:395–405

Jia Y, Ding Y, Shi Y et al (2016) The *cbfs* triple mutants reveal the essential functions of CBFs in cold acclimation and allow the definition of CBF regulons in *Arabidopsis*. New Phytol 212:345–353

Kakimoto T (2003) Perception and signal transduction of cytokinins. Annu Rev Plant Biol 54:605–627

Kanwar P, Sanyal S, Tokas I et al (2014) Comprehensive structural, interaction and expression analysis of CBL and CIPK complement during abiotic stresses and development in rice. Cell Calcium 56:81–95

Kim MH, Sasaki K, Imai R (2009) Cold shock domain protein 3 regulates freezing tolerance in *Arabidopsis thaliana*. J Biol Chem 284:23454–23460

Kim YS, Lee M, Lee JH et al (2015) The unified ICE-CBF pathway provides a transcriptional feedback control of freezing tolerance during cold acclimation in *Arabidopsis*. Plant Mol Biol 89:187–201

Knight H, Knight MR (2000) Imaging spatial and cellular characteristics of low temperature calcium signature after cold acclimation in *Arabidopsis*. J Exp Bot 51:1679–1686

Lee H, Seo P (2015) The MYB96-HHP module integrates cold and abscisic acid signaling to activate the CBF-COR pathway in *Arabidopsis*. Plant J 82:962–977

Lee H, Xiong L, Gong Z et al (2001) The *Arabidopsis HOS1* gene negatively regulates cold signal transduction and encodes a RING finger protein that displays cold-regulated nucleo-cytoplasmic partitioning. Genes Dev 15:912–924

Lee BH, Lee H, Xiong L et al (2002) A mitochondrial complex I defect impairs cold-regulated nuclear gene expression. Plant Cell 14:1235–1251

Lee BH, Henderson DA, Zhu JK (2005) The *Arabidopsis* cold-responsive transcriptome and its regulation by ICE1. Plant Cell 17:3155–3175

Liang Z, Xi D, Li S, Zheng G, Zhao S, Shi J, Wu C, Guo X (2011) A cotton group C MAP kinase gene, GhMPK2, positively regulates salt and drought tolerance in tobacco. Plant Mol Biol 77(1–2):17–31

Li H, Ye K, Shi Y et al (2017) *BZR1* positively regulates freezing tolerance via CBF-dependent and CBF-independent pathways in *Arabidopsis*. Mol Plant:S1674–S2052

Liu Q, Ksauga M, Sakuma Y et al (1998) Two transcription factors, DREB1 and DREB2, with an EREBP/AP2 DNA binding domain separate two cellular signal transduction pathways in drought and low temperature-responsive gene expression, respectively, in *Arabidopsis*. Plant Cell 10:1391–1406

Liu Z, Jia Y, Ding Y et al (2017) Plasma membrane CRPK1-mediated phosphorylation of 14-3-3 proteins induces their nuclear import to fine-tune CBF signaling during cold response. Mol Cell 66:117–128

Los DA, Murata N (2004) Membrane fluidity and its roles in the perception of environmental signals. Biochim Biophys Acta 1666:142–157

Ma Y, Dai X, Xu Y et al (2015) COLD1 confers chilling tolerance in rice. Cell 160:1209–1221

Mahajan S, Tuteja N (2005) Cold, salinity and drought stresses: an overview. Arch Biochem Biophys 444:139–158

McAinsh MR, Pittman JK (2009) Shaping the calcium signature. New Phytol 181:275–294

Miquel M, James D Jr, Dooner H et al (1993) *Arabidopsis* requires polyunsaturated lipids for low-temperature survival. Proc Natl Acad Sci U S A 90:6208–6212

Miura K, Jin JB, Lee J et al (2007) SIZ1-mediated sumoylation of ICE1 controls CBF3/DREB1A expression and freezing tolerance in *Arabidopsis*. Plant Cell 19:1403–1414

Miura K, Ohta M, Nakazawa M et al (2011) ICE1 Ser403 is necessary for protein stabilization and regulation of cold signaling and tolerance. Plant J 67:269–279

Novillo F, Alonso JM, Ecker JR et al (2004) CBF2/DREB1C is a negative regulator of CBF1/DREB1B and CBF3/DREB1A expression and plays a central role in stress tolerance in *Arabidopsis*. Proc Natl Acad Sci U S A 101:3985–3990

Oh SJ, Song SI, Kim YS et al (2005) *Arabidopsis* CBF3/DREB1A and ABF3 in transgenic rice increased tolerance to abiotic stress without stunting growth. Plant Physiol 138:341–351

Osbornea T, Roseb G, Wheeler T (2013) Variation in the global-scale impacts of climate change on crop productivity due to climate model uncertainty and adaptation. Agric For Meteorol 170:183–194

Orvar BL, Sangwan V, Omann F, Dhindsa RS (2000) Early steps in cold sensing by plant cells: the role of actin cytoskeleton and membrane fluidity. Plant J 23(6):785–794

Park S, Lee C, Doherty C et al (2015) Regulation of the *Arabidopsis* CBF regulon by a complex low-temperature regulatory network. Plant J 82:193–207

Pellegrineschi A, Reynolds M, Pacheco M et al (2004) Stress-induced expression in wheat of the *Arabidopsis thaliana* DREB1A gene delays water stress symptoms under greenhouse conditions. Genome 47:493–500

Ramankutty N, Foley JA, Norman J et al (2002) The global distribution of cultivable lands: current patterns and sensitivity to possible climate change. Glob Ecol Biogeogr 11:377–392

Ruell E, Cantrel C, Gawer M et al (2002) Activation of phospholipases-C and -D is an early response to a cold exposure in *Arabidopsis* suspension cells. Plant Physiol 130:999–1007

Sangwan V, Orvar BL, Beyerly J, Hirt H, Dhindsa RS (2002) Opposite changes in membrane fluidity mimic cold and heat stress activation of distinct plant MAP kinase pathways. Plant J 31(5):629–638

Sasaki K, Kim M-H, Imai R (2007) Arabidopsis COLD SHOCK DOMAIN PROTEIN2 is a RNA chaperone that is regulated by cold and developmental signals. Biochem Biophys Res Commun 364(3):633–638

Sharma P, Sharma N, Deswal R (2005) The molecular biology of the low-temperature response in plants. BioEssays 27:1048–1059

Shi H, Chan ZL (2014) *AtHAP5A* modulates freezing stress resistance in *Arabidopsis* independent of the CBF pathway. Plant Signal Behav 9:e29109

Shi H, Ye T, Zhong B et al (2014) *AtHAP5A* modulates freezing stress resistance in *Arabidopsis* through binding to CCAAT motif of *AtXTH21*. New Phytol 203:554–567

Shi Y, Ding Y, Yang S (2015) Cold signal transduction and its interplay with phytohormones during cold acclimation. Plant Cell Physiol 56:7–15

Sinha S, Raxwal VK, Joshi B, Jagannath A, Katiyar-Agarwal S, Goel S, Kumar A, Agarwal M (2015) De novo transcriptome profiling of cold-stressed siliques during pod filling stages in Indian mustard (Brassica juncea L.) Front Plant Sci 6:932

Solanke AK, Sharma AK (2008) Signal transduction during cold stress in plants. Physiol Mol Biol Plants 14:69–79

Steponkus PL (1984) Role of the plasma membrane in freezing injury and cold acclimation. Annu Rev Plant Physiol 35:543–584

Steponkus PL, Uemura M, Webb MS (1993) A contrast of the cryostability of the plasma membrane of winter rye and spring oat-two species that widely differ in their freezing tolerance and plasma membrane lipid composition. In: Steponkus PL (ed) Advances in low-temperature biology. JAI Press, London, pp 211–312

Tahtiharju S, Palva T (2001) Antisense inhibition of protein phosphatase 2C accelerates cold acclimation in *Arabidopsis thaliana*. Plant J 26:461–470

Tamminen I, Makela P, Heino P et al (2001) Ectopic expression of *ABI3* gene enhances freezing tolerance in response to abscisic acid and low temperature in *Arabidopsis thaliana*. Plant J 25:1–8

Townley HE, Knight MR (2002) Calmodulin as a potential negative regulator of *Arabidopsis* COR gene expression. Plant Physiol 128:1169–1172

Urao T, Miyata S, Yamaguchi-Shinozaki K et al (2000) Possible his to asp phosphorelay signalling in an *Arabidopsis* two-component system. FEBS Lett 478:227–232

Vannini C, Locatelli F, Bracale M et al (2004) Overexpression of the rice *Osmyb4* gene increases chilling and freezing tolerance of *Arabidopsis thaliana* plants. Plant J 37:115–127

Vogel JT, Zarka DG, Van Buskirk HA et al (2005) Roles of the CBF2 and ZAT12 transcription factors in configuring the low temperature transcriptome of *Arabidopsis*. Plant J 41:195–211

Wang X-C, Zhao Q-Y, Ma C-L, Zhang Z-H, Cao H-L, Kong Y-M, Yue C, Hao X-Y, Liang C, Ma J-Q, Jin J-Q, Li X, Yang Y-J (2013) Global transcriptome profiles of Camellia sinensis during cold acclimation. BMC Genomics 14(1):415

Wulfetange K, Lomin SN, Romanov GA et al (2011) The cytokinin receptors of *Arabidopsis* are located mainly to the endoplasmic reticulum. Plant Physiol 156:1808–1818

Xin Z, Mandaokar A, Chen J et al (2007) Arabidopsis ESK1 encodes a novel regulator of freezing tolerance. Plant J 49:786–799

Yamaguchi-Shinozaki K, Shinozaki K (2005) Organization of cis-acting regulatory elements in osmotic- and cold stress-responsive promoters. Trends Plant Sci 10:88–94

Yamaguchi-Shinozaki K, Shinozaki K (2006) Transcriptional regulatory networks in cellular responses and tolerance to dehydration and cold stresses. Annu Rev Plant Biol 57(1):781–803

Zhai H, Bai X, Zhu Y et al (2010) A single-repeat R3-MYB transcription factor MYBC1 negatively regulates freezing tolerance in *Arabidopsis*. Biochem Biophys Res Commun 394:1018–1023

Zhan X, Zhu JK, Lang Z (2015) Increasing freezing tolerance: kinase regulation of ICE1. Dev Cell 32:257–258

Zhao C, Zhang Z, Xie S et al (2016) Mutational evidence for the critical role of CBF transcription factors in cold acclimation in *Arabidopsis*. Plant Physiol 171:2744–2759

Zhu J, Shi H, Lee BH et al (2004) An *Arabidopsis* homeodomain transcription factor gene, *HOS9*, mediates cold tolerance through a CBF-independent pathway. Proc Natl Acad Sci U S A 101:9873–9878

Zhu J, Verslues PE, Zheng X et al (2005) *HOS10* encodes an R2R3-type MYB transcription factor essential for cold acclimation in plants. Proc Natl Acad Sci U S A 102:9966–9971

Zhu B, Xiong AS, Peng RH et al (2010) Over-expression of *ThpI* from *Choristoneura fumiferana* enhances tolerance to cold in *Arabidopsis*. Mol Biol Rep 37:961–966

Zhu X, Feng Y, Liang G et al (2013) Aequorin based luminescence imaging reveals stimulus and tissue specific Ca^{2+} dynamics in *Arabidopsis* plants. Mol Plant 6:444–455

Zuther E, Juszczak I, Lee YP et al (2015) Time-dependent deacclimation after cold acclimation in *Arabidopsis thaliana* accessions. Sci Rep 5:12199

12 Cross Talk Between Phytohormone Signaling Pathways Under Abiotic Stress Conditions and Their Metabolic Engineering for Conferring Abiotic Stress Tolerance

Sheezan Rasool, Uneeb Urwat, Muslima Nazir, Sajad Majeed Zargar, and M. Y. Zargar

Abstract

The environmental stresses, both biotic and abiotic, reduce crop harvests dramatically. However, abiotic stresses are responsible for the lion's share of harvest losses. We really need to devise strategies to prevent crop losses, so that we could meet the food demands of ever-growing human population. So, the need of the hour is to identify and understand the mechanisms deployed by plants to counteract abiotic stresses. Plants perceive and react to environmental stresses in a highly coordinated and interactive manner. Being sessile, plasticity enables them to adapt to harsh changing environmental conditions, mediated by elaborate signaling networks. The perception of abiotic stress triggers the activation of signal transduction cascades that interact with the baseline pathways transduced by phytohormones. The convergence points among hormone signal transduction cascades are considered cross talk, and together they form a signaling network. Through this mechanism, hormones interact by activating either a common second messenger or through a phosphorylation cascade. This chapter reviews the possible roles of phytohormones in abiotic stress tolerance and cross talk between phytohormone signaling and also about the metabolic engineering of phytohormones for conferring abiotic stress tolerance on crop plants which can prove an excellent target to prevent crop losses and mitigate the problem of feeding to increasing human population.

S. Rasool · U. Urwat · M. Nazir · S. M. Zargar (✉)
Division of Plant Biotechnology, Sher-e-Kashmir University of Agricultural Sciences & Technology of Kashmir, Srinagar, Jammu & Kashmir, India

M. Y. Zargar
Sher-e-Kashmir University of Agricultural Sciences & Technology of Kashmir, Srinagar, Jammu & Kashmir, India

S. M. Zargar, M. Y. Zargar (eds.), *Abiotic Stress-Mediated Sensing and Signaling in Plants: An Omics Perspective*, https://doi.org/10.1007/978-981-10-7479-0_12

Keywords
Abiotic stress · Phytohormones · Signal transduction · Metabolic engineering

12.1 Introduction

The emerging ecological impacts of climate change (Bellard et al. 2012) have led to various abiotic stresses affecting plant growth and development, thereby resulting in reduction of crop productivity. The ever-increasing human population is another major constraint to crop production (Wallace et al. 2003). Moreover, we need a substantial increase in agricultural productivity to feed the rapidly increasing world population. These constraints to global food supply and a balanced environment demand research and development of climate-smart agriculture which can be flexible to climate change (Wheeler and Von Braun 2013).

Plants grow in an environment with extreme levels of eliciting abiotic stress and have to acclimate to the changing environmental conditions. A critical aspect in abiotic stress biology is to identify how plants perceive the different stressors, how the early signals are received and converted within the plant, how diversified the response pathways are, and how they are genetically regulated (Yoshida et al. 2014). The abiotic stresses like drought, salinity, and extreme temperatures are most widespread and significant (Wani et al. 2013). As the stress tolerance traits are highly complex, the conventional breeding techniques have scarcely proven successful. In this direction, we need to device some novel and potent strategies. Engineering of phytohormones could be one such strategy to produce the climate-resilient and high-yielding crops.

Phytohormones are required by the plants in very low concentrations for regulation of their cellular processes. They act as chemical messengers to communicate cellular activities in higher plants (Vob et al. 2014). Phytohormones regulate internal as well as external stimuli and coordinate various signal transduction pathways during abiotic stress tolerance (Kazan 2015). Phytohormone engineering could prove a good platform for biotechnologists to improve crops both nutritionally and economically. This chapter presents an overview about phytohormones and their role in plant growth, development, and abiotic stress response and cross talk between phytohormone signaling and their metabolic engineering for conferring abiotic stress tolerance on crop plants to improve food quantity as well as quality.

12.2 Abiotic Stresses: Big Challenge

Abiotic stress factors, especially cold (low temperatures), drought, and salinity, greatly affect crop growth and development and thus are proving to be the major constraints in reducing crop yield and hindering agricultural production. Therefore, in order to increase the crop productivity, we need to explore how plant growth and development are affected by abiotic stresses not only at physiological levels but also at biochemical and molecular levels as well (Kazan 2015).

When a plant is exposed to abiotic stress, its metabolism gets disrupted affecting its physiology. Plants sense changes in their environment and respond to them in order to prevent damage of their bodies. During plant defense mechanism, early perception of stress and rapid and efficient response to it are a pivotal step in plant defense. After recognition, the plants' constitutive basal defense mechanisms (Andreasson and Ellis 2010) lead to an activation of complex signaling cascades of defense varying from one stress to another (Abou Qamar et al. 2009; Chinnusamy et al. 2004). Once plants are exposed to abiotic and/or biotic stress, specific ion channels and kinase cascades (Fraire-Velázquez et al. 2011) are activated; reactive oxygen species (ROS) (Laloi et al. 2004) and phytohormones like abscisic acid (ABA), jasmonic acid (JA), salicylic acid (SA), and ethylene (ET) accumulate; and the genetic machinery is rearranged leading to sufficient defense reactions and an increase in plant tolerance in order to minimize the biological damage caused by the stress (Fujita et al. 2006).

In recent years, most of the research has been done to understand the plant responses to individual abiotic or biotic stresses. However, conditions differ in the natural environments. In natural environment, crops never face a single stress but combinations of stresses. However, maximum molecular studies are carried out in laboratory or greenhouses, that does not simulate actual field conditions. Therefore, the experiments should be conducted to evaluate the effects of combination of stress factors instead of an individual stress which can prove more beneficial. Interestingly, when plants are exposed to multiple stresses, the result is that plants defend themselves not only against one stress but become more resistant to other stresses as well. This phenomenon is called cross-tolerance, which shows that plants have a strong regulatory mechanism that helps them to adapt to changing environment (Bowler and Fluhr 2000; Capiati et al. 2006; Suzuki et al. 2012). For instance, salt tolerance of tomato plants is increased by wounding (Capiati et al. 2006).

12.3 Phytohormones

12.3.1 Key Mediators of Plant Responses to Abiotic Stresses

Phytohormones are required by the plants in small quantities in regulating growth and development. Phytohormones act either at their site of synthesis or elsewhere in plants during their transport. Although there are various factors on which plants depend to respond to abiotic stresses, phytohormones being one of the most important substances to modulate physiological and molecular responses are considered as a requirement for survival of plants. Phytohormones include auxin (IAA), gibberellins (GAs), cytokinins (CKs), abscisic acid (ABA), ethylene (ET), brassinosteroids (BRs), salicylic acid (SA), and jasmonates (JAs). The strigolactones (SL) constitute a new class of phytohormones. Figure 12.1 shows the chemical structures of major phytohormones.

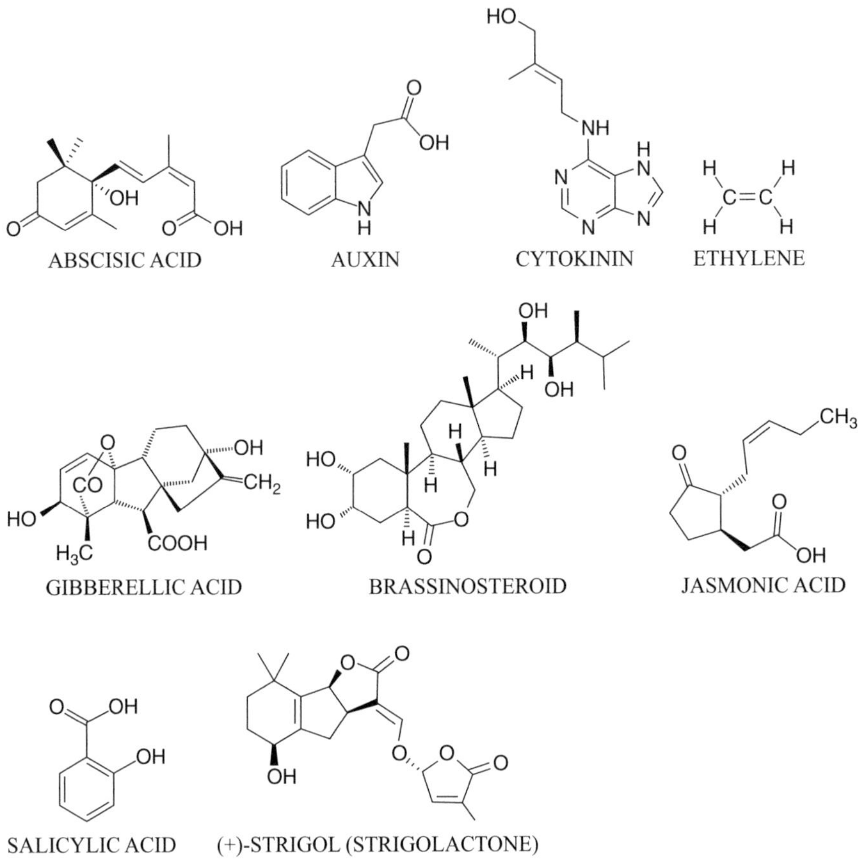

Fig. 12.1 Structures of major classes of phytohormones involved in abiotic stress tolerance in plants

12.3.1.1 Abscisic Acid (ABA): The Abiotic Stress Hormone

Abscisic acid (ABA) got its name because of its role in abscission of plant leaves, and perhaps it is the most studied phytohormones for its major role in plant adaptation to abiotic stresses and is, therefore, termed a "stress hormone." It is one of the important isoprenoid phytohormones and is synthesized in the plastidal 2-C methyl-D-erythritol 4-phosphate pathway. ABA plays an important role in various plant physiological and developmental processes like seed dormancy and development, stomatal opening, synthesis of storage proteins and lipids, embryo morphogenesis, etc. (Sreenivasulu et al. 2010).

ABA is considered as an essential messenger in adaptive response of plants to abiotic stress. Accumulation of ABA in response to cold stress has been reported in earlier studies (Lalk and Dörffling 1985). Higher levels of ABA have also been reported in salt-stressed tobacco cells and alfalfa seedlings (Singh et al. 1987; Luo et al. 1992). It has been observed that external application of ABA has increased drought tolerance in some plant species (Wei et al. 2015). ABA treatments could increase cold resistance in cucumber (Flores et al. 1988) and alfalfa (Mohapatra et al. 1988). It is also observed that ABA can increase salt stress in common bean and potato (Khadri et al. 2006; Etehadnia et al. 2008).

Its primary role is in regulating stomatal aperture, which is required to limit water loss from leaves under drought conditions. ABA also induces the expression of many genes whose products are important for stress responses and tolerance such as enzymes for osmoprotectant synthesis (Fujita et al. 2011). Transcriptome studies have shown that over 50% of the genes regulated by ABA are also controlled by drought or salinity, whereas cold-regulated transcriptome is not induced by other stresses (Sah et al. 2016).

12.3.1.2 Auxins (IAA)

IAA (indole-3-acetic acid) is one of the most multifunctional phytohormones and is important not only for plant growth and development but also for regulating plant growth under stress conditions (Kazan 2013). There are significant evidences that IAA plays an essential part in plants for adaptation to salt stress. It increases root and shoot growth of plants growing under salinity (Egamberdieva 2009). Auxin stimulates the transcription of a large number of genes called primary auxin response genes, and these genes have been analyzed and described in various plant species including *Arabidopsis*, rice, and soybean (Javid et al. 2011). However, much work is needed to be done in order to identify novel genes involved in abiotic stress tolerance so that we could engineer them to develop stress-tolerant crops.

12.3.1.3 Cytokinins (CKs)

Cytokinin hormones are well known for its role in growth and development. They got their name as they are the substances which promote cytokinesis in plants. There are abundant evidences indicating that cytokinins function in stress responses as well. Abiotic stress leads to the changes in the endogenous levels of CKs which indicates their vital roles during abiotic stress including drought (Kang et al. 2012) and salinity (Nishiyama et al. 2011).

Unlike ABA which is involved in inhibition of seed germination, CKs release seeds from dormancy. That is why CKs are often considered ABA antagonists. In water-deficit plants, decreased CK content and accumulation of ABA lead to an increased ABA/CK ratio. The decreased levels of CKs promote apical dominance, which along with the increased levels of ABA regulates opening of stomatal aperture and thus helps plants to get adapted to drought stress (O'Brien and Benkova 2013).

12.3.1.4 Ethylene (ET)

Ethylene is a gaseous hormone that regulates plant growth and development. Ethylene has long been regarded as a stress-related hormone. Morgan and Drew (1997) have reported that salinity and other abiotic stresses induce the synthesis of ACC (direct precursor of ethylene) and ethylene in many plant species. In two halophytes, *Cakile maritima* and *Thellungiella salsuginea*, ACC is concentrated more in leaves and roots under high salt stress, compared to glycophyte *Arabidopsis thaliana* (Ellouz et al. 2014). ET also provides heat tolerance to plants (Larkindale et al. 2005). Jasmonates (JAs) and ethylene (ET), often acting conjointly, help in regulating plant defense against pests and pathogens. During cold stress, the C-repeat binding factor (CBF) pathway is differentially regulated by JAs and ET. Major JA and ET signaling hubs such as JAZ proteins, CTR1, MYC2, components of the mediator complex, EIN2, EIN3, and several members of the AP2/ERF transcription factor gene family all have complex regulatory roles during abiotic stress adaptation (Kazan 2015).We need to understand the role of these phytohormones in plant abiotic stress tolerance that can help us to develop the crops tolerant to various abiotic stresses.

12.3.1.5 Gibberellins (GAs)

GAs constitute a group of tetracyclic diterpenes which greatly influence the processes of seed germination, leaf expansion, stem elongation, flower and trichome initiation, and fruit development. It is an important plant growth bioregulator that can increase the stress tolerance of many crop plants. When plants are exposed to abiotic stresses, GA accumulates rapidly. At certain concentrations during abiotic stress, GA has been observed to be important for the physiology and metabolism of various plants, as it imparts a mechanism for the regulation of metabolic processes as a function of sugar signaling and antioxidant enzymes (Iqbal et al. 2011). Against various stresses, one group of GA compounds protects plants, and therefore, the triazoles have been referred to as plant "multiprotectants" (Fletcher et al. 2000). It has been verified that the morphological and stress-protective effects of triazoles are inverted by GA (Gilley and Flecher 2007), thereby indicating an intimate relationship between GA and plant stress protection.

Recently, an auxin that promotes GA biosynthesis has been reported in different plant species (Wolbang et al. 2004). It has also been reported that application of GA leads to the enhanced catabolism of ABA (Gonai et al. 2004). We can conclude that modification of GA can prove an alluring approach for conferring the stress tolerance to plants. At the same time, we can integrate the different approaches to modulate GA in order to frame crop protection strategies with global implications.

12.3.1.6 Brassinosteroids (BRs)

BRs are a group of naturally occurring novel steroidal plant hormones that regulate plant growth and development by producing an array of physiological changes (Khripach et al. 2000). They show various activities including seed germination, cell growth, reproductive growth, vascular formation, and production of flowers and fruit (Khripach et al. 2000). To date, numerous studies have highlighted BR-induced stress tolerance to various environmental extremes such as high temperature (Li et al. 2007a, b), chilling (Wang et al. 2014), drought (Krishna 2003), salinity (Zhu 2002), and heavy metals (Bajguz 2010) in a range of plant species. Some studies have revealed that BRs significantly increased the dry mass accumulation and activities of antioxidant enzymes in lucerne under salinity (Zhang et al. 2007).

There are number of fields where further research can prove constructive, for instance, their sites of biosynthesis, pathways and enzymes involved in their biosynthesis, source–sink relationships, interactions with microorganisms, fungi and animals, developmental and stress physiology, and the recognition of applications.

12.3.1.7 Jasmonates (JAs)

Methyl jasmonate and its free acid, JA, jointly referred to as jasmonates, are vital cellular regulators involved in diverse plant developmental processes, such as seed germination, callus growth, primary root growth, flowering, formation of gum and bulb, and senescence (Pedranzani et al. 2007). In addition, JA is involved in various biotic and abiotic stress responses, for instance, plant defense responses to insect wounding, attack by various pathogens, and various environmental stresses like drought, cold, and salt stress (Creelman and Mullet 1995; Mei et al. 2006). The post-application of exogenously applied JA is reported to mitigate the salt stress in rice seedlings. In addition, exogenous JA application dramatically decreased sodium concentration. After salt treatment, Kang et al. (2005) have reported exogenous application of JA can change the endogenous hormone balance, such as ABA, which can serve as an important link for understanding the protection mechanism involved in salt stress tolerance. These results clearly indicate that exogenous application of JA not only is engaged in the defense against wounding and pathogen stress but also plays a significant role during drought and salt stress.

12.3.1.8 Salicylic Acid (SA)

SA is a phenolic compound produced by plants and functions as plant hormone and is widely used in organic synthesis. It has been found to play a critical role in the regulation of plant growth, development, and interaction with other organisms (Senaratna et al. 2000). Its role is evident in seed germination, glycolysis, flowering, fruit yield, ion uptake and transport, photosynthetic rate, stomatal conductance, and transpiration (Khan et al. 2003). Most genes that respond positively to acute SA treatment are related to stress and signaling pathways which eventually led to cell death. This includes the genes encoding heat shock proteins, chaperone, and antioxidants and genes that are responsible for the biosynthesis of secondary metabolite, such as sinapyl alcohol dehydrogenase, cytochrome P450, and cinnamyl alcohol dehydrogenase (Jumali et al. 2011).

The major role of SA in plant is thought to be the regulation of responses to biotic stresses; however, a large body of literature now suggests that SA is also involved in responses to several abiotic stresses like drought (Senaratna et al. 2000), salt stress (Khodary 2004; Fahad and Bano 2012), low temperature (Szalai et al. 2002; Tasgin et al. 2003), heavy metal stress (Yang et al. 2003), and heat (Larkindale and Huang 2004; Chakraborty and Tongden 2005). In maize plants under salinity, it was reported that SA increased the level of IAA and decreased the level of ABA (Fahad and Bano 2012). Furthermore, during stress condition, treatment of SA led to an increase in root growth. In maize, the activities of SOD, POD, and ascorbate peroxidase lead to the reduced catalase activity as compared to the plants treated with salt.

In the future, this plant hormone can be used as a management tool for providing tolerance to our agricultural crops against the aforesaid constraints and as a result aiding to accelerate our potential crop yield. However, a lot of work is still needed to elucidate the exact pathway of SA biosynthesis, so that we can explain the detailed molecular mechanism of SAs' role in abiotic stress.

12.3.1.9 Strigolactones (SLs)

Strigolactones (SLs) are a relatively new class of carotenoid-derived plant hormones known to regulate several developmental processes, particularly the regulation of shoot branching as well as the regulation of genes encoding LHC proteins and other photosynthetic units (Mayzlish-Gati et al. 2010; Waldie et al. 2014; Mashiguchi et al. 2009). Few recent studies have indicated a possible involvement of SLs in plants' responses to drought stress. Ectopic expression of the rice cystatin in *Glycine max* and *Arabidopsis* resulted in an increased accumulation of chlorophyll, enhanced shoot branching, and reduced mRNA abundance of SL-related enzymes (Quain et al. 2014). A comparative study involving wild-type and mutant *Arabidopsis* plants indicates their role in the development of root system architecture. Application of GR24, a synthetic and biologically active SL (Gomez-Rolden et al. 2008; Umehara et al. 2008), repressed lateral root formation in wild-type seedlings and SL synthesis mutants (max3 and max4) but not in the strigolactone response mutant (max2), suggesting the MAX2-dependent negative effect of strigolactone on lateral root formation (Kapulnik et al. 2011). They stimulate nodulation in the legume–rhizobium interaction process (Soto et al. 2010; Foo and Davies 2011). Overall, it can be concluded that SLs constitute an important group of signaling molecules and are key regulators of plants' developmental adaptations to changing environmental conditions. They have the potential to be used in agriculture for various purposes including as inducers of suicidal seed germination of parasitic plants (Vurro and Yoneyama 2012).

12.3.2 Role of Phytohormones in Abiotic Stress Tolerance

Phytohormones are essential for ability of plants to adapt to abiotic stresses by mediating a wide range of adaptive responses. Phytohormones help to alter the gene expression by preventing the degradation of transcriptional regulators via ubiquitin–proteasome system (Santner and Estelle 2010). Different phytohormones play significant roles in developing tolerance against various abiotic stresses like drought, cold, heat (high temperature), salinity, etc. One of the most studied topics in response to abiotic stress, especially water stress, is ABA signaling and ABA responsive genes. ABA synthesis is one of the fastest responses of plants to abiotic stress, triggering ABA-inducible gene expression (Yamaguchi-Shinozaki and Shinozaki 2006) and causing stomatal closure, thereby reducing the water loss via transpiration (Wilkinson and Davies 2010).

CK has an important role toward productivity and increased tolerance to various stresses (Zalabák et al. 2013). Reguera et al. (2013) reported that phytohormone engineering, i.e., cytokinin, leads to increased drought tolerance in rice. Similarly, Klay et al. (2014) reported that ethylene response factor leads to enhance salt and cold tolerance in tomato. Colebrook et al. (2014) reported that gibberellins (GA) have a significant role in response to abiotic stress. Increasing the GA biosynthesis and signaling induces tolerance toward submergence, whereas the reduction of GA levels restricts plant growth and development to many stresses like cold, salt, and osmotic stress. BR was reported (mostly the exogenous application) to induce the expression of stress-related genes, leading to the maintenance of photosynthetic activity, the activation of antioxidant enzymes, the accumulation of osmoprotectants, and the induction of other hormone responses (Divi and Krishna 2009). The overlap between hormone-regulated gene suites during the adaptive responses of plants to environmental stresses suggests the existence of complex network with extensive cross talk between the different hormone signaling pathways.

12.4 Cross Talk Between Phytohormone Signaling

In plants, the perception of abiotic stress triggers the activation of signal transduction cascades that interact with the baseline pathways transduced by phytohormones. The convergence points among hormone signal transduction cascades are considered cross talk, and together they form a signaling network. By virtue of this mechanism, hormones interact by triggering either a common second messenger or a phosphorylation cascade. These transduction cascades lead to the regulation of gene expression that directly affects the biosynthesis or action of other hormones, a process that represents an additional layer of hormonal cross talk.

ABA regulates the stomatal opening during stress; however, recent studies suggest that other hormones such as CK, ethylene, BR, JA, SA, and NO also affect stomatal function (Acharya and Assmann 2009). While ABA, BR, SA, JA, and NO induce stomatal closure, CK and IAA promote stomatal opening. NO operates as a key intermediate in the ABA-mediated signaling network that regulates stomatal closure (Ribeiro et al. 2009). ABA is also a regulator of strigolactone biosynthesis, as shown using tomato ABA-deficient mutants of different steps in the ABA biosynthetic pathway and specific inhibitors for different carotenoid cleaving enzymes (López-Ráez et al. 2010). Recently our own work has shown that expression of IPT (isopentenyl transferase, a gene encoding a key step in the biosynthesis of CK) under the control of a drought-inducible and senescence-inducible promoter (PSARK) in tobacco (*Nicotiana tabacum*) and rice results in a significant alteration of gene expression associated with hormone biosynthesis, response, and regulation (Peleg et al. 2011; Rivero et al. 2010). Transgenic tomato (*Solanum lycopersicum*) rootstocks expressing IPT had enhanced root CK synthesis that was shown to modify shoot hormonal balance under salinity stress (Ghanem et al. 2011). The role of auxins in drought tolerance was postulated; TLD1/OsGH3.13, encoding indole-3-acetic acid (IAA)-amido synthetase, was shown to enhance the expression of LEA (late embryogenesis abundant) genes, which correlated with the increased drought tolerance of rice seedlings (Zhang et al. 2009), and auxin was found to affect ethylene biosynthesis. Several members of the 1-aminocyclopropane-1-carboxylate synthase (ACS) gene family, encoding rate-limiting enzymes in ethylene biosynthesis, were shown to be regulated by auxin treatment (Tsuchisaka and Theologis 2004). CK was also shown to be a positive regulator of auxin biosynthesis, and it was postulated that a homeostatic feedback regulatory loop involving both CK and IAA signaling acts to maintain appropriate CK and IAA concentrations in developing root and shoot tissues (Jones et al. 2010). GA and BR regulate many common physiological processes. OsGSR1, a member of the GAST (GA-stimulated transcript) gene family, was found to play key roles in both BR and GA signaling pathways and to mediate the interaction between them (Wang et al. 2009b). GA is also associated with SA. The exogenous application of GA (GA3) induced increased expression levels of ICS1 (isochorismate synthase 1) and NPR1 (non-expressor of pathogenesis-related genes 1), genes involved in SA biosynthesis and SA action, respectively (Alonso-Ramirez et al. 2009).

It has been reported that reciprocation among environmentally activated ABA, ET, GA, and CK signals can prove crucial in determining the responses to different plant stresses (Qin et al. 2011; Iqbal et al. 2011; Krouk et al. 2011). The potential roles played by phytohormones in abiotic stress tolerance and cross talk between phytohormone signaling are illustrated in Fig. 12.2.

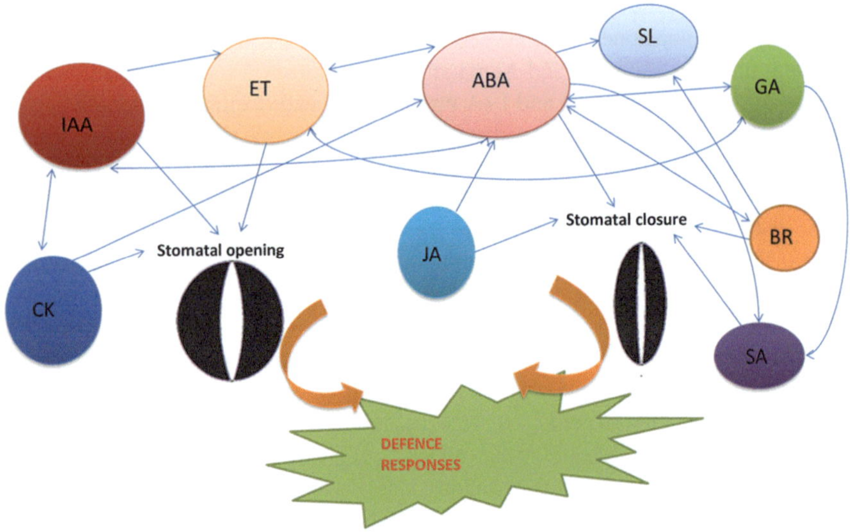

Fig. 12.2 The possible roles of phytohormones in abiotic stress tolerance and cross talk between phytohormone signaling. *IAA* indole-3-acetic acid, *CK* cytokinins, *ET* ethylene, *ABA* abscisic acid, *SL* strigolactones, *GA* gibberellic acid, *BR* brassinosteroids, *JA* jasmonic acid, *SA* salicylic acid

12.5 Metabolic Engineering of Phytohormones for Conferring Abiotic Stress Tolerance on Crop Plants

As discussed in detail in previous sections, phytohormones are considered as the key regulators of growth and development as well as the mediators of the response to environmental stress; engineering of these phytohormones and their signaling processes can prove excellent in enhancing the abiotic stress tolerance in plants and therefore can lead to an increase in agricultural production. Nonetheless, maintenance of hormonal balance is a critical affair and should be dealt with utmost responsibility.

ABA, also known as stress hormone, is perhaps the most sought-after hormone for engineering abiotic stress tolerance in crop plants, as it regulates various physiological processes ranging from stomatal opening to protein storage and provides adaptation to many stresses like drought, salt, and cold stresses. As a result, many of the key ABA biosynthetic pathway enzymes have been investigated transgenically for improved abiotic stress tolerance (Jewell et al. 2010).

NCED encodes 9-cis-epoxy carotenoid dioxygenase, an enzyme that catalyzes the conversion of neoxanthin to xanthoxin, a rate-limiting reaction in the synthesis of ABA. In *A. thaliana*, AtNCED3 is an important gene involved in drought stress-inducible ABA biosynthesis. NCED mutants were found to have defects in ABA accumulation under drought stress. Tobacco plants constitutively overexpressing SgNCED1 (from *Stylosanthes guianensis*) displayed a 51–77% increase in leaf ABA accumulation, which resulted in enhanced tolerance of the transgenic plants to drought and salinity (Zhang et al. 2008). The constitutive overexpression of LeNCED1 in tomato also resulted in increased ABA accumulation in the transgenic plants (Thompson et al. 2007).

CK is an antagonist to ABA; when plants get exposed to drought, it leads to the decrease in the levels of CK. Higher levels of CK promoted survival during drought, repressed leaf senescence, and induced higher levels of proline (Alvarez et al. 2008). Manipulation of endogenous CK levels was effective in delaying senescence. The IPT gene has been overexpressed in several plant species under different promoters, and the transgenic plants were tested for tolerance to various environmental stresses (reviewed by Ma 2008). The constitutive overexpression of IPT increased endogenous CK concentrations up to 150-fold and resulted in decreased root growth and in water stress (Smigocki and Owens 1989). For the conditional expression of hormone biosynthetic genes, the use of inducible promoters has made it possible to regulate the hormone levels without ill effects on growth and development which otherwise may be produced by large changes in hormone levels. The senescence-induced promoter PSAG12 (Gan and Amasino 1995) has been used to drive the IPT expression, resulting in a significant delay in plant senescence. However, a significant delay in flowering and reduced yield were also observed (reviewed by Ma 2008), probably due to altered source/sink relationships brought about by the lack of chlorophyll and protein degradation in source leaves (Rivero et al. 2007). The use of maturation-induced and stress-induced promoters (SARK, senescence-associated receptor-like kinase (Rivero et al. 2007)) to drive IPT expression in both dicots and monocots provided an alternative approach for the induction of IPT and the concomitant biosynthesis of CK, without the negative effects of constitutively high CK content on plant phenology (i.e., flowering time, plant architecture, etc.) (Peleg et al. 2011; Rivero et al. 2007, 2009, 2010). IPT was expressed in the whole plant, its maximal expression was attained during the drought episode, and the transgenic plants displayed enhanced drought tolerance and superior yields (Peleg et al. 2011).

Exogenous application of BRs was reported in diverse plant species to induce drought tolerance (Divi et al. 2010). The overexpression of AtDWF4, a gene involved in BR biosynthetic, under the control of a seed-specific oleosin promoter, resulted in improved germination of seeds that were previously treated with ABA, which suggests an antagonistic effect of BR on ABA-regulated processes. Furthermore transgenic seedlings were more tolerant to cold stress than wild-type seedlings (Divi and Krishna 2010). Antagonism between BR and ABA was recently demonstrated in transgenic PSARK<IPT rice plants, where the increase in CK induced BR-associated genes and repressed ABA-related processes (Peleg et al. 2011).

Sakamoto et al. (2003) modified GA levels by overexpression of OsGA2ox1, a gene encoding GA2-oxidase. OsGA2ox1 ectopic expression at the site of bioactive GA synthesis in shoots under the control of the promoter of a GA biosynthesis gene, OsGA3ox2 (D18), resulted in a semidwarf phenotype showing normal flowering and grain development. Attempts at engineering phytohormones for enhanced abiotic stress tolerance of plants are listed in Table 12.1

12.6 Conclusion

It can be concluded that phytohormone engineering represents an important platform for producing high-yielding and abiotic stress-tolerant crops which provides new opportunities to maintain sustainable crop production to feed the whole world under changing environmental conditions. Despite the rapid development of genomic technology, significant efforts have been made in recent years in elucidating the plant abiotic stress responses and tolerance; there are still various challenges that need to be decoded to bring to light the intricacies of stress signal transduction pathways. More comprehensive work is needed to be done at the genetic level of biosynthetic pathway of hormones like IAA, elucidating the pathway of various hormones like SA, upregulation of genes of ABA biosynthesis by abiotic stress, and hormone homeostasis like GA. Phytohormones show direct or indirect involvement in a wide spectrum of abiotic stresses, and credible research has confirmed that play significant roles in plant defense against abiotic stress. Undoubtedly, the phytohormone engineering can prove an encouraging field for plant biologist, but significant measures are needed to be taken to reach its pinnacle. Among the greatest challenges that remain to be addressed is the development of stable phytohormone-engineered abiotic stress-tolerant crops that can fulfill the demand of agricultural production to feed the world in the coming decades. As different stresses are most likely to occur simultaneously under field conditions, we should focus our study on

Table 12.1 Successful attempts made by phytohormone engineering for enhanced abiotic stress tolerance of plants

Plant hormone	Gene	Function of gene	Expression/ knockout	Phenotype of transgenics	References
ABA	*MoCo sulfurase*	Regulation of the last step of ABA biosynthesis	↑	Transgenic soybean showed higher biomass, yield, and overall enhanced drought tolerance	Li et al. (2013)
	LOS5	Key regulator of ABA biosynthesis	↑	Transgenic maize with enhanced ABA accumulation and increased drought tolerance	Lu et al. (2013)
	NCED	Important role in rate-limiting step of ABA biosynthesis for feedback control	↑	Increased levels of endogenous ABA, decreased stomatal conductance, and increased drought tolerance	Estrada-Melo et al. (2015)
	AtLOS5	Key regulator of ABA biosynthesis	↑	Increased salt tolerance attributed to enhanced Na^+ efflux and H^+ influx	Zhang et al. (2016)
	SnRK2.4	Important serine/threonine protein kinase in ABA signaling network	↑	Transgenic *Arabidopsis* exhibited enhanced tolerance to drought, salt, and freezing stresses associated with decreased water loss, improved photosynthesis, and osmotic potential	Mao et al. (2010)
	OsPIN3t	Auxin efflux carrier, important in polar auxin transport	↑	Increased drought tolerance in rice	Zhang et al. (2012)
	MsZEP	Important role in ABA biosynthesis	↑	Heterologous expression of gene resulted in better salt and drought tolerance	Zhang et al. (2015)
	ERA1	Encodes the b-subunit of farnesyltransferase, an enzyme associated with ABA-dependent signal transduction	↓	Transgenic canola (*Brassica napus* L.) carrying an ERA1 antisense construct controlled by the drought-inducible rd29A promoter from *A. thaliana* showed enhanced yield during drought stress	Wang et al. (2005)
	MLP43	Major latex protein-like protein 43 (MLP43) is a positive regulator of ABA response	↑	Overexpression of MLP43 conferred drought tolerance in *Arabidopsis*	Wang et al. (2016)

(continued)

Table 12.1 (continued)

Plant hormone	Gene	Function of gene	Expression/ knockout	Phenotype of transgenics	References
	PYR/PYL/ RCAR	ABA receptors		Drought tolerance in tomato	Gonzalez-Guzman et al. (2014)
	OsPYL3 and OsPYL9	ABA receptors	↑	Drought and cold resistance in rice	Tian et al. (2015)
	GhNAC2	ABA-induced leaf senescence	↑	Improved root growth in *Arabidopsis*, better vegetative growth and productivity	Gunapati et al. (2016)
	HVA1	ABA-responsive late embryogenesis abundant protein	↑	Drought tolerance and improved biomass in wheat	Sivamani et al. (2000)
	BnFTA	Farnesyltransferase, an enzyme associated with ABA-dependent signal transduction	↑	Drought tolerance in canola and increase in yield	Wang et al. (2009)
Auxin	*YUCCA6*	Important gene in auxin/IPA biosynthesis	↑	Phenotypes of potato with higher auxin content and enhanced drought tolerance	Kim et al. (2013)
	OsIAA6	A member of rice auxin/IAA gene family	↑	Better drought tolerance of transgenic rice plants via auxin biosynthesis regulation	Jung et al. (2015)
Cytokinins	*IPT*	Cytokinin biosynthesis	↑	Transgenic tomato showed enhanced growth and yield under salt stress	Ghanem et al. (2011)
			↑	Transgenic tobacco showed enhanced tolerance to salinity	Qiu et al. (2012)
	SlIPT3		↑	Transgenic tomato showed enhanced salinity stress tolerance	Žižkov et al. 2015
	AtCKX1	Cytokinin dehydrogenase	↑	Transgenic barley plants showed better drought tolerance via better dehydration avoidance	Pospíšilová et al. (2016)
	CKX	Cytokinin dehydrogenase	↑	Transgenic *Arabidopsis* plants overexpressing cytokinin oxidase/dehydrogenase gene showed enhanced drought tolerance	Werner et al. (2010)
	ERF-1 (JERF1)	Response factors for ethylene as well as jasmonates	↑	Rice plants showed increased drought tolerance	Zhang et al. (2010)

Ethylene	*ACC synthase*	Catalyzes rate-limiting step in ethylene biosynthesis	Gene silencing	Reduced ethylene levels with better drought tolerance in transgenic maize plants	Habben et al. (2014)
	ZmARGOS	Negative regulators of ethylene signal transduction	↑	Improved drought tolerance of transgenic *Arabidopsis* and maize plants	Shi et al. (2015)
	ETOL1		↑	Increased tolerance to drought and submergence	Du et al. (2014)
	NTHK1	Type II ethylene receptor homolog gene	↑	Transgenic *Arabidopsis*, with NTHK1, showed mRNA and protein expression and were salt sensitive	Wan-Hong et al. (2007)
	OsGSK1	BR negative regulator	Knockout of *OsGSK1*	Increased tolerance of knockout mutants to cold, heat, salt, and drought stresses	Koh et al. (2007)
Brassinosteroids	*AtHSD1*	Role in BR biosynthesis	↑	Overproduction of BR increased growth rate and seed yield, increased salinity tolerance	Li et al. (2007a, b)
	AtDWF4	Encodes a C-22 hydroxylase that catalyzes a rate-determining step in BR biosynthesis	↑	Overexpression in *B. napus* led to an increase in seed yield, oil content biomass, and root length, during drought and high temperature	Sahni et al. (2016)
	BdBRI1	BR receptor gene	↓	Improved drought tolerance with dwarf phenotypes of purple false brome	Feng et al. (2015)

↑: Overexpression

combination of stress responses which could act as cornerstone toward the goal of feeding the ever-increasing world population.

References

Abou Qamar S, Luo H, Laluk K, Mickelbart VM, Mengiste T (2009) Crosstalk between biotic and abiotic stress responses in tomato is mediated by AIM1 transcription factor. Plant J 58:1–13.14

Acharya B, Assmann S (2009) Hormone interactions in stomatal function. Plant Mol Biol 69:451–462

Alonso-Ramirez A, Rodriguez D, Reyes D, Jimenez JA, Nicolas G, Lopez-Climent M, Gomez-Cadenas A, Nicolas C (2009) Evidence for a role of gibberellins in salicylic acid-modulated early plant responses to abiotic stress in *Arabidopsis* seeds. Plant Physiol 150:1335–1344

Alvarez S, Marsh EL, Schroeder SG, Schachtman DP (2008) Metabolomic and proteomic changes in the xylem sap of maize under drought. Plant Cell Environ 31:325–340

Andreasson E, Ellis B (2010) Convergence and specificity in the *Arabidopsis* MAPK nexus. Trends Plant Sci 15:106–113

Bajguz A (2010) An enhancing effect of exogenous brassinolide on the growth and antioxidant activity in *Chlorella vulgaris* cultures under heavy metal stress. Environ Exp Bot 68:175–179

Bellard C, Bertelsmeier C, Leadley P, Thuiller W, Courchamp F (2012) Impacts of climate change on the future of biodiversity. Ecol Lett 15:365–377. https://doi.org/10.1111/j.1461-0248.2011.01736.x

Bowler C, Fluhr R (2000) The role of calcium and activated oxygen as signals for controlling cross-tolerance. Trends Plant Sci 5:241–246

Capiati DA, Pais SM, Tellez-Iñon MT (2006) Wounding increases salt tolerance in tomato plants: evidence on the participation of calmodulin-like activities in cross-tolerance signaling. J Exp Bot 57:2391–2400

Chakraborty U, Tongden C (2005) Evaluation of heat acclimation and salicylic acid treatments as potent inducers of thermotolerance in Cicer arietinum L. Curr Sci 89:384–389

Chinnusamy V, Schumaker K, Zhu JK (2004) Molecular genetics perspectives on cross-talk and specificity in abiotic stress signalling in plants. J Exp Bot 55:225–236

Colebrook EH, Thomas SG, Phillips AL, Hedden P (2014) The role of gibberellin signalling in plant responses to abiotic stress. J Exp Biol 217:67–75

Creelman RA, Mullet JE (1995) Jasmonic acid distribution and action in plants: regulation during development and response to biotic and abiotic stress. Proc Natl Acad Sci U S A 92:4114–4119

Divi U, Krishna P (2009) Brassinosteriod: a biotechnological target for enhancing crop yield and stress tolerance. New Biotechnol 26:131–136

Divi U, Krishna P (2010) Overexpression of the brassinosteroid biosynthetic gene AtDWF4 in *Arabidopsis* seeds 0vercomes abscisic acid-induced inhibition of germination and increases cold tolerance in transgenic seedlings. J Plant Growth Regul 29:385–393

Divi U, Rahman T, Krishna P (2010) Brassinosteroid-mediated stress tolerance in *Arabidopsis* shows interactions with abscisic acid, ethylene and salicylic acid pathways. BMC Plant Biol 10:151

Du H, Wu N, Cui F, You L, Li XH, Xiong LZ (2014) A homolog of ETHYLENE OVER-PRODUCER, OsETOL1, differentially modulates drought and submergence tolerance in rice. Plant J 78:834–849

Egamberdieva D (2009) Alleviation of salt stress by plant growth regulators and IAA producing bacteria in wheat. Acta Physiol Plant 31:861–864

Ellouz H, Hamed KB, Hernandez I, Cela J, Muller M, Magne C et al (2014) A comparative study of the early osmotic, ionic, redox and hormonal signaling response in leaves and roots of two halophytes and a glycophyte to salinity. Planta 240:1299–1317. https://doi.org/10.1007/s00425-014-2154-7

Estrada-Melo AC, Ma C, Reid MS, Jiang CZ (2015) Overexpression of an ABA biosynthesis gene using a stress-inducible promoter enhances drought resistance in petunia. Hortic Res 2:15013. https://doi.org/10.1038/hortres.2015.13

Etehadnia M, Waterer DR, Tanino KK (2008) The method of ABA application affects salt stress responses in resistant and sensitive potato lines. J Plant Growth Regul 27:331–341. https://doi.org/10.1007/s00344-008-9060-9

Fahad S, Bano A (2012) Effect of salicylic acid on physiological and biochemical characterization of maize grown in saline area. Pak J Bot 44:1433–1438

Feng Y, Yin YH, Fei SZ (2015) Downregulation of BdBRI1, a putative brassinosteroid receptor gene produces a dwarf phenotype with enhanced drought tolerance in Brachypodium distachyon. Plant Sci 234:163–173

Fletcher RA, Gill A, Davis TD, Sankhla N (2000) Triazoles as plant growth regulators and stress protectants. Hortic Rev 24:55–138

Flores A, Grau A, Laurich F, Dorffling K (1988) Effect of new terpenoid analogues of abscisic acid on chilling and freezing resistance. J Plant Physiol 132:362–369. https://doi.org/10.1016/S0176-1617(88)80121-4

Foo E, Davies NW (2011) Strigolactones promote nodulation in pea. Planta 234:1073–1081

Fraire-Velázquez S, Rodríguez-Guerra R, Sánchez-Calderón L (2011) In: Shanker A (ed) Abiotic and biotic stress response crosstalk in plants-physiological, biochemical and genetic perspectives. InTech Open Access Company, Rijeka, pp 1–26

Fujita M, Fijita Y, Noutoshi Y, Takahashi F, Narusaka Y, Yamaguchi-Shinozaki K, Shinozaki K (2006) Crosstalk between abiotic and biotic stress responses: a current view from the points of convergence in the stress signaling networks. Curr Opin Plant Biol 9:436–442.19

Fujita Y, Fujita M, Shinozaki K, Yamaguchi-Shinozaki K (2011) ABA mediated transcriptional regulation in response to osmotic stress in plants. J Plant Res 124:509–525. https://doi.org/10.1007/s10265-011-0412-3

Gan S, Amasino RM (1995) Inhibition of leaf senescence by auto-regulated production of cytokinin. Science 270:1986–1988

Ghanem ME, Albacete A, Smigocki AC, Frébort I, Pospíšilová H, Martínez-Andújar C, Acosta M, Sánchez-Bravo J, Lutts S, Dodd IC et al (2011) Root-synthesized cytokinins improve shoot growth and fruit yield in salinized tomato. J Exp Bot 62:125–140

Gilley A, Flecher RA (2007) Gibberellin antagonizes paclobutrazol induced stress protection in wheat seedlings. J Plant Physiol 103:200–207

Gomez-Rolden V, Fermas S, Brewer PB, Puech-Pages V, Adun E, Pillot J, Letisse F, Matusova R, Danoun S, Portais JC, Bouwmeester H, Becard G, Beveridge CA, Rameau C, Rochange SF (2008) Strigolactone inhibition of shoot branching. Nature 455:189–194

Gonai T, Kawahara S, Tougou M, Satoh S, Hashiba T, Hirai N, Kawaide H, Kamiya Y, Yoshioka T (2004) Abscisic acid in the thermoinhibition of lettuce seed germination and enhancement of its catabolism by gibberellin. J Exp Bot 55:111–118

Gonzalez-Guzman M, Rodriguez L, Lorenzo-Orts L, Pons C, Sarrion Perdigones A, Fernandez MA et al (2014) Tomato PYR/PYL/RCAR abscisic acid receptors show high expression in root, differential sensitivity to the abscisic acid agonist quinabactin, and the capability to enhance plant drought resistance. J Exp Bot 65:4451–4464. https://doi.org/10.1093/jxb/ eru219

Gunapati S, Naresh R, Ranjan S, Nigam D, Hans A, Verma PC, Gadre R, Pathre UV, Sane AP, Sane VA (2016) Expression of GhNAC2 from G. herbaceum, improves root growth and imparts tolerance to drought in transgenic cotton and *Arabidopsis*. Sci Rep 6:24978. https://doi.org/10.1038/srep24978

Habben JE, Bao XM, Bate NJ, De Bruin JL, Dolan D, Hasegawa D, Helentjaris TG, Lafitte RH, Lovan N, Mo H, Reimann K, Schussler JR (2014) Transgenic alteration of ethylene biosynthesis increases grain yield in maize under field drought-stress conditions. Plant Biotechnol J 12:685–693

Iqbal N, Nazar R, Iqbal MRK, Masood A, Nafees AK (2011) Role of gibberellins in regulation of source sink relations under optimal and limiting environmental conditions. Curr Sci 100:998–1007

Javid MG, Sorooshzadeh A, Moradi F, Sanavy SAMM, Allahdadi I (2011) The role of phytohormones in alleviating salt stress in crop plants. Aust J Crop Sci 5:726–734

Jewell MC, Campbell BC, Godwin ID (2010) Transgenic plants for abiotic stress resistance. In: Kole C, Michler C, Abbott AG, Hall TC (eds) Transgenic crop plants, Vol. 2: utilization and biosafety. Springer-Verlag, Berlin, pp 67–131

Jones B, Gunneras SA, Petersson SV, Tarkowski P, Graham N, May S, Dolezal K, Sandberg G, Ljung K (2010) Cytokinin regulation of auxin synthesis in *Arabidopsis* involves a homeostatic feedback loop regulated via auxin and cytokinin signal transduction. Plant Cell 22:2956–2969

Jumali SS, Said IM, Ismail I, Zainal Z (2011) Genes induced by high concentration of salicylic acid in Mitragyna speciosa. Aust J Crop Sci 5:296–303

Jung H, Lee DK, Choi DY, Kim JK (2015) OsIAA6, a member of the rice Aux/IAA gene family, is involved in drought tolerance and tiller outgrowth. Plant Sci 236:304–312

Kang DJ, Seo YJ, Lee JD, Ishii R, Kim KU, Shin DH, Park SK, Jang SW, Lee IJ (2005) Jasmonic acid differentially affects growth, ion uptake and abscisic acid concentration in salt-tolerant and salt-sensitive rice cultivars. J Agron Crop Sci 191:273–282

Kang NY, Cho C, Kim NY, Kim J (2012) Cytokinin receptor-dependent and receptor-independent path ways in the dehydration response of *Arabidopsis* thaliana. J Plant Physiol 169:1382–1391

Kapulnik Y, Delaux PM, Resnick N, Mayzlish-Gsti E, Wininger S, Bhattarcharya C, Sejalon-Delmas N, Combier JP, Bécard G, Belausov E, Beeckman T, Dor E, Hershenhorn J, Koltai H (2011) Strigolactones affect lateral root formation and root-hair elongation in *Arabidopsis*. Planta 233:209–216

Kazan K (2013) Auxin and the integration of environmental signals into plant root development. Ann Bot 112:1655–1665

Kazan K (2015) Diverse roles of jasmonates and ethylene in abiotic stress tolerance. Trends Plant Sci 20:219–229

Khadri M, Tejera NA, Lluch C (2006) Alleviation of salt stress in common bean (Phaseolus vulgaris) by exogenous abscisic acid supply. J Plant Growth Regul 25:110–119. https://doi.org/10.1007/s00344-005-0004-3

Khan W, Prithiviraj B, Smith D (2003) Photosynthetic response of corn and soybean to foliar application of salicylates. J Plant Physiol 160:485–492

Khodary SEA (2004) Effect of salicylic acid on growth, photosynthesis and carbohydrate metabolism in salt stressed maize plants. Int J Agric Biol 6:5–8

Khripach V, Zhabinskii VN, De Groot AE (2000) Twenty years of brassinosteroids: steroidal plant hormones warrant better crops for the XXI century. Ann Bot 86:441–447

Kim IJ, Baek D, Park HC, Chun HJ, Oh DH, Lee MK, Cha JY, Kim WY, Kim MC, Chung WS, Bohnert HJ (2013) Overexpression of *Arabidopsis* YUCCA6 in potato results in high-auxin developmental phenotypes and enhanced resistance to water deficit. Mol Plant 6:337–349

Klay I, Pirrello J, Riahi L, Bernadac A, Cherif A, Bouzayen M, Bouzid S (2014) Ethylene response factors Sl-ERFB3 is responsive to abiotic stresses and mediates salt and cold stress response regulation in tomato. Sci World J 2014:1–12

Koh S, Lee SC, Kim MK, Koh JH, Lee S, An G, Choe S, Kim SR (2007) T-DNA tagged knockout mutation of rice OsGSK1, an orthologue of *Arabidopsis* BIN2, with enhanced tolerance to various abiotic stresses. Plant Mol Biol 65:1158–1164

Krishna P (2003) Brassinosteroid-mediated stress responses. J Plant Growth Regul 22:289–297

Krouk G, Ruffel S, Gutierrez RA, Gojon A, Crawford NM, Coruzzi GM, Lacombe B (2011) A framework integrating plant growth with hormones and nutrients. Trends Plant Sci 16:178–182

Lalk I, Dörffling K (1985) Hardening, abscisic acid, proline and freezing resistance in two winter wheat varieties. Physiol Plant 63:287–292. https://doi.org/10.1111/j.1399-3054.1985.tb04267.x

Laloi C, Appel K, Danon A (2004) Reactive oxygen signalling: the latest news. Curr Opin Plant Biol 7:323–328

Larkindale J, Huang B (2004) Thermotolerance and antioxidant systems in Agrostisstolonifera: involvement of salicylic acid, abscisic acid, calcium, hydrogen peroxide, and ethylene. J Plant Physiol 161:405–413

Larkindale J, Hall DJ, Knight MR, Vierling E (2005) Heat stress phenotypes of *Arabidopsis* mutants implicate multiple signaling pathways in the acquisition of thermo-tolerance. Plant Physiol 138:882–897

Li FL, Asami T, Wu XZ, Tsang EWT, Cutler AJ (2007a) A putative hydroxysteroid dehydrogenase involved in regulating plant growth and development. Plant Physiol 145:87–97

Li KR, Wang HH, Han G, Wang QJ, Fan J (2007b) Effects of brassinolide on the survival, growth and drought resistance of Robinia pseudoacacia seedlings under water stress. New For 35:255–266

Li YJ, Zhang JC, Zhang J, Hao L, Hua JP, Duan LS, Zhang MC, Li ZH (2013) Expression of an *Arabidopsis* molybdenum factor sulphurase gene in soybean enhances drought tolerance and increases yield under field conditions. Plant Biotechnol J 11:747–758

López-Ráez JA, Kohlen W, Charnikhova T, Mulder P, Undas AK, Sergeant MJ, Verstappen F, Bugg TDH, Thompson AJ, Ruyter-Spira C et al (2010) Does abscisic acid affect strigolactone biosynthesis? New Phytol 187:343–354

Lu Y, Li Y, Zhang J, Xiao Y, Yue Y, Duan L, Li Z (2013) Overexpression of *Arabidopsis* molybdenum cofactor sulfurase gene confers drought tolerance in maize (Zea mays L.) PLoS One 8:e52126. https://doi.org/10.1371/journal.pone.0052126

Luo M, Liu JH, Mahapatra S, Hiu RD, Hill RD, Mahapatra SS (1992) Characterization of a gene family encoding abscisic acid and environmental stress-inducible proteins of alfalfa. J Biol Chem 267:15367–15374

Ma Q-H (2008) Genetic engineering of cytokinins and their application to agriculture. Crit Rev Biotechnol 28:213–232

Mao X, Zhang H, Tian S, Chang X, Jing R (2010) TaSnRK2.4, an SNF1-type serine/threonine protein kinase of wheat (Triticum aestivum L.), confers enhanced multistress tolerance in *Arabidopsis*. J Exp Bot 61:683–696

Mashiguchi K, Sasaki E, Shimada Y, Nagae M, Ueno K, Nakano T, Yoneyama K, Suzuki Y, Asami T (2009) Feedback-regulation of strigolactone biosynthetic genes and strigolactone-regulated genes in *Arabidopsis*. Biosci Biotechnol Biochem 73:2460–2465

Mayzlish-Gati E, LekKala SP, Resnick N, Wininger S, Bhattacharya C, Lemcoff JH, Kapulnik Y, Koltai H (2010) Strigolactones are positive regulators of light-harvesting genes in tomato. J Exp Bot 61:3129–3136

Mei C, Qi M, Sheng G, Yang Y (2006) Inducible over expression of a rice allene oxide synthase gene increases the endogenous jasmonic acid level, PR gene expression, and host resistance to fungal infection. Mol Plant-Microbe Interact 19:1127–1137

Mohapatra SS, Poole RJ, Dhindsa RS (1988) Abscisic acid-regulated gene expression in relation to freezing tolerance in alfalfa. Plant Physiol 87:468–473. https://doi.org/10.1104/pp.87.2.468

Morgan PW, Drew MC (1997) Ethylene and plant responses to stress. Physiol Plant 100:620–630. https://doi.org/10.1111/j.1399-3054.1997.tb03068.x

Nishiyama R, Watanabe Y, Fujita Y, Tien LD, Kojima M, Werner T, Vankova R, Yamaguchi-Shinozaki K, Shinozaki K, Kakimoto T, Sakakibara H, Schmuelling T, Lam-Son PT (2011) Analysis of cytokinin mutants and regulation of cytokinin metabolic genes reveals important regulatory roles of cytokinins in drought, salt and abscisic acid responses, and abscisic acid biosynthesis. Plant Cell 23:2169–2183

O'Brien JA, Benkova E (2013) Cytokinin cross-talking during biotic and abiotic stress responses. Front Plant Sci 4:451

Pedranzani H, Sierra-de-Grado R, Vigliocco A, Miersch O, Abdala G (2007) Cold and water stresses produce change in endogenous Jasmonates in two populations of *Pinus pinaster* Ait. Plant Growth Regul l52:111–116

Peleg Z, Reguera M, Tumimbang E, Walia H, Blumwald E (2011) Cytokinin-mediated source/sink modifications improve drought tolerance and increases grain yield in rice under water-stress. Plant Biotechnol. https://doi.org/10.1111/ j.1467-7652.2010.00584.x. in press

Pospíšilová H, Jiskrová E, Vojta P, Mrízová K, Kokáš F, Čudejková MM, Bergougnoux V, Plíhal O, Klimešová J, Novák O, Dzurová L (2016) Transgenic barley overexpressing a cytokinin dehydrogenase gene shows greater tolerance to drought stress. New Biotechnol 33:692–705. https://doi.org/10.1016/j.nbt.2015.12.005

Qin F, Kazuo S, Kazuo YS (2011) Achievements and challenges in understanding plant abiotic stress responses and tolerance. Plant Cell Physiol 52:1569–1582

Qiu WM, Liu MY, Qiao GR, Jiang J, Xie LH, Zhuo RY (2012) An isopentenyl transferase gene driven by the stress-inducible rd29A promoter improves salinity stress tolerance in transgenic tobacco. Plant Mol Biol Report 30:519–528

Quain MD, Makgopa ME, Márquez-García B, Comadira G, Fernandez-Garcia N, Olmos E, Schnaubelt D, Kunert KJ, Foyer CH (2014) Ectopic phytocystatin expression leads to enhanced drought stress tolerance in soybean (*Glycine max*) and *Arabidopsis* thaliana through effects on strigolactone pathways and can also result in improved seed traits. Plant Biotechnol J 12:903–913

Reguera M, Peleg Z, Abdel-Tawab YM, Tumimbang EB, Delatorre CA, Blumwald E (2013) Stress- induced cytokinin synthesis increases drought tolerance through the coordinated regulation of carbon and nitrogen assimilation in rice. Plant Physiol 163(4):1609–1622

Ribeiro DM, Desikan R, Bright JO, Confraria ANA, Harrison J, Hancock JT, Barros RS, Neill SJ, Wilson ID (2009) Differential requirement for NO during ABA-induced stomatal closure in turgid and wilted leaves. Plant Cell Environ 32:46–57

Rivero RM, Kojima M, Gepstein A, Sakakibara H, Mittler R, Gepstein S, Blumwald E (2007) Delayed leaf senescence induces extreme drought tolerance in a flowering plant. Proc Natl Acad Sci U S A 104:19631–19636

Rivero RM, Shulaev V, Blumwald E (2009) Cytokinin-dependent photorespiration and the protection of photosynthesis during water deficit. Plant Physiol 150:1530–1540

Rivero RM, Gimeno J, Van Deynze A, Walia H, Blumwald E (2010) Enhanced cytokinin synthesis in tobacco plants expressing PSARK<IPT prevents the degradation of photosynthetic protein complexes during drought. Plant Cell Physiol 51:1929–1941

Sah SK, Reddy KR, Li J (2016) Abscisic acid and abiotic stress tolerance in crop plants. Front Plant Sci 7:571. https://doi.org/10.3389/fpls.2016.00571

Sahni S, Prasad BD, Liu Q, Grbic V, Sharpe A, Singh SP, Krishana P (2016) Overexpression of brassinosteroid biosynthesis gene DWF4 in Brassica napus simultaneously increases seed yield and stress tolerance. Sci Rep 6:28298. https://doi.org/10.1038/srep28298

Sakamoto T, Yoichi M, Kanako I, Masatomo K, Hironori I, Toshiaki K, Shuichi I, Makoto M, Hiroshi T (2003) Genetic manipulation of gibberellin metabolism in transgenic rice. Nat Biotechnol 21:909–913

Santner A, Estelle M (2010) The ubiquitin-proteasome system regulates plant hormone signaling. Plant J 61:1029–1040

Senaratna T, Touchell D, Bumm E, Dixon K (2000) Acetylsalicylic acid (aspirin) and salicylic acid induce multiple stress tolerance in bean and tomato plants. Plant Growth Regul 30:157–161

Shi J, Habben JE, Archibald RL, Drummond BJ, Chamberlin MA, Williams RW, Lafitte HR, Weers BP (2015) Overexpression of ARGOS genes modifies plant sensitivity to ethylene, leading to improved drought tolerance in both *Arabidopsis* and maize. Plant Physiol 169:266–282

Singh N, La Rosa K, Handa PC, Hasegawa AKPM, Bressan RA (1987) Hormonal regulation of protein synthesis associated with salt tolerance in plant cells. Proc Natl Acad Sci U S A 84:739–743. https://doi.org/10.1073/pnas.84.3.739

Sivamani E, Bahieldin A, Wraith JM, Al-Niemi T, Dyer WE, Ho TD et al (2000) Improved biomass productivity and water use efficiency under water deficit conditions in transgenic wheat constitutively expressing the barley HVA1 gene. Plant Sci 155:1–9. https://doi.org/10.1016/S0168-9452(99) 00247-2

Smigocki AC, Owens LD (1989) Cytokinin-to-auxin ratios and morphology of shoots and tissues transformed by a chimeric isopentenyl transferase gene. Plant Physiol 91:808–811

Soto MJ, Fernandez-Aparicio MN, Castellanos-Morales V, Garcia-Garrido JM, Ocampo JA, Delgado MJ, Vierheilig H (2010) First indications for the involvement of strigolactones on nodule formation in alfalfa (Medicago sativa). Soil Biol Biochem 42:383–385

Sreenivasulu N, Radchuk V, Alawady A, Borisjuk L, Weier D, Staroske N, Fuchs J, Miersch O, Strickert M, Usadel B, Wobus U, Grimm B, Weber H, Weschke W (2010) De-regulation of abscisic acid contents causes abnormal endosperm development in the barley mutant seg8. Plant J 64:6489–6603

Suzuki N, Koussevitzky S, Mittler R, Miller G (2012) ROS and redox signalling in the response of plants to abiotic stress. Plant Cell Environ 35:259–270.36

Szalai G, Tari I, Janda T, Pestenacz A, Paldi E (2002) Effects of cold acclimation and salicylic acid on changes in ACC and MACC contents in maize during chilling. Biol Plant 43:637–640

Tasgin E, Atici O, Nalbantoglu B (2003) Effects of salicylic acid and cold on freezing tolerance in winter wheat leaves. Plant Growth Regul 41:231–236

Thompson AJ, Andrews J, Mulholland BJ, McKee JMT, Hilton HW, Horridge JS, Farquhar GD, Smeeton RC, Smillie IRA, Black CR et al (2007) Overproduction of abscisic acid in tomato increases transpiration efficiency and root hydraulic conductivity and influences leaf expansion. Plant Physiol 143:1905–1917

Tian X, Wang Z, Li X, Lv T, Liu H, Wang L et al (2015) Characterization and functional analysis of pyrabactin resistance-like abscisic acid receptor family in rice. Rice 8:28. https://doi.org/10.1186/s12284-015-0061-6

Tsuchisaka A, Theologis A (2004) Unique and overlapping expression patterns among the *Arabidopsis* 1-amino cyclopropane-1- carboxylate synthase gene family members. Plant Physiol 136:2982–3000

Umehara M, Hanada A, Yoshid S, Akiyam K, Arite T, Takeda N, Kamiya N, Magome H, Kamiya Y, Shirasu K, Yoneyama K, Kyozuka J, Yamaguchi S (2008) Inhibition of shoot branching by new terpenoid plant hormones. Nature 455:195–200

Vob U, Bishopp A, Farcot E, Bennett MJ (2014) Modelling hormonal response and development. Trends Plant Sci 19:311–319

Vurro M, Yoneyama K (2012) Strigolactones-intriguing biologically active compounds: perspectives for deciphering their biological role and for proposing practical application. Pest Manag Sci 68:664–668

Waldie T, McCulloch H, Leyser O (2014) Strigolactones and the control of plant development: lessons from shoot branching. Plant J 79:607–622.201

Wallace JS, Acreman MC, Sullivan CA (2003) The sharing of water between society and ecosystems: from conflict to catchment–based co– management. Philos Trans R Soc B Biol Sci 358:2011–2026. https://doi.org/10.1098/rstb.2003.1383

Wang Y, Ying J, Kuzma M, Chalifoux M, Sample A, McArthur C, Uchacz T, Sarvas C, Wan J, Dennis DT et al (2005) Molecular tailoring of farnesylation for plant drought tolerance and yield protection. Plant J 43:413–424

Wang Y, Beaith M, Chalifoux M, Ying J, Uchacz T, Sarvas C et al (2009a) Shoot-specific down-regulation of protein farnesyltransferase (α-subunit) for yield protection against drought in canola. Mol Plant 2:191–200. https://doi.org/10.1093/mp/ssn088

Wang L, Wang Z, Xu Y, Joo S-H, Kim S-K, Xue Z, Xu Z, Wang Z, Chong K (2009b) OsGSR1 is involved in crosstalk between gibberellins and brassinosteroids in rice. Plant J 57:498–510

Wang XH, Shu C, Li HY, Hu XQ, Wang YX (2014) Effects of 0.01% brassinolide solution application on yield of rice and its resistance to autumn low-temperature damage. Acta Agric Jiangxi 26:36–38. in Chinese with English abstract

Wang Y, Yang L, Chen X, Ye T, Zhong B, Liu R et al (2016) Major latex protein-like protein 43 (MLP43) functions as a positive regulator during abscisic acid responses and confers drought tolerance in *Arabidopsis thaliana*. J Exp Bot 67:421–434. https://doi.org/10.1093/jxb/erv477

Wan-Hong C, Jun L, Xin-Jian H, Rui-Ling M, Hua-Lin Z, Shou-Yi C, Jin-Song Z (2007) Modulation of ethylene responses affects plant salt-stress responses. Plant Physiol 143:707–719

Wani SH, Singh NB, Haribhushan A, Mir JA (2013) Compatible solute engineering in plants for abiotic stress tolerance- the role of glycine betaine. Curr Genomics 14:157–165

Wei L, Wang L, Yang Y, Wang P, Guo T, Kang G (2015) Abscisic acid enhances tolerance of wheat seedlings to drought and regulates transcript levels ABA and abiotic stress tolerance of genes encoding ascorbate-glutathione biosynthesis. Front Plant Sci 6:458. https://doi.org/10.3389/fpls.2015.00458

Werner T, Nehnevajova E, Köllmer I, Novák O, Strnad M, Krämer U, Schmulling T (2010) Root-specific reduction of cytokinin causes enhanced root growth, drought tolerance, and leaf mineral enrichment in *Arabidopsis* and tobacco. Plant Cell 22:3905–3920

Wheeler T, Von Braun J (2013) Climate change impacts on global food security. Science 341:508–513. https://doi.org/10.1126/science.1239402

Wilkinson S, Davies WJ (2010) Drought ozone, ABA and ethylene: new insights from cell to plant to community. Plant Cell Environ 33:510–525

Wolbang CM, Chandler PM, Smith JJ, Ross JJ (2004) Auxin from the developing inflorescence is required for the biosynthesis of active gibberellins in barley stems. Plant Physiol 134:769–776

Yamaguchi-Shinozaki K, Shinozaki K (2006) Transcriptional regulatory networks in cellular responses and tolerance to dehydration and cold stresses. Annu Rev Plant Biol 57:781–803

Yang ZM, Wang J, Wang SH, Xu LL (2003) Salicylic acid induced aluminium tolerance by modulation of citrate efflux from roots of Cassia tora L. Planta 217:168–174

Yoshida T, Mogami J, Yamaguchi-Shinozaki K (2014) ABA-dependent and ABA-independent signaling in response to osmotic stress in plants. Curr Opin Plant Biol 21:133–139. https://doi.org/10.1016/j.pbi.2014.07.009

Zalabák D, Pospíšilová H, Šmehilová M, Mrízová K, Frébort I, Galuszka P (2013) Genetic engineering of cytokinin metabolism: prospective way to improve agricultural traits of crop plants. Biotechnol Adv 31:97–117

Zhang S, Hu J, Zhang Y, Xie XJ, Knapp A (2007) Seed priming with brassinolide improves lucerne (Medicago sativa L.) seed germination and seedling growth in relation to physiological changes under salinity stress. Aust J Agric Res 58:811–815

Zhang Y, Yang J, Lu S, Cai J, Guo Z (2008) Overexpressing SgNCED1 in tobacco increases ABA level, antioxidant enzyme activities, and stress tolerance. J Plant Growth Regul 27:151–158

Zhang S-W, Li C-H, Cao J, Zhang Y-C, Zhang S-Q, Xia Y-F, Sun D-Y, Sun Y (2009) Altered architecture and enhanced drought tolerance in rice via the down-regulation of indole-3-Acetic Acid by TLD1/OsGH3.13 activation. Plant Physiol 151:1889–1901

Zhang ZJ, Li F, Li DJ, Zhang HW, Huang RF (2010) Expre ssion of ethylene response factor JERF1 in rice improves tolerance to drought. Planta 232:765–774

Zhang Q, Li J, Zhang W, Yan S, Wang R, Zhao J et al (2012) The putative auxin effl ux carrier OsPIN3t is involved in the drought stress response and drought tolerance. Plant J 72:805–816

Zhang ZQ, Wang YF, Chang LQ, Zhang T, An J, Liu YS, Cao YM, Zhao X, Sha XY, Hu TM, Yang PZ, Zep MS (2015) A novel zeaxanthin epoxidase gene from alfalfa (Medicago sativa),confers drought and salt tolerance in transgenic tobacco. Plant Cell Rep 14:1–5

Zhang J, Yu HY, Zhang YS, Wang YB, Li MY, Zhang JC, Duan LS, Zhang MC, Li ZH (2016) Increased abscisic acid levels in transgenic maize overexpressing AtLOS5 mediated root ion fluxes and leaf water status under salt stress. J Exp Bot 67:1339–1355. 528. https://doi.org/10.1093/jxb/erv528

Zhu JK (2002) Salt and drought stress signal transduction in plants. Annu Rev Plant Physiol Plant Mol Biol 53:247–273

Žižková E, Dobrev PI, Muhovski Y, Hošek P, Hoyerová K, Haisel D, Hichri I (2015) Tomato (Solanum lycopersicum L.) SlIPT3 and SlIPT4 isopentenyltransferases mediate salt stress response in tomato. BMC Plant Biol 15:85. https://doi.org/10.1186/s12870-015-0415-7

MIX
Papier aus verantwortungsvollen Quellen
Paper from responsible sources
FSC® C105338

If you have any concerns about our products,
you can contact us on
ProductSafety@springernature.com

In case Publisher is established outside the EU,
the EU authorized representative is:
Springer Nature Customer Service Center GmbH
Europaplatz 3, 69115 Heidelberg, Germany

Printed by Libri Plureos GmbH
in Hamburg, Germany